Digital Therapeutics for Mental Health and Addiction

Digital Therapeutics for Mental Health and Addiction

The State of the Science and Vision for the Future

Edited by

Nicholas C. Jacobson

Center for Technology and Behavioral Health, Geisel School of Medicine at Dartmouth College, Lebanon, New Hampshire, United States; Department of Biomedical Data Science, Geisel School of Medicine at Dartmouth College, Lebanon, New Hampshire, United States; Department of Psychiatry, Geisel School of Medicine at Dartmouth College, Lebanon, New Hampshire, United States; Department of Computer Science, Dartmouth College, Hanover, New Hampshire, United States

Tobias Kowatsch

Future Health Technologies Programme, Campus for Research Excellence and Technological Enterprise, Singapore-ETH Centre, Singapore, Singapore; Centre for Digital Health Interventions, Department of Management, Technology, and Economics, ETH Zurich, Zurich, Switzerland; Centre for Digital Health Interventions, Institute of Technology Management, University of St. Gallen, St. Gallen, Switzerland; Institute for Implementation Science in Health Care, University of Zurich, Zurich, Switzerland; School of Medicine, University of St Gallen, St Gallen, Switzerland

Lisa A. Marsch

Center for Technology and Behavioral Health, Geisel School of Medicine at Dartmouth College, Lebanon, New Hampshire, United States; Department of Biomedical Data Science, Geisel School of Medicine at Dartmouth College, Lebanon, New Hampshire, United States; Department of Psychiatry, Geisel School of Medicine at Dartmouth College, Lebanon, New Hampshire, United States

ELSEVIER

ACADEMIC PRESS
An imprint of Elsevier
elsevier.com/books-and-journals

For Information on all Academic Press publications visit our website at https://www.elsevier.com/books-and-journals

Publisher: Mara Conner
Acquisitions Editor: Carrie Bolger
Editorial Project Manager: Mariana L. Kuhl
Production Project Manager: Prasanna Kalyanaraman
Cover Designer: Matthew Limbert

Typeset by Aptara, New Delhi, India

Working together to grow libraries in developing countries

www.elsevier.com • www.bookaid.org

Contents

Contributors

Ryan Allred, University of Washington Department of Psychiatry and Behavioral Sciences, CREATIV Lab, Seattle, WA, U.S.A.

Patricia A. Areán, University of Washington Department of Psychiatry and Behavioral Sciences, CREATIV Lab, Seattle, WA, U.S.A.

Christina M. Armstrong, Office of Connected Care, U.S. Department of Veterans Affairs, Washington D.C.

Timothy Bickmore, Khoury College of Computer Sciences, Northeastern University, Boston, MA, USA

Christina A. Brezing, Columbia University Irving Medical Center, Department of Psychiatry and New York State Psychiatric Institute, Division on Substance Use Disorders, New York, NY, United States

Alan Jeffrey Budney, Geisel School of Medicine at Dartmouth, Center for Technology and Behavioral Health, Dartmouth College, Hanover, NH, USA

Caterina Bérubé, Centre for Digital Health Interventions, Department of Management, Technology and Economics, ETH Zurich, Zurich, Switzerland

Aimee N.C. Campbell, Columbia University Irving Medical Center, Department of Psychiatry and New York State Psychiatric Institute, Division on Substance Use Disorders, New York, NY, United States

Megan Coder, PharmD, MBA, Digital Therapeutics Alliance, Arlington, Virginia, USA

Gavin Doherty, School of Computer Science and Statistics, Trinity College Dublin, Ireland

Amanda Edwards-Stewart, AES Psychological Service, LLC, Tukwila, Washington

Elgar Fleisch, Centre for Digital Health Interventions, Department of Management, Technology and Economics, ETH Zurich, Zurich, Switzerland; Centre for Digital Health Interventions, Institute of Technology Management, University of St. Gallen (ITEM-HSG), St. Gallen, Switzerland

Milena Heinsch, School of Medicine and Public Health, College of Health, Medicine, and Wellbeing, the University of Newcastle, Callaghan, NSW, Australia

Donald M. Hilty, Northern California Veterans Administration Health Care System Professor, Department of Psychiatry & Behavioral Sciences, UC Davis 10535 Hospital Way, Mather, CA

Marcello Ienca, Department of Health Sciences and Technology, ETH Zurich

Nicholas C. Jacobson, Center for Technology and Behavioral Health, Geisel School of Medicine at Dartmouth College, Lebanon, New Hampshire, United States; Department of Biomedical Data Science, Geisel School of Medicine at Dartmouth College, Lebanon, New Hampshire, United States; Department of Psychiatry, Geisel School of Medicine at Dartmouth College, Lebanon, New Hampshire, United States; Department of Computer Science, Dartmouth College, Hanover, New Hampshire, United States

Benjamin Kaveladze, Department of Psychological Science, University of California, Irvine, Irvine, CA, USA

Frances Kay-Lambkin, Hunter Medical Research Institute, Newcastle, NSW, Australia

Olivia Clare Keller, Centre for Digital Health Interventions, Department of Management, Technology and Economics, ETH Zürich, Zürich, Switzerland

Roman Keller, Future Health Technologies Programme, Campus for Research Excellence and Technological Enterprise, Singapore-ETH Centre, Singapore, Singapore; Saw Swee Hock School of Public Health, National University of Singapore, Singapore, Singapore

Predrag Klasnja, School of Information, University of Michigan, Ann Arbor, MI, United States; Kaiser Permanente Washington Health Research Institute, Seattle, WA, United States

Tobias Kowatsch, Future Health Technologies Programme, Campus for Research Excellence and Technological Enterprise, Singapore-ETH Centre, Singapore, Singapore; Centre for Digital Health Interventions, Department of Management, Technology, and Economics, ETH Zurich, Zurich, Switzerland; Institute for Implementation Science in Health Care, University of Zurich, Zurich, Switzerland; School of Medicine, University of St Gallen, St Gallen, Switzerland

Constantin Landers, Department of Health Sciences and Technology, ETH Zurich

Camilla M. Lee, Center for Technology and Behavioral Health, Geisel School of Medicine at Dartmouth College, Lebanon, New Hampshire, United States

David D. Luxton, Department of Psychiatry and Behavioral Sciences, University of Washington School of Medicine, Seattle, Washington

Jacqueline Mair, Future Health Technologies Programme, Campus for Research Excellence and Technological Enterprise, Singapore-ETH Centre, Singapore, Singapore; Saw Swee Hock School of Public Health, National University of Singapore, Singapore, Singapore

Lisa A. Marsch, Center for Technology and Behavioral Health, Geisel School of Medicine at Dartmouth College, Lebanon, New Hampshire, United States; Department of Biomedical Data Science, Geisel School of Medicine at Dartmouth College, Lebanon, New Hampshire, United States; Department of Psychiatry, Geisel School of Medicine at Dartmouth College, Lebanon, New Hampshire, United States

Susan A. Murphy, Department of Statistics, Computer Science, John A. Paulson School of Engineering and Applied Sciences, Harvard University, Boston, Massachusetts

Inbal Nahum-Shani, Data-Science for Dynamic Decision-making Center (d3c), Institute for Social Research, University of Michigan, Ann Arbor, Michigan

John A. Naslund, Department of Global Health and Social Medicine, Harvard Medical School, Boston, MA, United States

Leysan Nurgalieva, Department of Computer Science, Aalto University, Espoo, Finland

Teresa O'Leary, Khoury College of Computer Sciences, Northeastern University, Boston, MA, USA

Rachel E. Quist, Center for Technology and Behavioral Health, Geisel School of Medicine at Dartmouth College, Lebanon, New Hampshire, United States

Dara Sampson, School of Medicine and Public Health, College of Health, Medicine, and Wellbeing, the University of Newcastle, Callaghan, NSW, Australia

Stephen M. Schueller, Department of Psychological Science, University of California, Irvine, Irvine, CA, USA; Department of Informatics, University of California, Irvine, Irvine, CA, USA

Nicholas J. Seewald, Department of Health Policy and Management, Johns Hopkins Bloomberg School of Public Health, Baltimore, MD, United States

Matisyahu Shulman, Columbia University Irving Medical Center, Department of Psychiatry and New York State Psychiatric Institute, Division on Substance Use Disorders, New York, NY, United States

Shawna N. Smith, Department of Health Management and Policy, University of Michigan School of Public Health, Ann Arbor, MI, United States; Department of Psychiatry, University of Michigan, Ann Arbor, MI, United States

Jessica Spagnolo, Département des sciences de la santé communautaire, Université de Sherbrooke, Sherbrooke, QC, Canada; Centre de recherche Charles-Le Moyne (CR-CLM), Campus de Longueuil, Université de Sherbrooke, QC, Canada

Cara Ann Struble, Geisel School of Medicine at Dartmouth, Center for Technology and Behavioral Health, Dartmouth College, Hanover, NH, USA

Gisbert Wilhelm Teepe, Centre for Digital Health Interventions, Department of Management, Technology and Economics, ETH Zürich, Zürich, Switzerland

Florian v. Wangenheim, Future Health Technologies Programme, Campus for Research Excellence and Technological Enterprise, Singapore-ETH Centre, Singapore, Singapore; Centre for Digital Health Interventions, Department of Management, Technology, and Economics, ETH Zurich, Zurich, Switzerland

David W. Wetter, Center for Health Outcomes and Population Equity (HOPE), Department of Population Health Sciences, Huntsman Cancer Institute, University of Utah, Salt Lake City, Utah

Blanche Wies, Department of Health Sciences and Technology, ETH Zurich

Foreword

This book brings together a range of topics across a broad continuum of digital therapeutics. Digital therapeutics—software designed to prevent, treat, or manage pathology—accessed via mobile devices or personal computers have been a growing part of the healthcare landscape over the last decade. The COVID-19 pandemic underscores the value of remotely delivered interventions, in that, traditional methods of providing treatment have limitations and can be difficult or even impossible to access. Therefore, additional treatment options are urgently needed.

Digital therapeutics represent a rapidly expanding opportunity to address significant public health challenges. While thousands of interventions are currently available to the public, many have never been tested for efficacy. However, the number of validated interventions is growing, with some now authorized by the Food and Drug Administration (FDA) and others being utilized by health organizations following rigorous testing. These technologies cover a broad range of therapeutic areas, including mental health, metabolic disease, substance use, and more. Their potential to promote public health and wellbeing is only beginning to be tapped. For example, these areas of priorities were recently identified by the White House Office of National Drug Control Policy (ONDCP) in making clear the need to expand access to evidence-based treatment including "…reimbursement for motivational incentives and digital treatment for addiction…."

As digital therapeutic interventions become more ubiquitous and the industry continues to see tremendous growth, academic investigators have a significant opportunity to create engaging, effective, and ultimately the wide adoption of new digital therapeutics. This book will cover the field of digital therapeutics including the extant literature on their efficacy, effectiveness, scalability, and cost-effectiveness. Additionally, digital therapeutics are currently being extended with novel directions, including personalized interventions delivered in real-time. Issues around broad adoption, including how digital therapeutics can be monetized and scaled, ethical issues, cultural adaptations, privacy and security concerns, and potential pitfalls must be considered. These seminal topics will elucidate the pathways, requirements, and capabilities needed to develop and sustain effective digital therapeutic interventions.

As a class of interventions, however, the rapidity of digital therapeutic development has eclipsed efforts to determine the extent to which they work. For example, there are hundreds of smoking cessation apps available in the market, yet very few follow standard clinical guidelines and evidence of clinical validation is unclear. The importance of validation in the target population is as important for digital therapeutics as more traditional medications. Yet the pathway leading to treatment in the clinic is much more varied for digital therapeutic interventions. All medications in the United States need to be approved by the FDA prior to being prescribed by physicians. Some digital interventions, on the other hand, can be given to patients without FDA review and authorization. It is important to build on foundational, peer-reviewed studies. Digital therapeutics research should evaluate clinical findings in appropriately designed randomized clinical trials, in comparison to treatment as usual or a relevant approved treatment, when available.

The development of safe and effective digital therapeutic options will be important to guide research and to add to the armamentarium of interventions. Digital therapeutics offer unique treatment options and the opportunity to deliver interventions with fidelity and state-of-the-art practices. This seminal book offers a comprehensive overview of the field of digital therapeutics to help guide (1) researchers and graduate students in the fields of biomedical engineering, health informatics, biomedical data science, software engineering, and computer science, (2) persons in industry within the field of digital health, as well as in the business of healthcare, and (3) policy makers and persons who study digital mental health from the legal and/or humanities fields. These areas of expertise and related fields will be important to coalescence to deliver highly efficacious and sustainable digital therapeutic treatment options to those who need them.

Will M. Aklin, Ph.D.
Director, Behavioral Therapy Development Program
Division of Therapeutics and Medical Consequences
National Institute on Drug Abuse (NIDA), National Institutes of Health (NIH)

Chapter 1

Introduction: A vision for the field of digital therapeutics

Nicholas C. Jacobson [a,b,c,d], Tobias Kowatsch [e,f,g] and Lisa A. Marsch [a,b,c]

[a] *Center for Technology and Behavioral Health, Geisel School of Medicine at Dartmouth College, Lebanon, New Hampshire, United States,* [b] *Department of Biomedical Data Science, Geisel School of Medicine at Dartmouth College, Lebanon, New Hampshire, United States,* [c] *Department of Psychiatry, Geisel School of Medicine at Dartmouth College, Lebanon, New Hampshire, United States,* [d] *Department of Computer Science, Dartmouth College, Hanover, New Hampshire, United States,* [e] *Centre for Digital Health Interventions, Department of Management, Technology, and Economics, ETH Zurich, Zürich, Switzerland,* [f] *Centre for Digital Health Interventions, Institute of Technology Management, University of St. Gallen, St. Gallen, Switzerland,* [g] *Future Health Technologies Programme, Campus for Research Excellence and Technological Enterprise, Singapore-ETH Centre, Singapore, Singapore*

1.1 Why read this book?

The purpose of this book is to introduce the topic of digital therapeutics to a broad audience. Traditional treatments for mental health and addiction have not been capable of scaling to meet the large number of persons who need or seek treatment. Digital therapeutics—software designed to prevent, treat, or manage pathology—may help to close a longstanding access to treatment barrier, providing scalable, low-cost interventions to a broad audience. Within the past decade, there has been a massive growth of digital therapeutics, but there is limited integration of the literature on the current state of the science. Mental health clinicians and researchers alike have difficulty in keep abreast of the advancement of these topics, including evidence for direct-to-consumer standalone digital therapeutics, the effectiveness of integrating digital treatments within traditional healthcare settings, as well as recent innovations which are currently and will continue to transform the field of digital therapeutics toward therapeutic models which are more personalized, adaptable, and engaging. The goal of this book is to introduce these topics, discuss cutting-edge advancements, and offer important considerations for implementation of digital therapeutics. The editors believe that this combination of features has not been offered in any other books to date.

1.2 Who is this book for?

The intended audience of this book includes: (1) researchers and graduate students in the fields of biomedical engineering, health informatics, biomedical data science, software engineering, and computer science, (2) persons in industry within the field of digital health, as well as in the business of healthcare, and (3) policy makers and persons who study digital mental health from the legal and/or humanities fields. We have designed this book as a seminal text introducing persons without a background in mental health and substance use to the problems faced within the fields of mental health and addiction, and how digital therapeutics may be leveraged to help treat these persons and fill gaps within the traditional care system. Although the book is tailored for audiences with broad technical backgrounds, this book could also be useful for readers with more mental health backgrounds (e.g., clinical psychology, counseling psychology, counseling, psychiatry, substance use counseling, and social work programs).

1.3 What topics are covered in this book?

First, we introduce a comprehensive overview of the field of digital therapeutics and research on their efficacy, effectiveness, scalability, and cost-effectiveness. *Second*, we introduce the novel directions in which digital therapeutics are currently being extended, including personalized interventions delivered in real-time. *Third*, we review important considerations

Digital Therapeutics for Mental Health and Addiction: The State of the Science and Vision for the Future. DOI: https://doi.org/10.1016/B978-0-323-90045-4.00008-3

surrounding digital therapeutics, including how they can be monetized and scaled, ethical issues, cultural adaptations, privacy and security concerns, and potential pitfalls.

1.3.1 First section: Introduction to digital therapeutics

The next chapter of this book (Chapter 2: Using digital therapeutics to target gaps and failures in traditional mental health and addiction treatments) provides an overview of the current gaps and failings of the traditional mental health and addiction care system, including how and why most persons needing mental health treatment do not receive it. The chapter then introduces and defines digital therapeutics, and discusses how digital therapeutics can be used to help remedy gaps in the traditional care system.

After introducing the problems faced within the traditional care systems, the subsequent chapter discusses the first wave of scalable digital therapeutics (Chapter 3: First wave of scalable digital therapeutics: internet-based programs for direct-to-consumer standalone care for mental health and addiction). This chapter introduces the first major wave of evidence-based scalable treatments for the masses, focusing on the use of internet-based treatments delivered directly to consumers (outside of healthcare systems). This chapter both introduces the seminal advances made by pioneers within digital therapeutics, as well as reviews the current evidence base.

After discussing the first wave of digital therapeutics, the following chapter (Chapter 4: Second wave of scalable digital therapeutics: mental health and addiction treatment apps for direct-to-consumer standalone care) discusses the emergence of smartphone applications that are used for symptom monitoring and standalone interventions and made available direct-to-consumers, highlighting how these tools are differentiated from web-based counterparts, and reviews their current evidence base.

With Chapters 3 and 4 most principally devoted to standalone care, the next chapter (Chapter 5: Blending digital therapeutics with traditional treatments within the healthcare system) reviews the integration of digital therapeutics and traditional treatments in order to extend and augment traditional treatments for mental health and addiction.

1.3.2 Second section: The new frontier

The second section of this book will discuss how digital therapeutics can leverage recent advancements in technology for more effective interventions. The first chapter of this section (Chapter 6: Receptivity to mobile health interventions) will introduce the importance of engagement in and receptivity toward mobile health interventions. Relevant receptivity studies are reviewed after describing an "ideal" mobile health intervention and the specific processes involved in receptivity, that is, receiving, processing, and using support. Then, pertinent factors that seem helpful for detecting and predicting receptive states are summarized. The chapter concludes with various technical challenges intervention authors face when developing receptive-capable mobile health interventions.

The following chapter (Chapter 7: Analytics for adapting interventions to context targeting vulnerability and receptivity) will discuss the design of effective just-in-time adaptive interventions (JITAIs). JITAIs leverage intensive longitudinal data from a target person and combine it with context information to deliver the most appropriate support at opportune moments. The authors clarify the definition and operationalization of vulnerability and receptivity in JITAIs, and offer specific research questions to support intervention authors in identifying the most suitable study designs such as observational studies or microrandomized trials.

Therapeutic alliance is an important relationship quality between patients and health care providers that is robustly linked to treatment success. Chapter 8 (Digital therapeutic alliance) will explain how this therapeutic relationship can also be developed between an individual and a given digital therapeutic. In particular, the authors discuss therapeutic alliance in digital mental health interventions that are either supported by humans or unsupported, that is, in purely technological interventions. The chapter also discusses how to measure digital therapeutic alliance, its impact on outcomes in digital mental health interventions, and what intervention authors can do to foster therapeutic alliance with digital therapeutics.

Chapter 9 (Conversational agents on smartphones and the web) will then introduce conversational agents and how these computer programs that imitate communication with human beings can be used for health service delivery. In particular, the authors discuss conversational agents deployed over the web and on smartphones that offer scalable interventions for mental health and addiction. Affordances of web- and smartphone-based conversational agents, such as low cost, reach or availability, as well as safety concerns, or benefits of conversational agents, are described, too. For example, individuals may feel less stigma when conversing about mental health or addiction with conversational agents, in contrast to human health coaches.

In complementing Chapter 9, the authors of Chapter 10 (Voice-based conversational agents for sensing and support: examples from academia and industry) will detail the rise of voice assistants as potential outlets for care delivery in ways that are both intuitive and engaging. The chapter will also discuss the potential of acoustic data streams that can be used by such technology for the detection of vulnerable states. Moreover, the authors present various examples of voice assistants from both academia and industry, and outline potentials for mental health and addiction treatment. The chapter concludes with a discussion on various challenges developers of voice assistants face, such as using vocal biomarkers via automated speech analysis or concerns around data and conversation privacy.

1.3.3   Third section: Structural considerations

Chapter 11 starts the section of this book focused on the broad array of structural considerations in the field of digital therapeutics. This chapter (Chapter 11: Design considerations for preparation, optimization, evaluation, and maintaining digital therapeutics) introduces a framework designed to guide developers of digital therapeutics across their lifespan, including their formative development, optimization, evaluation, implementation, and maintenance. This framework may support the development of maximally effective, engaging, implementable, and sustainable digital therapeutics.

Chapter 12 (Chapter 12: Cultural adaptations of digital therapeutics) then embraces the important topic of adapting digital therapeutics for diverse cultures, languages, and contexts. The chapter provides an overview of frameworks for guiding cultural adaptation of evidence-based interventions, the evidence supporting such as adaptation, and recommendations for advancing this field.

The following chapter (Chapter 13: Implementation, business models, and regulation) provides an overview of the progression of the digital therapeutics industry. This chapter reviews business models of digital therapeutics, including how they are being commercialized, scaled, and regulated across the world.

Next, Chapter 14 (Chapter 14: Lessons learned and potential pitfalls) reviews the major lessons that have emerged from the past decade of research and development of digital therapeutics as applied to mental health and substance use. It additionally reviews challenges and suggests strategies to increase utilization, engagement, and accessibility of clinically validated digital therapeutics worldwide.

The following chapter (Chapter 15: Privacy and security) discusses how the growth in digital therapeutics presents an unprecedented and important need to concurrently address the privacy and security vulnerabilities inherent to some of these interventions. This chapter reviews common risks related to patient data in digital therapeutics, suggests strategies for navigating the relevant regulatory landscape, and discusses methods for evaluating and addressing security and privacy considerations with digital therapeutics.

Chapter 16 (Ethical considerations of digital therapeutics for mental health) introduces the ethical challenges faced by researchers and clinicians utilizing digital therapeutics, including issues related to transparency, autonomy, fairness, and quality. The chapter provides an overview of the role health care providers can play in navigating these ethical considerations.

The final chapter of this book (Chapter 17: A look forward to digital therapeutics in 2040 and how clinicians and institutions get there) presents a vision for how digital therapeutics may evolve and transform over the next 10–20 years, discussing how the current tools will adjust as new technological innovations mature and advance.

1.4   What will readers learn?

After reading this book, readers will have a strong foundation to understand both the current utility of digital therapeutics in addressing systemic inadequacies in the current care system for mental health and substance use. Additionally, readers will understand the current evidence base of digital therapeutics, both as standalone interventions as well as interventions that are integrated within the care system. Readers will also be introduced to recent and ongoing advances in the field of digital therapeutics, including how therapeutics can be dynamically adapted to place and time to provide care in the moment when persons are most likely to benefit and in their times of greatest need, and the connections formed between persons and their digital therapeutics. Readers will learn about digital therapeutics that are developed through linguistic and verbal interactions mimicking conversational patterns across a series of devices. Readers will understand study designs used to test and optimize digital therapeutics, tailoring digital therapeutics to integrate cultural differences, and potential areas of concern and pitfalls including privacy, security, and ethical considerations. Lastly, readers will be introduced to the marked growth within the field of digital therapeutics and how they are projected to evolve over time and fundamentally improve models of mental health and substance use care worldwide.

Chapter 2

Using digital therapeutics to target gaps and failures in traditional mental health and addiction treatments

Nicholas C. Jacobson [a,b,c,d], Rachel E. Quist [a], Camilla M. Lee [a] and Lisa A. Marsch [a,b,c]

[a] Center for Technology and Behavioral Health, Geisel School of Medicine at Dartmouth College, Lebanon, New Hampshire, United States, [b] Department of Biomedical Data Science, Geisel School of Medicine at Dartmouth College, Lebanon, New Hampshire, United States, [c] Department of Psychiatry, Geisel School of Medicine at Dartmouth College, Lebanon, New Hampshire, United States, [d] Department of Computer Science, Dartmouth College, Hanover, New Hampshire, United States

2.1 Prevalence and impact of mental health and substance use disorders

Globally, one in five people suffer from a mood, anxiety, or substance abuse disorder (Steel et al., 2014) and one in seven children meet criteria for a mental disorder (Ferrari et al., 2022). These rates are higher in certain populations: lesbian, gay, bisexual, and transgender (LGBT) people are more than twice as likely to have a mental health disorder in their lifetime (Semlyen, King, Varney, & Hagger-Johnson, 2016) and about a third LGBT youth meet criteria for a disorder (Mustanski, Garofalo, & Emerson, 2010). Income inequality also factors into mental health: the likelihood of having a mental health issue is twice as high among low-income people as high-income people (Patel, Araya, de Lima, Ludermir, & Todd, 1999). The first onset for any mental health disorder begins in early childhood or adolescence; half of all lifetime cases start by age 14 and three quarters by age 24 (Kessler et al., 2005). Up to 79% of people experiencing chronic illness will experience a mental health disorder in their lifetime (Solano, Gomes, & Higginson, 2006). Furthermore, an estimated 8.9 million people in the United States have a comorbid mental health and substance use disorder (Substance Abuse and Mental Health Services Administration [SAMHSA], 2016). Therefore, mental health disorders are a significant and pertinent problem worldwide and some groups have even higher rates of mental health issues.

The impact of these disorders worldwide is significant in many ways. In the United States, estimates for national spending on outpatient treatment for depression alone are US$ 17 billion (Hockenberry, Joski, Yarbrough, & Druss, 2019), with an estimated 200 million days lost from work each year (Conti & Burton, 1994). Lost productive time among US workers costs US$44 billion to employers every year, in contrast to US$13 billion lost from peers without depression (Stewart et al., 2003). In 2010, mental health and substance use disorders accounted for 183.9 million Disability Adjusted Life Years (DALYs), or 7.4% of the worldwide rate (Whiteford et al., 2013). This measure calculates the global burden of disease by combining years of life lost due to premature mortality with years of healthy life lost due to disability. However, the actual number may be closer to 13% of all DALYs worldwide; on average, mental health disorders contributed to a 6.8-day decrease in healthy days per month (Vigo, Thornicroft, & Atun, 2016).

In line with loss of healthy life, those with mental health disorders ranked their health-related quality of life (HRQoL) lower than those without mental health disorders (Spitzer et al., 1995). The HRQoL for those with mental health disorders was also significantly lower than those with other chronic conditions, such as back pain, diabetes, or hypertension (Cook & Harman, 2008). Additionally, chronic mental health disorders such as social anxiety and dysthymia have been associated

Digital Therapeutics for Mental Health and Addiction: The State of the Science and Vision for the Future. DOI: https://doi.org/10.1016/B978-0-323-90045-4.00005-8

with worse HRQoL than episodic disorders (Saarni et al., 2007). Mortality rates in general are higher among people with mental health disorders, resulting in a median 10-year decrease in life expectancy (Walker, McGee, & Druss, 2015). Worldwide, substance use accounts for 4% of all deaths and 5.4% of the global burden of disease (World Health Organization, 2010). Therefore, the global prevalence of mental health disorders, including substance use disorders, is an economic and global health burden.

2.2   Lack of treatment receipt in the traditional care system

Despite the high prevalence of these disorders and the related human and economic outcomes, only an estimated 41% of adults and 45% of adolescents worldwide have received any form of care for their mental health disorders, and a substantial subset of these people do not receive minimally adequate care for their disorders, where minimally adequate mental health care is defined as at least 2 months of treatment with an appropriate medication as well as at least four visits with any physician or, without medication, at least eight visits with a mental health professional. Lack of access to treatment is associated with demographic factors such as race, gender, and socioeconomic status (Wang, Demler, & Kessler, 2002). These demographic factors can interfere with protective factors against developing mental health issues, such as having autonomy when responding to severe life events, having access to resources that allow choices when confronting severe life events, and having support from family, friends, and health providers when confronting severe life events (World Health Organization, 2004).

One study in California found that men, Latinos, Asians, young people, older adults, people with less education, uninsured adults, and individuals with limited English proficiency were significantly more likely to have unmet need (Tran & Ponce, 2017). In the United States, Latino and African-American parents are more likely to seek familial or community-based help than professional help for children with mental health issues (McMiller & Weisz, 1996). People living in rural populations are less likely to seek out treatment than nonrural populations (Crumb, Mingo, & Crowe, 2019) despite the fact that rural populations experience mental illness at equal or greater rates than nonrural populations, especially when it comes to women (Hauenstein, Boyd, Submission, & H., 1994). LGBT individuals are more likely than their non-LGBT counterparts to have unmet need (Whaibeh, Mahmoud, & Vogt, 2020). Worldwide, both attitudinal and structural factors are barriers to receiving care for mental health issues, though some research has indicated that treatment for mild or moderate mental illness can be more impacted by attitudinal behaviors while serious mental illness may be more impacted by structural barriers (Andrade et al., 2014). Low perceived need for treatment was listed as a common reason for respondents in the National Comorbidity Survey Replication study for not seeking treatment, with attitudinal reasons being more common than structural barriers among those with perceived need (Mojtabai et al., 2011). Serious mental illness is defined as a mental illness diagnosis that typically leads to extensive inpatient and outpatient treatment and results in significant disability in one or more major life domains (Parabiaghi, Bonetto, Ruggeri, Lasalvia, & Leese, 2006).

2.3   Barriers to traditional treatment: stigma, personal beliefs, and cultural competence

Despite the fact that up to 57% of individuals around the world report that they are in daily contact with an individual who they think suffers from a mental illness, including addiction (Seeman, Tang, Brown, & Ing, 2016), one of the biggest attitudinal barriers to care is the stigma surrounding treatment for mental health issues. Cultures across the world have relatively stable negative prejudices toward people with mental illness over the past 20 years despite efforts to publicize a model of mental illness as a biological, medical disease (Schomerus et al., 2012). In LGBT populations, stigma surrounding mental illness can intersect with stigma surrounding sexual or gender identity, which leads to discrimination and internalized shame, making access to care even more difficult (Wainberg et al., 2017). Men and non-White individuals are more likely to report stigmatized attitudes toward people with mental illness or substance use disorders; however, increased reported stigma by people of color may be confounded by socioeconomic status and discrimination due to identity (Corrigan & Watson, 2007).

Individuals with substance use disorders face higher levels of stigma and discrimination than those with other psychiatric disorders. Only 11% of people with substance use disorders received professional help in the past year (Substance Abuse and Mental Health Services Administration [SAMHSA], 2019). In the public eye, individuals with substance use disorders are likely to be seen as dangerous and unpredictable, unable to make decisions about treatment or finances, and to be blamed for their own condition (Yang, Wong, Grivel, & Hasin, 2017). Even healthcare professionals have a more negative attitude toward individuals with substance use disorders than individuals with other mental health issues (van Boekel, Brouwers, van Weeghel, & Garretsen, 2013).

In addition to stigma from family, friends, and society, stigma can also stem from personal beliefs of an individual with mental illness. Despite the finding that religious participation can be a protective factor against mental illness (Levin & Chatters, 1998), mental illness can be seen by both the individual and their family and friends as a lack of religious connection (Crumb et al., 2019). Worldwide, low perceived need is the most common reason for not initiating treatment among those with mild or moderate cases. A desire to handle the problem on one's own was the most common reason for not seeking treatment among those who perceived a need for treatment (Andrade et al., 2014). Another reason is due to perceived inadequacy of treatment. Of individuals who received treatment for their serious mental illness, only 38.9% received treatment that was "minimally adequate" as defined earlier in this chapter (Wang et al., 2002). Similarly, a large number of individuals with depression received inadequate treatment for their depression; those who had recently attempted suicide received equally inadequate care as those who had not (Oquendo, Malone, Ellis, Sackeim, & Mann, 1999). Worldwide, the most common reason for dropping out of mental health treatment is perceived ineffectiveness of care (Andrade et al., 2014).

2.4 Barriers to traditional treatment: cultural competence

The perceived cultural competency of a mental health provider is also a barrier to treatment for some groups. Cultural competency is defined as "providing health care that meets the needs of a population diverse in gender, race, ethnicity, sexual orientation, age, religion, [dis]ability, language, national origin, immigration status and socioeconomic status" (Bassey & Melluish, 2013). Cross-cultural attitudinal barriers to mental health treatment include cognitive barriers, affective barriers, and value orientation barriers, and effective practitioners must be able to understand how these barriers relate to their patient (Leong & Kalibatseva, 2011). A lack of understanding in these areas can lead to underutilization or low retention for mental health services (Whaibeh et al., 2020).

Even when an individual with mental illness overcomes attitudinal barriers to seek out treatment, there are many structural barriers that can limit access to care. One of the most frequently cited structural barriers to care is ability to pay (Andrade et al., 2014). Like most areas of medical care, assessment and treatment for mental illness is expensive. In 2006, inpatient treatment—often required for more severe mental illness cases and for substance use disorders—ranged from $766 per day to over $1000 per day ($1060 and $1833 respectively when adjusted for inflation) (Stensland, Watson, & Grazier, 2012) while average household income in 2006 was $48,450 (Statistical Abstract of the United States, 2006). In 2013, the average annual cost for managing depression in the United States ranged from $8662 to $16,375 depending on the severity of symptoms (Chow et al., 2019) and the average annual cost for managing schizophrenia hovered around $21,672 (Fitch, Iwasaki, & Villa, 2014). For comparison, the average household income in 2013 was $52,250 (Noss, 2012). This means that simply managing the care for mental illness could cost over 40% of the average household's income without insurance.

2.5 Barriers to traditional treatment: high cost and lack of insurance coverage

Because of the financial costs, in most developed or developing countries insurance coverage is one of the most significant deciding factors for whether or not an individual can receive care (Mechanic, 2002). In the United States, 28 million people did not have any insurance and millions more were uninsured for part of the year (Keisler-Starkey & Bunch, 2021). A 2010 comparison of physicians and psychiatrists revealed that about 90% of physicians accept private insurance in comparison to only 55% of psychiatrists (Bishop, Press, Keyhani, & Pincus, 2014). This is largely due to how managed care determines reimbursement rates; practitioners are underpaid when they accept insurance as payment in comparison to accepting out-of-pocket fees. Therapists are also disincentivized by the limitations imposed by managed care, which include decreased flexibility for clinical judgment, restrictions on who can receive treatment, and increased pressure to make referrals for prescriptions/medication (Cohen, Marecek, & Gillham, 2006). Additionally, adults with severe mental illness have the lowest rates of receiving care, while adults with public insurance have the highest rates (Rowan, McAlpine, & Blewett, 2013). Thirty-seven percent of adults with severe mental illness lack insurance for at least part of the year, compared to 28% of adults without a severe mental illness. Out-of-pocket expenses and premiums exceed 5% of income for at least 2 years for 40% of adults with at least 3 chronic mental health conditions and 20% of adults with one chronic condition (Cunningham, 2009).

Even when individuals are willing to seek out care and have insurance to pay for it, government-sponsored health insurance plans are often lacking in their mental health coverage. For example, in the United States, Medicare recipients' restriction of access to care depends on availability of providers who will accept Medicare. About 48.6% of rural counselors referred an existing client to a different counselor due to Medicare ineligibility. These mid-treatment referrals disrupt the therapeutic alliance and contribute to the problem of limited access and long waitlists for treatment (Fullen, Brossoie, Dolbin-MacNab, Lawson, & Wiley, 2020). Finding in-network practitioners is easier with private insurance, but still challenging. Some private insurance plans have limits on the amount of inpatient days they will cover, which can be detrimental to those with severe mental or substance use disorders that need inpatient care (Cohen et al., 2006).

Lack of health insurance coverage is not specific to the United States; while nearly 50 countries have attained universal coverage, many other countries have insurance rates close to 0. In Africa, out-of-pocket health expenses account for nearly 45% of total health care expenditures. In Asia, countries like Korea and China have nearly universal coverage, while India's coverage level is under 10% of the population. More well off Latin American countries like Chile and Costa Rica have almost universal coverage, while poorer countries like Colombia have coverage rates around 30%. Europe boasts the highest number of countries with universal coverage, but effective access to the system remains an issue for many countries in the region (International Labour Office, 2008).

2.6 Barriers to traditional treatment: transportation and appointment time availability

Another structural barrier to accessing mental health treatment is whether or not the individual seeking care has the ability to access the provider. Lack of available transportation has long been established as a significant barrier to mental health and substance use care (Sommers, 1989). This lack of access is exacerbated by rural living situations; Australian people living in rural regions were nearly 20 times more likely to delay seeking treatment by over 1 year in comparison to those living in urban locations (Green, Hunt, & Stain, 2012). When transportation is provided free of charge, distance to care does not decrease utilization of mental health treatment programs (Whetten et al., 2006).

Frequently, the lack of public transportation in rural areas couples with the fact that driving distances tend to be much longer. Patients might not be able to drive themselves to receive mental health care as there is often a comorbidity between mental health conditions and physical health conditions that would interfere with and could even prevent a patient's ability to drive. The prevalence of comorbid mental and physical conditions has increased significantly within the past decades (Sartorious, 2013). Comorbid diseases are also rapidly increasing at younger ages. Although comorbid diseases are often overlooked, it is recommended that psychiatrists and other clinicians promote the idea that all conditions have both psychological and somatic components (Sartorious, 2013). Mental health disorders are a risk factor for developing a chronic condition and vice versa. The combination is associated with higher costs of treatment, decreased quality of life, increased symptom burden (Sartorius, 2018).

Additionally, the location of appointments and the times appointments are available may interfere with the patient's responsibilities. Many mental health care workers have hours that fall within the working day, which may require clients to take off work, find childcare, or sacrifice family time (Harvey & Gumport, 2015). The inconvenience and inability to get an appointment are additional structural barriers that prevent access (Sareen et al., 2007). A study that surveyed respondents in the National Comorbidity Survey Replication with common 1-year DSM-IV disorders revealed that those that are married or cohabiting with a partner report higher levels of structural barriers, but only among milder mental health cases. This finding may reflect how marriage correlates with increased family responsibilities and consequently, higher demands on time and financial resources (Mojtabai et al., 2011). The barrier to seeking treatment might only be overcome when cases are more serious.

2.7 Barriers to traditional treatment: inadequate number of mental health providers

One in five people means that of the 5.2 billion adults currently alive in the world, approximately 1.3 billion are suffering from a mental health disorder, and 762 million have not received any form of care. The previously discussed attitudinal and structural barriers to receiving care are responsible in many ways for the number of people with untreated mental illness. However, even if all the problems surrounding insurance, access, and stigma were solved, there would still be millions of people without adequate mental health care because the number of providers of mental health care is too low. On average across the globe, there are 10.7 mental health workers per 100,000 people (Morris, Lora, McBain, & Saxena, 2012). If 20,000 (or one in five) of those people have a mental health disorder, each mental health worker would need to treat 1869 patients each year. Although this is clearly not plausible, a few calculations will determine just how implausible it is.

Let's say a psychologist works a typical 40-hour week and spends all of their time in-session with clients for 45 minutes at a time. At the end of 1 week, this person would be able to help about 53 people. It is important to note that this number is not plausible for most providers. In fact, most psychotherapists average around 25–30 clients in their caseload because they are not solely practicing therapy; psychotherapists also manage clinical notes, treatment reports, telephone calls, and billing among other responsibilities. Utilizing a large dataset ($N = 18,322$) from a counseling center, the average client caseload (referring to a rolling estimate of unique individual clients seen by a clinician in the past 30 days) in 2016 was determined to be 26.45, the highest the average client caseload seen between 2009 and 2017 (Bailey, Erekson, Goates-Jones, Andes, & Snell, 2021). The average number of hours of direct client interaction was found to be 22 hours in 2008 (Michalski, Mulvey, & Kohout, 2008).

However, let's imagine a scenario involving the maximum client caseload of a therapist at 53 for now. Let's assume that this person is incredibly good at their job, and therefore each client requires only ten sessions before they no longer need treatment. This psychologist then helps 53 individuals every 10 weeks. In 1 year, a single professional working as hard as they can for 50 out of the 52 weeks of the year can help approximately 265 patients—or just under 12% of the estimated number of patients they would need to serve in the current system in order to meet the deficit of care. These statistics are all estimates, and therefore the numbers derived here are not exact. However, the case presented here is a sobering example of the challenge of meeting the demands of global mental health care.

Both the length of treatment and suggested caseload for the mental health practitioner mentioned above are quite unrealistic. For instance, the National Institute for Health and Care Excellence (NICE) guidelines recommend 16–20 cognitive-behavioral therapy (CBT) sessions for those struggling with depression. Additionally, reviews of the length of therapy approximate that half of patients recover after an average of 15–20 sessions. There are a specific number of treatments of 12–16 weeks in duration that lead to clinically significant improvements. Therapists and patients sometimes prefer to extend sessions over a 6-month period to maintain goals or increase remission. Comorbid conditions and personality disorders may require 12–18 months of treatment (American Psychological Association, Division 12, 2017). As far as caseload, even meeting with 20–35 individual clients in a week may be too much for a single mental health practitioner. In one survey of 29 directors of community mental health centers in a rural area, two thirds reported feelings of high emotional exhaustion and low personal accomplishment (Rohland, 2000) and 54% of community mental health workers in northern California reported feelings of high emotional exhaustion (Webster & Hackett, 1999). These feelings of exhaustion and low accomplishment correlate with worse interactions between practitioners and patients, including negative attitudes toward patients in mental health treatment facilities (Holmqvist & Jeanneau, 2006). Altogether, this research suggests that despite the sobering lack of providers to handle the need we present above, the estimates are likely to be far worse as they present unrealistic overoptimistic estimates about both caseload, therapeutic efficacy, and therapeutic efficiency.

Even for those who are eventually able to access care, help often comes after a significant delay from providers. On average, the average wait time to an initial assessment can be up to 110 days, depending on the presenting problem (Kowalewski, McLennan, & McGrath, 2011). The extensive waiting times for care leads to an increased no-show rate and an increase in client symptom severity, including an increased risk of suicide (DiMino & Blau, 2012; Williams, Latta, & Conversano, 2008). Waiting times are also linked to a reduced likelihood of responding to treatment once it is eventually started. Clinically significant and reliable patient outcome deterioration is amplified by increased waiting times, and effects are significant between 3 and 12 months of waiting time (Reichert & Jacobs, 2018).

The lack of access to mental health care is not due to underemployment of providers. In fact, in the United States, psychologists and mental health counselors have an unemployment rate of 1.5%, significantly below the national unemployment rate of 5.2% (Bureau of Labor Statistics, 2021). There have been several suggestions for how to close the gap between mental health disorders and providers of mental health care. Perhaps the most obvious and most frequently suggested solution is to train more mental health care providers, which is the main focus of the Improving Access to Psychological Treatments program being implemented across England. However, because of the sheer number of providers needed to adequately address the gap in mental health care, increasing the number of mental health care workers would require a vast advertisement, incentivization, and training effort that would likely take several generations to implement and take effect.

Some solutions have argued for training programs that are offered online, making them more accessible for those interested in working in mental health (Fairburn & Patel, 2014). While an internet-based training program is more scalable than a traditional training program, it still requires an increased interest in employment in mental health fields. An increase of the necessary magnitude would require a massive incentivization program, as mental health fields are often rejected as potential career fields due to cost, years of schooling, or high workload and low compensation compared to all other areas of medicine (Dial, Haviland, & Pincus, 2005).

Even if it was feasible to sufficiently increase the number of mental health workers worldwide, there are other significant barriers which block individuals from receiving the care they need. Some of these barriers are personal. Seeking out care takes a high level of initiative and requires clients to come to terms with uncomfortable ideas, such as readiness to change and admit they could benefit from external help. In fact, referring individuals to care too early can actually have adverse effects, including increased risk of active suicidal ideation (Jacobson et al., 2020). Often, people with mental health disorders do not seek out treatment because they believe their problems are not severe enough or believe they can handle the problems on their own (Andrade et al., 2014). On the other hand, those with severe mental health disorders are also unlikely to receive in-person treatment. For instance, people with agoraphobia are unlikely to leave their homes, and people with severe depression are unlikely to get out of bed (American Psychiatric Association, 2013).

The effects of delayed or inaccessible mental health care are long-term and significant for both adolescents and adults. In adolescents, behavioral disorders such as attention deficit-hyperactivity disorder (ADHD) are treated more frequently than other disorders, such as anxiety. About 72% of children with ADHD receive some form of treatment, in contrast with 41% of children with anxiety (Wang et al., 2005). There are a wide variety of personal risks that can stem from delayed or nonexistent access to care for adolescents. Many of these are long-term consequences that can persist into adulthood, such as sleep issues, academic difficulties, and an increase in delinquent behavior (Jolliffe et al., 2019; Mazzone et al., 2007). There is also evidence that a lack of care for one mental health disorder can lead to comorbidity with other disorders, such as untreated anxiety leading to depression or vice versa (Jacobson & Newman, 2017). In addition, a lack of access to care as an adolescent may decrease the likelihood that the same person will seek care as an adult (Sheppard, Deane, & Ciarrochi, 2018). Perhaps most concerningly, unaddressed mental health problems in adolescents are directly linked to increased rates of suicide (Kessler et al., 2005), even at subthreshold levels of depression and anxiety (Balázs et al., 2013). Globally, suicide is the fourth-leading cause of death among people aged 15–19.

One of the most significant areas impacted by mental health disorders for adults is employment. 14% of years lived with disability globally are known to be caused by depression, anxiety and other mental illnesses (James et al., 2018)). In addition to disability, untreated mental illness contributes to higher rates of unemployment for adults, and this unemployment in turn contributes to more severe psychiatric symptoms and cognitive functioning (McGurk & Mueser, 2003; Yoo et al., 2016). Unemployment leads to poverty, which in turn leads to a higher likelihood for mental health disorders. Globally, those with the lowest incomes in a community suffer 1.5–3 times more frequently from depression, anxiety, and other common mental illnesses than those with the highest incomes (Lund et al., 2010). Untreated mental health also leads to difficulty in interpersonal relationships and an increase in suicide (Davidson, Wingate, Grant, Judah, & Mills, 2011; Huh, Kim, Yu, & Chae, 2014). Approximately 500,000 suicides annually are documented worldwide, or 1.4% of premature deaths; underreporting of suicides may range anywhere from 20% to 100% of the reported total due to religious beliefs, stigma, or legislation (Bachmann, 2018).

2.8   Barriers to traditional treatment: patient narratives

To convey the levels of difficulty in finding treatment, narratives of three different people trying to receive mental health care are provided as follows:

John is upper-class with insurance and disposable income. He can afford to pay out-pocket fees for therapy. Because he does not have to worry about in-network care, his primary challenge is finding a therapist that can take on a new patient. He may need to call several providers before one can take him, and/or he may have to be put onto a wait-list before he can receive help. Practices frequently turn patients away as they are already so full. However, he will likely be able to begin therapy within a few weeks, as he has a wider range of mental health providers he can access.

Mary is middle-class with insurance. She has the same challenge as John in terms of finding an available therapist, but she is further limited since out-of-pocket fees are not feasible. She must find a therapist who not only accepts insurance, but accepts her particular brand of insurance. The therapists in her area who are in-network have longer waitlists than the therapists John has access to, and Mary may have to decide if it is worth it for her to travel further in order to access an in-network therapist. While Mary would prefer in-network treatment, she does have several other options, including copays (the client pays a portion of the fees), reimbursement for out-of-network care (insurance pays a portion), or sliding scale fees (fees based on client income).

Henry is low-income and is insured through government-sponsored insurance. As Henry cannot afford the high cost of therapy, receiving treatment is nearly impossible. Henry must look into other options, such as crisis lines and community support groups. The nearest community support group is 15 miles from Henry's house, and there is no direct access to the group by the bus, Henry's primary mode of transportation. Because of all the roadblocks he encounters in accessing care, Henry decides it is not worth it to continue trying. His condition will likely worsen as he is not receiving the care he needs.

2.9   Leveraging digital therapeutics to transform treatment models for mental health and substance use disorders

The confluence of factors reviewed above underscores the urgent need for solutions to scale-up access to effective treatments for mental health and substance use disorders. Digital therapeutics are a key part of the solution to tackling these challenges and are redefining the future of mental health care. Digital therapeutics are software used to prevent, treat, or manage a medical disorder or disease (Digital Therapeutics Alliance, 2022). They package an entire model of care into a unified, seamless digital delivery system.

Although digital therapeutics have been applied to many health domains, they have been especially transformative to the field of mental health and substance use. A large and growing literature (including many chapters in this book) demonstrate the robust and replicable clinical effectiveness of digital therapeutics for mental health and/or substance use (Deady et al., 2017). These include digitally delivered treatments such as cognitive behavioral therapy, contingency management, problem-solving therapy and behavior activation (Budney, Borodovsky, Marsch, & Lord, 2019; Deady et al., 2017; Naslund et al., 2017). And there are many features of these digital interventions that address the numerous challenges in accessing effective care endemic in our current models of mental health care delivery.

First, digital therapeutics offer anytime/anywhere mental health care. They are accessible 24/7 on mobile devices for on-demand access to therapeutic support. The majority of the world's population has a mobile device, and it is typically within arm's reach (Pew Research Center, 2019b). And global access to mobile devices continues to grow (BankMyCell, 2022) exponentially even in many traditionally underserved populations, including among persons with mental health and substance use disorders and in low and middle income countries (Collins et al., 2016; Naslund et al., 2017). There are now over 10 billion mobile subscriptions worldwide (BankMyCell, 2022). Given that there are more mobile phone subscriptions in the world than there are people in the world, digital therapeutics offer the potential to transform access to mental health care.

This also means digital therapeutics offer the ability to scale-up access to mental health care at a population level. This is exceptionally important given the large and growing mental health needs globally and the insufficient mental health care workforce to meet those needs. Digital therapeutics can thus extend the reach and the impact of the mental health workforce (as additional therapeutic resources offered to persons in care), and they can also function as stand-alone tools and provide care to individuals who may not otherwise have access. When digital therapeutics are offered in blended care models along with clinician-delivered care, they may make the therapeutic processes more efficient and effective. They may also free up providers' time to allow them to treat more individuals and reduce long waiting lists for treatment.

Moreover, given that digital therapeutics transcend geographic boundaries and provide on-demand therapeutic support, they can overcome the many transportation barriers associated with traveling to traditional treatment programs. Additionally, given that they do not require synchronous communication with a clinician, they overcome the many time constraints associated with participating in care that is offered only at specific appointment times. This benefit is particularly realized when digital therapeutics are offered as stand-alone tools and not along with clinician-delivered models of care.

Additionally, therapeutic support offered on digital platforms may be much less stigmatizing than the stigma associated with many of our existing models of mental health care (Muñoz, 2010). They can be utilized anonymously and in the privacy of one's own home. This may be particularly appealing to individuals when addressing sensitive topics such as substance use and other risk behavior.

Another appealing feature of digital therapeutics is that they often have a lower barrier to entry than traditional models of care which typically require extensive clinical assessments, specific appointment times, proper insurance coverage and long waitlist time. This may not only lead to more individuals seeking care for mental health problems, but it may also mean that they enter care at an earlier stage. Thus, individuals may be able to prevent escalation of a mental health challenge and/or reduce prolonged suffering from mental health problems. Additionally, because individuals may readily access a wide array of digital therapeutics, they may try out more than one to find the best digital therapeutic for them. This process of exploring and self-selecting a digital therapeutic may lead to higher levels of continued engagement with care, addressing the high dropout typically evident in care (Yardley et al., 2016).

Digital therapeutics are also well-poised to tackle the cost barrier that many people experience in traditional models of care. Although clinically validated digital therapeutics may have a notable up-front development and evaluation cost, the cost to deploy and maintain them is far less than the cost of clinician-delivered care in health care settings. This is not to suggest that digital therapeutics should replace traditional models of mental health care which are already in short supply but rather to underscore their utility in enhancing the scope of care. Their cost-effectiveness in the mental health care arena is increasingly being documented (Aspvall et al., 2021; Velez et al., 2021; Wang, Gellings Lowe, Jalali, & Murphy, 2021) and cost-effective care is a central hallmark of increasingly popular pay-for-performance or value-based reimbursement payer models of healthcare (Stewart, Lareef, Hadley, & Mandell, 2017). A recent study conducted by McKinsey in partnership with the German Managed Care Association estimated that up to €4.3 billion in potential healthcare savings could have been realized in 2018 if the German healthcare system had adopted all of the available digital solutions for patient self-treatment and patient self-care (Hehner, Biesdorf, & Möller, 2018).

Further, digital therapeutics ensure the delivery of mental health care with fidelity to the most evidence-based practices. This is significant given the widespread variability in the quality of mental health care (which often includes delivery of ineffective care) across the globe. And as the state of the science continues to evolve, best practices can be readily incorporated into the content and functionality of digital therapeutics and can accelerate the pace at which scientific innovations can be widely accessible and impact people's daily lives.

Additionally, digital therapeutics can provide ongoing and adaptive treatment that is responsive to individuals' changing clinical trajectories over time. This is clinically important given that individuals have distinct mental health and substance use experiences, risk factors, and treatment needs. And they may need therapeutic support at specific points in time and for differing durations of time. Digital therapeutics do not need to be static and always work the same way across individuals or even within an individual, but rather they can enable dynamic computational models to predict and respond to people's changing needs, goals, and clinical trajectories over time (Carpenter, Menictas, Nahum-Shani, Wetter, & Murphy, 2020; Goldstein et al., 2017; Nahum-Shani, Hekler, & Spruijt-Metz, 2015). Digital devices can measure individuals' clinical status and needs in real time (e.g., conduct passive, unobtrusive ecological sensing that provides continuous measurement of individuals' behavior and physiology, such as sleep, social interactions, physical activity, electrodermal activity, and/or cardiac activity or prompt individuals to respond to brief queries about their clinical status and functioning). These data can then inform personalized digital therapeutics that deliver therapeutic support in response to individuals needs and in the moment. These interventions, sometimes referred to as ecological momentary interventions (EMIs) or Just-in-Time Adaptive Interventions (JITAIs) are designed to be directly responsive to the health needs of an individual, such as when they are craving a substance of abuse or when they may be at risk of self-harm.

Further, digital therapeutics can provide culturally competent care. Indeed, they can be readily adapted for different populations, languages, and contexts to offer more engaging and relevant care to individuals (refer to Chapter 12 in this book on *Cultural adaptations of digital therapeutics*).

Digital therapeutics also offer great promise to extend beyond siloed models of care targeting a single health domain to address the co-occurring needs of individuals. For example, substance use disorders often co-occur with other mental health conditions such as anxiety or depression needs (e.g., almost 4% of US adults or 9.2 million individuals experience both a substance use and another mental health disorder) (Substance Abuse and Mental Health Services Administration [SAMHSA], 2016), but most clinicians who are trained to treat substance use disorder are often not also trained to treat other mental health conditions. Digital therapeutics can be transdiagnostic and can arguably embrace whatever combination of needs and goals an individual may have.

Importantly, digital therapeutics can also greatly reduce some of the striking disparities in treatment access and treatment quality evident in healthcare settings across the globe. For example, in the United States, it is well-documented that Black and Latino/a individuals with substance use disorders have greater barriers to accessing, completing, and having satisfactory experiences with treatment relative to their White counterparts (Marsh, Cao, Guerrero, & Shin, 2009; Mennis & Stahler, 2016). Indeed, Black Americans and Latino/a individuals are 69% and 75% as likely to complete a substance use treatment episode, respectively, as White Americans (Mennis & Stahler, 2016). And perceived stigma and lack of social support in substance use treatment models of care have been shown to impact Blacks more than Whites (Pinedo, Zemore, & Mulia, 2020). Additionally, rural regions have lower capacity for substance use treatment, including fewer clinicians and treatment settings, and individuals in rural communities have lengthier commutes to treatment, relative to those in urban and suburban regions.

Although White Americans (82%) are more likely to have a desktop or laptop computer at home compared to 58% of Blacks and 57% of Latino/a individuals, Black and Latino/a individuals are just as likely as Whites to have access to a smartphone and more likely than Whites to seek health information on their phones (Pew Research Center, 2019b). Individuals in the rural U.S. are also showing marked increases in access to mobile devices (Pew Research Center, 2019a). Thus, offering interventions on mobile platforms has great potential to overcome the disparities endemic to many traditional care models (Gibbons, 2008; Zhang et al., 2017, 2019). Digital therapeutics are also arguably more timely and significant than ever, given the marked shift to remote models of care as a result of the COVID-19 global pandemic (Guthrie, 2022).

2.10 A time of opportunity to transform models of mental health and substance use care via digital therapeutics

This is an exciting moment in time in the history of digital therapeutics. As reviewed in this chapter, digital therapeutics effectively tackle many of the numerous barriers to accessing effective mental health and substance use care across the globe. And there is now an unprecedented surge of interest and activity in embedding digital therapeutics into new models of health and healthcare.

Investment in the realm of digital therapeutics has exploded globally in recent years. The number of technology start-ups focused on digital therapeutics has grown tremendously in just the past 3 years, and large global pharmaceutical companies are also increasingly launching digital therapeutics initiatives (Hendrickson, 2019). A forecast by Insider Intelligence last year projected that the digital therapeutics market was expected to grow dramatically within the next 3–5 years to a projected market valuation of $9 billion in 2025 (up from a $3 billion valuation in 2019 (Burrone, Graham, & Bevan, 2022). However, strikingly, the long-term effects of the global COVID-19 pandemic as well as an increasing prevalence of mental health and chronic conditions have now led to a massively increased projected valuation of the digital therapeutics market to $56 billion by 2025 (Insider Intelligence, 2020). Among the growth in investment and demand for digital therapeutics, investment in digital therapeutics focused on expanding access to mental health and substance use disorder treatment resources has also been surging (Stotz & Zweig, 2018). And according to a recent international survey conducted by McKinsey and Company, more than 75% of all patients expect to use digital services (Park, Garcia-Palacios, Cohen, & Varga, 2019).

Additionally, a growing array of paths for deploying clinically validated digital therapeutics to individuals has emerged. Such paths range from prescriptions for digital therapeutics from qualified healthcare providers to activation codes for digital therapeutics supported by a private or governmental payor, employer, or pharmacy benefit manager. And, several countries support national coverage frameworks for digital therapeutics (see Chapter 11 in this book on *Building the digital therapeutics industry: regulation, evaluation, and implementation*).

Additionally, digital health research has been expanding in the academic community across a broad array of health domains. And funding for research on digital therapeutics has been markedly increasing. For example, the US National Institutes of Health (NIH) funded approximately 200–250 grants per year on digital treatments in the 2000s, but they now fund almost double that number of grants per year on digital treatments (particularly since smartphones have become nearly ubiquitous (Riley, Oh, Aklin, & Wolff-Hughes, 2019). It is noteworthy that this increase in funding of digital health grants occurred during a period when the total number of competitive awards from NIH remained relatively flat.

Finally, a growing number of industry, scientific, governmental, clinical, payor, and community stakeholders are engaging in collaborative activities to implement best practices in digital therapeutics. We are now poised to transform mental health and substance use care to create sustainable paths to digital therapeutics to improve individuals' lives across the globe.

References

American Psychiatric Association, A. P. A. (2013). Diagnostic and statistical manual of mental disorders (DSM-5). American Psychiatric Association, Washington, DC.

American Psychological Association, Division 12 (2017). How long will it take for treatment to work? Retrieved January 26, 2022, from https://www.apa.org/ptsd-guideline/patients-and-families/length-treatment.

Andrade, L. H., Alonso, J., Mneimneh, Z., Wells, J. E., Al-Hamzawi, A., Borges, G., et al. (2014). Barriers to mental health treatment: Results from the WHO World Mental Health surveys. *Psychological Medicine, 44*(6), 1303–1317. https://doi.org/10.1017/S0033291713001943.

Aspvall, K., Sampaio, F., Lenhard, F., Melin, K., Norlin, L., Serlachius, E., et al. (2021). Cost-effectiveness of internet-delivered vs in-person cognitive behavioral therapy for children and adolescents with obsessive-compulsive disorder. *JAMA Network Open, 4*(7), e2118516. https://doi.org/10.1001/jamanetworkopen.2021.18516.

Bachmann, S. (2018). Epidemiology of suicide and the psychiatric perspective. *International Journal of Environmental Research and Public Health, 15*(7), 1425. https://doi.org/10.3390/ijerph15071425.

Bailey, R. J., Erekson, D. M., Goates-Jones, M., Andes, R. M., & Snell, A. N. (2021). Busy therapists: Examining caseload as a potential factor in outcome. *Psychological Services, 18*(4), 574–583. https://doi.org/10.1037/ser0000462.

Balázs, J., Miklósi, M., Keresztény, Á., Hoven, C. W., Carli, V., Wasserman, C., et al. (2013). Adolescent subthreshold-depression and anxiety: Psychopathology, functional impairment and increased suicide risk. *Journal of Child Psychology and Psychiatry, 54*(6), 670–677. https://doi.org/10.1111/jcpp.12016.

BankMyCell (2022). How many people have smartphones worldwide. Retrieved January 27, 2022, from https://www.bankmycell.com/blog/how-many-phones-are-in-the-world.

Bassey, S., & Melluish, S. (2013). Cultural competency for mental health practitioners: A selective narrative review. *Counselling Psychology Quarterly, 26*(2), 151–173. https://doi.org/10.1080/09515070.2013.792995.

Bishop, T. F., Press, M. J., Keyhani, S., & Pincus, H. A. (2014). Acceptance of insurance by psychiatrists and the implications for access to mental health care. *JAMA Psychiatry, 71*(2), 176–181. https://doi.org/10.1001/jamapsychiatry.2013.2862.

Budney, A. J., Borodovsky, J. T., Marsch, L. A., & Lord, S. E. (2019). Technological innovations in addiction treatment. The assessment and treatment of addiction (pp. 75–90). Elsevier. https://doi.org/10.1016/B978-0-323-54856-4.00005-5.

Bureau of Labor Statistics (2021). Unemployment rate drops to 5.2 percent in August 2021: *The Economics Daily: U.S. Bureau of Labor Statistics*. Retrieved July 21, 2022, from https://www.bls.gov/opub/ted/2021/unemployment-rate-drops-to-5-2-percent-in-august-2021.htm.

Burrone, V., Graham, L., & Bevan, A. (2022). Digital therapeutics: past trends and future prospects. Evidera. Retrieved January 27, from https://www.evidera.com/digital-therapeutics-past-trends-and-future-prospects/.

Carpenter, S. M., Menictas, M., Nahum-Shani, I., Wetter, D. W., & Murphy, S. A. (2020). Developments in mobile health just-in-time adaptive interventions for addiction science. *Current Addiction Reports, 7*(3), 280–290. https://doi.org/10.1007/s40429-020-00322-y.

Chow, W., Doane, M. J., Sheehan, J., Alphs, L., & Le, H. (2019). Economic burden among patients with major depressive disorder: an analysis of healthcare resource use, work productivity, and direct and indirect costs by depression severity. 4.

Cohen, J., Marecek, J., & Gillham, J. (2006). Is three a crowd? Clients, clinicians, and managed care. *The American Journal of Orthopsychiatry, 76*(2), 251–259. https://doi.org/10.1037/0002-9432.76.2.251.

Collins, K. M., Armenta, R. F., Cuevas-Mota, J., Liu, L., Strathdee, S. A., & Garfein, R. S. (2016). Factors associated with patterns of mobile technology use among persons who inject drugs. *Substance Abuse, 37*(4), 606–612. https://doi.org/10.1080/08897077.2016.1176980.

Conti, D. J., & Burton, W. N. (1994). The Economic Impact of Depression in a Workplace. *Journal of Occupational and Environmental Medicine, 36*(9), 983–988.

Cook, E. L., & Harman, J. S. (2008). A comparison of health-related quality of life for individuals with mental health disorders and common chronic medical conditions. *Public Health Reports, 123*(1), 45–51. https://doi.org/10.1177/003335490812300107.

Corrigan, P. W., & Watson, A. C. (2007). The stigma of psychiatric disorders and the gender, ethnicity, and education of the perceiver. *Community Mental Health Journal, 43*(5), 439–458. https://doi.org/10.1007/s10597-007-9084-9.

Crumb, L., Mingo, T. M., & Crowe, A. (2019). "Get over it and move on": The impact of mental illness stigma in rural, low-income United States populations. *Mental Health & Prevention, 13*, 143–148. https://doi.org/10.1016/j.mhp.2019.01.010.

Cunningham, P. J. (2009). Chronic burdens: The persistently high out-of-pocket health care expenses faced by many Americans with chronic conditions. *Issue Brief (Commonwealth Fund), 63*, 1–14.

Davidson, C. L., Wingate, L. R., Grant, D. M., Judah, M. R., & Mills, A. C. (2011). Interpersonal suicide risk and ideation: The influence of depression and social anxiety. *Journal of Social and Clinical Psychology, 30*(8), 842–855. https://doi.org/10.1521/jscp.2011.30.8.842.

Deady, M., Choi, I., Calvo, R. A., Glozier, N., Christensen, H., & Harvey, S. B. (2017). eHealth interventions for the prevention of depression and anxiety in the general population: A systematic review and meta-analysis. *BMC Psychiatry [Electronic Resource], 17*(1), 310. https://doi.org/10.1186/s12888-017-1473-1.

Dial, T. H., Haviland, M. G., & Pincus, H. A. (2005). Datapoints: M.D. faculty salaries in psychiatry and all faculty departments, 1980-2001. *Psychiatric Services, 56*(2), 142. https://doi.org/10.1176/appi.ps.56.2.142. 142.

Digital Therapeutics Alliance. (2022). Retrieved January 27, 2022, from https://dtxalliance.org/

DiMino, J., & Blau, G. (2012). The relationship between wait time after triage and show rate for intake in a nonurgent student population. *Journal of College Student Psychotherapy, 26*(3), 241–247. https://doi.org/10.1080/87568225.2012.685857.

Fairburn, C. G., & Patel, V. (2014). The global dissemination of psychological treatments: a road map for research and practice. *American Journal of Psychiatry, 171*(5), 495–498. https://doi.org/10.1176/appi.ajp.2013.13111546.

Ferrari, A. J., Santomauro, D. F., Herrera, A. M. M., Shadid, J., Ashbaugh, C., Erskine, H. E., et al. (2022). Global, regional, and national burden of 12 mental disorders in 204 countries and territories, 1990-2019: A systematic analysis for the Global Burden of Disease Study 2019. *The Lancet Psychiatry, 9*, 137–150. https://doi.org/10.1016/S2215-0366(21)00395-3.

Fitch, K., Iwasaki, K., & Villa, K. F. (2014). Resource utilization and cost in a commercially insured population with schizophrenia. *American Health & Drug Benefits, 7*(1), 18–26.

Fullen, M. C., Brossoie, N., Dolbin-MacNab, M. L., Lawson, G., & Wiley, J. D. (2020). The impact of the Medicare mental health coverage gap on rural mental health care access. *Journal of Rural Mental Health, 44*(4), 243–251. https://doi.org/10.1037/rmh0000161.

Gibbons, M. C. (2008). An overview of healthcare disparities. In M. C. Gibbons (Ed.), Ehealth solutions for healthcare disparities (pp. 3–10). Baltimore, Maryland: Springer. https://doi.org/10.1007/978-0-387-72815-5_1.

Goldstein, S. P., Evans, B. C., Flack, D., Juarascio, A., Manasse, S., Zhang, F., et al. (2017). Return of the JITAI: Applying a just-in-time adaptive intervention framework to the development of m-health solutions for addictive behaviors. *International Journal of Behavioral Medicine, 24*(5), 673–682. https://doi.org/10.1007/s12529-016-9627-y.

Green, A. C., Hunt, C., & Stain, H. J. (2012). The delay between symptom onset and seeking professional treatment for anxiety and depressive disorders in a rural Australian sample. *Social Psychiatry and Psychiatric Epidemiology, 47*(9), 1475–1487. https://doi.org/10.1007/s00127-011-0453-x.

Guthrie, T. (2022).(n.d.). How COVID-19 will accelerate a digital therapeutics revolution. Forbes. Retrieved January 27, from https://www.forbes.com/sites/columbiabusinessschool/2020/04/21/how-covid-19-will-accelerate-a-digital-therapeutics-revolution/.

Harvey, A. G., & Gumport, N. B. (2015). Evidence-based psychological treatments for mental disorders: Modifiable barriers to access and possible solutions. *Behaviour Research and Therapy, 68*, 1–12. https://doi.org/10.1016/j.brat.2015.02.004.

Hauenstein, E. J., Boyd, M. E., & Submission, Features, H. C. (1994). Depressive symptoms in young women of the piedmont. *Women & Health, 21*(2–3), 105–123. https://doi.org/10.1300/J013v21n02_07.

Hehner, S., Biesdorf, S., & Möller, M. (2018). Digitizing healthcare—Opportunities for Germany | McKinsey. Retrieved January 27, 2022, from https://www.mckinsey.com/industries/healthcare-systems-and-services/our-insights/digitizing-healthcare-opportunities-for-germany.

Hendrickson, Z. (2019). Big pharma is breaking into the $3 billion digital therapeutics market. *Business Insider*. https://www.businessinsider.com/pharma-edging-into-digital-therapeutics-2019-9.

Hockenberry, J. M., Joski, P., Yarbrough, C., & Druss, B. G. (2019). Trends in treatment and spending for patients receiving outpatient treatment of depression in the United States, 1998-2015. *JAMA Psychiatry, 76*(8), 810–817 https://doi.org/10/ggxb7n.

Holmqvist, R., & Jeanneau, M. (2006). Burnout and psychiatric staff's feelings towards patients. *Psychiatry Research, 145*(2), 207–213. https://doi.org/10.1016/j.psychres.2004.08.012.

Huh, H. J., Kim, S.-Y., Yu, J. J., & Chae, J.-H. (2014). Childhood trauma and adult interpersonal relationship problems in patients with depression and anxiety disorders. *Annals of General Psychiatry, 13*(1), 26. https://doi.org/10.1186/s12991-014-0026-y.

Intelligence, Insider (2020). The digital theraputics report. Business Insider.

International Labour Office, International Labour Office, & Social Security Department. (2008). Social health protection: An ILO strategy towards universal access to health care. Geneva, Switzerland: ILO.

Jacobson, N. C., & Newman, M. G. (2017). Anxiety and depression as bidirectional risk factors for one another: A meta-analysis of longitudinal studies. *Psychological Bulletin, 143*(11), 1155–1200. https://doi.org/10.1037/bul0000111.

Jacobson, N. C., Yom-Tov, E., Lekkas, D., Heinz, M., Liu, L., & Barr, P. J. (2020). Impact of online mental health screening tools on help-seeking, care receipt, and suicidal ideation and suicidal intent: Evidence from internet search behavior in a large U.S. cohort. *Journal of Psychiatric Research, 145,* 276–283. S0022395620310694 https://doi.org/10.1016/j.jpsychires.2020.11.010.

James, S. L., Abate, D., Abate, K. H., Abay, S. M., Abbafati, C., Abbasi, N., et al. (2018). Global, regional, and national incidence, prevalence, and years lived with disability for 354 diseases and injuries for 195 countries and territories, 1990–2017: A systematic analysis for the Global Burden of Disease Study 2017. *The Lancet, 392*(10159), 1789–1858. https://doi.org/10.1016/S0140-6736(18)32279-7.

Jolliffe, D., Farrington, D. P., Brunton-Smith, I., Loeber, R., Ahonen, L., & Palacios, A. P. (2019). Depression, anxiety and delinquency: Results from the Pittsburgh Youth Study. *Journal of Criminal Justice, 62,* 42–49. https://doi.org/10.1016/j.jcrimjus.2018.08.004.

Keisler-Starkey, K., & Bunch, L. N. (2021). Health insurance coverage in the United States: 2020. 40.

Kessler, R. C., Berglund, P., Demler, O., Jin, R., Merikangas, K. R., & Walters, E. E. (2005). Lifetime prevalence and age-of-onset distributions of DSM-IV disorders in the national comorbidity survey replication. *Archives of General Psychiatry, 62*(6), 593. https://doi.org/10.1001/archpsyc.62.6.593.

Kowalewski, K., McLennan, J. D., & McGrath, P. J. (2011). A preliminary investigation of wait times for child and adolescent mental health services in Canada. *Journal of the Canadian Academy of Child and Adolescent Psychiatry, 20*(2), 112–119.

Leong, F. T. L., & Kalibatseva, Z. (2011). Cross-cultural barriers to mental health services in the United States. *Cerebrum: The Dana Forum on Brain Science, 2011,* 5.

Levin, J. S., & Chatters, L. M. (1998). Religion, health, and psychological well-being in older adults: Findings from three national surveys. *Journal of Aging and Health, 10*(4), 504–531. https://doi.org/10.1177/089826439801000406.

Lund, C., Breen, A., Flisher, A. J., Kakuma, R., Corrigall, J., Joska, J. A., et al. (2010). Poverty and common mental disorders in low and middle income countries: A systematic review. *Social Science & Medicine (1982), 71*(3), 517–528. https://doi.org/10.1016/j.socscimed.2010.04.027.

Marsh, J. C., Cao, D., Guerrero, E., & Shin, H.-C. (2009). Need-service matching in substance abuse treatment: Racial/ethnic differences. *Evaluation and Program Planning, 32*(1), 43–51. https://doi.org/10.1016/j.evalprogplan.2008.09.003.

Mazzone, L., Ducci, F., Scoto, M. C., Passaniti, E., D'Arrigo, V. G., & Vitiello, B. (2007). The role of anxiety symptoms in school performance in a community sample of children and adolescents. *BMC Public Health [Electronic Resource], 7*(1), 347. https://doi.org/10.1186/1471-2458-7-347.

McGurk, S. R., & Mueser, K. T. (2003). Cognitive functioning and employment in severe mental illness. *The Journal of Nervous and Mental Disease, 191*(12), 789–798. https://doi.org/10.1097/01.nmd.0000100921.31489.5a.

McMiller, W. P., & Weisz, J. R. (1996). Help-seeking preceding mental health clinic intake among African-American, Latino, and Caucasian Youths. *Journal of the American Academy of Child & Adolescent Psychiatry, 35*(8), 1086–1094. https://doi.org/10.1097/00004583-199608000-00020.

Mechanic, D. (2002). Removing barriers to care among persons with psychiatric symptoms. *Health Affairs, 21*(3), 137–147. https://doi.org/10.1377/hlthaff.21.3.137.

Mennis, J., & Stahler, G. J. (2016). Racial and ethnic disparities in outpatient substance use disorder treatment episode completion for different substances. *Journal of Substance Abuse Treatment, 63,* 25–33. https://doi.org/10.1016/j.jsat.2015.12.007.

Michalski, D., Mulvey, T., & Kohout, J. (2010). *2008 Apa Survey of Psychology Health Service Providers.* https://www.apa.org/workforce/publications/08-hsp/report.pdf.

Mojtabai, R., Olfson, M., Sampson, N. A., Jin, R., Druss, B., Wang, P. S., et al. (2011). Barriers to mental health treatment: Results from the National Comorbidity Survey Replication. *Psychological Medicine, 41*(8), 1751–1761. https://doi.org/10.1017/S0033291710002291.

Morris, J., Lora, A., McBain, R., & Saxena, S. (2012). Global mental health resources and services: A WHO survey of 184 countries. *Public Health Reviews, 34*(2), 3. https://doi.org/10.1007/BF03391671.

Muñoz, R. F. (2010). Using evidence-based internet interventions to reduce health disparities worldwide. *Journal of Medical Internet Research, 12*(5), e1463. https://doi.org/10.2196/jmir.1463.

Mustanski, B. S., Garofalo, R., & Emerson, E. M. (2010). Mental health disorders, psychological distress, and suicidality in a diverse sample of lesbian, gay, bisexual, and transgender youths. *American Journal of Public Health, 100*(12), 2426–2432. https://doi.org/10.2105/AJPH.2009.178319.

Nahum-Shani, I., Hekler, E. B., & Spruijt-Metz, D. (2015). Building health behavior models to guide the development of just-in-time adaptive interventions: A pragmatic framework. *Health Psychology, 34S,* 1209–1219. https://doi.org/10.1037/hea0000306.

Naslund, J. A., Aschbrenner, K. A., Araya, R., Marsch, L. A., Unützer, J., Patel, V., & Bartels, S. J. (2017). Digital technology for treating and preventing mental disorders in low-income and middle-income countries: A narrative review of the literature. *The Lancet Psychiatry, 4*(6), 486–500. https://doi.org/10.1016/S2215-0366(17)30096-2.

Noss, A. (2012). *Household Income: 2013.* 6.

Oquendo, M. A., Malone, K. M., Ellis, S. P., Sackeim, H. A., & Mann, J. J. (1999). Inadequacy of antidepressant treatment for patients with major depression who are at risk for suicidal behavior. *American Journal of Psychiatry, 156*(2), 190–194. https://doi.org/10.1176/ajp.156.2.190.

Parabiaghi, A., Bonetto, C., Ruggeri, M., Lasalvia, A., & Leese, M. (2006). Severe and persistent mental illness: A useful definition for prioritizing community-based mental health service interventions. *Social Psychiatry and Psychiatric Epidemiology, 41*(6), 457–463. https://doi.org/10.1007/s00127-006-0048-0.

Park, S., Garcia-Palacios, J., Cohen, A., & Varga, Z. (2019). From treatment to prevention: The evolution of digital healthcare. *Nature, 573*(7775). https://www.cheric.org/research/tech/periodicals/view.php?seq=1774536.

Patel, V., Araya, R., de Lima, M., Ludermir, A., & Todd, C. (1999). Women, poverty and common mental disorders in four restructuring societies. *Social Science & Medicine (1982), 49*(11), 1461–1471. https://doi.org/10.1016/s0277-9536(99)00208-7.

Pew Research Center (2019). Some digital divides between rural, urban, suburban America persist. Retrieved January 27, 2022, from https://www.pewresearch.org/fact-tank/2021/08/19/some-digital-divides-persist-between-rural-urban-and-suburban-america/.

Pew Research Center (2019). Smartphone ownership is growing rapidly around the world, but not always equally. Retrieved January 27, 2022, from https://www.pewresearch.org/global/2019/02/05/smartphone-ownership-is-growing-rapidly-around-the-world-but-not-always-equally/.

Pew Research Center (2021). Black, Hispanic adults less likely to have broadband or traditional PC than White adults. Retrieved January 27, 2022, from https://www.pewresearch.org/fact-tank/2021/07/16/home-broadband-adoption-computer-ownership-vary-by-race-ethnicity-in-the-u-s/.

Pinedo, M., Zemore, S., & Mulia, N. (2020). Black-White differences in barriers to specialty alcohol and drug treatment: Findings from a qualitative study. *Journal of Ethnicity in Substance Abuse, 21*, 112–126. https://doi.org/10.1080/15332640.2020.1713954.

Reichert, A., & Jacobs, R. (2018). The impact of waiting time on patient outcomes: Evidence from early intervention in psychosis services in England. *Health Economics, 27*(11), 1772–1787. https://doi.org/10.1002/hec.3800.

Riley, W. T., Oh, A., Aklin, W. M., & Wolff-Hughes, D. L. (2019). National Institutes of Health Support of digital health behavior research. *Health Education & Behavior, 46*(2_suppl.), 12S–19S. https://doi.org/10.1177/1090198119866644.

Rohland, B. M. (2000). A survey of burnout among mental health center directors in a rural state. *Administration and Policy in Mental Health and Mental Health Services Research, 27*(4), 221–237. https://doi.org/10.1023/A:1021361419155.

Rowan, K., McAlpine, D. D., & Blewett, L. A. (2013). Access and cost barriers to mental health care, by insurance status, 1999–2010. *Health Affairs, 32*(10), 1723–1730. https://doi.org/10.1377/hlthaff.2013.0133.

Saarni, S. I., Suvisaari, J., Sintonen, H., Pirkola, S., Koskinen, S., Aromaa, A., et al. (2007). Impact of psychiatric disorders on health-related quality of life: General population survey. *The British Journal of Psychiatry, 190*(4), 326–332. https://doi.org/10.1192/bjp.bp.106.025106.

Sareen, J., Jagdeo, A., Cox, B. J., Clara, I., ten Have, M., Belik, S.-L., et al. (2007). Perceived barriers to mental health service utilization in the United States, Ontario, and the Netherlands. *Psychiatric Services (Washington, D.C.), 58*(3), 357–364. https://doi.org/10.1176/ps.2007.58.3.357.

Sartorious, N. (2013). Comorbidity of mental and physical diseases: A main challenge for medicine of the 21st century. *Shanghai Archives of Psychiatry, 25*(2), 68–69. https://doi.org/10.3969/j.issn.1002-0829.2013.02.002.

Sartorius, N. (2018). Comorbidity of mental and physical disorders: A key problem for medicine in the 21st century. *Acta Psychiatrica Scandinavica, 137*(5), 369–370. https://doi.org/10.1111/acps.12888.

Schomerus, G., Schwahn, C., Holzinger, A., Corrigan, P. W., Grabe, H. J., Carta, M. G., et al. (2012). Evolution of public attitudes about mental illness: A systematic review and meta-analysis. *Acta Psychiatrica Scandinavica, 125*(6), 440–452. https://doi.org/10.1111/j.1600-0447.2012.01826.x.

Seeman, N., Tang, S., Brown, A. D., & Ing, A. (2016). World survey of mental illness stigma. *Journal of Affective Disorders, 190*, 115–121. https://doi.org/10.1016/j.jad.2015.10.011.

Semlyen, J., King, M., Varney, J., & Hagger-Johnson, G. (2016). Sexual orientation and symptoms of common mental disorder or low wellbeing: Combined meta-analysis of 12 UK population health surveys. *BMC Psychiatry [Electronic Resource], 16*(1), 67. https://doi.org/10.1186/s12888-016-0767-z.

Sheppard, R., Deane, F. P., & Ciarrochi, J. (2018). Unmet need for professional mental health care among adolescents with high psychological distress. *Australian & New Zealand Journal of Psychiatry, 52*(1), 59–67. https://doi.org/10.1177/0004867417707818.

Solano, J. P., Gomes, B., & Higginson, I. J. (2006). A comparison of symptom prevalence in far advanced cancer, AIDS, heart disease, chronic obstructive pulmonary disease and renal disease. *Journal of Pain and Symptom Management, 31*(1), 58–69. https://doi.org/10.1016/j.jpainsymman.2005.06.007.

Sommers, I. (1989). Geographic location and mental health services utilization among the chronically mentally Ill. *Community Mental Health Journal, 25*(2), 132–144. https://doi.org/10.1007/BF00755385.

Spitzer, R. L., Kroenke, K., Linzer, M., Hahn, S. R., Williams, J. B. W., deGruy, F. V., III, et al. (1995). Health-related quality of life in primary care patients with mental disorders: results from the PRIME-MD 1000 study. *JAMA, 274*(19), 1511–1517. https://doi.org/10.1001/jama.1995.03530190025030.

Statistical Abstract of the United States: (2006). 59. https://www.census.gov/library/publications/2005/compendia/statab/125ed.html.

Steel, Z., Marnane, C., Iranpour, C., Chey, T., Jackson, J. W., Patel, V., & Silove, D. (2014). The global prevalence of common mental disorders: A systematic review and meta-analysis 1980-2013. *International Journal of Epidemiology, 43*(2), 476–493. https://doi.org/10.1093/ije/dyu038.

Stensland, M., Watson, P. R., & Grazier, K. L. (2012). An examination of costs, charges, and payments for inpatient psychiatric treatment in community hospitals. *Psychiatric Services, 63*(7), 666–671. https://doi.org/10.1176/appi.ps.201100402.

Stewart, R. E., Lareef, I., Hadley, T. R., & Mandell, D. S. (2017). Can we pay for performance in behavioral health care? *Psychiatric Services (Washington, D.C.), 68*(2), 109–111. https://doi.org/10.1176/appi.ps.201600475.

Stewart, W. F., Ricci, J. A., Chee, E., Hahn, S. R., & Morganstein, D. (2003). Cost of lost productive work time among US workers with depression. *JAMA, 289*(23), 3135–3144. https://doi.org/10.1001/jama.289.23.3135.

Stotz, C., & Zweig, J. (2018, November 5). Going digital to disrupt the addiction epidemic | Rock Health. https://rockhealth.com/insights/going-digital-to-disrupt-the-addiction-epidemic/.

Substance Abuse and Mental Health Services Administration [SAMHSA]. (2016). Key substance use and mental health indicators in the United States: Results from the 2015 National Survey on Drug Use and Health. Center for Behavioral Health Statistics and Quality, HHS Publication No. SMA 16-4984, NSDUH Series H-51. Retrieved from https://www.samhsa.gov/data/

Substance Abuse and Mental Health Services Administration [SAMHSA]. (2019). Key substance use and mental health indicators in the United States: Results from the 2018 National Survey on Drug Use and Health (HHS Publication No. PEP19-5068, NSDUH Series H-54). Rockville, MD: Center for Behavioral Health Statistics and Quality, Substance Abuse and Mental Health Services Administration.

Tran, L. D., & Ponce, N. A. (2017). Who gets needed mental health care? Use of mental health services among adults with mental health need in California. *Californian Journal of Health Promotion, 15*(1), 36–45.

van Boekel, L. C., Brouwers, E. P. M., van Weeghel, J., & Garretsen, H. F. L. (2013). Stigma among health professionals towards patients with substance use disorders and its consequences for healthcare delivery: Systematic review. *Drug and Alcohol Dependence, 131*(1), 23–35. https://doi.org/10.1016/j.drugalcdep.2013.02.018.

Velez, F. F., Luderer, H. F., Gerwien, R., Parcher, B., Mezzio, D., & Malone, D. C. (2021). Evaluation of the cost-utility of a prescription digital therapeutic for the treatment of opioid use disorder. *Postgraduate Medicine, 133*(4), 421–427. https://doi.org/10.1080/00325481.2021.1884471.

Vigo, D., Thornicroft, G., & Atun, R. (2016). Estimating the true global burden of mental illness. *The Lancet Psychiatry, 3*(2), 171–178. https://doi.org/10.1016/S2215-0366(15)00505-2.

Wainberg, M. L., Scorza, P., Shultz, J. M., Helpman, L., Mootz, J. J., Johnson, K. A., et al. (2017). Challenges and opportunities in global mental health: a research-to-practice perspective. *Current Psychiatry Reports, 19*(5), 28. https://doi.org/10.1007/s11920-017-0780-z.

Walker, E. R., McGee, R. E., & Druss, B. G. (2015). Mortality in mental disorders and global disease burden implications: A systematic review and meta-analysis. *JAMA Psychiatry, 72*(4), 334–341. https://doi.org/10.1001/jamapsychiatry.2014.2502.

Wang, P. S., Berglund, P., Olfson, M., Pincus, H. A., Wells, K. B., & Kessler, R. C. (2005). Failure and delay in initial treatment contact after first onset of mental disorders in the National Comorbidity Survey Replication. *Archives of General Psychiatry, 62*(6), 603–613. https://doi.org/10.1001/archpsyc.62.6.603.

Wang, P. S., Demler, O., & Kessler, R. C. (2002). Adequacy of treatment for serious mental illness in the United States. *American Journal of Public Health, 92*(1), 92–98. https://doi.org/10.2105/AJPH.92.1.92.

Wang, W., Gellings Lowe, N., Jalali, A., & Murphy, S. M. (2021). Economic modeling of reSET-O, a prescription digital therapeutic for patients with opioid use disorder. *Journal of Medical Economics, 24*(1), 61–68. https://doi.org/10.1080/13696998.2020.1858581.

Webster, L., & Hackett, R. K. (1999). Burnout and leadership in community mental health systems. *Administration and Policy in Mental Health and Mental Health Services Research, 26*(6), 387–399. https://doi.org/10.1023/A:1021382806009.

Whaibeh, E., Mahmoud, H., & Vogt, E. L. (2020). Reducing the treatment gap for LGBT mental health needs: the potential of telepsychiatry. *The Journal of Behavioral Health Services & Research, 47*(3), 424–431. https://doi.org/10.1007/s11414-019-09677-1.

Whetten, R., Whetten, K., Pence, B. w., Reif, S., Conover, C., & Bouis, S. (2006). Does distance affect utilization of substance abuse and mental health services in the presence of transportation services? *Aids Care, 18*, 27–34. https://doi.org/10.1080/09540120600839397.

Whiteford, H. A., Degenhardt, L., Rehm, J., Baxter, A. J., Ferrari, A. J., Erskine, H. E., et al. (2013). Global burden of disease attributable to mental and substance use disorders: Findings from the Global Burden of Disease Study 2010. *The Lancet, 382*(9904), 1575–1586. https://doi.org/10.1016/S0140-6736(13)61611-6.

Williams, M. E., Latta, J., & Conversano, P. (2008). Eliminating the wait for mental health services. *The Journal of Behavioral Health Services & Research, 35*(1), 107–114. https://doi.org/10.1007/s11414-007-9091-1.

World Health Organization. (2004). Prevention of mental disorders. Geneva, Switzerland: World Health Organization. https://public.ebookcentral.proquest.com/choice/publicfullrecord.aspx?p=4978589.

World Health Organization, (2010). Atlas on substance use (2010): Resources for the prevention and treatment of substance use disorders (p. 138). Geneva, Switzerland.

Yang, L., Wong, L. Y., Grivel, M. M., & Hasin, D. S. (2017). Stigma and substance use disorders: An international phenomenon. *Current Opinion in Psychiatry, 30*(5), 378–388. https://doi.org/10.1097/YCO.0000000000000351.

Yardley, L., Spring, B. J., Riper, H., Morrison, L. G., Crane, D. H., Curtis, K., et al. (2016). Understanding and promoting effective engagement with digital behavior change interventions. *American Journal of Preventive Medicine, 51*(5), 833–842. https://doi.org/10.1016/j.amepre.2016.06.015.

Yoo, K.-B., Park, E.-C., Jang, S.-Y., Kwon, J. A., Kim, S. J., Cho, K., et al. (2016). Association between employment status change and depression in Korean adults. *BMJ Open, 6*(3), e008570. https://doi.org/10.1136/bmjopen-2015-008570.

Zhang, X., Hailu, B., Tabor, D. C., Gold, R., Sayre, M. H., Sim, I., et al. (2019). Role of health information technology in addressing health disparities: Patient, clinician, and system perspectives. *Medical Care, 57*(Suppl. 2), S115–S120. https://doi.org/10.1097/MLR.0000000000001092.

Zhang, X., Pérez-Stable, E. J., Bourne, P. E., Peprah, E., Duru, O. K., Breen, N., et al. (2017). Big data science: Opportunities and challenges to address minority health and health disparities in the 21st century. *Ethnicity & Disease, 27*(2), 95–106. https://doi.org/10.18865/ed.27.2.95.

Chapter 3

First wave of scalable digital therapeutics: Internet-based programs for direct-to-consumer standalone care for mental health and addiction

Aimee N.C. Campbell, Christina A. Brezing and Matisyahu Shulman

Columbia University Irving Medical Center, Department of Psychiatry and New York State Psychiatric Institute, Division on Substance Use Disorders, New York, NY, United States

3.1 Development of internet-based programs

Internet-based programs (IBPs) to treat substance use disorders (SUD) and other mental health disorders originated in the latter part of the 1990s with full emergence of randomized controlled research trials of IBPs in the early 2000s. The rise in IBPs coincided with the widespread use of the internet globally which increased from 16 million users in 1995 (0.4% of the world's population) to over a billion users by the end of 2005 (15.7% of the world's population) (Internet World Stats, 2021). In 2021, global internet use was 4.9 billion users representing 63% of the world's population (International Telecommunication Union, 2021).

The original promise of IBPs centered upon their capacity to increase access to treatment and improve fidelity of the delivery of evidence-based treatments that may produce similar outcomes to clinician delivered interventions. A major impetus for pursuing IBPs was (and is) the large gap between the need for behavioral health care to treat substance use and other mental health disorders and the availability, acceptability, and access to evidence-based services.

The gap between need for treatment and receipt of treatment is well documented with little change over the last three decades. Data from the 2019 National Survey on Drug and Health administered by SAMHSA showed that 19.3 million Americans (7.7%) reported a substance use disorder in the prior year; 51.5 million (20.6%) reported a mental health disorder (with 9.5 million reporting co-occurring substance use and mental health disorders) (SAMHSA, 2020). Societal costs of substance use are reported to be $740 billion annually (NIDA, nd), while mental health disorder estimates of societal cost are over $193 billion annually (Kessler et al., 2008). Despite the significant need, only about 12% of people with a SUD received treatment at a specialty facility in the prior year, with an even lower percentage for younger individuals (SAMHSA, 2020). About 26% of people with a mental health disorder perceived an unmet need for services in the past year (SAMHSA, 2020). The top two reasons reported for this unmet mental health care need were cost (43.9%) and not knowing where to go (33.1%) (SAMHSA, 2020). Recent provisional data on drug overdose deaths show a 30% increase in 2020 compared to 2019, equating to 93,398 fatalities (Ahmad, Rossen, & Sutton, 2021). This suggests the COVID-19 pandemic, beginning in early 2020 may have exacerbated the treatment gap, further highlighting the need to improve access and effectiveness of behavioral health interventions.

The treatment gap for behavioral health continues for a number of key reasons, many of which IBPs have the capacity to effectively address. These include: stigma related to substance use and mental health disorders; impeded access to care due to low information, geographic barriers/transportation, timing of services (i.e., accommodating work schedules), low integration within the larger healthcare system, and cost; lack of perceived need for specialty care; concerns regarding privacy and confidentiality; inconsistent delivery of services and lack of standardization of care; and the need for more personalized treatment planning (Campbell, Muench, & Nunes, 2014). In brief, IBPs are capable of providing confidential, personalized assessment, education, and treatment with fidelity (and without need for ongoing training and supervision of clinical staff), when and where individuals might be ready, motivated, or at a heightened level of concern but ambivalent about seeking formal services for behavioral health issues. Research has demonstrated that IBPs can be delivered cost-effectively and produce results similar to face-to-face care (Barak, Hen, Boneil-Nissim, & Shapira, 2008; Carlbring, Andersson, Cuijpers, Riper, & Hedman-Lagerlöf, 2018; Chebli, Blaszczynski, & Gainsbury, 2016; Hedman, Ljótsson, & Lindefors, 2012; Mitchell, Joshi, Patel, Lu, & Naslund, 2021; Olmstead, Ostrow, & Carroll, 2010).

In addition to IBPs addressing the long-standing treatment gap, evolving features of the psychiatric and substance use disorder treatment fields contributed to the advent of mHealth. Andersson (2018) describes three key features conducive to IBP development in behavioral health: (1) emergence of evidence-based interventions as standard of care (i.e., the cumulative research knowledge beginning to clearly demonstrate efficacious interventions with core features); (2) burgeoning literature on guided self-help practices, such as skills training, breathing techniques, mindfulness, and relaxation that could more easily be translated onto an internet-based platform; and (3) longer-term use of computerized testing and other types of interventions creating a context for which IBPs could be seen as viable alternatives to face-to-face care. Common core components of evidence-based behavioral health practices are found in early IBPs. The manualization of skills-based interventions such as cognitive behavioral therapy, relapse prevention, and motivational interviewing provided accessible content to translate onto internet-based platforms. For example, brief alcohol interventions (e.g., Cunningham, Wild, Cordingley, van Mierlo, & Humphreys, 2009; Hester, Squires, & Delaney, 2005) were designed to assess for high risk drinking, offer objective, normative feedback to end users, and include elements of motivational enhancement therapy to elicit behavior change (e.g., confidence in and likelihood to make changes). The majority of mental health IBPs offer a form of cognitive behavioral therapy (i.e., iCBT) to include psychoeducation, behavioral activation, cognitive restructuring, goal setting, coping skills, and homework assignments (e.g., Andersson, Bergstrom, Carlbring, Kaldo, & Ekselius, 2005; Kumar, Sattar, Bseiso, Khan, & Rutkofsky, 2017).

The rise of the internet and advances in technology are of course essential to the use and promise of IBPs. As previously noted, the internet globally has 4.9 billion users representing 63% of the world's population (International Telecommunication Union, 2021). In particular, there was a 17% increase from 2019 to 2021, likely brought about by a "COVID connectivity boost" as the need for remote communication became more essential (International Telecommunication Union, 2021). The percentage of adults in the United States who report using the internet has increased from 52% in 2000 to 76% in 2010 and, in 2021, stands at 93% (Pew Research Center, 2021a). Thus, IBPs are potentially accessible to all but 7% of the US adult population. Connectivity also can vary greatly by region (e.g., country development status; rural vs. urban areas), however, and an important factor to consider when developing and disseminating IBPs. For example, in least developed countries, almost three-quarters of the population have never connected to the internet (International Telecommunication Union, 2021).

The advent of smartphones and in particular, user friendly iPhones in 2007, opened up additional options for digital therapeutics, including 24/7 portable access. Smartphone ownership has increased from 35% of the adult US population in 2011 to 85% in 2021 (Pew Research Center, 2021b). Differences in smartphone ownership do exist, including by age group (e.g., only 61% of those 65+ own smartphones), economic status (e.g., 76% of those making less than 30K own smartphones), and geographic location (e.g., 80% of those in rural areas own smartphones) (Pew Research Center, 2021b). As a whole, however, the vast majority of US adults have ready access to the internet and mobile devices to utilize the internet. Recent research also suggests that people with substance use disorders also have high rates of mobile technology use, although less access among certain subgroups, including older people and those with unstable housing (Collins et al., 2016), as well as common challenges maintaining uninterrupted access to data plans.

3.1.1 Internet-based programs versus mobile applications

Similarities and differences between IBPs and mobile applications result in pros and cons (see Chapter 5 for information on the rise of mobile applications). IBPs or mobile websites (used via an internet browser) can be accessed from any device, but are not stored on the device. Thus, IBPs do not require the end user to manually update the programs like downloadable mobile apps which are housed on individual smartphones (Turner-McGrievy et al., 2007). Similarly, IBPs do not require specific operating systems (i.e., for android or iOS) which can make some mobile apps obsolete as operating systems are upgraded or earlier versions are no longer supported. IBPs (and websites) are less expensive and require less time to develop and pilot test compared to mobile apps (Turner-McGrievy et al., 2007). Drawbacks to IBPs include: needing to actively go to a specific website to access the program, whereas mobile apps are on a mobile device and available immediately; mobile apps can use phone functionality like cameras, data collection mechanisms like wearable sensors, and easy push notifications; and mobile apps may work offline while mobile websites require an internet connection (Turner-McGrievy et al., 2007). According to Turner-McGrievy et al. (2007) data security remains a risk for both internet and mobile programs with common issues like back end data security. Accessing IBPs does introduce an opportunity for internet hacking (see later chapter on Privacy and Security for more information).

3.2 Internet-based programs for substance use disorders

Several automated psychosocial interventions for substance use disorders have been developed, tested, and are available for broader use. As the internet and a digitally connected ecosystem developed, many of these interventions have been repackaged for wider dissemination as applications. Overall, reviews of IBPs found these interventions to be effective compared to no treatment or similar in comparison to standard in-person treatment (Chebli et al., 2016; Gainsbury & Blaszczynski, 2011). Studies vary widely in terms of choice of psychosocial interventions, with the most common interventions based in CBT and motivational interviewing, and include both automated programs and remote treatment delivered via the internet. These studies also included a variety of populations and targeted a range of substances.

Reviews (Chebli et al., 2016; Gainsbury & Blaszczynski, 2011) of IBPs for substance use disorders did, however, identify consistent methodological issues across trials. Although most studies included random allocation and control or comparison groups, several did not. In addition, many of the studies included relatively small sample sizes (less than 100 individuals) drawn from a specific population, such as specific schools or colleges. Sampling bias was also noted as potential methodological concern, with primarily individuals interested in online treatment who may not reflect the general population. Newer research may overcome some of these limitations in testing IBPs across more diverse samples and across geographic areas. The following section reviews three IBPs with the broadest research base, all of which have now been transitioned into applications and disseminated for more general clinical use.

3.2.1 Therapeutic education system

The therapeutic education system (TES) was among the first developed and extensively researched automated treatment intervention for substance use disorders (Bickel, Marsch, Buchhalter, & Badger, 2008; Campbell et al., 2014). TES was modeled on the Community Reinforcement Approach (CRA) and engineered to deliver neurobehavioral therapy for patients with substance use disorders in conjunction with standard treatment. TES delivers CRA therapy as a series of interactive lessons accessed through an online portal. Therapy lessons are comprised of a cognitive behavioral therapy component and skill-building exercises and take 10–15 minutes to complete. Therapy lesson content is delivered primarily via didactic content and may include videos, animations, and graphics and covers topics in cognitive behavioral relapse prevention and lifestyle skills to support abstinence and recovery. A subset of core lessons are completed first and in order, followed by access to all lessons with the recommendation to complete at least four lessons per week. Clinicians can engage in the treatment by encouraging lessons to complete (or repeat) based on individual patient needs.

The program also includes contingency management with rewards (i.e., draws from a virtual prize bowl which can be redeemed for varying levels of prizes) contingent upon: (1) completing therapy lessons, and (2) toxicology screens that are negative for substance use. In most trials the draws are used to obtain prizes based on Petry's "prize bowl" method (Stitzer & Petry, 2006), which implements contingency management at relatively low cost by having most of the rewards be an encouraging message (e.g., "nice job," "keep up the good work") and some of the time a prize, mainly small prizes (approximately \$1–5) and occasionally larger prizes of approximately \$20 or approximately \$100. In some trials, a more sophisticated reward schedule is used, using escalating rewards to more strongly reinforce positive behaviors, with higher total rewards. Other trials have shown benefit to the CRA modules without the addition of contingency management (Marsch et al., 2014).

Research base on TES. TES is among the best studied automated interventions for substance use disorder. Several initial studies showed promise that the intervention was effective in improving outcomes with substance use disorders (Bickel et al., 2008; Marsch et al., 2011). A large multisite trial using this intervention in eight community-based substance use disorder clinics showed that the TES program improved outcomes on drug use at 12 weeks after individuals enter treatment for nonopioid (alcohol and illicit substance use) use disorders (Campbell et al., 2014).

Although the TES program did not show benefit in individuals with opioid use disorder not on medications indicated for opioid use disorder, several trials have shown benefits for modified versions of the system in conjunction with medications for opioid use disorder (methadone [Marsch et al., 2014] and buprenorphine [Christensen et al., 2014]).

Transition to commercially available application. The rights to the TES program were purchased by the technology start-up Pear Therapeutics. Based on the above trial information (Campbell et al., 2014), the data were submitted to the FDA and the product was cleared to be marketed as a prescription digital therapeutic in 2017. TES is currently marketed as the reSET and reSET-o applications and is available via prescription for nonopioid substance use disorders and opioid use disorder, respectively. Disseminating contingency management in the community has historically been perceived as a challenge due to the direct costs of providing rewards. The current marketed version of the TES program has been able to provide contingencies in the community using coverage from some private and Medicaid insurance plans.

3.2.2 The CHESS–ACHESS program

ACHESS is a second early digital therapeutic that has been translated into an application and widely disseminated. First developed in 1989, CHESS was among the earliest digital therapeutics and has been studied and found efficacious for a variety of substance use disorders and populations (Gustafson, Bosworth, Hawkins, Boberg, & Bricker, 1992, 2002, 2014; Hochstatter et al., 2021; Johnson et al., 2016). The earliest iteration of CHESS included nine services delivered via a software program and local university network. These included information delivered in various text-based formats and a discussion group (Gustafson, Bosworth, Hawkins, Boberg, & Bricker, 1992). The program was designed to address a range of health conditions including substance use disorders. The program was expanded and used as a smartphone based intervention and shown to be effective for alcohol and other substance use disorders.

This intervention is based on self-determination theory (SDT), which holds that meeting three needs—for autonomy, competence, and relatedness—improves a person's adaptive functioning. *ACHESS'* services provide antecedent-appropriate intervention(s) that: (1) boost autonomy (intrinsic motivation) by selecting from multiple services those most likely to be most personally meaningful to the participant; (2) offer information, monitoring, and tools to increase competence; and/or (3) increase relatedness.

3.2.3 CBT4CBT

A third well-studied early IBP is the CBT4CBT (Computer Based Training for Cognitive Behavioral Therapy) program. The program includes video, quizzes, and interactive exercises to teach recovery skills across seven content areas that each take about 1 hour to complete. The seven lessons of CBT4CBT are: Recognize the Triggers, Deal with Cravings, Stand Up for Yourself, Plan Don't Panic, Stop and Think, Go Against the Flow, and Stay Safe. Each lesson can be repeated and each one offers practice exercises which can be reviewed any time after completing the lesson. CBT4CBT has been tested and found to be effective across multiple randomized controlled trials (Carroll et al., 2008, 2014a, 2014b; Kiluk et al., 2016; Shi, Henry, Dwy, Orazietti, & Carroll, 2019). CBT4CBT has also been studied with regards to mechanism of action with several trials demonstrating improvement in coping skills after completion of the program (Kiluk et al., 2017; Sugarman, Nich, & Carroll, 2010). CBT4 CBT has been packaged for general commercial clinical dissemination and is included in the connections app marketed by CHESS health (see https://www.chess.health/solutions/).

3.2.4 Dual diagnosis patients

Several digital therapeutics were designed for individuals with dual diagnoses and have evidence for benefit from one or more trials. Most of these include text, video, and audio delivered CBT modules. These include SHADE (Self-Help for Alcohol and Other Drug Use and Depression) (Kay-Lambkin, Baker, Lewin, & Carr, 2009; 2011) and Breaking Free Online (Elison, Davies, & Ward, 2015; Elison, Humphreys, Ward, & Davies, 2014, 2017).

3.2.5 Prevention and screening

IBPs are ideal tools in the prevention and screening for substance use disorders because of the broad need across the general population and in diverse settings (e.g., primary care, emergency departments, schools, places of employment, etc.). Several interventions have been developed and tested for this purpose, many with good evidence of efficacy from clinical trials. Many of these target specific populations such as adolescents or college students (Ganz et al., 2018; Hausheer, Doumas, & Esp, 2018; Hustad, Barnett, Borsari, & Jackson, 2010; Jander, Crutzen, Mercken, Candel, & de Vries, 2016; Martinez-Montilla et al., 2020) or adults with high risk or problem alcohol use (Postel et al., 2010).

3.3   Internet-based programs for mental health (nonsubstance related disorders)

As with substance use disorders, evidence-based treatments for nonsubstance related psychiatric disorders that can be delivered via in person care with clear and structured protocols for cognitive techniques, provide patient education, and support behavioral activation were the first psychotherapies to be trialed through internet-based delivery methods. These psychotherapies include mindfulness exercises, cognitive behavioral therapy (CBT) adapted across disorders via internet-based CBT (iCBT), dialectal behavioral therapy (DBT), interpersonal psychotherapy (IPT), and acceptance commitment therapy (ACT) (Cuijpers et al., 2021; Washburn, Yu, Rubin, & Zhou, 2021). These internet-based psychotherapies have been studied for diverse mental health conditions that focus on mood, anxiety, eating, personality, psychosis, trauma, neurodevelopmental, and cognition with good supporting evidence for their utilization and effectiveness (Simon, Robertson, & Lewis, 2021; Taylor, Graham, Flatt, Waldherr, & Fitzsimmons-Craft, 2021).

Many clinical trials have examined these IBPs for mental health disorders in special populations, including pregnant women (Bright, Mughal, & Wajid, 2019), individuals with significant medical illness, such as HIV, cancer, and COVID-19, demonstrating comparable effects to in-person treatment-as-usual (Nissen, O'Connor, & Kaldo, 2020). Internet delivered psychotherapeutic interventions for psychiatric disorders are helpful across the lifecycle, with studies supporting their use in children (Aspvall, Andersson, & Melin, 2021), as well as the elderly (Chen et al., 2020), a population that historically has limited technological fluency and thereby was thought to need additional guidance and support (Barnard, Bradley, Hodgson, & Lloyd, 2013). With small modifications, these IBPs have been tailored to support the unique needs of historically marginalized populations struggling with behavioral health disorders such as people who are undomiciled, LGBTQ+ youth (Strauss et al., 2019), and individuals with severe and persistent mental illness (Beentjes, van Gaal, Goossens, & Schoonhoven, 2016). These interventions for psychiatric disorders have been tailored to be socially and culturally competent to improve relevance and engagement for people across diverse ethnic and racial backgrounds (Bansa et al., 2018). Various IBPs have focused on not only treating the client or patient, but also providing platforms for support networks, such as the therapist, physician, family member, peer, or other caregiver. For the scope of this chapter, we will focus on IBPs for mood and anxiety disorders which have been the most widely studied and have the broadest applications for use.

3.3.1   Mood and anxiety disorders

Mood and anxiety disorders have a significant negative impact on individuals' functioning, are associated with increased risk of medical morbidity and mortality, and overall diminish quality of life (Donohue & Pincus, 2007). There are numerous effective evidence-based treatments for mood and anxiety disorders; however, there is a dearth of trained providers in these modalities who are accessible to patients, particularly in rural areas, resulting in the majority of people with depression not receiving treatment, as noted earlier in this chapter (Collins, Westra, Dozois, & Burns, 2004). Alternatively, IBPs for depression and anxiety may serve a role as part of stepped-care, allowing for an initial trial of psychotherapy via the internet, followed by assessment and referral for additional treatment if the patient does not achieve full remission of symptoms (Aspvall et al., 2021). IBPs for mood and anxiety disorders are seen as one possible solution to addressing the gap that exists in the current mental health care system.

Most IBP studies for mood and anxiety disorders have focused on disorder-specific interventions, such as major depressive disorder or panic disorder, though greater than 25% of studies completed have transdiagnostic applications (Dear et al., 2015). Few studies of IBP for mood and anxiety disorders present "blended treatments," or treatments that are combining more than one psychotherapeutic modality, particularly in the routine care setting (Erbe, Eichert, Riper, & Ebert, 2017). On average, interventions utilizing CBT, the most common psychotherapy adapted for an internet-based intervention, include eight "sessions," mirroring modules provided in traditional clinical practice. Many studies include guidance from study research staff that range from administrative to motivational in supporting treatment; thus, it is unclear how many IBPs might stand-alone. Supplementary modes of contact include text messaging reminders, email, telephone calls, and face-to-face contact as part of study participation (Etzelmueller, Vis, & Karyotaki, 2020). With IBPs for depression, many studies included systems for assessing and monitoring safety concerns, such as suicidality, or if there was significant clinical worsening, prompting contact of the participant, follow-up with structured risk assessments, and possible referrals for additional treatment if warranted (Ebert & Baumeister, 2017).

Numerous research studies have evaluated the feasibility of providing IBP for anxiety and depression, including in routine care. Feasibility has included factors such as enrollment, adherence, assessment of participant characteristics, and participant satisfaction with different elements of the internet-delivered treatment. In one meta-analysis pooling depression and anxiety interventions, with the majority of participants women and a mean age of 38 years, the average participant completed 61.2% of the online sessions and 61.3% of participants achieved full adherence (i.e., completing all treatment components)

(Etzelmueller et al., 2020). However, studies have found conflicting results for adherence to online interventions for depression and anxiety, with some finding substantial attrition (Christensen, Griffiths, & Farrer, 2009; Kelders, Kok, Ossebaard, & Van Gemert-Pijnen, 2012; Lintvedt et al., 2011; Richards et al., 2015) and others with a median of 80% of participants demonstrating full adherence (Andrews, Cuijpers, Craske, McEvoy, & Titov, 2010). One particular study explored predictors of adherence to IBP for depression and found age, perception of health, and change in depression severity from baseline to be positively associated with adherence (Castro, López-del-Hoyo, & Peake, 2018). A separate study found that the overall "acceptability," of an IBP predicted engagement and adherence (Gulliver et al., 2021), reflective of the quality of the internet-based format from a user perspective. Individual characteristics of patients also impact acceptability. These include agreeableness, lower levels of stigma toward their diagnosis and treatment, and more positive help-seeking attitudes (Gulliver et al., 2021).

Another important factor to consider is how engagement and adherence is defined and measured. Adherence to IBPs for behavioral health is different than adherence to a medication and must take into account the concept of intended use, defined as the amount of intervention content experienced by the participant to derive maximum benefit, as defined by its developers (Kelders et al., 2012). While participants may not complete all modules or even the study recommended dose, lower rates of adherence may result in significantly improved psychiatric symptoms (Sieverink, Kelders, & van Gemert-Pijnen, 2017). It is often difficult to predict the intended usage early in development of IBPs and therefore this construct may not be clearly defined or characterized in studies. A greater understanding of factors that contribute to engagement in IBPs for mental health is needed.

Patient satisfaction is regularly assessed in studies of IBPs for anxiety and depression (>58% of all studies in meta-analyses [Josephine, Josefine, Philipp, David, & Harald, 2017]); however, there is little consensus or consistent instrumentation utilized, making results difficult to interpret. Generally, studies that assess patient satisfaction find that participants generally rate their treatment "high" to "very high" in satisfaction (Nguyen, Attkisson, & Stegner, 1983). However, in some studies, there is notable selection bias in terms of which participants engage long enough in the studies to provide feedback potentially inflating satisfaction (Rothwell, 2005).

There are numerous studies examining the effectiveness of IBPs for mood and anxiety disorders. A meta-analysis of 19 studies of iCBT for depression and anxiety discovered pooled effect sizes of 1.78 for depression and 0.94 for anxiety studies (Etzelmueller et al., 2020), with a wide range of effect sizes from 0.42 to 1.88 (Etzelmueller et al., 2020). Other studies have found that IBPs without therapist support have lower effect sizes than those that include at least some therapist guidance. The studies with additional therapist support are thought to be more similar with traditional in-person psychotherapy for mood and anxiety disorders (Andersson et al., 2006). These findings suggest that the degree and type of human contact in studies of IBPs for mood and anxiety disorders impact the effectiveness of the treatment, with less contact leading to a lower effect size and greater contact increasing effect size (Johansson & Andersson, 2012). Other studies have demonstrated no treatment effect differences for face-to-face treatment as compared to self-guided IBPs for phobias (Andersson, Cuijpers, Carlbring, Riper, & Hedman, 2014; Andrews, Davies, & Titov, 2011) and panic disorder (Bergrstom et al., 2010). It is possible that the specific mood and/or anxiety disorder diagnosis is more or less amenable to treatment with an IBP that does not have substantial human support during the treatment. However, it is generally accepted that for widespread treatment, IBP's are noninferior to face-to-face CBT and demonstrate more consistent fidelity to the evidence-based protocol throughout the duration of treatment (Wagner, Horn, & Maercker, 2014).

IBP's for depression and anxiety are scalable and cost-efficient due to their ease of access through any internet enabled device. The ability to access evidence-based psychotherapy modules through the internet has paved the way for development of web-based applications and commercialization of these treatment protocols for widespread access. iCBT, in particular, has seen adoption across a number of different platforms and settings targeting transdiagnostic mental health disorders though most notably for depression and anxiety (Rosso et al., 2017; Carl et al., 2020). Utilization of IBPs to target large populations in municipalities, such as schools or prisons, are becoming more common to address the demand for high quality treatment of depression and anxiety in the absence of sufficient in-person providers (Teesson et al., 2020; Woodhouse et al., 2016).

In summary, the literature to date presents a vast landscape of heterogenous, yet effective, IBPs for mood and anxiety disorders. Given the numerous options in this space, clinicians should recommend IBPs that have been tested in randomized controlled trials to their patients. Future directions in the field point toward utilizing advanced analysis of patient-level data. These type of patient-level data analyses over the course of treatment has the potential to reveal helpful, less helpful, and potentially harmful components of IBPs, and may provide insight for how to tailor and individualize these interventions for specific patients and thereby enhance their utilization, engagement and ultimate effectiveness in producing positive mood and anxiety outcomes (Furukawa, Suganuma, & Ostinelli, 2021).

3.4 Summary and next steps for internet-based programs for substance use and mental health disorders

There is now an extensive evidence base for the efficacy of IBPs for substance use and mental health disorders. IBPs retain a number of strengths that make evidence-based behavioral health interventions available to broad audiences, including low cost development, less burden to end users to monitor changes to the program and to smartphone operating systems, and access to the service across devices with internet capacity.

There are also clear directions for additional research to address important outstanding questions. First, although multiple studies have demonstrated the potential for comparability between internet-delivered and clinician-delivered interventions (Bickel et al., 2008; Chaple et al., 2016), as well as acting as clinician extenders—whereby the IBP could replace some face-to-face service provision with superior results (Campbell et al., 2014), future studies are needed to understand when and where this comparability exists. For example, a recent systematic review of the noninferiority claim among cognitive behavioral interventions for anxiety suggests there may be limited evidence from randomized controlled trials to support this claim with recommendations for studies specifically designed to test this question (O'Kearney, Kim, Dawson, & Calear, 2019). Similarly, cost effectiveness of IBPs has been demonstrated across studies with a recent systematic review of IBPs for anxiety and depression showing that 81% of 27 trials reported cost-effectiveness (Mitchell et al., 2021). However, the authors cautioned that additional research with adequate controls and well specified cost components were needed, as well as understanding cost in the context of lower resource settings and with newer technologies. This recommendation has been echoed in several other reviews conducted within the last 5 years (Erbe et al., 2017; Jimenez-Molina et al., 2019; Sin et al., 2020).

As with other digital therapeutics, a third area in need of additional research is behavioral health IBPs with evidence of effectiveness in diverse populations. A number of studies suggest that IBPs may be equally or even more effective among historically marginalized racial and ethnic end users. For example, a randomized control trial of computer-assisted CBT demonstrated significant reductions in anxiety and depression among Black participants but not White participants (Jonassaint et al., 2020). Campbell et al. (2017) reported that a computer-assisted psychoeducation intervention for SUDs appeared to be equally effective across a number of substance use outcomes among non-Hispanic Black, non-Hispanic White, and Hispanic/Latinx treatment seekers. However, there are fewer IBPs developed specifically with diverse populations or that contain adequate numbers of diverse participants to perform sufficiently powered secondary analyses and this work is needed, along with systematic cultural adaptation of IBPs originally developed for mainstream/general population. In a systematic review of participation of Black participants in eHealth and mHealth studies, James and colleagues (James, Harville, Sears, Efunbumi, & Bondoc, 2017) concluded that there was low representation of Black participants overall and that tailoring of incentives, recruitment, and retention strategies are needed that is specific to target condition, population, and community. See later chapter for content on cultural adaptations of digital therapeutics.

The proliferation of downloadable apps also offers key characteristics that may lessen the attraction to IBPs, especially the ability to use smartphone functionality as part of the intervention. Engagement with and adherence to both web apps and phone apps is a substantial concern (Batterham et al., 2021; Birnbaum, Lewiw, Rosen, & Ranney, 2015); phone apps may have greater options to address this compared to IBPs. For example, including "gaming" aspects to programs is seen as one strategy for improving engagement and incentivizing ongoing use. This is likely easier to achieve with a smartphone app that can offer direct reminders and is readily available in one tap, with or without internet access. A recent systematic review of 35 studies produced a checklist of mHealth design features that increase end user experience and thus engagement, including personalization, reinforcement, communication, navigation, credibility, message presentation, and interface aesthetics (Wei et al., 2020). This highlights the importance of engaging the end user in the development of IBPs and understanding mechanisms which directly lead to greater adherence.

Finally, scaling up evidence-based IBPs, especially for substance use disorders, is an ongoing issue. Two critical challenges remain: (1) finding a mechanism or business plan to keep costs low to end users while maintaining the IBP; and (2) understanding how best to disseminate effective IBPs to people who might benefit, especially those who are out of care and may not have access to a provider that can readily refer them to appropriate, acceptable, effective and available IBPs (e.g., Grady, Yoong, Sutherland, Hopin, & Nathan, 2018). Systematic planning for implementation during the design and testing of IBPs is important to address some of these challenges, including reduction of program build-out costs, keeping pace with the shifting technology landscape, and designing supports to sustain the intervention for wider dissemination (Li et al., 2019). Understanding the strengths and challenges associated with behavioral health IBPs will result in the retention of these interventions in the digital therapeutic toolbox and should be considered as viable options for ongoing research and dissemination.

References

Ahmad, F. B., Rossen, L. M., & Sutton, P. (2021). Provisional drug overdose death counts. National Center for Health Statistics. https://www.cdc.gov/nchs/nvss/vsrr/drug-overdose-data.htm.

Andersson, G. (2018). Internet interventions: Past, present, and future. *Internet Intervention, 12*, 181–188.

Andersson, G., Bergstrom, J., Carlbring, P., Kaldo, V., & Ekselius, L. (2005). Internet-based self-help for depression: Randomised controlled trial. *British Journal of Psychiatry, 187*, 456–461.

Andersson, G., Carlbring, P., Holmström, A., Sparthan, E., Furmark, T., Nilsson-Ihrfelt, E., et al. (2006). Internet-based self-help with therapist feedback and in vivo group exposure for social phobia: A randomized controlled trial. *Journal of Consulting and Clinical Psychology, 74*(4), 677–686.

Andersson, G., Cuijpers, P., Carlbring, P., Riper, H., & Hedman, E. (2014). Guided internet-based vs. face-to-face cognitive behavior therapy for psychiatric and somatic disorders: A systematic review and meta-analysis. *World Psychiatry, 13*(3), 288–295.

Andrews, G., Cuijpers, P., Craske, M. G., McEvoy, P., & Titov, N. (2010). Computer therapy for the anxiety and depressive disorders is effective, acceptable and practical health care: A meta-analysis. *Plos One, 5*, e13196.

Andrews, G., Davies, M., & Titov, N. (2011). Effectiveness randomized controlled trial of face to face versus Internet cognitive behaviour therapy for social phobia. *Australian New Zealand Journal of Psychiatry, 45*(4), 337–340.

Aspvall, K., Andersson, E., Melin, K., Norlin, L., Eriksson, V., Vigerland, S., et al. (2021). Effect of an internet-delivered stepped-care program vs in-person cognitive behavioral therapy on obsessive-compulsive disorder symptoms in children and adolescents: A randomized clinical trial. *JAMA, 325*(18), 1863–1873.

Bansa, M., Brown, D., DeFrino, D., Mahoney, N., Saulsberry, A., Marko-Holguin, M., et al. (2018). A little effort can withstand the hardship: Fielding an internet-based intervention to prevent depression among urban racial/ethnic minority adolescents in a primary care setting. *Journal of the National Medical Association, 110*(2), 130–142.

Barak, A., Hen, L., Boneil-Nissim, M., & Shapira, N. A. (2008). Comprehensive review and a meta-analysis of the effectiveness of internet-based psychotherapeutic interventions. *Journal of Technology in Human Services, 26*(2-4), 109–160.

Barnard, Y., Bradley, M. D., Hodgson, F., & Lloyd, A. D. (2013). Learning to use new technologies by older adults: Perceived difficulties, experimentation behaviour and usability. *Computational Human Behavior, 29*(4), 1715–1724.

Batterham, P. J., Calear, A. L., Sunderland, M., Kay-Lambkin, F., Farrer, L. M., Christensen, H., et al. (2021). A brief intervention to increase uptake and adherence of an internet-based program for depression and anxiety (enhancing engagement with psychosocial interventions): Randomized controlled trial. *Journal of Medical Internet Research, 23*(7), e23029.

Beentjes, T. A. A., van Gaal, B. G. I., Goossens, P. J. J., & Schoonhoven, L. (2016). Development of an e-supported illness management and recovery programme for consumers with severe mental illness using intervention mapping, and design of an early cluster randomized controlled trial. *BMC Health Services Research, 19*(16), 20.

Bergrstom, J., Andersson, G., Ljotsson, B., Ruck, C., Andreewitch, S., Karlsson, A., et al. (2010). Internet-versus group-administered cognitive beahvior therapy for panic disorder in a psychiatric setting: A randomized trial. *BMC Psychiatry [Electronic Resource], 2*(10), 54.

Bickel, W. K., Marsch, L. A., Buchhalter, A., & Badger, G. (2008). Computerized behavior therapy for opioid dependent outpatients: A randomized, controlled trial. *Experimental and Clinical Psychopharmacology, 16*, 132–143.

Birnbaum, F., Lewiw, D. M., Rosen, R., & Ranney, M. L. (2015). Patient engagement and the design of digital health. *Academy of Emergency Medicine, 22*(6), 754–756.

Bright, K. S., Mughal, M. K., Wajid, A., Lane-Smith, M., Murray, L., Roy, N., et al. (2019). Internet-based interpersonal psychotherapy for stress, anxiety, and depression in prenatal women: Study protocol for a pilot randomized controlled trial. *Trials, 20*, 814.

Campbell, A. N., Mongomery, L., Sanchez, K., Pavlicova, M., Hu, M., Newville, L., et al. (2017). Racial/ethnic subgroup differences in outcomes and acceptability of an internet-delivered intervention for substance use disorders. *Journal of Ethnicity in Substance Abuse, 16*(4), 460–478.

Campbell, A. N., Nunes, E. V., Matthews, A. G., Stitzer, M., Miele, G. M., Polsky, D., et al. (2014). Internet-delivered treatment for substance abuse: A multisite randomized controlled trial. *American Journal of Psychiatry, 171*(6), 683–690.

Campbell, A. N. C., Muench, F., & Nunes, E. V. (2014). Technology-based behavioral interventions for alohol and drug use problems. In L. Marsch, S. Lord, & J. Dallery (Eds.), *Behavioral healthcare and technology: Using science-based innovations to transform practice*. New York: Oxford University Press, Inc.

Carl, J. R., Miller, C. B., Henry, A. L., Davis, M. L., Stott, R., Smits, J. A. J., et al. (2020). Efficacy of digital cognitive behavioral therapy for moderate-to-severe symptoms of generalized anxiety disorder: A randomized controlled trial. *Depression and Anxiety, 37*(12), 1168–1178.

Carlbring, P., Andersson, G., Cuijpers, P., Riper, H., & Hedman-Lagerlöf, E. (2018). Internet-based vs. face-to-face cognitive behavior therapy for psychiatric and somatic disorders: An updated systematic review and meta-analysis. *Cognitive Behaviour Therapy, 47*(1), 1–18.

Carroll, K. M., Ball, S. A., Martino, S., Nich, C., Babuscio, T. A., Nuro, K. F., et al. (2008). Computer-assisted delivery of cognitive-behavioral therapy for addiction: A randomized trial of CBT4CBT. *American Journal of Psychiatry, 165*(7), 881–888.

Carroll, K. M., Kiluk, B. D., Nich, C., DeVito, E. E., Decker, S., LaPaglia, D., et al. (2014a). Toward empirical identification of a clinically meaningful indicator of treatment outcome: Features of candidate indicators and evaluation of sensitivity to treatment effects and relationship to one year follow up cocaine use outcomes. *Drug and Alcohol Dependence, 137*, 3–19.

Carroll, K. M., Kiluk, B. D., Nich, C., Gordon, M. A., Portnoy, G. A., Marino, D. R., et al. (2014b). Computer-assisted delivery of cognitive-behavioral therapy: Efficacy and durability of CBT4CBT among cocaine-dependent individuals maintained on methadone. *American Journal of Psychiatry, 171*(4), 436–444.

Castro, A., López-del-Hoyo, Y., Peake, C., Mayoral, F., Botella, C., García-Campayo, J., et al. (2018). Adherence predictors in an Internet-based Intervention program for depression. *Cognivtive Behavioral Therapy, 47*(3), 246–261.

Chaple, M., Sacks, S., McKendrick, K., Marsch, L. A., Belenko, S., Leukefeld, C., et al. (2016). A comparative study of the therapeutic education system for incarcerated substance-abusing offenders. *The Prison Journal, 96*(3), 485–508.

Chebli, J., Blaszczynski, A., & Gainsbury, S. M. (2016). Internet-based interventions for addictive behaviours: A systematic review. *Journal of Gambling Studies, 32*(4), 1279–1304.

Chen, A. T., Slattery, K., Tomasino, K. N., Rubanovich, C. K., Bardsley, L. R., & Mohr, D. C. (2020). Challenges and benefits of an internet-based intervention with a peer support component for older adults with depression: qualitative analysis of textual data. *JMIR, 22*(6), e17586.

Christensen, H., Griffiths, K. M., & Farrer, L. (2009). Adherence in internet interventions for anxiety and depression. *JMIR, 11*, e13.

Christensen, D. R., Landes, R. D., Jackson, L., Marsch, L. A., Mancino, M. J., Chopra, M. P., et al. (2014). Adding an internet-delivered treatment to an efficacious treatment package for opioid dependence. *Journal of Consulting and Clinical Psychology, 82*(6), 964.

Collins, K. A., Westra, H. A., Dozois, D. J., & Burns, D. D. (2004). Gaps in accessing treatment for anxiety and depression: Challenges for the delivery of care. *Clinical Psychology Review, 24*, 583–616.

Collins, K. M., Armenta, R. F., Cuevas-Mota, J., Liu, L., Strathdee, S. A., & Garfein, R. S. (2016). Factors associated with patterns of mobile technology use among persons who inject drugs. *Substance Abuse, 37*(4), 606–612.

Cuijpers, P., Pineda, B. S., Quero, S., Karyotaki, E., Struijs, S. Y., Figueroa, C. A., et al. (2021). Psychological interventions to prevent the onset of depressive disorders: A meta-analysis of randomized controlled trials. *Clinical Psychology Review, 83*, 101955.

Cunningham, J. A., Wild, T. C., Cordingley, J., van Mierlo, T., & Humphreys, K. (2009). A randomized controlled trial of an internet-based intervention for alcohol abusers. *Addiction, 104*, 2023–2032.

Dear, B. F., Staples, L. G., Terides, M. D., Karin, E., Zou, J., Johnston, L., et al. (2015). Transdiagnostic versus disorder-specific and clinician-guided versus self-guided internet-delivered treatment for generalized anxiety disorder and comorbid disorders: A randomized controlled trial. *Journal of Anxiety Disorders, 36*, 63–77.

Donohue, J. M., & Pincus, H. A. (2007). Reducing the societal burden of depression: A review of economic costs, quality of care and effects of treatment. *Pharmacoeconomics, 25*(1), 7–24.

Ebert, D. D., & Baumeister, H. (2017). Internet-based self-help interventions for depression in routine care. *JAMA Psychiatry, 74*(8), 852–853.

Elison, S., Davies, G., & Ward, J. (2015). An outcomes evaluation of computerized treatment for problem drinking using Breaking Free Online. *Alcoholism Treatment Quarterly, 33*(2), 185–196.

Elison, S., Humphreys, L., Ward, J., & Davies, G. (2014). A pilot outcomes evaluation for computer assisted therapy for substance misuse–an evaluation of Breaking Free Online. *Journal of Substance Use, 19*(4), 313–318.

Elison, S., Jones, A., Ward, J., Davies, G., & Dugdale, S. (2017). Examining effectiveness of tailorable computer-assisted therapy programmes for substance misuse: Programme usage and clinical outcomes data from Breaking Free Online. *Addictive Behaviors, 74*, 140–147.

Erbe, D., Eichert, H., Riper, H., & Ebert, D. D. (2017). Blending face-to-face and internet-based interventions for the treatment of mental disorders in adults: Systematic review. *JMIR, 19*(9), e306.

Etzelmueller, A., Vis, C., Karyotaki, E., Baumeister, H., Titov, N., Berking, M., et al. (2020). Effects of internet-based cognitive behavioral therapy in routine care for adults in treatment for depression and anxiety: Systematic review and meta-analysis. *JMIR, 22*(8), e18100.

Furukawa, T. A., Suganuma, A., Ostinelli, E. G., Andersson, G., Beevers, C. G., Shumake, J., et al. (2021). Dismantling, optimising, and personalising internet cognitive behavioural therapy for depression: A systematic review and component network meta-analysis using individual participant data. *Lancet Psychiatry, 8*(6), 500–511.

Gainsbury, S., & Blaszczynski, A. (2011). A systematic review of internet-based therapy for the treatment of addictions. *Clinical Psychology Review, 31*(3), 490–498.

Ganz, T., Braun, M., Laging, M., Schermelleh-Engel, K., Michalak, J., & Heidenreich, T. (2018). Effects of a stand-alone web-based electronic screening and brief intervention targeting alcohol use in university students of legal drinking age: A randomized controlled trial. *Addictive Behaviors, 77*, 81–88.

Grady, A., Yoong, S., Sutherland, R., Hopin, L., & Nathan, N. (2018). Improving the public health impact of eHealth and mHealth interventions. *New Zealand Journal of Public Health, 42*(2), 118–119.

Gulliver, A., Calear, A. L., Sunderland, M., Kay-Lambkin, F., Farrer, L. M., & Batterham, P. J. (2021). Predictors of acceptability and engagement in a self-guided online program for depression and anxiety. *Internet Interventions, 25*, 100400.

Gustafson, D. H., Bosworth, K., Hawkins, R. P., Boberg, E. W., & Bricker, E. (1992). CHESS: A computer-based system for providing information, referrals, decision support and social support to people facing medical and other health-related crises. Proceedings/the … annual symposium on computer applications in medical care (pp. 161–165). Bethesda: American Medical Informatics Association.

Gustafson, D. H., Hawkins, R. P., Boberg, E. W., McTavish, F., Owens, B., Wise, M., et al. (2002). CHESS: 10 years of research and development in consumer health informatics for broad populations, including the underserved. *International Journal of Medical Informatics, 65*(3), 169–177.

Gustafson, D. H., McTavish, F. M., Chih, M-Y., Atwood, A. K., Johnson, R. A., Boyle, M. G., et al. (2014). A smartphone application to support recovery from alcoholism: A randomized clinical trial. *JAMA Psychiatry, 71*(5), 566–572.

Hausheer, R., Doumas, D. M., & Esp, S. (2018). Evaluation of a web-based alcohol program alone and in combination with a parent campaign for ninth-grade students. *Journal of Addictions & Offender Counseling, 39*(1), 15–30.

Hedman, E., Ljótsson, B., & Lindefors, N. (2012). Cognitive behavior therapy via the Internet: A systematic review of applications, clinical efficacy and cost–effectiveness. *Expert Review of Pharmacoeconomics & Outcomes Research, 12*(6), 745–764.

Hester, R. K., Squires, D. D., & Delaney, H. D. (2005). The Drinker's check-up: 12-month outcomes of a controlled clinical trial of a stand-alone software program for problem drinkers. *Journal of Substance Abuse Treatment, 28*, 159–169.

Hochstatter, K. R., Gustafson, D. H., Sr, Landucci, G., Pe-Romashko, K., Cody, O., Maus, A., et al. (2021). Effect of an mHealth intervention on hepatitis C testing uptake among people with opioid use disorder: randomized controlled trial. *JMIR mHealth and uHealth, 9*(2), e23080.

Hustad, J. T., Barnett, N. P., Borsari, B., & Jackson, K. M. (2010). Web-based alcohol prevention for incoming college students: A randomized controlled trial. *Addictive Behaviors, 35*(3), 183–189.

International Telecommunications Union (ITU). (2021, November 30). 2.9 billion people still offline. Accessed January 25, 2022 from https://www.itu.int/en/ITU-D/Statistics/Pages/stat/default.aspx.

Internet World Stats (2021, July 3). Internet growth statistics. Accessed January 25, 2022 from https://www.internetworldstats.com/emarketing.htm.

James, D. C., Harville, I. I. C., Sears, C., Efunbumi, O., & Bondoc, I. (2017). Participation of African Americans in eHealth and mHealth studies: A systematic review. *Telemedicine and e-Health, 23*(5), 351–364.

Jander, A., Crutzen, R., Mercken, L., Candel, M., & de Vries, H. (2016). Effects of a web-based computer-tailored game to reduce binge drinking among Dutch adolescents: A cluster randomized controlled trial. *JMIR, 18*(2), e4708.

Jimenez-Molina, A., Fanco, P., Martinez, V., Martinez, P., Rojas, G., & Araya, R. (2019). Internet-based interventions for the prevention and treatment of mental disorders in Latin America: A scoping review. *Frontiers in Psychiatry, 10*, 664.

Johansson, R., & Andersson, G. (2012). Internet-based psychological treatments for depression. *Expert Review of Neurotherapeutics, 12*, 861–870.

Johnson, K., Richards, S., Chih, M-Y., Moon, T. J., Curtis, H., & Gustafson, D. H. (2016). A pilot test of a mobile app for drug court participants. *Substance Abuse: Research and Treatment, 10*(SART), S33390.

Jonassaint, C. R., Belnap, B. H., Huang, Y., Karp, J. F., Abebe, K. Z., & Rollman, B. L. (2020). Racial differences in the effectiveness of internet-delivered mental health care. *Journal of General Internal Medicine, 35*(2), 490–497.

Josephine, K., Josefine, L., Philipp, D., David, E., & Harald, B. (2017). Internet- and mobile-based depression interventions for people with diagnosed depression: A systematic review and meta-analysis. *Journal of Affective Disorders, 223*, 28–40.

Kay-Lambkin, F. J., Baker, A. L., Kelly, B., & Lewin, T. J. (2011). Clinician-assisted computerised versus therapist-delivered treatment for depressive and addictive disorders: A randomised controlled trial. *Medical Journal of Australia, 195*, S44–S50.

Kay-Lambkin, F. J., Baker, A. L., Lewin, T. J., & Carr, V. J. (2009). Computer-based psychological treatment for comorbid depression and problematic alcohol and/or cannabis use: A randomized controlled trial of clinical efficacy. *Addiction, 104*(3), 378–388.

Kelders, S. M., Kok, R. N., Ossebaard, H. C., & Van Gemert-Pijnen, J. E. W. C. (2012). Persuasive system design does matter: A systematic review of adherence to web-based interventions. *JMIR, 14*, e152.

Kessler, R. C., Heeringa, S., Lakoma, M. D., Petukhova, M., Rupp, A. E., Schoenbaum, M., et al. (2008). Individual and societal effects of mental disorders on earnings in the United States: Results from the national comorbidity survey replication. *American Journal of Psychiatry, 165*(6), 703–711.

Kiluk, B. D., DeVito, E. E., Buck, M. B., Hunkele, K., Nich, C., & Carroll, K. M. (2017). Effect of computerized cognitive behavioral therapy on acquisition of coping skills among cocaine-dependent individuals enrolled in methadone maintenance. *Journal of Substance Abuse Treatment, 82*, 87–92.

Kiluk, B. D., Devore, K. A., Buck, M. B., Nich, C., Frankforter, T. L., LaPaglia, D. M., et al. (2016). Randomized trial of computerized cognitive behavioral therapy for alcohol use disorders: Efficacy as a virtual stand-alone and treatment add-on compared with standard outpatient treatment. *Alcoholism: Clinical and Experimental Research, 40*(9), 1991–2000.

Kumar, V., Sattar, Y., Bseiso, A., Khan, S., & Rutkofsky, I. H. (2017). The effectiveness of internet-based cognitive behavioral therapy in treatment of psychiatric disorders. *Cureus, 9*(8), e1626.

Li, D. H., Brown, C. H., Gallo, C., Morgan, E., Sullivan, P. S., Young, S. D., et al. (2019). Design considerations for implementing eHealth behavioral interventions for HIV prevention in evolving sociotechnical landscapes. *Current HIV/AIDS Report, 16*(4), 335–348.

Lintvedt, O. K., Griffiths, K. M., Sorensen, K., Ostvik, A. R., Wang, C. E. A., Eisemann, M., et al. (2011). Evaluating the effectiveness and efficacy of unguided internet-based self-help intervention for the prevention of depression: A randomized controlled trial. *Clinical Psychology and Psychotherapy, 20*, 10–27.

Marsch, L. A., Grabinski, M. J., Bickel, W. K., Desrosiers, A., Guarino, H., Muehlbach, B., et al. (2011). Computer-assisted HIV prevention for youth with substance use disorders. *Substance Use & Misuse, 46*(1), 46–56.

Marsch, L. A., Guarino, H., Acosta, M., Aponte-Melendez, Y., Cleland, C., Grabinski, M., et al. (2014). Web-based behavioral treatment for substance use disorders as a partial replacement of standard methadone maintenance treatment. *Journal of Substance Abuse Treatment, 46*(1), 43–51.

Martinez-Montilla, J. M., Mercken, L., de Vries, H., Candel, M., Lima-Rodríguez, J. S., & Lima-Serrano, M. (2020). A web-based, computer-tailored intervention to reduce alcohol consumption and binge drinking among Spanish adolescents: Cluster randomized controlled trial. *JMIR, 22*(1), e15438.

Mitchell, L. M., Joshi, U., Patel, V., Lu, C., & Naslund, J. A. (2021). Ecomonic evaluations of internet-based psychological interventions for anxiety disorders and depression: A systematic review. *Journal of Affective Disorders, 284*, 157–182.

National Institute on Drug Abuse (NIDA). (nd). Cost of substance abuse. Accessed at https://archives.drugabuse.gov/trends-statistics/costs-substance-abuse.

Nguyen, T. D., Attkisson, C., & Stegner, B. L. (1983). Assessment of patient satisfaction: Development and refinement of a service evaluation questionnaire. *Evaluation and Program Planning, 6*(3-4), 299–313.

Nissen, E. R., O'Connor, M., Kaldo, V., Højris, I., Borre, M., Zachariae, R., et al. (2020). Internet-delivered mindfulness-based cognitive therapy for anxiety and depression in cancer survivors: A randomized controlled trial. *Psycho-Oncology, 29*(1), 68–75.

O'Kearney, R., Kim, S., Dawson, R. L., & Calear, A. L. (2019). Are claims of non-inferiority of Internet and computer-based cognitive-behavioural therapy compared with in-person cognitive-behavioural therapy for adults with anxiety disorders supported by the evidence from head-to-head randomised controlled trials? A systematic review. *Australian and New Zealand Journal of Psychiatry, 53*(9), 851–865.

Olmstead, T. A., Ostrow, C. D., & Carroll, K. M. (2010). Cost-effectiveness of computer-assisted training in cognitive-behavioral therapy as an adjunct to standard care for addiction. *Drug and Alcohol Dependence, 110*(3), 200–207.

Pew Research Center. (April, 2021a). Internet/broadband fact sheet. Accessed from https://www.pewresearch.org/internet/fact-sheet/internet-broadband/. Washington, D.C.

Pew Research Center. (April, 2021b). Mobile fact sheet. Accessed from https://www.pewresearch.org/internet/fact-sheet/mobile/. Washington, D.C.

Postel, M. G., de Haan, H. A., Huurne Ter, E. D., Becker, E. S., & de Jong, C. A. (2010). Effectiveness of a web-based intervention for problem drinkers and reasons for dropout: Randomized controlled trial. *JMIR, 12*(4), e68.

Richards, D., Timulak, L., O'Brien, E., Hayes, C., Vigano, N., Sharry, J., et al. (2015). A randomized controlled trial of an internet-delivered treatment: Its potential as a low-intensity community intervention for adults with symptoms of depression. *Behaviour Research and Therapy, 75*, 20–31.

Rosso, I. M., Killgore, W. D., Olson, E. A., Webb, C. A., Fukunaga, R., Auerbach, R. P., et al. (2017). Internet-based cognitive behavior therapy for major depressive disorder: A randomized controlled trial. *Depression and Anxiety, 34*(3), 236–245.

Rothwell, P. M. (2005). External validity of randomised controlled trials: 'to whom do the results of this trial apply?' *Lancet, 365*(9453), 82--93.

Shi, J. M., Henry, S. P., Dwy, S. L., Orazietti, S. A., & Carroll, K. M. (2019). Randomized pilot trial of web-based cognitive-behavioral therapy adapted for use in office-based buprenorphine maintenance. *Substance Abuse, 40*(2), 132–135.

Sieverink, F., Kelders, S. M., & van Gemert-Pijnen, J. E. W. C. (2017). Clarifying the concept of adherence to eHealth technology: Systematic review on when usage becomes adherence. *JMIR, 19*(12), e402.

Simon, N., Robertson, L., Lewis, C., Roberts, N. P., Bethell, A., Dawson, S., et al. (2021). Internet-based cognitive and behavioural therapies for post-traumatic stress disorder (PTSD) in adults. *Cochrane Database Systematic Review, 5*(5), 1–108.

Sin, J., Galeazzi, G., McGregor, E., Collom, J., Taylor, A., Barrett, B., et al. (2020). Digital interventions for screening and treating common mental disorders or symptoms of common mental illness in adults: Systematic review and meta-analysis. *JMIR, 22*(9), E20581.

Stitzer, M., & Petry, N. (2006). Contingency management for treatment of substance abuse. *Annual Review of Clinical Psychology, 2*, 411–434.

Strauss, P., Morgan, H., Wright, Toussaint, D., Lin, A., Winter, S., et al. (2019). Trans and gender diverse young people's attitudes towards game-based digital mental health interventions: A qualitative investigation. *Internet Interventions, 18*, 100280.

Substance Abuse and Mental Health Services Administration (SAMHSA). (2020). *Key substance use and mental health indicators in the United States: Results from the 2019 National Survey on Drug Use and Health (HHS Publication No. PEP20-07-01-001, NSDUH Series H-55).* Rockville, MD: Center for Behavioral Health Statistics and Quality, Substance Abuse and Mental Health Services Administration Retrieved from https://www.samhsa.gov/data/.

Sugarman, D. E., Nich, C., & Carroll, K. M. (2010). Coping strategy use following computerized cognitive-behavioral therapy for substance use disorders. *Psychology of Addictive Behaviors, 24*(4), 689.

Taylor, C. B., Graham, A. K., Flatt, R. E., Waldherr, K., & Fitzsimmons-Craft, E. E. (2021). Current state of scientific evidence on internet-based interventions for the treatment of depression, anxiety, eating disorders and substance abuse: An overview of systematic reviews and meta-analyses. *European Journal of Public Health, 31*(S1), i3–i10.

Teesson, M., Newton, N. C., Slade, T., Chapman, C., Birrell, L., Mewton, L., et al. (2020). Combined prevention for substance use, depression, and anxiety in adolescence: A cluster-randomised controlled trial of a digital online intervention. *Lancet Digit Health, 2*(2), e74–e84.

Turner-McGrievy, G. M., Hales, S. B., Schoffman, D. E., Valafar, H., Brazendale, K., Weaver, R. G., et al. (2007). Choosing between responsive-design websites versus mobile apps for your mobile behavioral intervention: Presenting four case studies. *Translational Behavioral Medicine, 7*(2), 224–232.

Wagner, B., Horn, A. B., & Maercker, A. (2014). Internet-based versus face-to-face cognitive-behavioral intervention for depression: A randomized controlled non-inferiority trial. *Journal of Affective Disorders, 152--154*, 113–121.

Washburn, M., Yu, M., Rubin, A., & Zhou, S. (2021). Web-based acceptance and commitment therapy (ACT) for symptoms of anxiety and depression: Within-group effect size benchmarks as tools for clinical practice. *Journal of Telemedicine and Telecare, 27*(5), 314–322.

Wei, Y., Zheng, P., Deng, H., Wang, X., Li, X., & Fu, H. (2020). Design features for improving mobile health interventions user engagement: Systematic review and thematic analysis. *JMIR, 22*(12), e21687.

Woodhouse, R., Neilson, M., Martyn-St James, M., Glanville, J., Hewitt, C., & Perry, A. E. (2016). Interventions for drug-using offenders with co-occurring mental health problems: A systematic review and economic appraisal. *Health Justice, 4*(1), 10.

Chapter 4

Second wave of scalable digital therapeutics: Mental health and addiction treatment apps for direct-to-consumer standalone care

Patricia A. Areán and Ryan Allred

University of Washington Department of Psychiatry and Behavioral Sciences, CREATIV Lab, Seattle, WA, U.S.A.

4.1 Overview of second-wave digital therapeutics

Second-wave digital mental and behavioral health care consists of numerous forms of technology with the aim of helping people manage their mood and behavior. These technologies take several forms, such as *mobile phone-based applications (apps)*, in which care is delivered either through SMS (text-based) technologies or through an app program where people have access to digital education and strategies for mood and behavioral management, *symptom trackers*, in which the consumer uses an app to track goals, symptoms, and side effects, *mindfulness applications*, which are based in meditation strategies, *online peer communities*, where people find support from others in managing their mental health challenges, and *serious games*, where people receive care using algorithmic-based programs that have the look and feel of a video game. Digital therapeutics such as these have been found to be highly effective in over 100 randomized controlled trials (RCTs) (Andrews et al., 2018; Firth et al., 2017a, 2017b), but there is considerable variability as to which types of digital therapeutics are the most effective (Arean et al., 2016; Karyotaki et al., 2017; Kenter et al., 2015; Weisel et al., 2019; Wright et al., 2005), and whether commercial versions of these therapeutics are as effective as the research-grade versions (Torous et al., 2018). The purpose of this chapter is to review the recent evidence supporting the use of second-wave digital therapeutics, the challenges encountered in using these tools, how to use these tools in practice, and finally future directions in the research and development of self-guided mental health tools.

4.2 Evidence base for second wave digital therapeutics

As of this writing, there are 10,000–20,000 (Clay, 2021) digital therapeutics for mental health and stress reduction. These apps purport to provide support for depression and anxiety (Andersson, Cuijpers, Carlbring, Riper, & Hedman, 2014; Andrews et al., 2018; Linardon, Cuijpers, Carlbring, Messer, & Fuller-Tyszkiewicz, 2019), PTSD (Olthuis et al., 2016), substance use disorders (Tofighi, Chemi, Ruiz-Valcarcel, Hein, & Hu, 2019), and for symptom-specific problems, such as sleep disturbance (Zachariae, Lyby, Ritterband, & O'Toole, 2016). In addition, these technologies appear to be quite helpful for managing mild, moderate, and severe behavioral and mental health problems and are effective across the lifespan (Anguera, Gunning, & Arean, 2017; Ebert et al., 2015; Pratap et al., 2018). These tools vary in the number of features they offer. Some are based in psychotherapeutic principles, offering education and self-guided strategies for symptom management and function, some are simple mood tracking systems that offer advice to help with the improvement of mood, some teach mindfulness strategies, others offer support through peer contact, and others are serious game technologies. These apps also vary in the degree with which there is evidence to support claims for mental health challenges they purport to

Digital Therapeutics for Mental Health and Addiction: The State of the Science and Vision for the Future. DOI: https://doi.org/10.1016/B978-0-323-90045-4.00011-3
31

	Depression & Anxiety	PTSD	Substance Use	Total
Coach-Guided Apps	66	8	30	104
Self-Guided Apps	18	5	2	25
Mindfulness Apps	16	4	2	22
Online Peer Communities	10	0	3	13
Serious Games	9	2	0	11
Tracking Apps*	0	0	0	0

*These counts only include basic symptom-monitoring apps and exclude tracking apps that also have integrated interventions or evidence-based strategies

FIGURE 4.1 Approximate number of trials finding that app use leads to improved clinical outcomes, grouped by app type and mental health condition.

manage, with the majority of evidence supporting digital therapeutics that include guidance from a mental health practitioner or coach (Mohr et al., 2021; Wang, Varma, & Prosperi, 2018) (see Fig. 4.1).

In this section, we discuss and review the evidence for each mental health app type to help inform the reader which digital therapeutics are worth investing in, based on the evidence available as of this writing. For each type, we will describe the components of research-grade digital therapeutics (apps that were used in the clinical trials) and then describe how known commercial examples may vary from the research-grade offerings. We note here that in two large consensus forums regarding standards for digital therapeutics, experts have suggested that not every commercial tool developed needs to undergo a randomized clinical trial, so long as the commercial offering possesses the features that were present in the research-grade versions (Mohr et al., 2021; National Institute of Mental Health, 2015). We cannot be exhaustive here regarding the nearly 20,000 second-wave digital therapeutics now available to consumers, but we will provide examples of some of the top tools and recommend search engines to help clinicians and consumers select commercial tools most like research-grade therapeutics.

Guided digital therapeutics. Guided digital therapeutics are mobile applications that are often based in known, evidence-based psychotherapies and behavioral interventions, and include access to a clinician or coach as part of the treatment package. Research-grade digital therapeutics are designed to be mobile phone-based applications that generally have an educational component (how to use this app, explanation of the evidence-based practice), in-app activities that are designed to help consumers learn evidence-based strategies for habit control or mood management, and finally support from an actual clinician or trained coach who provides support in using the in-app activities or actual counseling through video chat or text messaging. Guided digital therapeutics include the following features (Bakker, Kazantzis, Rickwood, & Rickard, 2016):

- Uses strategies from evidence-based behavioral interventions—in particular, cognitive-behavioral therapy, dialectical behavioral therapy, problem-solving treatment, behavioral activation, or exposure therapy
- Activity/goal setting and tracking
- Behavior/mood tracking
- Education about coping methods
- Access to a clinician or coach
- Real-time engagement

Access to a coach or therapist is critical for these tools to be effective. Research finds that having a clinician or coach as part of the tool enhances consumer engagement, accountability, and potentially assures better safety monitoring (Mohr et al., 2021). Recent evidence suggests that access to either interventions or coaches in real-time optimizes skill acquisition for the consumer, as the advice and tools can be applied in the very moment they are needed (Depp et al., 2010; Donker et al., 2013). It is unclear to what degree the in-app activities contribute to improvement in clinical and behavioral outcomes, given that most studies find use of the in-app tools without clinician support are not nearly as effective as use of the tools in combination with clinical support (Wright et al., 2019), and only very small, proof-of-concept studies exist to date which find no difference in outcomes between traditional, face-to-face counseling and counseling augmented with digital technologies (Broglia, Millings, & Barkham, 2019). This would seem to suggest that the powerful component in guided digital therapeutics is the clinical interaction with a coach or counselor. More research is needed to disentangle which of the research-grade features are critical to outcomes.

As for commercial examples, as many as 81 companies globally offer guided digital therapeutics, either as part of a national healthcare system (Duffy, Enrique, Connell, Connolly, & Richards, 2019; National Institute for Health and Clinical Excellence (NICE), 2006, 2009; Richards et al., 2020; Titov et al., 2017), as part of employee assistance programs, as contracts through health plans, or as fee-for-service models (Mohr et al., 2021). Commercial offerings do vary in the degree to which they rely on research-grade features. Some companies rely on the in-app therapeutic tools, with access to coaching to support the use of the tool, to the service being primarily a technology-based communication platform between consumers and providers, where consumers have access to a library of tools, but these tools are not central to care. We describe five examples of such companies.

Two large US-based companies, Talkspace and Betterhelp, offer on-demand text messaging-based care with a licensed clinician, with the option of adding on virtual visits that are not unlike telepsychology visits. For these companies, message-based care is central to the tool. While tele-psychotherapy has a very strong evidence base, stand-alone message-based care does not (Hoermann, McCabe, Milne, & Calvo, 2017), although it could be assumed that the opportunity to receive intervention in the moment of need may be a potent intervention (Hull, Malgaroli, Connolly, Feuerstein, & Simon, 2020). However, this feature has not been rigorously investigated to date, although Talkspace is currently funded by the National Institute of Mental Health to conduct a Sequential Multiple Assignment Randomized Trial comparing message-based care to tele-psychotherapy [trial registration number: NCT04513080].

Three other companies, two US based and one UK based, offer consumers a combination of services similar to research-grade mental health apps. Sanvello, Ginger.io, and SilverCloud offer self-guided tools and access to a therapist, where consumers utilize both the self-guided tools and regular telepsychotherapy, which includes (for Ginger.io and Sanvello) on-demand messaging. Sanvello and Ginger.io also include additional support tools, where Ginger.io's on-demand services are managed by lay coaches, and Sanvello offers peer support with or without access to a licensed clinician.

Evidence base for guided digital therapeutics. The study of digital therapeutics in the management of depression and anxiety has by far the largest evidence base. As of this writing, at least 66 well-designed RCTs find that guided digital therapeutics that contain the features described above are effective for depression and anxiety disorders and are as good as traditional face-to-face therapies (Andersson et al., 2014; Andrews et al., 2018; Linardon et al., 2019). A majority of research-grade digital therapeutics studied are based in cognitive-behavioral or behavioral activation therapies. Most of the evidence in existence is short-term impact, with most studies following participants for no more than 12 weeks. The few studies that have longer follow-up periods find the effectiveness compared to control conditions tends to be very small for both depression and anxiety (Linardon et al., 2019).

PTSD is a complicated condition and treatments for this disorder require regular contact with a clinician and the completion of exposure-based therapy that can either be in vivo or written. Exposure therapy exercises require considerable reflection on the part of the consumer, with specific guidance from a clinician to help manage both the symptoms of PTSD and the cognitive reprocessing that needs to take place for exposure to work. As one can imagine, the creation of mobile health tools to emulate the therapeutic experience consumers encounter in person is complicated, and a guided digital therapeutic seems necessary. Self-guided tools do exist, but are meant to be used in conjunction with face-to-face treatment (Owen et al., 2018). An advantage of digital therapeutics when applied to PTSD treatment is the opportunity for more frequent contact with the provider. Interventions such as cognitive processing therapy and exposure therapy were originally designed to be offered more than once a week, a practice that is no longer widely available (Vinograd & Craske, 2020). Although there are fewer studies of guided digital therapeutics for PTSD, the studies available are promising. As of this writing, there have been eight clinical trials that have shown these therapeutics to be effective (Olthuis et al., 2016). A recent systematic review, however, found that most research-grade PTSD apps were not designed to be well integrated with traditional clinical care, and suffered from significant issues with consumer engagement and app acceptability (Rodriguez-Paras et al., 2017).

The use of digital therapeutics was first developed for habit control conditions, health enhancement, and alcohol misuse (Marsch et al., 2020). One recent systematic review of digital therapeutics identified 30 clinical trials in the substance abuse field and found guided therapeutics to be highly effective in habit control and craving management (Steinkamp et al., 2019). However, for therapeutics geared toward opioid use disorders, there are very few on the market. Many are clinician-facing and focused on supporting conversations about medication titration, and none with any evidence to support these technologies (Nuamah, Mehta, & Sasangohar, 2020). Although this area of research has been active for decades, the use of guided smartphone applications in the context of substance use disorders is still in development.

Self-guided second-wave digital therapeutics. Self-guided digital therapeutics are stand-alone mobile applications that provide in-app exercises based on evidence-based practices without access to a coach or clinician. In that regard, research-grade versions of these tools are similar to the guided versions. As stated above, self-guided digital therapeutics are not as effective for moderate to severe mental and behavioral health problems, but do have an effect on mild mental health challenges (Arean et al., 2016; Cuijpers et al., 2011; Drissi, Ouhbi, Janati Idrissi, & Ghogho, 2020; Pratap et al., 2018;

Rathbone & Prescott, 2017). However, symptom severity alone should not be used to determine who should use self-guided digital therapeutics, as some trials have shown that digital mental health treatments can be effective across all severities (Mohr, Kwasny, Meyerhoff, Graham, & Lattie, 2021).

Research-grade self-guided tools that have an evidence base, like with their guided counterparts, have a basis in evidence-based practices, in particular cognitive-behavioral therapy (Cuijpers et al., 2011), problem-solving therapy (Arean et al., 2016), dialectical behavior therapy (O'Grady et al., 2020), and there is one example for acceptance and commitment therapy (Järvelä-Reijonen et al., 2018). Self-guided apps with an evidence base have the following features:

- Uses strategies from evidence-based behavioral interventions
- Activity/goal setting and tracking
- Behavior/mood tracking
- Education about coping methods

Compared to research-grade digital therapeutics, commercial offerings vary from having all of the features noted above to none of them. In one review of commercial apps based in cognitive-behavioral therapy or behavioral activation, the authors identified one hundred and seventeen apps on App and Play stores making claims to be based in these therapies, yet only 12 met the quality standards mentioned above (Huguet et al., 2016). Further, self-guided apps are poorly regulated (Parker, Bero, Gillies, Raven, & Grundy, 2018; Torous et al., 2019). However, this is true for all second-wave digital therapeutics and is a problem that may need to be remedied in the future. Another challenge with self-guided digital therapeutics is that, without the guidance of a coach, they are not used by consumers as intended by developers. One recent study found that, when compared to research participant engagement with self-guided digital therapeutics, consumers who access these tools tend to engage differently, using the therapeutic repeatedly in the first few weeks of download, but then not using the tool after (Baumel & Kane, 2018). It should be noted here that this study was not able to collect clinical outcomes, and thus it is unclear if this type of engagement is optimal or suboptimal. Although consumers of self-guided digital therapeutics have found non–evidence-based tools to provide in the moment relief for symptoms, these tools generally have lower consumer satisfaction rating on the App and Play stores (Baumel, Torous, Edan, & Kane, 2020). We discuss the issue of engagement later in the chapter, but raise it here because engagement with self-guided digital therapeutics is markedly different than for guided digital therapeutics and may be the reason that they are not as effective as their guided counterparts (Mohr et al., 2019). Finally, although the evidence for these digital therapeutics is promising, a number of studies have found that people do not see these tools as replacing traditional treatments (Chan & Honey, 2021), and many feel these approaches to care would be marginally effective because of the lack of accountability (Renn, Hoeft, Lee, Bauer, & Arean, 2019).

Evidence base for self-guided digital therapeutics. There are very few randomized clinical trials of self-guided digital therapeutics, compared to guided tools, with many publications in this area focused largely on proof-of-concept engagement strategies and app design (Miralles et al., 2020). A majority of the research conducted to date has focused on depression and anxiety management (Miralles et al., 2020) and finds that the application of self-guided digital therapeutics to be more effective than no treatment or to controls that do not provide evidence-based treatment (Park et al., 2020). In the area of PTSD, there are no less than 201 apps for the assessment and treatment of PTSD, of which 45 are specific to evidence-based PTSD treatment (Rodriguez-Paras et al., 2017). However, a recent systematic review has found that there are very few clinical trials of apps for PTSD and the tools that are based in evidence-based treatments are not meant to be stand-alone interventions (Rodriguez-Paras et al., 2017). Finally, there is a small yet emerging evidence base supporting the use of self-guided digital therapeutics for alcohol use disorders (though less so for other substance use), where self-guided tools based on self-determination theory are showing promise in the reduction of risky drinking behaviors (Tofighi, Abrantes, & Stein, 2018).

Mood tracking apps. As their name suggests, mood tracking apps are meant to increase a consumers' awareness of their mood and related symptoms through the regular monitoring and tracking of these symptoms. While mood tracking apps do not typically offer a treatment or intervention, they can often lead to helpful information-seeking for mood disorders (Scherr & Goering, 2020). Mood tracking is a hallmark feature of many evidence-based treatments, and in CBT is considered an important therapeutic element. Mood tracking apps commonly include features such as symptom logging, journaling, and the regular collection of activity and mood data via smartphone sensors and ecological momentary assessments (EMA's), which are brief and usually automated survey questions that prompt users to note how they are feeling in the moment during frequent intervals. Modern mood tracking apps are often used in conjunction with "wearables"—devices such as FitBits or Apple watches that can track physical activity through various built-in sensors. Wearables offer a streamlined method of mood tracking through their ability to collect data from users in their natural environment without requiring any time or effort from the user, as data is automatically collected in the background.

Top-rated mood tracking apps like Happify, myStrength, PTSD Coach, and Daylio typically include additional activities and resources beyond simple symptom tracking (MindTools.io, 2021; One Mind PsyberGuide, 2021). Happify uses short questionnaires to track one's mood and uses the responses to suggest mood-boosting activities. Similarly, myStrength uses questionnaire responses to offer personalized activities and a library of video clips, audio clips, and other helpful information. Like Happify and myStrength, PTSD Coach allows users to track symptoms and also includes a robust library of resources like a diagnosis checklist, skills to help improve symptoms, and information on obtaining supports. Apps like Daylio, which serves as a more basic digital mood journal, and other tracking apps also provide a "progress" section where users can see how their symptoms have changed over time.

4.2.1   Evidence base for mood tracking apps

Like many digital therapeutic apps, mood tracking apps focus on the most common mental health conditions: depression, anxiety, and PTSD. However, since most mood tracking apps focus on symptom tracking instead of treatment, very few trials have studied clinical outcomes for these conditions. Instead, most studies have focused on the usability and feasibility of tracking apps (Sequeira et al., 2020). Studies looking at tracking apps for depression and anxiety largely show high feasibility (McCall, Ali, Yu, Fontelo, & Khairat, 2021; Sequeira et al., 2020), with the most feasible app features being wellness trackers, brief personalized interventions (Hetrick et al., 2018), prompts for completing self-reports, and encouraging messages of support (Shrier & Spalding, 2017). Similarly, the evidence shows that mood tracking apps related to PTSD are largely acceptable and feasible. In one study, 79% of veterans found an app called Cogito Companion to be acceptable (Betthauser et al., 2020), while another study examining an app called T2 Mood Tracker found that 100% of participants rated the app as "somewhat" or "very" easy to use (Bush, Ouellette, & Kınn, 2014).

The evidence base has also reviewed the impact of mood tracking apps on resource-seeking. This evidence shows that the self-monitoring of one's depressive symptoms in a mood tracking app can be associated with higher depression-related resource-seeking (Scherr & Goering, 2020).

Mindfulness apps. Mindfulness apps are designed to help consumers learn the principles of mindful meditation. Research has demonstrated that a regular practice in meditation, which usually involves 10–20 minutes of focused attention on body awareness, can improve perceived stress, anxiety, depression, and psychological well-being (Gál, Ştefan, & Cristea, 2021). Effective mindfulness apps typically include guided meditation with instruction, timers, and reminders to meditate. Second-wave digital therapeutic tools to support training in mindfulness practice abound, with 2500 such apps having been launched since 2015 (Gál et al., 2021).

The most highly rated commercial mindfulness apps are Headspace and Calm, both of which include guided meditation exercises (Gál et al., 2021). These two apps, however, target different aspects of wellbeing. Headspace includes modules specific for depression, anxiety, and habit control, where the guided meditation will focus more specifically on the types of distractions and thoughts that impact these disorders. Calm, however, does not have specific modules for mental health conditions, but rather focuses on behaviors that mediate mental and behavioral health, such as sleep.

Evidence base for mindfulness apps. The evidence base for these second-wave digital therapeutics is relatively small, with Headspace being one of the most commonly studied interventions (Howells, Ivtzan, & Eiroa-Orosa, 2016; Rosen, Paniagua, Kazanis, Jones, & Potter, 2018). The most common outcome measured in these studies is either depression or well-being, with anxiety, PTSD, and substance use disorder being in the minority of existing studies. Mindfulness can have a positive impact on stress (Goyal et al., 2014), depression, anxiety, and substance use disorder (Goldberg et al., 2018), with somewhat limited support for PTSD (Hilton et al., 2017). A recent meta-analysis of research on these apps has found them to be effective for the same conditions traditional meditation practice is indicated for, with 34 randomized clinical trials as of this writing. However, this study cautions that there is considerable heterogeneity in effect sizes of these apps as well as sample sizes, thus interpretation of these data should be made with caution (Gál et al., 2021). Finally, engagement, as with all second-wave digital therapeutics, is less than desired by the developers. However, one study found that while all app types are effective, consumers much preferred apps that included guided meditation (Mani, Kavanagh, Hides, & Stoyanov, 2015).

Online peer communities. Online peer communities, also known as peer-to-peer supports, are online communities that allow individuals to virtually connect with one another, discuss their experiences, and share resources and emotional support with one another. Research shows that these online peer communities can increase access to mental health treatment while providing a helpful space to reduce stigma (Ali, Farrer, Gulliver, & Griffiths, 2015; Naslund, Aschbrenner, Marsch, & Bartels, 2016). With an ever-growing percentage of people who have access to the internet (about 93% of young adults in the United States (Lenhart, Purcell, Smith, & Zickuhr, 2010)), online peer communities have become a particularly scalable digital therapeutic that is very popular among this demographic.

Peer communities exist primarily within social networking and group forum sites, with a smaller portion of communities existing through smartphone apps. Among social networking sites, Facebook groups are the most popular for peer communities, as Facebook is one of the largest social networking sites and hosts more open-access support groups that are specific to health issues (Prescott, Rathbone, & Brown, 2020; Ridout & Campbell, 2018). Another type of networking site, called moderated online social therapy (MOST), is a platform built more intentionally for peer-to-peer support and typically focuses on specific mental health conditions like depression or anxiety (Ridout & Campbell, 2018). Among smartphone apps for peer-to-peer support, seven cups of tea (7Cups) is one of the most commonly used. 7Cups offers peer support through several avenues including one-on-one chat with a trained volunteer, group support chats, and community forums (MindTools.io, 2021).

4.2.2 Evidence base for online peer communities

Much of the recent evidence base for online peer communities, which include several randomized clinical trials, looks at social media platforms like Facebook. This aligns with findings that Facebook groups serve as one of the most popular avenues for peer-to-peer support (Prescott et al., 2020). In terms of mental health outcomes, depression and anxiety have been studied the most, with at least ten trials showing improved depression symptoms (Griffiths, Calear, & Banfield, 2009) and at least one trial showing improved anxiety symptoms (Ellis, Campbell, Sethi, & O'Dea, 2011). Peer-to-peer supports also show some preliminary effectiveness for substance use, especially for short-term improvements with tobacco use (Woodruff, Edwards, Conway, & Elliott, 2001, 2007). For PTSD, however, there are no known trials showing that online peer communities lead to improved outcomes.

Despite the small evidence base related to clinical outcomes for online peer communities, they have still proven to be beneficial for those with mental health difficulties—especially in young people (Ali et al., 2015; Pretorius, Chambers, & Coyle, 2019; Ridout & Campbell, 2018). These types of supports can increase feelings of connectedness, provide helpful resources, and reduce stigma around mental health conditions, particularly for those experiencing depression and anxiety (Ali et al., 2015; Naslund et al., 2016; Pretorius et al., 2019; Ridout & Campbell, 2018). This is illustrated by one cross-sectional study looking at online depression communities, which found that heavy users of online depression groups rated a mean score of 4.24 out of 5 for level of agreement with the statement "I gain knowledge about various treatments," and a mean score of 4.35 out of 5 for the statement "I feel connected with others" (Nimrod, 2013).

Serious games. Serious games are video games that are designed to serve a purpose above and beyond play. In the context of this chapter, serious games are games designed to improve the mood or cognitive performance of users. Early serious games were all played on a computer, but several serious games are now available as mobile apps (Fleming et al., 2017; Lau, Smit, Fleming, & Riper, 2016). As of this writing, serious games are still novel as a digital therapeutic for mental health and the research on them is limited (Johnson et al., 2016). However, serious games do exist that use cognitive training to improve general mental well-being and a few also target mood disorders like depression. Successful serious games employ two important features for effective cognitive remediation: adaptive algorithms and multitasking paradigms. Serious games with the most effect in proof-of-concept trials use a benchmark called productive struggle, where the game is set to allow users to succeed in 80% of the trials they play (Anguera et al., 2017). As users improve (or tire), the game adapts to ensure users are experiencing optimal productive struggle. The second feature, multitasking, consists of the user completing two distracting tasks. In a recent FDA-approved serious game, EndeavourRx, users must navigate through an obstacle course while simultaneously selecting predetermined objects from a series of similarly looking objects.

Top-rated serious games for cognitive training include CogniFit, HAPPYneuron Pro, and Lumosity, all of which are smartphone apps. These apps all use games to target cognitive functions like working memory, visual processing, speed, and attention (One Mind PsyberGuide, 2021). Games targeted more specifically at mood disorders or other mental health conditions include Sparx, a CBT-based video game for depression and anxiety, Journey of the Wild Divine and EndeavorRx, games designed to help children diagnosed with ADHD, and Personal Zen, a mobile app-based game aimed at reducing stress and anxiety.

4.2.3 Evidence base for serious games

The evidence base for serious games remains sparse. Although there is ample research on serious games related to cognitive performance, few randomized clinical trials exist for clinical outcomes of mental health disorders. To the authors' knowledge, there is not yet enough evidence to show that serious games can improve outcomes for PTSD, anxiety, or substance use. Although, it should be noted that veterans with PTSD may be more likely to play video games, making serious games a viable option for future research on PTSD treatment (Grant, Spears, & Pedersen, 2018). Some preliminary evidence does exist showing that serious games can be effective in treating depression (Anguera et al., 2017; Fleming et al., 2017; Gunning, Anguera, Victoria, & Arean, 2021; Lau et al., 2016; Li, Theng, & Foo, 2014). However, this evidence is still in its early stages and more research is needed to support the use of serious games for depression.

4.3 Cautions and limitations with second-generation digital therapeutics

Direct-to-consumer, stand-alone mental health apps are not suited for everyone who may be seeking care for mental health challenges. The development and rise of these tools have been quite rapid and it is only as they have been used in the wild that we are beginning to learn what the challenges are with using these digital therapeutics. Challenges can be categorized into these key themes: engagement, accessibility, data security and privacy, and risk management.

Engagement. Engagement is the degree to which consumers use a digital therapeutic in the way it was designed to be used for an optimal length of time (treatment dosage) to ensure long-term behavioral and emotional change (Cole-Lewis, Ezeanochie, & Turgiss, 2019). Several studies find that people do not engage as developers intended, and long-term engagement is poor (Hatch, Hoffman, Ross, & Docherty, 2018; Lattie et al., 2019; Ng, Firth, Minen, & Torous, 2019; Yeager & Benight, 2018). Indeed, a recent comparison of research engagement with research-grade digital therapeutics to real-world engagement of digital therapeutics found that research participants were four times more likely to engage with the same technological offering than real-world consumers (Baumel, Edan, & Kane, 2019; Kelders, Kok, Ossebaard, & Van Gemert-Pijnen, 2012). Although engagement tends to be better when consumers have access to clinicians and coaches, less than 1% of consumers who access guided digital therapeutics complete all sessions (Gilbody et al., 2015).

However, the association between engagement and outcome is not well understood. Evidence suggests that many people may benefit from 2 weeks of DMH use (Arean et al., 2016) and in the informatics field, engagement with digital health technology has been found to be cyclical, with people having many lapses and re-engagements over time depending on the fit of the tool for the person's goals and the contexts in which they have engaged with the tool (Epstein, Ping, Fogarty, & Munson, 2015; Lin, Althoff, & Leskovec, 2018). Hence, optimal engagement is not a one-size/context-fits-all phenomenon. Indeed, our own research points to the fact that when consumers are asked about their preferences for digital therapeutics versus traditional therapies, many indicate that they are unclear as to how such offerings would be helpful for behavioral or mental health challenges (Renn et al., 2019), and while digital therapeutic companies saw a substantial increase in the number of people requesting services during COVID-19, we found that only 16% of those in need during the pandemic ever looked to these tools as a means of coping (Arean et al., 2021). Thus, although these technologies are effective, consumers may prefer to use these tools in ways developers have not considered and may not even feel that technology is an ideal method of care delivery.

Accessibility. Another problem developers of second-wave digital therapeutics have not well considered is whether their tools are accessible to everyone. Those from lower incomes, those who live in areas where broadband is inaccessible, or those who live with disabilities are often overlooked in the development of these therapeutics. For instance, people with visual and hearing disabilities often find such devices to be unusable. Although smart phones and laptops have methods for making accommodations for vision and hearing impairment, many apps are designed in such a way that operating system tools designed to help with accommodations cannot work (Sayal, Subbalakhmi, & Saini, 2020). Those with lower incomes often use older generation phones that are incompatible with newer generation applications. The use of new apps on older phones creates a wealth of problems, including battery drain and increased cost owing to data plan limitations (Liu et al., 2020). Finally, people who live in rural areas often struggle with consistent connectivity, making therapeutics that provide on-demand services untenable (Liu et al., 2020). However, these challenges are not insurmountable, and future developers should consider how to create tools that mitigate these access barriers.

Privacy standards. Although the EU and United Kingdom have created regulations to protect consumer smartphone data, in the United States such regulation is not as strict (Huckvale, Torous, & Larsen, 2019). This problem is changing, particularly in the wake of data breach scandals involving Facebook and Cambridge Analytica. Privacy rules and standards can be hard to follow, but to protect one's privacy and minimize risks, it is imperative to read privacy rules before selecting a second-wave digital therapeutic.

Risk management. As of this writing, there is no evidence that second-wave digital therapeutics are harmful (Ebert et al., 2016). However, some commercial tools may make claims about the rigor of their tool or the potential therapeutic benefit without ample evidence in support of those claims. Indeed, a serious game for cognition, Lumosity, was ordered by the FCC to pay a two million dollar fine for making unsubstantiated claims (Federal Trade Commission, 2016). Care should be taken in selecting tools based on company claims.

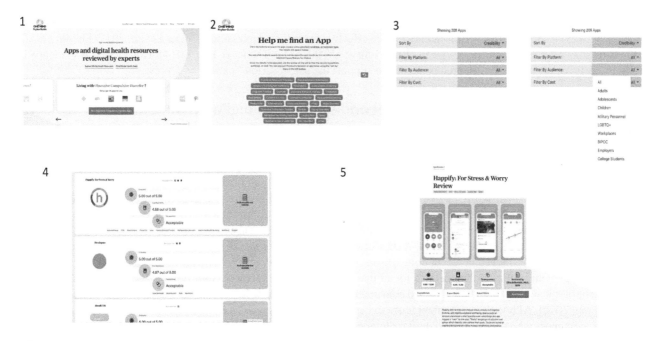

FIGURE 4.2   OneMind/Psyberguide platform. (1) select the condition of interest; (2) select tool focus/purpose; (3) sort by criteria you feel is most important; (4) list of tools are provided, with (5) a independent review of the tool.

4.4   How to find effective second-wave digital therapeutics

The universe of second-wave digital therapeutics is an ever-changing one, where apps come and go rapidly and are changing to keep up with the latest advances in mobile technology. Although we were able to provide a primer as to what features should be present in these technologies to be considered evidence-based, it can be very difficult for the average consumer or clinician to determine which of the tens of thousands of apps available have the best features. In addition to the need for second-generation digital therapeutics to be evidence-based, issues such as engagement, accessibility, privacy, and risk management, which we reviewed above as potential limitations, need to be taken into consideration. Fortunately, companies have begun to surface that will do this work for the consumer and clinician.

Organizations like One Mind PsyberGuide, the Organization for the Review of Care and Health Applications (ORCHA), Mindapps and Mindtools.io are helpful and reliable resources for learning about evidence-based digital therapeutics. These tools use a systematic review of mental health apps to find which ones are most credible and backed by evidence-based treatments. Although each platform has a unique presentation of data and different names for their rating categories, they all assess similar app characteristics and constructs (Carlo, Hosseini Ghomi, Renn, & Arean, 2019). Each organization's website allows a user to search for apps by a specific condition (e.g., depression, anxiety, PTSD, etc.) or treatment type (e.g., cognitive behavioral therapy, dialectical behavior therapy, etc.). The search results show the top-rated apps in a selected category based on ratings like credibility, user experience, and privacy transparency. Users can read about the purpose of each app and will see a summary of its features. Some of these tools (i.e., One Mind PsyberGuide) will also link any of the app's supporting research for the user to review. Given their ease of access, extensive app databases, and reliance on evidence-based results, these sites are useful tools for anyone looking to use digital therapeutics (see Fig. 4.2).

4.5   How to use second-wave digital therapeutics in practice

Clinicians and consumers can potentially benefit from using these technologies in the management of behavioral and mental health care. While there are a number of opinion pieces to help clinicians incorporate second-wave digital therapeutics into their practice, in truth there is little evidence-based guidance on the best way to use these tools as part of one's clinical practice. Realistically, guided digital therapeutics will be less likely to be of benefit to clinicians since these tools already offer counseling in some form, with the exception of peer support tools, which are helpful to consumers who would benefit from the support of others going through similar experiences.

Second-wave digital therapeutics can be used successfully as an electronic method for assessing and monitoring consumer outcomes, akin to asking consumers to track their mood using paper-pencil form. The symptom tracking tools

Should you use a digital tool in your practice?

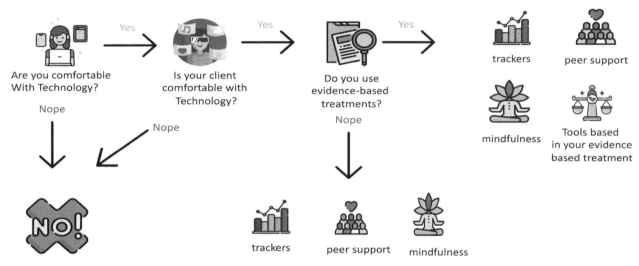

FIGURE 4.3 Digital Therapeutic Decision Tree.

we reviewed above can be handy methods for consumers to track their mood on a daily basis, however, a real challenge for clinicians is the fact that very few of these tools are compatible with existing electronic health systems, making data access—and protection—complicated (Connolly, Kuhn, Possemato, & Torous, 2021). If clinicians select tools and there is no means for integrating or sharing consumer data safely and securely into the medical record, then clinicians will need to spend time looking over the electronic data, much as they would with a paper–pencil form.

Self-guided digital therapeutics may be particularly helpful for consumers who feel they need on-going motivation and real-time access to therapeutic strategies. Clinicians will be somewhat limited to selecting apps, unless they offer cognitive-behavioral therapy as part of their care. Having an app make recommendations based on one therapeutic model that is not in-line with the therapeutic methods clinicians are using may cause some confusion for the consumer. As we discussed above, some second-wave digital therapeutics were designed specifically to be adjuncts of evidence-based treatments, for instance apps for PTSD, and therefore are designed to support clinicians in real-time monitoring of consumer engagement with out-of-session work as well as daily tracking (see Fig. 4.3).

4.6 Summary and future directions

In summary, second-wave digital therapeutics have enough evidence to be deemed effective for managing mental and behavioral health conditions (Andrews et al., 2018; Anguera et al., 2017; Fleming et al., 2017; Gál et al., 2021; Goldberg et al., 2018; Griffiths et al., 2009; Gunning et al., 2021; Lau et al., 2016; Li et al., 2014; Linardon et al., 2019; Olthuis et al., 2016; Park et al., 2020; Steinkamp et al., 2019), but the evidence varies based on what features are available. Guided apps have the greatest evidence, but self-guided tools are beginning to emerge as a viable option for people with mild presentations of mental and behavioral conditions. Peer support and serious games still need greater study and scrutiny, but the evidence, albeit small, is showing promise as well. Apps can be effectively used as adjuncts to traditional mental health treatment, but clinicians should be aware that clients will have preferences for how these tools play a role in their recovery. It appears that clients may prefer to use these tools as supports in goal setting and symptom tracking, and once the goal is reached to set these tools aside until they feel they need them again. Engagement with these tools is still poorly understood, and more work is needed to determine what is optimal engagement for improved clinical outcomes. Of major importance is the need to include consumers in the decision and development of second-wave digital therapeutics. Based on the extant research on consumer preferences for these tools, many still prefer access to a clinician over access to a digital therapeutic when it comes to receiving treatment, and the role they see technology playing in their recovery is still unclear. As a result, future research should look into what the typical consumer is looking for in a digital therapeutic, the role they feel it should play in their recovery, and how to define optimal engagement with these tools.

References

Ali, K., Farrer, L., Gulliver, A., & Griffiths, K. M. (2015). Online peer-to-peer support for young people with mental health problems: A systematic review. *JMIR Mental Health, 2*(2), e19. https://doi.org/10.2196/mental.4418.

Andersson, G., Cuijpers, P., Carlbring, P., Riper, H., & Hedman, E. (2014). Guided internet-based vs. face-to-face cognitive behavior therapy for psychiatric and somatic disorders: A systematic review and meta-analysis. *World Psychiatry, 13*(3), 288–295. https://doi.org/10.1002/wps.20151.

Andrews, G., Basu, A., Cuijpers, P., Craske, M. G., McEvoy, P., English, C. L., et al. (2018). Computer therapy for the anxiety and depression disorders is effective, acceptable and practical health care: An updated meta-analysis. *Journal of Anxiety Disorders, 55*, 70–78. https://doi.org/10.1016/j.janxdis.2018.01.001.

Anguera, J. A., Gunning, F. M., & Arean, P. A. (2017). Improving late life depression and cognitive control through the use of therapeutic video game technology: A proof-of-concept randomized trial. *Depression and Anxiety, 34*(6), 508–517. https://doi.org/10.1002/da.22588.

Arean, P. A., Hallgren, K. A., Jordan, J. T., Gazzaley, A., Atkins, D. C., Heagerty, P. J., et al. (2016). The use and effectiveness of mobile apps for depression: Results from a fully remote clinical trial. *Journal of Medical Internet Research, 18*(12), e330. https://doi.org/10.2196/jmir.6482.

Arean, P. A., Mata-Greve, F., Johnson, M., Pullmann, M., Griffith Fillipo, I., Comtois, K., et al. (2021). Mental health and perceived usability of digital mental health tools among essential workers and unemployed during COVID-19: A remote survey study. *JMIR Mental Health*. https://doi.org/10.2196/28360.

Bakker, D., Kazantzis, N., Rickwood, D., & Rickard, N. (2016). Mental health smartphone apps: Review and evidence-based recommendations for future developments. *JMIR Mental Health, 3*(1), e7. https://doi.org/10.2196/mental.4984.

Baumel, A., Edan, S., & Kane, J. M. (2019). Is there a trial bias impacting user engagement with unguided e-mental health interventions? A systematic comparison of published reports and real-world usage of the same programs. *Translational Behavioral Medicine, 9*(6), 1020–1033. https://doi.org/10.1093/tbm/ibz147.

Baumel, A., & Kane, J. M. (2018). Examining predictors of real-world user engagement with self-guided eHealth interventions: Analysis of mobile apps and websites using a novel dataset. *Journal of Medical Internet Research, 20*(12), e11491. https://doi.org/10.2196/11491.

Baumel, A., Torous, J., Edan, S., & Kane, J. M. (2020). There is a non-evidence-based app for that: A systematic review and mixed methods analysis of depression- and anxiety-related apps that incorporate unrecognized techniques. *Journal of Affective Disorders, 273*, 410–421. https://doi.org/10.1016/j.jad.2020.05.011.

Betthauser, L. M., Stearns-Yoder, K. A., McGarity, S., Smith, V., Place, S., & Brenner, L. A. (2020). Mobile app for mental health monitoring and clinical outreach in veterans: Mixed methods feasibility and acceptability study. *Journal of Medical Internet Research, 22*(8), e15506. https://doi.org/10.2196/15506.

Broglia, E., Millings, A., & Barkham, M. (2019). Counseling with guided use of a mobile well-being app for students experiencing anxiety or depression: Clinical outcomes of a feasibility trial embedded in a student counseling service. *JMIR MHealth and UHealth, 7*(8), e14318. https://doi.org/10.2196/14318.

Bush, N. E., Ouellette, G., & Kinn, J. (2014). Utility of the T2 Mood Tracker mobile application among army warrior transition unit service members. *Military Medicine, 179*(12), 1453–1457. https://doi.org/10.7205/MILMED-D-14-00271.

Carlo, A. D., Hosseini Ghomi, R., Renn, B. N., & Arean, P. A. (2019). By the numbers: Ratings and utilization of behavioral health mobile applications. *NPJ Digital Medicine, 2*, 54. https://doi.org/10.1038/s41746-019-0129-6.

Chan, A. H. Y., & Honey, M. L. L. (2021). User perceptions of mobile digital apps for mental health: Acceptability and usability – An integrative review. *Journal of Psychiatric and Mental Health Nursing, 29*(1), 147–168. https://doi.org/10.1111/jpm.12744.

Clay, R. A. (2021). Mental health apps are gaining traction. *Monitor on Psychology, 52*(1), 55.

Cole-Lewis, H., Ezeanochie, N., & Turgiss, J. (2019). Understanding health behavior technology engagement: Pathway to measuring digital behavior change interventions. *JMIR Formative Research, 3*(4), e14052. https://doi.org/10.2196/14052.

Connolly, S. L., Kuhn, E., Possemato, K., & Torous, J. (2021). Digital clinics and mobile technology implementation for mental health care. *Current Psychiatry Reports, 23*(7), 38. https://doi.org/10.1007/s11920-021-01254-8.

Cuijpers, P., Donker, T., Johansson, R., Mohr, D. C., van Straten, A., & Andersson, G. (2011). Self-guided psychological treatment for depressive symptoms: A meta-analysis. *Plos One, 6*(6), e21274. https://doi.org/10.1371/journal.pone.0021274.

Depp, C. A., Mausbach, B., Granholm, E., Cardenas, V., Ben-Zeev, D., Patterson, T. L., et al. (2010). Mobile interventions for severe mental illness: Design and preliminary data from three approaches. *The Journal of Nervous and Mental Disease, 198*(10), 715–721. https://doi.org/10.1097/NMD.0b013e3181f49ea3.

Donker, T., Petrie, K., Proudfoot, J., Clarke, J., Birch, M.-R., & Christensen, H. (2013). Smartphones for smarter delivery of mental health programs: A systematic review. *Journal of Medical Internet Research, 15*(11), e247. https://doi.org/10.2196/jmir.2791.

Drissi, N., Ouhbi, S., Janati Idrissi, M. A., & Ghogho, M. (2020). An analysis on self-management and treatment-related functionality and characteristics of highly rated anxiety apps. *International Journal of Medical Informatics, 141*, 104243. https://doi.org/10.1016/j.ijmedinf.2020.104243.

Duffy, D., Enrique, A., Connell, S., Connolly, C., & Richards, D. (2019). Internet-delivered cognitive behavior therapy as a prequel to face-to-face therapy for depression and anxiety: A naturalistic observation. *Frontiers in Psychiatry, 10*, 902. https://doi.org/10.3389/fpsyt.2019.00902.

Ebert, D. D., Donkin, L., Andersson, G., Andrews, G., Berger, T., Carlbring, P., et al. (2016). Does internet-based guided-self-help for depression cause harm? An individual participant data meta-analysis on deterioration rates and its moderators in randomized controlled trials. *Psychological Medicine, 46*(13), 2679–2693. https://doi.org/10.1017/S0033291716001562.

Ebert, D. D., Zarski, A.-C., Christensen, H., Stikkelbroek, Y., Cuijpers, P., Berking, M., et al. (2015). Internet and computer-based cognitive behavioral therapy for anxiety and depression in youth: A meta-analysis of randomized controlled outcome trials. *Plos One, 10*(3), e0119895. https://doi.org/10.1371/journal.pone.0119895.

Ellis, L., Campbell, A., Sethi, S., & O'Dea, B. (2011). Comparative randomized trial of an online cognitive-behavioral therapy program and an online support group for depression and anxiety. *Journal of CyberTherapy and Rehabilitation, 4*, 461–467.

Epstein, D. A., Ping, A., Fogarty, J., & Munson, S. A. (2015). A lived informatics model of personal informatics. In *Proceedings of the 2015 ACM international joint conference on pervasive and ubiquitous computing* (pp. 731–742). https://doi.org/10.1145/2750858.2804250.

Federal Trade Commission. (2016, January 5). *Lumosity to pay $2 million to settle FTC deceptive advertising charges for its "Brain Training" program.* Federal Trade Commission. https://www.ftc.gov/news-events/press-releases/2016/01/lumosity-pay-2-million-settle-ftc-deceptive-advertising-charges.

Firth, J., Torous, J., Nicholas, J., Carney, R., Pratap, A., Rosenbaum, S., et al. (2017a). The efficacy of smartphone-based mental health interventions for depressive symptoms: A meta-analysis of randomized controlled trials. *World Psychiatry, 16*(3), 287–298. https://doi.org/10.1002/wps.20472.

Firth, J., Torous, J., Nicholas, J., Carney, R., Rosenbaum, S., & Sarris, J. (2017b). Can smartphone mental health interventions reduce symptoms of anxiety? A meta-analysis of randomized controlled trials. *Journal of Affective Disorders, 218*, 15–22. https://doi.org/10.1016/j.jad.2017.04.046.

Fleming, T. M., Bavin, L., Stasiak, K., Hermansson-Webb, E., Merry, S. N., Cheek, C., et al. (2017). Serious games and gamification for mental health: Current status and promising directions. *Frontiers in Psychiatry, 7*, 215. https://doi.org/10.3389/fpsyt.2016.00215.

Gál, É., Ştefan, S., & Cristea, I. A. (2021). The efficacy of mindfulness meditation apps in enhancing users' well-being and mental health related outcomes: A meta-analysis of randomized controlled trials. *Journal of Affective Disorders, 279*, 131–142. https://doi.org/10.1016/j.jad.2020.09.134.

Gilbody, S., Littlewood, E., Hewitt, C., Brierley, G., Tharmanathan, P., Araya, R., et al. (2015). Computerised cognitive behaviour therapy (cCBT) as treatment for depression in primary care (REEACT trial): Large scale pragmatic randomised controlled trial. *BMJ (Clinical Research Ed.), 351*, h5627. https://doi.org/10.1136/bmj.h5627.

Goldberg, S. B., Tucker, R. P., Greene, P. A., Davidson, R. J., Wampold, B. E., Kearney, D. J., et al. (2018). Mindfulness-based interventions for psychiatric disorders: A systematic review and meta-analysis. *Clinical Psychology Review, 59*, 52–60. https://doi.org/10.1016/j.cpr.2017.10.011.

Goyal, M., Singh, S., Sibinga, E. M. S., Gould, N. F., Rowland-Seymour, A., Sharma, R., et al. (2014). Meditation programs for psychological stress and well-being: A systematic review and meta-analysis. *JAMA Internal Medicine, 174*(3), 357–368. https://doi.org/10.1001/jamainternmed.2013.13018.

Grant, S., Spears, A., & Pedersen, E. R. (2018). Video games as a potential modality for behavioral health services for young adult veterans: Exploratory analysis. *JMIR Serious Games, 6*(3), e15. https://doi.org/10.2196/games.9327.

Griffiths, K. M., Calear, A. L., & Banfield, M. (2009). Systematic review on internet support groups (ISGs) and depression (1): Do ISGs reduce depressive symptoms? *Journal of Medical Internet Research, 11*(3), e40. https://doi.org/10.2196/jmir.1270.

Gunning, F. M., Anguera, J. A., Victoria, L. W., & Arean, P. A. (2021). A digital intervention targeting cognitive control network dysfunction in middle age and older adults with major depression. *Translational Psychiatry, 11*(1), 1–9. https://doi.org/10.1038/s41398-021-01386-8.

Hatch, A., Hoffman, J. E., Ross, R., & Docherty, J. P. (2018). Expert consensus survey on digital health tools for patients with serious mental illness: Optimizing for user characteristics and user support. *JMIR Mental Health, 5*(2), e46. https://doi.org/10.2196/mental.9777.

Hetrick, S. E., Robinson, J., Burge, E., Blandon, R., Mobilio, B., Rice, S. M., et al. (2018). Youth codesign of a mobile phone app to facilitate self-monitoring and management of mood symptoms in young people with major depression, suicidal ideation, and self-harm. *JMIR Mental Health, 5*(1), e9. https://doi.org/10.2196/mental.9041.

Hilton, L., Maher, A. R., Colaiaco, B., Apaydin, E., Sorbero, M. E., Booth, M., et al. (2017). Meditation for posttraumatic stress: Systematic review and meta-analysis. *Psychological Trauma: Theory, Research, Practice and Policy, 9*(4), 453–460. https://doi.org/10.1037/tra0000180.

Hoermann, S., McCabe, K. L., Milne, D. N., & Calvo, R. A. (2017). Application of synchronous text-based dialogue systems in mental health interventions: Systematic review. *Journal of Medical Internet Research, 19*(8), e267. https://doi.org/10.2196/jmir.7023.

Howells, A., Ivtzan, I., & Eiroa-Orosa, F. J. (2016). Putting the 'app' in happiness: A randomised controlled trial of a smartphone-based mindfulness intervention to enhance wellbeing. *Journal of Happiness Studies, 17*(1), 163–185 https://doi.org/10/f7788d.

Huckvale, K., Torous, J., & Larsen, M. E. (2019). Assessment of the data sharing and privacy practices of smartphone apps for depression and smoking cessation. *JAMA Network Open, 2*(4), e192542. https://doi.org/10.1001/jamanetworkopen.2019.2542.

Huguet, A., Rao, S., McGrath, P. J., Wozney, L., Wheaton, M., Conrod, J., et al. (2016). A systematic review of cognitive behavioral therapy and behavioral activation apps for depression. *Plos One, 11*(5), e0154248. https://doi.org/10.1371/journal.pone.0154248.

Hull, T. D., Malgaroli, M., Connolly, P. S., Feuerstein, S., & Simon, N. M. (2020). Two-way messaging therapy for depression and anxiety: Longitudinal response trajectories. *BMC Psychiatry [Electronic Resource], 20*, 297. https://doi.org/10.1186/s12888-020-02721-x.

Järvelä-Reijonen, E., Karhunen, L., Sairanen, E., Muotka, J., Lindroos, S., Laitinen, J., et al. (2018). The effects of acceptance and commitment therapy on eating behavior and diet delivered through face-to-face contact and a mobile app: A randomized controlled trial. *The International Journal of Behavioral Nutrition and Physical Activity, 15*(1), 22. https://doi.org/10.1186/s12966-018-0654-8.

Johnson, D., Deterding, S., Kuhn, K.-A., Staneva, A., Stoyanov, S., & Hides, L. (2016). Gamification for health and wellbeing: A systematic review of the literature. *Internet Interventions, 6*, 89–106. https://doi.org/10.1016/j.invent.2016.10.002.

Karyotaki, E., Riper, H., Twisk, J., Hoogendoorn, A., Kleiboer, A., Mira, A., et al. (2017). Efficacy of self-guided internet-based cognitive behavioral therapy in the treatment of depressive symptoms: A meta-analysis of individual participant data. *JAMA Psychiatry, 74*(4), 351. https://doi.org/10.1001/jamapsychiatry.2017.0044.

Kelders, S. M., Kok, R. N., Ossebaard, H. C., & Van Gemert-Pijnen, J. E. W. C. (2012). Persuasive system design does matter: A systematic review of adherence to web-based interventions. *Journal of Medical Internet Research, 14*(6), e152. https://doi.org/10.2196/jmir.2104.

Kenter, R. M. F., van de Ven, P. M., Cuijpers, P., Koole, G., Niamat, S., Gerrits, R. S., et al. (2015). Costs and effects of internet cognitive behavioral treatment blended with face-to-face treatment: Results from a naturalistic study. *Internet Interventions, 2*(1), 77–83. https://doi.org/10.1016/j.invent.2015.01.001.

Lattie, E. G., Adkins, E. C., Winquist, N., Stiles-Shields, C., Wafford, Q. E., & Graham, A. K. (2019). Digital mental health interventions for depression, anxiety, and enhancement of psychological well-being among college students: Systematic review. *Journal of Medical Internet Research, 21*(7), e12869. https://doi.org/10.2196/12869.

Lau, H. M., Smit, J. H., Fleming, T. M., & Riper, H. (2016). Serious games for mental health: Are they accessible, feasible, and effective? A systematic review and meta-analysis. *Frontiers in Psychiatry, 7*, 209. https://doi.org/10.3389/fpsyt.2016.00209.

Lenhart, A., Purcell, K., Smith, A., & Zickuhr, K. (2010). *Social media & mobile internet use among teens and young adults*. Millenials: Pew Research Center.

Li, J., Theng, Y.-L., & Foo, S. (2014). Game-based digital interventions for depression therapy: A systematic review and meta-analysis. *Cyberpsychology, Behavior and Social Networking, 17*(8), 519–527. https://doi.org/10.1089/cyber.2013.0481.

Lin, Z., Althoff, T., & Leskovec, J. (2018). I'll be back: On the multiple lives of users of a mobile activity tracking application. In *Proceedings of the 2018 world wide web conference* (pp. 1501–1511). https://doi.org/10.1145/3178876.3186062.

Linardon, J., Cuijpers, P., Carlbring, P., Messer, M., & Fuller-Tyszkiewicz, M. (2019). The efficacy of app-supported smartphone interventions for mental health problems: A meta-analysis of randomized controlled trials. *World Psychiatry, 18*(3), 325–336. https://doi.org/10.1002/wps.20673.

Liu, P., Astudillo, K., Velez, D., Kelley, L., Cobbs-Lomax, D., & Spatz, E. S. (2020). Use of mobile health applications in low-income populations. *Circulation: Cardiovascular Quality and Outcomes, 13*(9), e007031. https://doi.org/10.1161/CIRCOUTCOMES.120.007031.

Mani, M., Kavanagh, D. J., Hides, L., & Stoyanov, S. R. (2015). Review and evaluation of mindfulness-based iPhone apps. *JMIR MHealth and UHealth, 3*(3), e82. https://doi.org/10.2196/mhealth.4328.

Marsch, L. A., Campbell, A., Campbell, C., Chen, C.-H., Ertin, E., Ghitza, U., et al. (2020). The application of digital health to the assessment and treatment of substance use disorders: The past, current, and future role of the National Drug Abuse Treatment Clinical Trials Network. *Journal of Substance Abuse Treatment, 112S*, 4–11. https://doi.org/10.1016/j.jsat.2020.02.005.

McCall, T., Ali, M. O., Yu, F., Fontelo, P., & Khairat, S. (2021). Development of a mobile app to support self-management of anxiety and depression in African American women: A usability study. *JMIR Formative Research, 17;5*(8), e24393. https://doi.org/10.2196/24393.

MindTools.io. (2021). *MindTools.io*. MindTools.Io. https://mindtools.io/.

Miralles, I., Granell, C., Díaz-Sanahuja, L., Van Woensel, W., Bretón-López, J., Mira, A., et al. (2020). Smartphone apps for the treatment of mental disorders: systematic review. *JMIR MHealth and UHealth, 8*(4), e14897. https://doi.org/10.2196/14897.

Mohr, D. C., Azocar, F., Bertagnolli, A., Choudhury, T., Chrisp, P., Frank, R., et al. (2021). Banbury forum consensus statement on the path forward for digital mental health treatment. *Psychiatric Services (Washington, D.C.), 72*(6), 677–683. https://doi.org/10.1176/appi.ps.202000561.

Mohr, D. C., Kwasny, M. J., Meyerhoff, J., Graham, A. K., & Lattie, E. G. (2021). The effect of depression and anxiety symptom severity on clinical outcomes and app use in digital mental health treatments: Meta-regression of three trials. *Behaviour Research and Therapy, 147*, 103972. https://doi.org/10.1016/j.brat.2021.103972.

Mohr, D. C., Schueller, S. M., Tomasino, K. N., Kaiser, S. M., Alam, N., Karr, C., et al. (2019). Comparison of the effects of coaching and receipt of app recommendations on depression, anxiety, and engagement in the IntelliCare platform: Factorial randomized controlled trial. *Journal of Medical Internet Research, 21*(8), e13609. https://doi.org/10.2196/13609.

Naslund, J. A., Aschbrenner, K. A., Marsch, L. A., & Bartels, S. J. (2016). The future of mental health care: Peer-to-peer support and social media. *Epidemiology and Psychiatric Sciences, 25*(2), 113–122. https://doi.org/10.1017/S2045796015001067.

National Institute for Health and Clinical Excellence (NICE). (2006). *Computerised cognitive behaviour therapy for depression and anxiety*. National Institute for Health and Clinical Excellence (NICE), London.

National Institute for Health and Care Excellence (NICE). (2009). *Depression in adults: Recognition and management*. National Institute for Health and Care Excellence (NICE), London, England https://www.nice.org.uk/guidance/CG90.

National Institute of Mental Health. (2015). *Opportunities and challenges of developing information technologies on behavioral and social science clinical research*. National Institute of Mental Health https://www.nimh.nih.gov/about/advisory-boards-and-groups/namhc/namhc-workgroups/namhc-bssr-workgroup-charge.

Ng, M. M., Firth, J., Minen, M., & Torous, J. (2019). User engagement in mental health apps: A review of measurement, reporting, and validity. *Psychiatric Services (Washington, D.C.), 70*(7), 538–544. https://doi.org/10.1176/appi.ps.201800519.

Nimrod, G. (2013). Challenging the internet paradox: Online depression communities and well-being. *International Journal of Internet Science, 8*, 30–48.

Nuamah, J., Mehta, R., & Sasangohar, F. (2020). Technologies for opioid use disorder management: Mobile app search and scoping review. *JMIR MHealth and UHealth, 8*(6), e15752. https://doi.org/10.2196/15752.

O'Grady, C., Melia, R., Bogue, J., O'Sullivan, M., Young, K., & Duggan, J. (2020). A mobile health approach for improving outcomes in suicide prevention (SafePlan). *Journal of Medical Internet Research, 22*(7), e17481. https://doi.org/10.2196/17481.

Olthuis, J. V., Wozney, L., Asmundson, G. J. G., Cramm, H., Lingley-Pottie, P., & McGrath, P. J. (2016). Distance-delivered interventions for PTSD: A systematic review and meta-analysis. *Journal of Anxiety Disorders, 44*, 9–26. https://doi.org/10.1016/j.janxdis.2016.09.010.

One Mind PsyberGuide. (2021). *One Mind PsyberGuide*. One Mind PsyberGuide https://onemindpsyberguide.org.

Owen, J. E., Kuhn, E., Jaworski, B. K., McGee-Vincent, P., Juhasz, K., Hoffman, J. E., et al. (2018). VA mobile apps for PTSD and related problems: Public health resources for veterans and those who care for them. *MHealth, 4*, 28. https://doi.org/10.21037/mhealth.2018.05.07.

Park, C., Zhu, J., Ho Chun Man, R., Rosenblat, J. D., Iacobucci, M., Gill, H., et al. (2020). Smartphone applications for the treatment of depressive symptoms: A meta-analysis and qualitative review. *Annals of Clinical Psychiatry, 32*(1), 48–68.

Parker, L., Bero, L., Gillies, D., Raven, M., & Grundy, Q. (2018). The "hot potato" of mental health app regulation: A critical case study of the Australian policy arena. *International Journal of Health Policy and Management, 8*(3), 168–176. https://doi.org/10.15171/ijhpm.2018.117.

Pratap, A., Renn, B. N., Volponi, J., Mooney, S. D., Gazzaley, A., Arean, P. A., et al. (2018). Using mobile apps to assess and treat depression in Hispanic and Latino populations: Fully remote randomized clinical trial. *Journal of Medical Internet Research, 20*(8), e10130. https://doi.org/10.2196/10130.

Prescott, J., Rathbone, A. L., & Brown, G. (2020). Online peer to peer support: Qualitative analysis of UK and US open mental health Facebook groups. *Digital Health, 6.* https://doi.org/10.1177/2055207620979209.

Pretorius, C., Chambers, D., & Coyle, D. (2019). Young people's online help-seeking and mental health difficulties: Systematic narrative review. *Journal of Medical Internet Research, 21*(11), e13873. https://doi.org/10.2196/13873.

Rathbone, A. L., & Prescott, J. (2017). The use of mobile apps and SMS messaging as physical and mental health interventions: Systematic review. *Journal of Medical Internet Research, 19*(8), e295. https://doi.org/10.2196/jmir.7740.

Renn, B. N., Hoeft, T. J., Lee, H. S., Bauer, A. M., & Arean, P. A. (2019). Preference for in-person psychotherapy versus digital psychotherapy options for depression: Survey of adults in the U.S. *NPJ Digital Medicine, 2,* 6. https://doi.org/10.1038/s41746-019-0077-1.

Richards, D., Enrique, A., Eilert, N., Franklin, M., Palacios, J., Duffy, D., et al. (2020). A pragmatic randomized waitlist-controlled effectiveness and cost-effectiveness trial of digital interventions for depression and anxiety. *NPJ Digital Medicine, 3*(1), 1–10. https://doi.org/10.1038/s41746-020-0293-8.

Ridout, B., & Campbell, A. (2018). The use of social networking sites in mental health interventions for young people: Systematic review. *Journal of Medical Internet Research, 20*(12), e12244. https://doi.org/10.2196/12244.

Rodriguez-Paras, C., Tippey, K., Brown, E., Sasangohar, F., Creech, S., Kum, H.-C., et al. (2017). Posttraumatic stress disorder and mobile health: App investigation and scoping literature review. *JMIR MHealth and UHealth, 5*(10), e156. https://doi.org/10.2196/mhealth.7318.

Rosen, K. D., Paniagua, S. M., Kazanis, W., Jones, S., & Potter, J. S. (2018). Quality of life among women diagnosed with breast cancer: A randomized waitlist controlled trial of commercially available mobile app-delivered mindfulness training. *Psycho-Oncology, 27*(8), 2023–2030. doi:10.2196/mhealth.7318.

Sayal, R., Subbalakhmi, C., & Saini, H. S. (2020). Mobile app accessibility for visually impaired. *International Journal of Advanced Trends in Computer Science and Engineering, 9*(1), 182–185. https://doi.org/10.30534/ijatcse/2020/27912020.

Scherr, S., & Goering, M. (2020). Is a self-monitoring app for depression a good place for additional mental health information? Ecological momentary assessment of mental help information seeking among smartphone users. *Health Communication, 35*(8), 1004–1012. https://doi.org/10.1080/10410236.2019.1606135.

Sequeira, L., Perrotta, S., LaGrassa, J., Merikangas, K., Kreindler, D., Kundur, D., et al. (2020). Mobile and wearable technology for monitoring depressive symptoms in children and adolescents: A scoping review. *Journal of Affective Disorders, 265,* 314–324. https://doi.org/10.1016/j.jad.2019.11.156.

Shrier, L. A., & Spalding, A. (2017). "Just take a moment and breathe and think": Young women with depression talk about the development of an ecological momentary intervention to reduce their sexual risk. *Journal of Pediatric and Adolescent Gynecology, 30*(1), 116–122. https://doi.org/10.1016/j.jpag.2016.08.009.

Steinkamp, J. M., Goldblatt, N., Borodovsky, J. T., LaVertu, A., Kronish, I. M., Marsch, L. A., et al. (2019). Technological interventions for medication adherence in adult mental health and substance use disorders: A systematic review. *JMIR Mental Health, 6*(3), e12493. https://doi.org/10.2196/12493.

Titov, N., Dear, B. F., Staples, L. G., Bennett-Levy, J., Klein, B., Rapee, R. M., et al. (2017). The first 30 months of the MindSpot clinic: Evaluation of a national e-mental health service against project objectives. *The Australian and New Zealand Journal of Psychiatry, 51*(12), 1227–1239. https://doi.org/10.1177/0004867416671598.

Tofighi, B., Abrantes, A., & Stein, M. D. (2018). The role of technology-based interventions for substance use disorders in primary care: A review of the literature. *The Medical Clinics of North America, 102*(4), 715–731. https://doi.org/10.1016/j.mcna.2018.02.011.

Tofighi, B., Chemi, C., Ruiz-Valcarcel, J., Hein, P., & Hu, L. (2019). Smartphone apps targeting alcohol and illicit substance use: Systematic search in commercial app stores and critical content analysis. *JMIR MHealth and UHealth, 7*(4), e11831. https://doi.org/10.2196/11831.

Torous, J., Andersson, G., Bertagnoli, A., Christensen, H., Cuijpers, P., Firth, J., et al. (2019). Towards a consensus around standards for smartphone apps and digital mental health. *World Psychiatry, 18*(1), 97–98. https://doi.org/10.1002/wps.20592.

Torous, J., Firth, J., Huckvale, K., Larsen, M. E., Cosco, T. D., Carney, R., et al. (2018). The emerging imperative for a consensus approach toward the rating and clinical recommendation of mental health apps. *The Journal of Nervous and Mental Disease, 206*(8), 662–666. https://doi.org/10.1097/NMD.0000000000000864.

Vinograd, M., & Craske, M. G. (2020). Chapter 1—History and theoretical underpinnings of exposure therapy. In T. S. Peris, E. A. Storch, & J. F. McGuire (Eds.), *Exposure therapy for children with anxiety and OCD* (pp. 3–20). Amsterdam, The Netherlands: Elsevier. https://doi.org/10.1016/B978-0-12-815915-6.00001-9.

Wang, K., Varma, D. S., & Prosperi, M. (2018). A systematic review of the effectiveness of mobile apps for monitoring and management of mental health symptoms or disorders. *Journal of Psychiatric Research, 107,* 73–78. https://doi.org/10.1016/j.jpsychires.2018.10.006.

Weisel, K. K., Fuhrmann, L. M., Berking, M., Baumeister, H., Cuijpers, P., & Ebert, D. D. (2019). Standalone smartphone apps for mental health—A systematic review and meta-analysis. *NPJ Digital Medicine, 2*(1), 1–10. https://doi.org/10.1038/s41746-019-0188-8.

Woodruff, S. I., Conway, T. L., Edwards, C. C., Elliott, S. P., & Crittenden, J. (2007). Evaluation of an internet virtual world chat room for adolescent smoking cessation. *Addictive Behaviors, 32*(9), 1769–1786. https://doi.org/10.1016/j.addbeh.2006.12.008.

Woodruff, S. I., Edwards, C. C., Conway, T. L., & Elliott, S. P. (2001). Pilot test of an internet virtual world chat room for rural teen smokers. *The Journal of Adolescent Health, 29*(4), 239–243. https://doi.org/10.1016/s1054-139x(01)00262-2.

Wright, J. H., Owen, J. J., Richards, D., Eells, T. D., Richardson, T., Brown, G. K., et al. (2019). Computer-assisted cognitive-behavior therapy for depression: A systematic review and meta-analysis. *The Journal of Clinical Psychiatry, 80*(2), 18r12188. https://doi.org/10.4088/JCP.18r12188.

Wright, J. H., Wright, A. S., Albano, A. M., Basco, M. R., Goldsmith, L. J., Raffield, T., et al. (2005). Computer-assisted cognitive therapy for depression: Maintaining efficacy while reducing therapist time. *The American Journal of Psychiatry, 162*(6), 1158–1164. https://doi.org/10.1176/appi.ajp.162.6.1158.

Yeager, C. M., & Benight, C. C. (2018). If we build it, will they come? Issues of engagement with digital health interventions for trauma recovery. *MHealth, 4*, 37. https://doi.org/10.21037/mhealth.2018.08.04.

Zachariae, R., Lyby, M. S., Ritterband, L. M., & O'Toole, M. S. (2016). Efficacy of internet-delivered cognitive-behavioral therapy for insomnia—A systematic review and meta-analysis of randomized controlled trials. *Sleep Medicine Reviews, 30*, 1–10. https://doi.org/10.1016/j.smrv.2015.10.004.

Chapter 5

Blending digital therapeutics within the healthcare system

Olivia Clare Keller[a], Alan Jeffrey Budney[b], Cara Ann Struble[b] and Gisbert Wilhelm Teepe[a]

[a] *Centre for Digital Health Interventions, Department of Management, Technology and Economics, ETH Zürich, Zürich, Switzerland,* [b] *Geisel School of Medicine at Dartmouth, Center for Technology and Behavioral Health, Dartmouth College, Hanover, NH, USA*

5.1 Introduction

This chapter provides insight into the history and current status of blended care (BC) for mental health disorders (MHD) and substance use disorders (SUD). BC combines elements of psychological therapy with digital therapeutics and can be seen as an alternative to person-to-person traditional psychotherapy. Few authors have attempted to define this type of care, one reason being the diverse range of BC interventions trialed in different contexts. Consequently, an important contribution of this chapter is to provide a working definition of BC.

Blended care relies heavily on advances in technology, which is why it is important to understand BC in the context of related developments in standalone apps and computerized self-help interventions. For a background on internet-based treatment programs, it is suggested to first read Chapter 3 in this volume, "First wave of scalable digital therapeutics: Internet-based programs for direct-to-consumer standalone care for mental health and addiction." Likewise, for context on standalone treatment apps, one may benefit from reading Chapter 4, "Second wave of scalable digital therapeutics: Mental health and addiction treatment apps for direct-to-consumer standalone care," before continuing with this chapter. The illustration (Fig. 5.1) puts BC into the context of standalone and person-to-person treatments.

One might ask why standalone apps and internet-based treatment programs do not constitute sufficient alternatives to traditional psychotherapy. Is it necessary for technology to become part of psychological therapy? While BC will likely never replace traditional psychotherapy, there have been many drivers of change encouraging an evolution and integration of technology into psychotherapy. Technological advancements, ubiquitous access to mobile applications, successful research evaluating online-delivered interventions, a need to increase access to care and reduce treatment costs, are just a few examples of cultural progressions that naturally led to the exploration of ways technology might improve psychotherapies.

It is also useful to look at the vessels that have worked on integrating technology into existing services. The Improving Access to Psychological Therapies (IAPT) program (National Health Service, 2021) in the United Kingdom (UK) can be credited with advancing computerized self-help by integrating this type of care into primary care services over the past 15 years. In the United States (US), some health insurers have been investing into digital health technologies, leading to standalone and BC offerings becoming visible and accessible to many clients. The systemic integration of BC into psychological therapy services, although sparse, has come with quality control and resulted in improved disorder-specific programs. This is visible in countries such as Germany, the UK, and Australia which offer a range of BC treatments free or at a reduced cost to clients.

What is the potential added value of BC? BC can provide effective and client-centered treatment, backed by technology and input from healthcare providers. While high-quality traditional psychotherapy can offer effective treatments, BC can additionally reduce the burden on the client and the provider, as well as facilitate the collection and monitoring of data that can guide therapy and potentially improve services. The IAPT program in the UK has further demonstrated how BC can be a cost-effective way of increasing access to care.

Digital Therapeutics for Mental Health and Addiction: The State of the Science and Vision for the Future. DOI: https://doi.org/10.1016/B978-0-323-90045-4.00016-2

FIGURE 5.1 Blended treatment in context.

What comprises BC has not been well defined, but is a topic of interest to researchers, service systems, therapists, and clients. There exists a significant gap between what research has evaluated and the practice of BC today—not least because of how fast technology has been evolving and because many BC interventions developed by private industry are proprietary. Examining evidence for the myriad of BC and digital therapies that have appeared in the market is well beyond the scope of this chapter. Rather, this chapter seeks to illustrate the status quo and the adoption of BC around the world by providing examples of how BC has been implemented, focusing on regulations in selected countries.

Although BC would appear to provide many potential benefits, the challenges that confront the adoption of and access to BC treatments are substantial and are a key area to address to advance care for MHD and SUD. Hence, we discuss the translation and uptake of BC from the client, provider, and health systems perspective. The last section of the chapter looks to the future and touches on technology advances, research directions, provider training, and evolving BC models that can advance the impact of BC on delivering quality care to those affected by MHD and SUD.

5.2 Blended care terminology

5.2.1 Internet-delivered treatments

This chapter aims to provide a working definition of BC while recognizing that this is a rapidly evolving treatment model for MHD and SUD. A number of umbrella terms exist to describe digitally enabled psychotherapy, including digital therapeutics, internet-delivered treatments, and internet- and mobile-based interventions. For example, Kenter et al. (2015) proposed the following: Internet-delivered treatments are "structured programs offered via the internet which are based on evidence-based therapies. Patients can work through these programs independently" or these programs "might be offered with coaching via telephone, email or face-to-face in order to keep patients motivated and to help them better understand the techniques and assignments." This definition combines "standalone" or what some call "self-guided" digital interventions (no healthcare provider support or guidance included; Christensen, Griffiths, Mackinnon, & Brittliffe, 2006), with what others have termed "guided" digital interventions that are supported by an e-coach or healthcare provider. Guided interventions can vary considerably in the amount and content of the support or guidance offered and the context in which the support is provided. The difficulty in the labeling of digital interventions becomes more confusing in that healthcare providers in primary care settings sometimes prescribe self-guided digital interventions, for example, iCBT (or computerized CBT [cCBT]) for depression or anxiety disorders, without offering additional support services (Weisel et al., 2019). To some extent this reflects an evolution of traditional bibliotherapy as it has been practiced by healthcare providers including psychotherapists (Marks, Cavanagh, & Gega, 2007). In contrast, a guided version of the same intervention, iCBT for anxiety or depressive disorders, can often be accessed through primary or secondary healthcare services; for example, it is widely available in the UK and offered by the IAPT framework (National Health Service, 2021). Moreover, in recognition of the different support needs of clients, some programs offer a compromise in the form of "on-demand" support which can be requested or booked by clients as they work through a digital treatment program. For all types of digital interventions, there is a trend toward clients receiving more targeted and personalized forms of support as a consequence of digital platforms offering technological features such as direct connection to the internet, passive sensing, and tracking of symptoms.

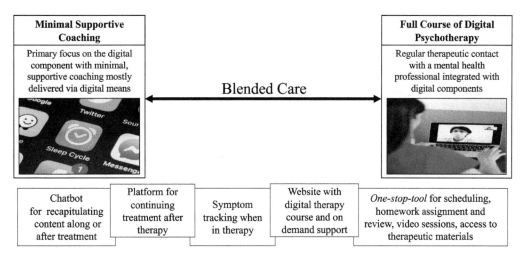

FIGURE 5.2 Continuum of blended care.

5.2.2 Blended care

Blended care (BC), which has also been labeled "blended treatment" (Erbe, Eichert, Riper, & Ebert, 2017; Wentzel, van der Vaart, Bohlmeijer, & van Gemert-Pijnen, 2016) or "blended interventions" (Erbe et al., 2017), integrates synchronous ("live" or "in real time") and asynchronous ("not occurring in real time") elements of care by leveraging technological components (i.e., a platform or app) and support from a healthcare provider in order to create a treatment package with added value. When comparing BC to traditional face-to-face psychotherapy, the idea is that "this approach aims at retaining the positive aspects associated with both forms of therapy while mitigating the disadvantages (Erbe et al., 2017)." Initially, BC was defined as blending face-to-face sessions with online therapy sessions (Kooistra et al., 2014). However, many other constellations of person-to-person (face-to-face or digitally enabled person-to-person) interventions have been developed and implemented. Some have included the integration of online components in a traditional treatment process, others have offered video, telephone, or text-based therapy appointments (Wentzel et al., 2016). Combinations of face-to-face appointments and digital components can include either the integrated or sequential use of each format (Erbe et al., 2017).

Unfortunately, not only is there a lack of comparative studies evaluating the outcomes of BC versus person-to-person, or of BC versus self-guided digital treatments, but little data exists to inform the optimal ratio of person-to-person and digital components (Erbe et al., 2017). Because BC continues to evolve, we propose BC best be understood as a continuum, with an idealized version of BC offering a high level of integration of the digital treatment components with regular therapy sessions delivered by a trained clinician (e.g., a licensed clinical psychologist or psychotherapist). However, BC can range from a primary focus on the digital component with minimal, supportive coaching to a full course of psychotherapy integrated with digital components designed to support and enhance the process and outcomes of the treatment. Fig. 5.2 illustrates this spectrum of BC. Therefore ideally, the type of BC offered would be determined by the clinical need of the individual seeking care. For instance, within the National Health Service in the UK, this is a key consideration of the "stepped care model" (Kendrick & Pilling, 2012) that seeks to match and adapt the level of treatment to a client's needs. To illustrate this, a client with mild to moderate symptoms of panic disorder might be offered a course of BC which involves working with a psychological wellbeing practitioner (a specialist mental health worker in the UK) and simultaneously working through an internet-based program between sessions. Alternatively, a client with severe post-traumatic stress disorder from a car accident may benefit from a higher level of support. Such a client may opt to work person-to-person with an accredited cognitive behavior therapist (a licensed type of psychotherapist in the UK) who integrates the use of a virtual reality exposure therapy headset to reduce the fear response to stimuli associated with the trauma. In both cases, the client receives dual support from the provider and technology, with the two components working closely together for effective results. Arguably BC for MHD and SUD remains in its nascent to early stage of development. It will be most interesting to observe how different healthcare systems around the world integrate BC approaches, and how it will evolve to serve different client populations.

5.3   Components and structure of blended care

5.3.1   Therapeutic support

How much support and personalization should be included in BC is a key question confronting optimal development of BC interventions and is of importance to scalable and effective translation. Clearly, it is of value to include a human support element from a healthcare provider, for instance to adequately assess risk and to deal with crisis situations (Berger & Andersson, 2009; Cuijpers, Riper, & Karyotaki, 2018). However, optimal types and levels of support are being tested and likely will depend on the client population and individual care needs, that is, type and severity of the primary issue. A recent meta-analysis on internet-based programs for depression found that not only was guided iCBT considerably more effective than self-guided iCBT, but the effect size was lowest for e-mail, intermediate for telephone, and largest when face-to-face support was provided, suggesting that more personal and synchronous support is associated with higher effects in this population (Wright et al., 2019).

Another key question looks at who should be delivering this professional support. Evaluative studies have involved a diverse range of healthcare providers (e.g., primary care workers, psychologists, coaches) to deliver support and guidance, but few have compared types of providers or levels of training they have received. One systematic review found no significant effects when comparing different levels of training for coaches used alongside guided iCBT for MHD (Baumeister, Reichler, Munzinger, & Lin, 2014). A controlled trial that evaluated a guided digital intervention for anxiety disorders found that apart from GAD-7 scores, the coach-led and the clinician-led group showed no significant differences post-treatment (Johnson, Titov, Andews, Spence, & Dear, 2011). It is quite possible that the efficacy of type of provider may depend on the severity and complexity of the individual care needs. Consequently, a client with a milder mental health issue might readily benefit from working with a coach (rather than a psychotherapist) with adequate training regarding the specific treatment protocol adapted to BC. This idea is implemented by IAPT in the UK with the "stepped care model" that includes digitally enabled psychotherapy (National Collaborating Centre for Mental Health, 2021). Overall, much more comparative research is needed to determine the differential impact of type and level of support needed to maximize outcomes across BC interventions.

5.3.2   Digital components

Digital symptom monitoring tools are frequently used by providers to support clients in psychotherapy. BC offers access to a range of digital components that can be tailored to each client's needs, such as online exercises, alerts, or messaging functions. These elements are ideally integrated into an individualized care package for the client. In traditional clinical practice, individualized care is important to tailor therapy to problems and goals—this premise extends to BC. One challenge of studying BC interventions is separating technological from human support. Advanced technology may influence adherence effects traditionally assigned to the role of guidance from a healthcare provider. For instance, Baumeister et al. (2014) point out that adherence facilitating enhancement might stem from automated email and text-message prompts, web-design, and interactive tasks. Of course, the opposite effect might also be observed, that is, digital components could hinder adherence. If for instance a client has an unstable internet connection which stops their BC platform from saving their entries in a digital diary (e.g., behavior diary), or if the platform fails to deliver expected reminders, the client may lose motivation. Providers working with BC interventions are therefore encouraged to regularly discuss the use of digital components with the client, and problem-solve any issues that arise (van der Vaart et al., 2014).

Fig. 5.3 provides examples of some of the therapeutic and digital components of BC, as well as of ways in which they can be used as a treatment package.

5.3.3   Structure of blended care

BC provides an opportunity to enhance and optimize personalization of care. For example, the healthcare provider can assist the client with selecting and adapting the online content that best matches their problem profile. The seamless integration of psychotherapy sessions and digital components into a personalized and evidence-based treatment plan offered to clients should be a key feature of all BC interventions, but doing so effectively can present some challenges (van der Vaart et al., 2014). For example, a prerequisite for this is the provider's familiarity with the digital components, both in terms of content and function. Hence, providers must set aside time to access the digital platform on a regular basis to both ensure that they understand how the digital features work and to assist clients in obtaining the maximal benefits from engaging with the platform.

The sequence of the treatment components delivered in BC can vary depending on several factors. Erbe et al. (2017) suggest the terms "integrated blended intervention" and "sequential blended intervention" to describe the structuring

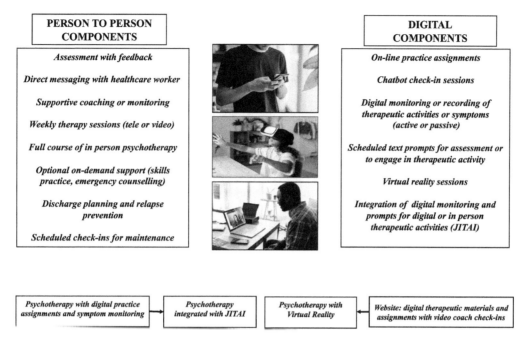

FIG. 5.3 Components of blended care.

of BC, depending on when therapist support is given over the course of treatment. Baumeister et al. (2014) propose that "one alternative approach might be spending more guidance at the beginning and transforming guidance to self-management over the course of the intervention." Sequential programs may be most valuable when different levels of care are provided, such as offering digital components while waiting for face-to-face therapy, using digital therapeutics as part of relapse prevention at the tail end of a face-to face intervention, or as an aftercare approach following a full course of face-to-face treatment. Wentzel et al. (2016) further suggest that BC should be approached as an opportunity to "integrate treatment modalities to reach a proper, tailored treatment plan," but acknowledged that "what type of blend works for whom, and why, is unclear" and that "the rationale for setting up BC is often lacking."

5.4 Blended care treatment approaches

BC interventions also differ in the type and quality of the psychological approach that guides the content of the person-to-person or digital components. As with any treatment for MHD and SUD, the type of approach used in BC should be based on high standards of care and adhere to best practices and guidelines for evidence-based therapies (Kendrick & Pilling, 2012). Which therapy approaches best lend themselves to these challenges? For many mental health issues, CBT is the most widely utilized treatment form and its efficacy for most common MHD and SUD is supported by numerous empirical studies.

Internet-based CBT (iCBT) emphasizes the practice of skills both during and between sessions. The technological component of therapy delivers psychoeducation and specific exercises, such as cognitive restructuring or prolonged exposure, while person-to-person therapy sessions help to tailor therapy, to carry out the exercises, and to problem-solve obstacles. Cuijpers et al. (2018) offer a rationale for why CBT is often the treatment approach of choice in BC: "CBT is a good starting point for such interventions, because it heavily relies on psychoeducation, homework, and other activities that the patient can do outside therapy sessions." Additionally, CBT emphasizes a collaborative approach to develop an individualized treatment plan. This is in line with increasing opportunities to tailor and personalize BC interventions. iCBT programs are being developed around the world and ongoing research into the effectiveness of iCBT contributes to increased insight into which clients most benefit from this type of care. Recently, there have been several studies evaluating the use of alternative therapy approaches in the context of BC such as acceptance and commitment therapy (ACT), transdiagnostic therapies, and virtual reality tools.

ACT is a "third-wave" evidence-based therapy with foundations in CBT, that focuses on acceptance-based emotional regulation, as well as on living according to personal values (Hayes, 2004). ACT has become more accepted and popular with behavioral psychologists over recent years and has begun to be studied in the context of BC. For instance, based on their controlled study, Witlox et al. (2021) confirmed that blended ACT (in this case, consisting of four face-to-face psychotherapy sessions and access to an online platform with ACT content) is a valuable treatment alternative to CBT for anxiety in later life.

Transdiagnostic approaches have generated great interest and hold much promise for the advancement of BC given the high comorbidity rates among MHD and SUD (Roca et al., 2009). These interventions target common underlying mechanisms among disorders (Păsărelu, Andersson, Bergman Nordgren, & Dobrean, 2017) and are therefore amenable to serving a broader clinical population, including the majority of clients who present with comorbid disorders. An initial controlled trial of a transdiagnostic-based BC (iCBT) for those with generalized anxiety disorder, social anxiety disorder, or panic disorder showed promising results (Johnson et al., 2011). The treatment included eight online modules with in-between session tasks, as well as weekly contact with a coach or clinician. Contact was offered in real time over the phone or via email or messaging functions. The coach and clinician roles were distinguished by the type of support they offered, with the coach redirecting to the online content rather than answering clinical questions. Receiving support from a coach or a clinician rendered superior clinical outcomes to the waitlist control group, with medium to large effects maintained at follow-up. While transdiagnostic approaches are not tailored to specific diagnoses, the advances in BC personalization with features such as virtual coaches, appointment reminders, responsive alerts with additional support or content, on-demand calls, and messaging functions make this an approach with great potential to create a personalized treatment plan via BC.

Virtual reality exposure therapy (VRET) offers a virtual immersive environment to conduct exposure-based exercises following the model of prolonged exposure in traditional face-to-face psychotherapy, often to treat anxiety disorders. VRET allows clients to experience fear-inducing situations without having to seek these out in real life. A scoping review identified that about a quarter of relevant studies using VRET to combat depression and anxiety disorders supplemented VRET with CBT (Baghaei et al., 2021). Therefore, it is important to recognize that VRET does not constitute an alternative to CBT, but rather, VRET is a tool to optimize CBT and facilitate practicing skills in a virtual environment rather than in real life (i.e., as an alternative to in vivo exposure). These virtual contexts can be easily modulated (Botella, Fernández-Álvarez, Guillén, García-Palacios, & Baños, 2017) and can facilitate a common understanding as VRET enables the therapist to take the client's view (Botella et al., 2017).

5.5 Development and value of blended care

5.5.1 Development of blended care treatments

The evolution and development of some types of BC interventions might be considered a natural process for many evidence-based healthcare providers. Most CBT-oriented interventions have traditionally integrated physical materials to supplement, reinforce, guide, and enhance a course of treatment. Common examples include psychoeducational self-help books (bibliotherapy) or handouts, worksheets that provide instructions for between-session practice assignments or that assist in the monitoring of key processes, mechanisms, or outcome targets. Examples include completing a functional analysis or tracking mood. Resources might also focus on a coping activity or on strategy reminders, such as stress management or relaxation audio tapes, or physical prompts to remind and trigger the use of therapeutic actions. As contemporary culture has more and more embraced technology in the form of smartphones, smartwatches, tablets, and personal computers, many of these therapeutic "supplements" have either been formally translated into digital formats, or providers and clients have intuitively exploited them to assess, monitor, and prompt therapeutic activities because of their practicality, efficiency, and ease of use. For example, providers might suggest: downloading mindfulness or relaxation apps; using sleep trackers or step counters on smartphones or watches; setting up alerts or alarms on phones or watches to prompt activities; monitoring and charting of behavior on formatted phone apps or computer programs; or provision of pdf versions of instructions or editable worksheets for cognitive or behavioral therapeutic practice assignments for those with PCs.

For healthcare treatment systems and for the leaders of the field who construct best-practice guidelines and policy, the development and path to BC is somewhat more complex. The appearance of what we refer to as *standalone* digital interventions, that is, applications that can be downloaded from the internet or computer software programs that can be purchased for use on personal computers or tablets, has made self-guided "treatment" programs readily available to the public. However, because most of these digital interventions do not have empirical studies supporting their efficacy, providers may be remiss in recommending these to individuals suffering from MHD and SUD. Academia and industry have sought to both develop and test existing standalone therapies, yet as will be discussed below, these efforts have typically led to the conclusion that BC, that is, inclusion of both a provider and the digital application, generally provide superior outcomes to standalone interventions and appear to typically yield at least equivalent outcomes to traditional, fully face-to-face psychotherapy (Carroll et al., 2008; Kay-Lambkin, Baker, Lewin, & Carr, 2009). That said, some standalone, self-help digital interventions may indeed provide benefits, as those who would have never sought out traditional therapy may obtain positive outcomes. Arguably, at this time, for many clients seeking care, BC options offer a greater probability for enhanced treatment outcomes. This will be further discussed when reviewing BC for depression below.

5.5.2 Value of blended care

The motivation to develop and implement BC interventions is multifaceted. For many evidence-based interventions, outcomes have much room for improvement that could potentially be enhanced with digital tools. Notably, access to treatment within the US and many other countries is poor. BC provides one possible solution to improving healthcare access for MHD and SUD across various phases of treatment and recovery. For instance, low-threshold iCBT has been implemented as an early step to bridge waiting times for subsequent face-to-face treatment (Erbe et al., 2017). BC combines strengths from traditional person-to-person psychotherapy and digital interventions while overcoming many limitations inherent with standalone models, and thus offers multiple potential benefits to clients, providers, and health systems.

5.5.2.1 Benefits to clients and healthcare providers

Clients: BC can offer clients up to 24/7 access to certain types of treatment support (Carlbring & Andersson, 2006) and reduces the impact of a variety of traditional treatment barriers (e.g., transportation, childcare, availability). The digital components of BC can also provide greater opportunity for clients to work at their own pace throughout the course of treatment (Erbe et al., 2017). Such interventions have been shown to lower client dropout (Campbell et al., 2014) and enhance motivation/engagement (Walters, Vader, & Harris, 2007), which may encourage self-mastery and empowerment, key factors in long-term positive outcomes (Wentzel et al., 2016). BC can expand treatment options for individuals with chronic mental health conditions by maintaining therapeutic relationships, contributing to more comprehensive aftercare or long-term support following more intensive face-to-face therapy (Erbe et al., 2017). Research suggests that clients are accepting of digital treatment components (Materia & Smyth, 2021; Wright & Wright, 1997). Clients benefit from the convenience of accessing portions of their treatment remotely and from tracking their progress. That said, the digital components of BC may not be attractive to all clients, with some concerns centering around data security, access to suitable technology devices, or simply the preference of traditional psychotherapy.

Healthcare providers: BC can increase the frequency, intensity, or "dose" of an intervention while offering substantial digital options and greater flexibility for providers. BC can potentially reduce the number of person-to-person therapy sessions by implementing some psychotherapeutic techniques digitally, increasing time and availability of providers to serve more clients (Budney, Borodovsky, Marsch, & Lord, 2019). Provider time can be more focused on clients with greater clinical needs that may require weekly person-to-person sessions, such as clients in crisis (e.g., suicidality, homicidality) and/or with complex comorbidities (e.g., cognitive impairments). Providers can adapt or augment digital components of treatment to meet the individual care needs of clients. For instance, providers might ask clients to monitor targeted thoughts, emotions, and/or behaviors between sessions via a smartphone app. Mobile technology allows for the collection of real-time client data (e.g., environmental, physiological) via passive sensing or ecological momentary assessments which could provide individualized just-in-time adaptive interventions (JITAIs) or skills to practice when high-risk situations or behaviors emerge (Nahum-Shani et al., 2018). Alternatively, supportive digital messaging could be utilized to enhance motivation or in response to detected stressors to assist with coping. Finally, BC can tailor to a wider variety of learning styles (e.g., visual/auditory aides), cultural considerations, and treatment preferences (Milward, Drummond, Fincham-Campbell, & Deluca, 2018). Since BC is a relatively new form of treatment, it is not surprising that there are also still multiple barriers for providers related to training, equipment, and processes, which are discussed in more detail later in this chapter.

5.5.2.2 Benefits to mental health services

While a wide range of BC models exist, one commonality is that BC provides the opportunity of collecting data as part of therapy. Better data in turn promises to help clients and providers in several ways. For instance, collecting outcome measures and documenting in-between session tasks allow clients and providers to better understand progress in therapy and can facilitate adjustments to processes or goals. For example, services may use aggregated data to alter the type of therapy offered to specific client populations, to offer additional training to providers, and more.

Advances in digital assessment, including performance tests and self-report measures, offer providers more options in BC models. Several standardized assessment protocols, often associated with high cost and training via face-to-face administration, have been programmed digitally (Bultler et al., 2001). Passive sensing and ecological momentary assessment options have also expanded with mobile advancement (e.g., smartphone apps, smartwatches) allowing for continuous physical and/or behavioral data collection that can facilitate JITAIs via less intrusive means. For example, digital methods can be used to increase the accuracy of the assessment of social anxiety symptoms, which tend to be underreported in traditional assessment (Jacobson, Summers, & Wilhelm, 2020) and other mental health conditions (Jacobson & Chung, 2020). A recent meta-analysis reported benefits of passive sensing including behavioral and client status change detection (Cornet & Holden, 2018). Further, virtual reality has shown promise as another novel approach to assessment (Parsons & Carlew, 2016; Parsons, Bowerly, Buckwalter, & Rizzo, 2007). Studies have found greater discloser of risky behaviors, including suicide and substance use, via digital tools (Butler, Villapiano, & Malinow, 2009; Proudfoot et al., 2003), lending support to their use in screening/monitoring.

Within health systems, reducing the number of face-to-face contacts lowers treatment costs (Budney et al., 2019; Erbe et al., 2017). BC may also enhance the quality of care delivered in that digital interventions validated in controlled trials can be delivered in places where such interventions are not available, including non-specialty clinics or primary care settings. An ancillary consequence of such availability is reduced burden on the health system by lessening the need for highly trained mental health providers within these settings. Supportive person-to-person contact from non-specialty providers in BC models provides benefit when evidence-based components are delivered digitally (Budney et al., 2019). Last, digital delivery of evidence-based interventions provides high treatment integrity and fidelity, which is not always the case with face-to-face delivery of these interventions (Moller et al., 2017).

In sum, by combining strengths from traditional face-to-face therapy and digital technology, BC holds promise of value. For clients, BC may strengthen outcomes of traditional psychotherapy while enhancing treatment access across different treatment phases for individuals with a variety of mental health concerns. BC has the potential to improve client motivation, engagement, and adherence (Figueroa, DeMasi, Hernandez-Ramos, & Aguilera, 2021) and feelings of self-mastery (Wentzel et al., 2016) while enhancing the frequency and intensity of high-fidelity evidence-based interventions. Providers can harness digital advances to augment therapy or assessment in a variety of ways to meet individualized client needs, while reducing client burden and cost of treatment delivery. Further, technology offers novel approaches to assessment that improve accuracy and frequency of status tracking over time. Last, BC models may offer a means to reduce treatment costs while improving access to high fidelity care, particularly in non-specialty clinics.

5.6 Illustrations

5.6.1 Examples of blended care interventions for MHD adapted for the use in the healthcare system

The following section provides the context of selected MHD along with examples of BC interventions to treat MHD from healthcare systems around the world.

5.6.1.1 Depression

Depression is a common and serious mental illness for which effective and accessible interventions are needed. Although, much evidence has accumulated for various clinical approaches to depression, there is not clear consensus about what treatment is most effective (Cuijpers, Cristea, Karyotaki, Reijnders, & Huibers, 2016; Cuijpers, Huibers, Ebert, Koole, & Andersson, 2013; Cuijpers, Van Straten, Andersson, & Van Oppen, 2008; Wampold & Imel, 2015), such that effectiveness at the client-level is highly heterogeneous (Hofmann, Asnaani, Vonk, Sawyer, & Fang, 2012; Loerinc et al., 2015), and by some reports only 30% of clients who receive treatment for depression achieve remission in their first treatment episode (Rush et al., 2006). Research on BC for depression as an alternative and perhaps more effective approach to treatment has been ongoing for more than 15 years, with early results showing strong potential of BC for both prevention and treatment. More recent studies found large effect sizes when comparing BC to self-guided interventions for depression (Andersson et al., 2009; Baumeister et al., 2014; Cowpertwait & Clarke, 2013; Richards & Richardson, 2012; Spek et al., 2007), and medium to large effects when BC was compared to psychoeducation for depression (Beiwinkel, Eißing, Telle, Siegmund-Schultze, & Rössler, 2017; Buntrock et al., 2015; Nobis et al., 2015).

A systematic review and meta-analysis of standalone digital interventions and BC for depression, however, found no significant difference between the two types of treatment, with both showing beneficial effects on depression symptoms and severity (Koenigbauer, Letsch, Doebler, Ebert, & Baumeister, 2017). As with BC for other MHD or SUD, some suggest that different symptom severity profiles may need unique BC approaches, or that stepped care models may prove useful. BC components and solutions for depression can be designed for and elaborated within the course of therapy or after the active treatment to help maintain treatment effects (Reins, Boß, Lehr, Berking, & Ebert, 2019). Improvement in cost-effectiveness was not investigated in either study (Koenigbauer et al., 2017), however, Buntrock et al. (2015) reported an acceptable cost–benefit ratio when addressing subthreshold depressive symptoms.

5.6.1.1.1 HelloBetter

One example of a publicly available BC is offered via courses from the German company HelloBetter. While one of their courses is a standalone treatment addressing stress and burnout, the remaining courses that address various mental health problems such as depression, panic disorder, and insomnia are designed as BC treatments (HelloBetter Online Classes Producer, 2022). Within the German statutory health insurance (SHI) system some insurances cover the BC courses, while the standalone course addressing stress and burnout can be prescribed as a pharmaceutical or medical device within the recently introduced Digital Health Care Act (Digitale-Versorgung-Gesetz, DVG). In total, 37 randomized controlled trials (RCTs) have been published investigating the effectiveness of the different available HelloBetter courses (HelloBetter Research Producer, 2022). Each BC course offered by HelloBetter consists of online support by qualified psychologists and psychotherapists and online content. Here we briefly describe courses for the prevention of depression and for acute depression and summarize the evidence of their effectiveness.

The BC course for depression prevention runs over a period of 6 weeks and is available in English and German (HelloBetter Depression Prevention Producer, 2022). In six 1-hour online course units, recommended for completion on a weekly basis, the users learn CBT-based strategies to improve mood, develop strategies to address problems step by step, and strengthen relationships. The supportive element of the treatment is individual support from a psychologist in the form of written feedback upon completion of each unit. The course is covered by some German health insurances or can be booked directly for a fee. Regarding the effectiveness of this online course, RCTs are cited by HelloBetter (HelloBetter Depression Prevention Producer, 2022).

Similar to the BC for depression prevention, the BC course for acute depression comprises six 1-hour courses designed to be completed on a weekly basis (HelloBetter Acute Depression Producer, 2022). Each course involves mood enhancement strategies derived from CBT approaches. The user's dashboard provides an overview of the main topics, the numbers of sessions completed, responses to diary questions, as well as a number of homework assignments to be completed and reviewed. Participants track their mood via a mood diary. The support component of the BC course consists of written feedback from a qualified psychologist on overall progress, submitted homework, and mood diary entries. Information about the psychologist ("View Profile") and messages from the psychologists are also provided on the user's dashboard. Video calls with the assigned psychologist are not available. This course is reimbursed by several insurances and or can be accessed for a fee. Empirical support of the efficacy of this course was documented in four controlled studies (HelloBetter Acute Depression Producer, 2022).

5.6.1.2 Anxiety disorders

Anxiety disorders are the second major MHD after depression affecting millions of people worldwide (Baxter, Scott, Vos, & Whiteford, 2013). Multiple types of anxiety disorders span a wide range of symptoms and severities. Here we provide examples of how BC has been employed to effectively treat a few of the more common types of anxiety disorders.

Generalized anxiety disorder (GAD) is characterized by an inability to contain hypothetical worries, often leading to physical symptoms of anxiety such as insomnia or panic. Various types of BC interventions have shown promise in helping reduce symptoms and the severity of GAD. For example, a BC that integrated face-to-face ACT with digital content delivered in web-based modules showed positive findings as a treatment alternative for older clients with anxiety symptoms (Witlox et al., 2021), and iCBT for GAD showed promising results when compared to face-to-face therapy across multiple trials (Andrews et al., 2018). In recent years, there have been an increasing number of studies using the unified protocol for face-to-face transdiagnostic treatment of emotional disorders (Barlow, Farchione, Fairholme, Ellard, Boisseau, Allen, & Ehrenreich-May, 2011; Richards, Richardson, Timulak, & McElvaney, 2015), which allows the targeting of characteristic symptoms of GAD.

Panic disorder with agoraphobia and social anxiety disorder (SAD) are both disorders that constrain the lives of clients significantly and cause anticipatory anxiety. Both conditions benefit greatly from exposure-based treatment protocols. Guided BC that involves exposure-based treatment and VRET have shown promising results for panic disorder with agoraphobia (Carl et al., 2019; Domhardt, Schröder, Geirhos, Steubl, & Baumeister, 2021). iCBT and several BC interventions with a focus on attention bias training have produce mixed results for SAD (Amir et al., 2009; Carlbring, Andersson, Cuijpers, Riper, & Hedman-Lagerlöf, 2018; Schmidt, Richey, Buckner, & Timpano, 2009; Williams, O'Moore, Mason, & Andrews, 2014). A meta-analysis of 37 trials concluded that self-guided and guided iCBT, as well as VRET were effective in treating SAD (Kampmann, Emmelkamp, & Morina, 2016), and a scoping review described several studies on SAD that integrated CBT and virtual reality with efficacy for reducing anxiety symptoms (Baghaei et al., 2021). A published clinical guide provides examples of some of the features of BC for SAD; for instance, social situations are recreated via a webcam to model and practice modifying self-focused attention and safety behaviors (Warnock-Parkes et al., 2020).

Specific phobias are some of the most common anxiety disorders, and typically develop during childhood. While not all specific phobias require treatment, a specific phobia can interfere with daily life. A specific phobia causes avoidance of situations and/or distress if a feared stimuli cannot be avoided. Ten years ago, few trials of internet-delivered treatments for specific phobias appeared in the literature, with examples of RCTs targeting spider phobia (Andersson et al., 2009) and snake phobia (Andersson et al., 2013). More recently, VRET demonstrated efficacy in extinguishing the response to a feared stimuli in phobias (e.g., Botella et al., 2017), with technological advancements enabling features such as a 360° immersive experience (Donker et al., 2019). Overall, research testing BCs for specific phobias remains relatively scarce (Mor et al., 2021).

5.6.1.2.1 The MindSpot Wellbeing course

In Australia, the MindSpot Wellbeing course (MindSpot Producer, 2022) is a web-based transdiagnostic BC program that targets underlying mechanisms of depression and anxiety disorders. MindSpot is a research center funded by the Australian government that offers evidence-based therapy services to Australian citizens at no cost. The Wellbeing course is one of several digital health programs offered by MindSpot. Wellbeing has five web-based modules on psychoeducation, tracking and challenging thoughts, avoidance, prolonged exposure, and relapse prevention. The program enables clients to have an initial assessment, as well as weekly scheduled support from a clinician, or on-demand support. The Wellbeing course is available for 8 weeks, with clients completing in-between session tasks. The Wellbeing transdiagnostic program has been evaluated in several RCTs showing promising results for SAD (Dear et al., 2016) and GAD (Dear et al., 2015).

5.6.1.2.2 Oxford VR (OVR)

Oxford VR (OVR Producer, 2022), a spinoff from the University of Oxford, specializes in delivering protocolized immersive experiences in virtual reality that are CBT-based. OVR bases its products on several scientific publications in collaboration with the University of Oxford that confirm the efficacy of VRET in treating anxiety disorders (Freeman et al., 2017, 2018). The conditions treated include specific phobias such as fear of heights, social anxiety, and psychosis. OVR works as a prescribed app with partners such as NHS services in the UK to offer automated treatments with a VR headset and a virtual coach, at times in conjunction with face-to-face therapy sessions (NHS Oxford Health Fudation Trust Producer, 2021). The programs are developed by engineers from the gaming industry to be engaging and realistic. Clients access 30-minute sessions to engage in graded exposure-based exercises in a safe environment. The user enters a virtual reality environment specific to their issue and is guided by a virtual coach to use CBT skills in order to gradually face a feared stimuli such as taking an elevator or moving in a busy crowd. One of the programs developed by OVR is "Social Engagement", which aims to help clients overcome anxious social avoidance is available through prescription to adults in the UK. through the Improving Access to Psychological Treatments NHS framework.

5.6.1.2.3 Ieso DigitalHealth

Ieso DigitalHealth (iesohealth Producer, 2022) is a company which offers CBT-based BC for a range of mental health conditions including specific phobias, anxiety disorders, post-traumatic stress disorder, and obsessive compulsive disorder. In many areas in the UK, Ieso Health appointments are available for free via NHS primary care services. Ieso base their treatments on a number of research studies conducted since 2009, including a RCT examining the effectiveness of chat-delivered iCBT for depression, which showed promising results for recovery (Kessler et al., 2009). Therapy appointments are delivered via video or via a chat function with additional features such as a messaging function and session transcripts supporting clients with reviewing therapy content and completing between-session tasks. To ensure that clients receive effective care, several aspects of care are addressed. First, therapy appointments include agenda-setting which is found in a face-to-face CBT session. Next, treatments are personalized collaboratively with the client. Third, the client's primary issue is matched to one of the CBT evidence-based manuals. Clients can choose between using a browser or the app to log in to their profile and connect to care. The app makes this type of BC available to adults without access to a personal computer. Additionally, it allows clients for whom face-to-face or video appointments present as a barrier, to access care by allowing clients to talk to their therapist via a chat function. Ieso is gathering data from transcripts and messages and analyses these data to further improve treatments.

5.6.2   Examples of BC interventions for SUD adapted for use in the healthcare system

Many innovative BC interventions for SUD have undergone efficacy testing with findings published in scientific journals. A much smaller number have demonstrated clinically significant results across multiple controlled trials including effectiveness trials conducted in community clinics. And only a handful of these have been translated to a format compatible for use in healthcare systems and have been made available to providers. The BC interventions for SUD that have made their way into healthcare systems have typically transformed existing evidenced-based, manualized behavioral therapies by delivering much of the active content (usually coping skills training or positive reinforcement) via a smartphone, tablet, or personal computer, with therapists serving to motivate, monitor, and provide the personal and emotional support needed to successfully adhere to and complete the course of BC. These BC offerings were designed for and tested with individuals seeking treatment for SUD or co-occurring disorders in an outpatient setting. Here, we provide examples of a few such BC interventions (ReSET/ReSET-O, SHADE, CBT4CBT/A-CHESS) that are currently available to mental health providers and those with SUD.

5.6.2.1   ReSET/ReSET-O (therapeutic education system)

ReSET@, a mobile app, was the first digital therapeutic authorized by the US Food and Drug Administration (FDA) as a prescription-only adjunct to psychosocial or medication SUD treatment. As such, providers can prescribe ReSET as part of a client's treatment program, and in many cases the client's health insurance will cover its cost. ReSET is based on multiple controlled clinical trials of a digital treatment, the treatment education system (TES), that had demonstrated positive treatment outcomes for opioid, cannabis, and methamphetamine use disorders, as well as reduction in HIV risk behaviors, with results comparable to those produced by clinician-delivered evidence-based treatment (Bickel, Marsch, Buchhalter, & Badger, 2008; Campbell et al., 2014; Marsch et al., 2015).

Drawing from the evidence-based community reinforcement approach (CRA) and CBT for treating SUD, TES embedded over 65 self-directed intervention modules designed to teach and enhance coping skills and improve psychosocial functioning. Multimedia features were integrated to facilitate interactive therapeutic exercises and to provide alternative learning styles and techniques. TES also included a system to support incentive-based contingency management programs (an important component of CRA), and provided a summary of client progress to the treatment provider to guide face-to-face sessions. Providers typically utilize check-ins to monitor, and problem solve any issues with completion of TES modules and to provide support, guidance, and assistance with acute problems and referrals.

ReSET is currently marketed as a clinical tool that can complement face-to-face or tele-therapy for SUD (Pear Therapeutics Producer, 2018; U.S. Food and Drug Administration FDA Producer, 2017). In response to the opioid crisis, a specialty version of ReSET was developed, ReSET-O. Both applications have been FDA-authorized, and these products are available by prescription, designated for use in concert with a licensed provider (not as a standalone), and are covered by some health insurances in the US.

5.6.2.2   SHADE (self-help for alcohol and other drug use and depression)

The Australian government has strongly embraced the potential of digital mental health (digital mHealth) tools and worked closely with academic clinical researchers and health systems across Australia to develop an online system that educates their population about and offers easily accessible pathways to mHealth services (Department of Health Australian Goverment Producer, 2021; Head to Health Producer, 2022). One BC program, SHADE, was developed and evaluated in controlled trials. Across multiple trials, SHADE accumulated evidence for significant reductions in symptoms of depression and in alcohol and cannabis use comparable to the effects of completely face-to-face interventions (Glasner et al., 2018; Kay-Lambkin et al., 2009; Kay-Lambkin, Simpson, Bowman, & Childs, 2014).

SHADE comprises 10 weekly 60-minute online modules that draw from CBT for SUD and Depression. CBT techniques include monitoring of mood, thoughts, and substance use, problem solving, identifying pros and cons, and drug refusal skills. Links between depressive symptoms and substance use are also explored. SHADE incorporates techniques from motivational enhancement therapy throughout the course of treatment. The first session begins with a person-to-person motivational enhancement session including assessment feedback. Education and planning related to subsequent CBT skills training (e.g., identifying goals, development of change plans) as well as a nonconfrontational discussion of behavior change and maintenance (Kay-Lambkin et al., 2009) are incorporated throughout the intervention. Weekly structured 10–15-minute provider check-ins are completed at the end of each session. Check-ins include a brief assessment, reviewing practice activities including mood monitoring and/or worksheets covering CBT skills introduced in the previous online module (i.e., homework review), and planning for homework completion through the next week (Glasner et al., 2018; Kay-Lambkin, Baker, Kelly, & Lewin, 2011).

SHADE is available at no cost in Australia through eCliPSE (Electronic Clinical Pathways to Service Excellence), an mHealth website managed by the Ministry of New South Wales. eCliPSE allows individuals to directly engage with SHADE and use it as a standalone program. After registering (Eclipse Producer, 2022), users complete a brief online assessment. eCLiPSE recommends treatment based upon questionnaire responses. All content (e.g., modules, worksheets, homework) are available online. Alternatively, providers can register for SHADE, allowing access to content and searching for clients. This BC option is strongly recommended. SHADE can be accessed commercially in the US and elsewhere through Cobalt Therapeutics, LLC's/Magellan Health's Computerized Cognitive Behavioral Therapy (CCBT) programs.

5.6.2.3 CBT4CBT (computer-based training for cognitive behavioral therapy)

CBT4CBT (Computer-based Training for Cognitive Behavioral Therapy) is a well-researched, efficacious BC developed for SUD treatment that can be web-delivered via personal computers, tablets, or mobile phones (Carroll et al., 2008). Multiple RCTs with diverse SUD samples have demonstrated that CBT4CBT can engender and maintain drug abstinence when integrated with face-to-face CBT for SUD (Carroll et al., 2008, 2014). More recently, CBT4CBT-buprenorphine, an adaptation that includes psychoeducation on buprenorphine therapy was found to increase drug abstinence when included as an adjunct to standard buprenorphine therapy (Shi, Henry, Dwy, Orazietti, & Carroll, 2019). CBT4CBT has also demonstrated positive effects on abstinence outcomes in a small, controlled trial for alcohol use disorder (Kiluk et al., 2016).

Similar to ReSET and SHADE, CBT4CBT is a self-guided program that comprises six modules with content primarily based on CBT for SUD. Psychoeducational and skills building materials in each module are presented in a variety of formats, such as engaging graphics, video, text, and audio, and each includes interactive assessments, and practice exercises. Techniques incorporated throughout CBT4CBT include identifying patterns of substance use and strengthening decision making, coping, problem solving, and drug refusal skills. Video demonstrations and practice assignments foster development of and reinforce use of the targeted therapeutic behaviors and skills (Carroll et al., 2008).

CBT4CBT for SUD is commercially available online (CBT4CBT LLC Producer, 2022) for direct purchase by licensed treatment providers in the US. CBT4CBT is not currently reimbursed through insurance but providers can prescribe CBT4CBT to appropriate clients. CBT4CBT (for SUD, alcohol, and buprenorphine) is also available via web or smartphone delivery through CHESS Health's Addiction Management "Connections" digital platform. Here, CBT4CBT is offered alongside a relapse prevention-based BC, A-CHESS, which is discussed below.

5.6.2.4 A-CHESS (addiction—comprehensive health enhancement support system)

A-CHESS is a mHealth smartphone application available in the US for clients aged 13 and older as an adjunct to face-to-face SUD treatment to improve adherence and reduce relapse. A-CHESS has been evaluated via multiple RCTs in a variety of SUD samples and treatment settings, including among clients engaged in medication-assisted treatment for opioid use disorder (OUD; Gustafson et al., 2016), women with SUD in a rural geographic location (Johnston, Mathews, Maus, & Gustafson, 2019), and drug court participants (Johnson et al., 2016). A RCT demonstrated short-term reduction in heavy drinking days, and higher abstinence rates up to 12-months post-intervention for A-CHESS users compared to a control (Gustafson et al., 2014). Greater use of the A-CHESS app has been associated with better outcomes, with some data suggesting that those with polysubstance use and without co-occurring MHD utilize A-CHESS (Gustafson et al., 2014; McTavish, Chih, Shah, & Gustafson, 2012).

A-CHESS connects users with 24/7 on-demand support and resources (e.g., audio-guided relaxation exercises) to cope with cravings, withdrawal symptoms, and high-risk situations to avoid relapse. The app incorporates social connections with peers, and motivational strategies. A-CHESS also includes adaptive monitoring such as global positioning system technology, which can track when an individual approaches an identified risky location and deliver an alert to the individual asking if they want to be in that location. A-CHESS also includes a panic button to contact social support during relapse-risk elevation. A 12-item weekly survey serves as a monitoring tool useful for predicting relapse (Chih et al., 2014). This tool assesses lifestyle balance, sleep quality, negative affect, and recent substance use. Users of A-CHESS can permit their providers to access this survey data and receive A-CHESS notifications if completion is below a specified threshold (Chih et al., 2014). A provider app, Companion, connects providers with clients and aides in integrative care facilitation.

A-CHESS is commercially available via the Connections smartphone app (Center for Health Care Strategies Producer, 2021). This evidence-based platform contains three distinct features including SUD provider searches and referrals (eIntervention), CBT4CBT as an adjunct to face-to-face treatment (eTherapy), and relapse prevention via A-CHESS (eRecovery).

Connections are available for purchase in the US by SUD treatment providers, state and local governments, hospitals, and health plans. The BC intervention format requires a provider code for individuals seeking treatment (ChessHealth Producer, 2022).

5.7 Challenges to the adoption of blended care

As described in Section 5.5 of this chapter, BC approaches to MHD and SUD provide many intriguing advantages to almost all stakeholders—clients, providers, and health systems. So why have BC treatments not become the first option offered to those seeking treatment for these health disorders? Here we discuss multiple factors that need to be addressed before healthcare systems can reap the full potential of digital therapeutics, in particular BC approaches.

Ironically, one of the major issues that has driven the development of digital therapeutics, that is, lack of access to evidence-based, efficacious interventions, is also a primary barrier to widespread use of BC for MHD and SUD. At the client level, although technology devices are seemingly ubiquitous, a substantial number of youth and adults either do not own or have reliable and easy access to the devices required for most BC offerings (smartphones, tablets, personal computers), or if they have access to a device, they may have limited access to the internet. Additionally, novel technologies often cannot be accessed with older personal computers or smartphones. Some technologies are also still being adapted for home use and scalable usage. For instance, the groundbreaking approach of VRET still largely relies on a costly VR headset (Baghaei et al., 2021). Unfortunately, disadvantaged populations—who are most vulnerable to MHD and SUD—are also those with the least access to the digital platforms needed to effectively engage with many BC programs. Moreover, specific groups, such as the elderly and those living in rural areas who face increased burden accessing quality healthcare, are less likely to own and be comfortable with technology devices (elderly) or have poor internet access (rural populations).

Access from the provider perspective also presents some difficulties. Awareness of BC interventions is low; how and where one purchases or gains access to BC models is not clear in many countries; training on how best to integrate BC models into practice is not readily available; research demonstrations and best practice guidelines are not widely published recommending BC options; and last, it is not clear how providers' time and effort integrating the digital components are paid for in many healthcare practice models. However, direct efforts to enhance access to and dissemination of BC is underway. For example, the American Psychological Association offers continuing education training for providers on the integration of technology into clinical practice (e.g., text and email, assessment and self-monitoring, pure and guided self-help programs), as well as ethical and privacy considerations (American Psychological Association, 2022). Similarly, the Center for m^{2}Health at Palo Alto University offers trainings and consultation on the use of technology in clinical practice (Center for m2Health & Palo Alto, 2022). From a health systems perspective, hindrances that must be addressed include: authorization, billing, and reimbursement structures for BC; personnel reorganization needs to consider modified caseloads, responsibilities, training, and required expertise to efficiently embrace and deliver BC; and the potential negative impact on morale or enthusiasm of clinical staff that can arise when asked to adopt digital procedures that to some extent replace them and minimize their perceived value. Futher, for many organizations, the initial cost of digital devices and tools and the infrastructure needed to integrate BC may impact already stressed operations that have outdated and low digital capacity.

Digital advancements have elevated privacy concerns for clients, providers, and health systems. This topic is discussed in detail in Chapter 15, titled "Privacy and security." Limits to confidentiality and problems with data security raise additional ethical concerns that pose a significant challenge in the adoption of BC. In terms of data privacy, information use and disclosure agreements are typically controlled by developers. Passive sensors may also inadvertently collect, store, or transmit private information to third party entities without knowledge to the user. Security risks include confidentiality breaches during collection or transmission of data, encryption issues, and other security flaws that increase vulnerability to malware attacks (Hall & McGraw, 2014). While not limited to BC, privacy, confidentiality, and data security are important to clients who are seeking or receiving SUD or MHD treatment. In BC models, clients should be informed of and consent to the additional limitations in privacy and security that are associated with some types of digital treatment components but as a result, may be concerned over the additional risks with the use of these tools (Ozair, Jamshed, Sharma, & Aggarwal, 2015). Providers similarly may lack trust in such technologies given their limited control over data privacy and security of protected client information. Another issue is that weakened regulations over security features might contribute to provider violations of the Health Insurance Portability and Accounting Act (HIPAA; Hale & Kvedar, 2014; Hall & McGraw, 2014). Within some health systems, the introduction of digital tools may not be feasible under current regulations or restrictions, further slowing BC adoption in these settings (Ramsey, Lord, Torrey, Marsch, & Lardiere, 2016).

While many digital treatment components (e.g., on-demand messaging, automated prompts for monitoring, or coping activities) have been suggested to improve adherence and retention outcomes in BC (e.g., Campbell et al., 2014), adherence to engaging with the digital components also presents an important challenge, particularly among populations that are difficult to engage in treatment (i.e., SUD; Ramos et al., 2021). For instance, studies of those with SUD suggest that after a brief initial period of high motivation, a substantial percentage disengage with digital tools (Attwood, Parke, Larsen, & Morton, 2017; Tait et al., 2014). Providers must also adhere consistently to their tasks for some digital tools to provide optimal impact, and some studies suggest that such adherence cannot be assumed. Thus, techniques to promote ongoing engagement with digital intervention components for clients and providers are needed to obtain positive outcomes that justify adoption of BC models.

5.8   Future directions and conclusions

BC models offer substantial opportunities to improve outcomes of, and access to, evidence-based treatments of MHD and SUD. Moreover, continued advances in technology and analytic approaches hold promise to expand further upon current BC possibilities and options. For example, JITAIs can be readily integrated into person-to-person behavioral therapies and have the potential to address some of the frequent barriers to adoption of BC as well as the potential for enhancing outcomes. JITAIs utilize mobile technology to collect real-time data from passive sensing or ecologically momentary assessments to facilitate personalization of the intervention based on a person's current environment and internal emotional or physiological state. Although JITAIs have shown promise in multiple experimental studies, continued advances in big data analytics, such as machine learning, and in passive sensing technology are needed to optimize and further validate specific applications that can be translated and integrated with person-to-person therapies to create feasible, appealing, and effective BC models.

To advance the general practice of BC, much attention and effort must be directed at provider uptake and adherence. As a rule, the application of new discoveries and innovations takes an inordinate amount of time (Morris, Wooding, & Grant, 2011; Rogers, 2010). As discussed above, we have identified several key factors that impede this process for digital treatments and many are similar to those that confront the adoption of evidence-based psychological interventions (Aarons, Wells, Zagursky, Fettes, & Palinkas, 2009; Lilienfeld, Ritschel, Lynn, Cautin, & Latzman, 2013; Lord, Moore, Ramsey, Dinauer, & Johnson, 2016; McGovern, Fox, Xie, & Drake, 2004). Adoption of BC will require multiple system- and provider-level efforts: easily accessible provider trainings, acceptance into and integration with existing academic and professional training programs, practice guidelines that include BC interventions, payor models for coverage of BC, healthcare system infrastructure to support BC implementation, and enhanced and standardized data security and privacy procedures with associated educational programs that can reassure providers and clients.

As various BC models appear in the scientific literature and the marketplace, each of which integrates digital components from a rapidly growing cadre of options, providers and clients need much more guidance to navigate digital applications that have been validated as effective in producing positive clinical outcomes. For payors to accept and approve BC, evidence of their value via health systems and health economic studies is sorely needed. How can such data be collected on a large scale to demonstrate efficacy? Controlled effectiveness and implementation clinical trials that identify effective treatment components of BC models are sorely lacking, and are needed to develop guidance in selection of BC treatment options. In addition, many health systems have yet to develop training models for providers that can motivate them to use BC and deliver them with integrity to optimize their application. However, much can be learnt from countries that are investing in digital treatment and training pathways. For instance, National Health Services such as in the UK and in Australia are collecting relevant data to help demonstrate efficacy and affordability of BC on a large scale. These data might be utilized to launch and adapt BC programs in other countries and healthcare settings.

At the client level, acceptance, and adherence to the digital components of BC provide a formidable challenge as well. In addition to allaying concerns about data privacy, technology availability and literacy, and acceptance of nontraditional or digital therapies for psychological issues, research and development efforts need to focus on testing strategies for maintaining engagement with the BC intervention over time, as most MHD and SUD are chronic conditions that require continued efforts at the client level to maintain progress and reduce the probability and severity of relapse. Personalized programs that develop healthy coping and monitoring programs that can be maintained over time with low associated client effort or burden may be most attractive for clients, providers, and health systems. Having relapse prevention check-ins and alerts via a BC platform or app may support clients to keep using therapy skills and reduce the number of clients returning for a new course of treatment. Data collection and monitoring in the background can help identify patterns of behavior, which in turn can inform personalized interventions. JITAIs help identify receptive states in order to deliver interventions at the right time. The future of digital features in BC will likely involve a lesser burden on the client through passive monitoring, as well as a higher degree of personalization.

Last, BC further holds the potential of increasing treatment options and specialist care. The distribution of trained psychotherapists varies geographically around the world. As in medicine, clients are not always able to find the right type of care for their mental health needs where they live as mental health care is scare and unequally distributed (World Health Organization, 2013). Negative experiences unique to minority groups such as microaggressions (Sue, Capodilupo, Torino, Bucceri, Holder, Nadal, & Esquilin, 2007) and implicit bias can additionally affect the quality of care received. BC holds the potential of creating content reflective of a specific minority group, as well as of connecting a client to specialist remote care, thereby increasing access and quality of care for underserved populations.

In conclusion, BC is still in an early stage of development and much guidance and support are still needed to see it reach its full potential. However, with some investment in research, provider training, and regulation, BC holds great promise in connecting more clients to high-quality care, while providing more treatment options at a lower burden, and reducing costs to clients, providers, and health systems. This has already been demonstrated by the NHS in the UK as well as by innovative BC programs for MHD and SUD coming to market in the US. Combining innovative technological features with data from passive tracking will give clients and providers a new level of insight into the journey of care to achieve a more personalized therapy experience that allows linking the traditionally distinct phases of assessment, treatment, and recovery.

References

Aarons, G. A., Wells, R. S., Zagursky, K., Fettes, D. L., & Palinkas, L. A. (2009). Implementing evidence-based practice in community mental health agencies: A multiple stakeholder analysis. *American Journal of Public Health, 99*(11), 2087–2095.

American Psychological Association. (2022). Continuing education: Incorporating technology into clinical practice. Retrieved January 11. https://www.apa.org/education-career/ce/1360496.

Amir, N., Beard, C., Taylor, C. T., Klumpp, H., Elias, J., Burns, M., et al. (2009). Attention training in individuals with generalized social phobia: A randomized controlled trial. *Journal of Consulting and Clinical Psychology, 77*(5), 961.

Andersson, G., Waara, J., Jonsson, U., Malmaeus, F., Carlbring, P., & Öst, L. G. (2009). Internet-based self-help versus one-session exposure in the treatment of spider phobia: A randomized controlled trial. *Cognitive Behaviour Therapy, 38*(2), 114–120.

Andersson, G., Waara, J., Jonsson, U., Malmaeus, F., Carlbring, P., & Öst, L.-G. (2013). Internet-based exposure treatment versus one-session exposure treatment of snake phobia: A randomized controlled trial. *Cognitive Behaviour Therapy, 42*(4), 284–291.

Andrews, G., Basu, A., Cuijpers, P., Craske, M., McEvoy, P., English, C., et al. (2018). Computer therapy for the anxiety and depression disorders is effective, acceptable and practical health care: An updated meta-analysis. *Journal of Anxiety Disorders, 55*, 70–78.

Attwood, S., Parke, H., Larsen, J., & Morton, K. L. (2017). Using a mobile health application to reduce alcohol consumption: A mixed-methods evaluation of the drinkaware track & calculate units application. *BMC Public Health [Electronic Resource], 17*(1), 1–21.

Baghaei, N., Chitale, V., Hlasnik, A., Stemmet, L., Liang, H.-N., & Porter, R. (2021). Virtual reality for supporting the treatment of depression and anxiety: Scoping review. *JMIR Mental Health, 8*(9), e29681.

Barlow, D. H., Farchione, T. J., Fairholme, C. P., Ellard, K. K., Boisseau, C. L., Allen, L. B., & Ehrenreich-May, J. (2011). Unified protocol for transdiagnostic treatment of emotional disorders: Therapist guide. Oxford University Press.

Baumeister, H., Reichler, L., Munzinger, M., & Lin, J. (2014). The impact of guidance on internet-based mental health interventions—A systematic review. *Internet Interventions, 1*(4), 205–215.

Baxter, A. J., Scott, K. M., Vos, T., & Whiteford, H. A. (2013). Global prevalence of anxiety disorders: A systematic review and meta-regression. *Psychological Medicine, 43*(5), 897–910.

Beiwinkel, T., Eißing, T., Telle, N.-T., Siegmund-Schultze, E., & Rössler, W. (2017). Effectiveness of a web-based intervention in reducing depression and sickness absence: Randomized controlled trial. *Journal of Medical Internet Research, 19*(6), e6546.

Berger, T., & Andersson, G. (2009). Internet-based psychotherapies: Characteristics and empirical evidence. *Psychotherapie, Psychosomatik, Medizinische Psychologie, 59*(3-4), 159–170.

Bickel, W. K., Marsch, L. A., Buchhalter, A. R., & Badger, G. J. (2008). Computerized behavior therapy for opioid-dependent outpatients: A randomized controlled trial. *Experimental and Clinical Psychopharmacology, 16*(2), 132–143.

Botella, C., Fernández-Álvarez, J., Guillén, V., García-Palacios, A., & Baños, R. (2017). Recent progress in virtual reality exposure therapy for phobias: A systematic review. *Current Psychiatry Reports, 19*(7), 1–13.

Budney, A. J., Borodovsky, J. T., Marsch, L. A., & Lord, S. E. (2019). Technological innovations in addiction treatment. In: I. Danovitch & L. J. Mooney (Eds.), The Assessment and treatment of addiction (pp. 75–90). St. Louis, Missouri: Elsevier.

Bultler, S. F., Budman, S. H., Goldman, R. J., Newman, F. J., Beckley, K. E., Trottier, D., et al. (2001). Initial validation of a computer-administered Addiction Severity Index: The ASI–MV. *Psychology of Addictive Behaviors, 15*(1), 4.

Buntrock, C., Ebert, D., Lehr, D., Riper, H., Smit, F., Cuijpers, P., et al. (2015). Effectiveness of a web-based cognitive behavioural intervention for subthreshold depression: Pragmatic randomised controlled trial. *Psychotherapy and Psychosomatics, 84*(6), 348–358.

Butler, S. F., Villapiano, A., & Malinow, A. (2009). The effect of computer-mediated administration on self-disclosure of problems on the Addiction Severity Index. *Journal of Addiction Medicine, 3*(4), 194.

Campbell, A. N., Nunes, E. V., Matthews, A. G., Stitzer, M., Miele, G. M., Polsky, D., et al. (2014). Internet-delivered treatment for substance abuse: A multisite randomized controlled trial. *American Journal of Psychiatry, 171*(6), 683–690.

Carl, E., Stein, A. T., Levihn-Coon, A., Pogue, J. R., Rothbaum, B., Emmelkamp, P., et al. (2019). Virtual reality exposure therapy for anxiety and related disorders: A meta-analysis of randomized controlled trials. *Journal of Anxiety Disorders, 61*, 27–36.

Carlbring, P., & Andersson, G. (2006). Internet and psychological treatment. How well can they be combined? *Computers in Human Behavior, 22*(3), 545–553.

Carlbring, P., Andersson, G., Cuijpers, P., Riper, H., & Hedman-Lagerlöf, E. (2018). Internet-based vs. face-to-face cognitive behavior therapy for psychiatric and somatic disorders: An updated systematic review and meta-analysis. *Cognitive Behaviour Therapy, 47*(1), 1–18.

Carroll, K. M., Ball, S. A., Martino, S., Nich, C., Babuscio, T. A., Nuro, K. F., et al. (2008). Computer-assisted delivery of cognitive-behavioral therapy for addiction: A randomized trial of CBT4CBT. *American Journal of Psychiatry, 165*(7), 881–888.

Carroll, K. M., Kiluk, B. D., Nich, C., Gordon, M. A., Portnoy, G. A., Marino, D. R., et al. (2014). Computer-assisted delivery of cognitive-behavioral therapy: Efficacy and durability of CBT4CBT among cocaine-dependent individuals maintained on methadone. *American Journal of Psychiatry, 171*(4), 436–444.

CBT4CBT LLC (Producer). (2022). Retrieved January 11. https://cbt4cbt.com/.

Center for Health Care Strategies (Producer). (2021). A-Chess. Retrieved 2022, January 11. https://www.chcs.org/digital-health-products/a-chess/.

Center for m2Health, Palo Alto University. (2022). Continuing education courses. Retrieved July 13. https://www.m2health.paloaltou.edu/continuing-education-courses.

ChessHealth (Producer). (2022). Customers. Retrieved January 11. https://www.chess.health/customers/.

Chih, M.-Y., Patton, T., McTavish, F. M., Isham, A. J., Judkins-Fisher, C. L., Atwood, A. K., et al. (2014). Predictive modeling of addiction lapses in a mobile health application. *Journal of Substance Abuse Treatment, 46*(1), 29–35.

Christensen, H., Griffiths, K., Mackinnon, A., & Brittliffe, K. (2006). Online randomized controlled trial of brief and full cognitive behaviour therapy for depression. *Psychological Medicine, 36*(12), 1737–1746.

Cornet, V. P., & Holden, R. J. (2018). Systematic review of smartphone-based passive sensing for health and wellbeing. *Journal of Biomedical Informatics, 77*, 120–132.

Cowpertwait, L., & Clarke, D. (2013). Effectiveness of web-based psychological interventions for depression: A meta-analysis. *International Journal of Mental Health and Addiction, 11*(2), 247–268.

Cuijpers, P., Cristea, I. A., Karyotaki, E., Reijnders, M., & Huibers, M. J. (2016). How effective are cognitive behavior therapies for major depression and anxiety disorders? A meta-analytic update of the evidence. *World Psychiatry, 15*(3), 245–258.

Cuijpers, P., Huibers, M., Ebert, D. D., Koole, S. L., & Andersson, G. (2013). How much psychotherapy is needed to treat depression? A metaregression analysis. *Journal of Affective Disorders, 149*(1-3), 1–13.

Cuijpers, P., Riper, H., & Karyotaki, E. (2018). Internet-based cognitive-behavioral therapy in the treatment of depression. *Focus: Journal of Life Long Learning in Psychiatry, 16*(4), 393.

Cuijpers, P., Van Straten, A., Andersson, G., & Van Oppen, P. (2008). Psychotherapy for depression in adults: A meta-analysis of comparative outcome studies. *Journal of Consulting and Clinical Psychology, 76*(6), 909.

Dear, B., Staples, L., Terides, M., Fogliati, V., Sheehan, J., Johnston, L., et al. (2016). Transdiagnostic versus disorder-specific and clinician-guided versus self-guided internet-delivered treatment for social anxiety disorder and comorbid disorders: A randomized controlled trial. *Journal of Anxiety Disorders, 42*, 30–44.

Dear, B., Staples, L., Terides, M., Karin, E., Zou, J., Johnston, L., et al. (2015). Transdiagnostic versus disorder-specific and clinician-guided versus self-guided internet-delivered treatment for generalized anxiety disorder and comorbid disorders: A randomized controlled trial. *Journal of Anxiety Disorders, 36*, 63–77.

Department of Health Australian Goverment (Producer). (2021). Digital mental health services. Retrieved March 3. https://www.health.gov.au/initiatives-and-programs/digital-mental-health-services.

Domhardt, M., Schröder, A., Geirhos, A., Steubl, L., & Baumeister, H. (2021). Efficacy of digital health interventions in youth with chronic medical conditions: A meta-analysis. *Internet Interventions, 24*, 1–12.

Donker, T., Cornelisz, I., Van Klaveren, C., Van Straten, A., Carlbring, P., Cuijpers, P., et al. (2019). Effectiveness of self-guided app-based virtual reality cognitive behavior therapy for acrophobia: A randomized clinical trial. *JAMA Psychiatry, 76*(7), 682–690.

Eclipse (Producer). (2022). Ehealth programs on eclipse. Retrieved January 11. https://eclipse.org.au/ehealth-programs-on-eclipse.

Erbe, D., Eichert, H.-C., Riper, H., & Ebert, D. D. (2017). Blending face-to-face and internet-based interventions for the treatment of mental disorders in adults: Systematic review. *Journal of Medical Internet Research, 19*(9), e306.

Figueroa, C. A., DeMasi, O., Hernandez-Ramos, R., & Aguilera, A. (2021). Who benefits most from adding technology to depression treatment and how? An analysis of engagement with a texting adjunct for psychotherapy. *Telemedicine and e-Health, 27*(1), 39–46.

Freeman, D., Haselton, P., Freeman, J., Spanlang, B., Kishore, S., Albery, E., et al. (2018). Automated psychological therapy using immersive virtual reality for treatment of fear of heights: A single-blind, parallel-group, randomised controlled trial. *The Lancet Psychiatry, 5*(8), 625–632.

Freeman, D., Reeve, S., Robinson, A., Ehlers, A., Clark, D., Spanlang, B., et al. (2017). Virtual reality in the assessment, understanding, and treatment of mental health disorders. *Psychological Medicine, 47*(14), 2393–2400.

Glasner, S., Kay-Lambkin, F., Budney, A. J., Gitlin, M., Kagan, B., Chokron-Garneau, H., et al. (2018). Preliminary outcomes of a computerized CBT/MET intervention for depressed cannabis users in psychiatry care. *Cannabis, 1*(2), 36–47.

Gustafson, D. H., Landucci, G., McTavish, F., Kornfield, R., Johnson, R. A., Mares, M.-L., et al. (2016). The effect of bundling medication-assisted treatment for opioid addiction with mHealth: Study protocol for a randomized clinical trial. *Trials, 17*(1), 1–12.

Gustafson, D. H., McTavish, F., Chih, M., Atwood, A., Johnson, R., & Boyle, M. (2014). A smartphone application to support recovery from alcoholism: A randomized clinical trial. *JAMA Psychiatry, 71*(5), 566–572. doi:10.1001/jamapsychiatry.2013.4642.

Hale, T. M., & Kvedar, J. C. (2014). Privacy and security concerns in telehealth. *AMA Journal of Ethics, 16*(12), 981–985.

Hall, J. L., & McGraw, D. (2014). For telehealth to succeed, privacy and security risks must be identified and addressed. *Health Affairs, 33*(2), 216–221.

Hayes, S. C. (2004). Acceptance and Commitment Therapy and the New Behavior Therapies: Mindfulness, Acceptance, and Relationship. In S. C. Hayes, V. M. Follette, & M. M. Linehan (Eds.), Mindfulness and acceptance: Expanding the cognitive-behavioral tradition (pp. 1–29). Guilford Press, New York.

Head to Health (Producer). (2022). Find digital mental health resources from trusted service providers. Retrieved January 11. https://www.headtohealth.gov.au/.

HelloBetter Research (Producer). (2022). Spitzenforschung von weltweit führenden Fachexperten seit 2008. Retrieved January 11. https://hellobetter.de/forschung/.

HelloBetter Acute Depression (Producer). (2022). Acute depression. Retrieved January 11. https://hellobetter.de/en/online-courses/acute-depression/.

HelloBetter Depression Prevention (Producer). (2022). Depression prevention. Retrieved January 11. https://hellobetter.de/en/online-courses/depression-prevention/.

HelloBetter Online Classes (Producer). (2022). Psychologische Online-Kurse – wirksam von HelloBetter. Retrieved January 11. https://hellobetter.de/online-kurse/.

Hofmann, S. G., Asnaani, A., Vonk, I. J., Sawyer, A. T., & Fang, A. (2012). *The efficacy of cognitive behavioral therapy: A review of meta-analyses Cognitive Therapy and Research, 36*, 427–440.

iesohealth (Producer). (2022). Beat anxiety and depression with online CBT on the NHS. Retrieved January 11. https://www.iesohealth.com/en-gb.

Jacobson, N. C., & Chung, Y. J. (2020). Passive sensing of prediction of moment-to-moment depressed mood among undergraduates with clinical levels of depression sample using smartphones. *Sensors, 20*(12), 3572.

Jacobson, N. C., Summers, D., & Wilhelm, S. (2020). Digital biomarkers of social anxiety severity: Digital phenotyping using passive smartphone sensors. *Journal of Medical Internet Research, 22*(5), e16875.

Johnson, K., Richards, S., Chih, M.-Y., Moon, T. J., Curtis, H., & Gustafson, D. H. (2016). A pilot test of a mobile app for drug court participants. *Substance Abuse: Research and Treatment, 10*, 1–7.

Johnson, L., Titov, N., Andews, G., Spence, J., & Dear, B. F. (2011). A RCT of a transdiagnostic internet-delivered treatment for three anxiety disorders: Examination of support roles and disorder-specific outcomes. *Plos One, 6*(11), e28079.

Johnston, D. C., Mathews, W. D., Maus, A., & Gustafson, D. H. (2019). Using smartphones to improve treatment retention among impoverished substance-using Appalachian women: A naturalistic study. *Substance Abuse: Research and Treatment, 13*, 1178221819861377.

Kampmann, I. L., Emmelkamp, P. M., & Morina, N. (2016). Meta-analysis of technology-assisted interventions for social anxiety disorder. *Journal of Anxiety Disorders, 42*, 71–84.

Kay-Lambkin, F. J., Baker, A. L., Kelly, B., & Lewin, T. J. (2011). Clinician-assisted computerised versus therapist-delivered treatment for depressive and addictive disorders: A randomised controlled trial. *Medical Journal of Australia, 195*, S44–S50.

Kay-Lambkin, F. J., Baker, A. L., Lewin, T. J., & Carr, V. J. (2009). Computer-based psychological treatment for comorbid depression and problematic alcohol and/or cannabis use: A randomized controlled trial of clinical efficacy. *Addiction, 104*(3), 378–388.

Kay-Lambkin, F. J., Simpson, A. L., Bowman, J., & Childs, S. (2014). Dissemination of a computer-based psychological treatment in a drug and alcohol clinical service: An observational study. *Addiction Science & Clinical Practice, 9*(1), 1–9.

Kendrick, T., & Pilling, S. (2012). Common mental health disorders—Identification and pathways to care: NICE clinical guideline. *British Journal of General Practice, 62*(594), 47–49.

Kenter, R. M., van de Ven, P. M., Cuijpers, P., Koole, G., Niamat, S., Gerrits, R. S., et al. (2015). Costs and effects of internet cognitive behavioral treatment blended with face-to-face treatment: Results from a naturalistic study. *Internet Interventions, 2*(1), 77–83.

Kessler, D., Lewis, G., Kaur, S., Wiles, N., King, M., Weich, S., et al. (2009). Therapist-delivered internet psychotherapy for depression in primary care: A randomised controlled trial. *The Lancet, 374*(9690), 628–634.

Kiluk, B. D., Devore, K. A., Buck, M. B., Nich, C., Frankforter, T. L., LaPaglia, D. M., et al. (2016). Randomized trial of computerized cognitive behavioral therapy for alcohol use disorders: Efficacy as a virtual stand-alone and treatment add-on compared with standard outpatient treatment. *Alcoholism: Clinical and Experimental Research, 40*(9), 1991–2000.

Koenigbauer, J., Letsch, J., Doebler, P., Ebert, D. D., & Baumeister, H. (2017). Internet-and mobile-based depression interventions for people with diagnosed depression: A systematic review and meta-analysis. *Journal of Affective Disorders, 223*, 28–40.

Kooistra, L. C., Wiersma, J. E., Ruwaard, J., van Oppen, P., Smit, F., Lokkerbol, J., et al. (2014). Blended vs. face-to-face cognitive behavioural treatment for major depression in specialized mental health care: Study protocol of a randomized controlled cost-effectiveness trial. *BMC Psychiatry [Electronic Resource], 14*(1), 1–11.

Lilienfeld, S. O., Ritschel, L. A., Lynn, S. J., Cautin, R. L., & Latzman, R. D. (2013). Why many clinical psychologists are resistant to evidence-based practice: Root causes and constructive remedies. *Clinical Psychology Review, 33*(7), 883–900.

Loerinc, A. G., Meuret, A. E., Twohig, M. P., Rosenfield, D., Bluett, E. J., & Craske, M. G. (2015). Response rates for CBT for anxiety disorders: Need for standardized criteria. *Clinical Psychology Review, 42*, 72–82.

Lord, S., Moore, S. K., Ramsey, A., Dinauer, S., & Johnson, K. (2016). Implementation of a substance use recovery support mobile phone app in community settings: Qualitative study of clinician and staff perspectives of facilitators and barriers. *JMIR Mental Health, 3*(2), e4927.

Marks, I. M., Cavanagh, K.,& Gega, L. (2007). Hands-on Help: Computer-aided Psychotherapy. Maudsley Monographs No. 49. Hove, United Kingdom: Psychology Press.

Marsch, L. A., Guarino, H., Grabinski, M. J., Syckes, C., Dillingham, E. T., Xie, H., et al. (2015). Comparative effectiveness of web-based vs. educator-delivered HIV prevention for adolescent substance users: A randomized, controlled trial. *Journal of Substance Abuse Treatment, 59*, 30–37.

Materia, F. T., & Smyth, J. M. (2021). Acceptability of intervention design factors in mHealth intervention research: experimental factorial study. *JMIR mHealth and uHealth, 9*(7), e23303.

McGovern, M. P., Fox, T. S., Xie, H., & Drake, R. E. (2004). A survey of clinical practices and readiness to adopt evidence-based practices: Dissemination research in an addiction treatment system. *Journal of Substance Abuse Treatment, 26*(4), 305–312.

McTavish, F. M., Chih, M.-Y., Shah, D., & Gustafson, D. H. (2012). How patients recovering from alcoholism use a smartphone intervention. *Journal of Dual Diagnosis, 8*(4), 294–304.

Milward, J., Drummond, C., Fincham-Campbell, S., & Deluca, P. (2018). What makes online substance-use interventions engaging? A systematic review and narrative synthesis. *Digital Health, 4*, 2055207617743354.

MindSpot (Producer). (2022). A digital mental health clinic for all Australians. Retrieved January 11. https://www.mindspot.org.au/.

Moller, A. C., Merchant, G., Conroy, D. E., West, R., Hekler, E., Kugler, K. C., et al. (2017). Applying and advancing behavior change theories and techniques in the context of a digital health revolution: Proposals for more effectively realizing untapped potential. *Journal of Behavioral Medicine, 40*(1), 85–98.

Mor, S., Grimaldos, J., Tur, C., Miguel, C., Cuijpers, P., Botella, C., et al. (2021). Internet-and mobile-based interventions for the treatment of specific phobia: A systematic review and preliminary meta-analysis. *Internet Interventions, 26*, 100462.

Morris, Z. S., Wooding, S., & Grant, J. (2011). The answer is 17 years, what is the question: Understanding time lags in translational research. *Journal of the Royal Society of Medicine, 104*(12), 510–520.

Nahum-Shani, I., Smith, S. N., Spring, B. J., Collins, L. M., Witkiewitz, K., Tewari, A., et al. (2018). Just-in-time adaptive interventions (JITAIs) in mobile health: Key components and design principles for ongoing health behavior support. *Annals of Behavioral Medicine, 52*(6), 446–462.

National Collaborating Centre for Mental Health. (2021). The improving access to psychological therapies manual. Retrieved January 11. https://www.england.nhs.uk/wp-content/uploads/2018/06/the-iapt-manual-v5.pdf.

National Health Service [NHS]. (2021). Adult improving access to psychological therapies programme. Retrieved December 16. https://www.england.nhs.uk/mental-health/adults/iapt/.

NHS Oxford Health Fudation Trust (Producer). (2021). OxfordVR Apps for health. Retrieved December 16. https://www.oxfordhealth.nhs.uk/apps/oxfordvr/.

Nobis, S., Lehr, D., Ebert, D. D., Baumeister, H., Snoek, F., Riper, H., et al. (2015). Efficacy of a web-based intervention with mobile phone support in treating depressive symptoms in adults with type 1 and type 2 diabetes: A randomized controlled trial. *Diabetes Care, 38*(5), 776–783.

OVR (Producer). (2022). Transforming mental health for millions with VR therapy. Retrieved January 11. https://ovrhealth.com/.

Ozair, F. F., Jamshed, N., Sharma, A., & Aggarwal, P. (2015). Ethical issues in electronic health records: A general overview. *Perspectives in Clinical Research, 6*(2), 73–76.

Păsărelu, C. R., Andersson, G., Bergman Nordgren, L., & Dobrean, A. (2017). Internet-delivered transdiagnostic and tailored cognitive behavioral therapy for anxiety and depression: A systematic review and meta-analysis of randomized controlled trials. *Cognitive Behaviour Therapy, 46*(1), 1–28.

Parsons, T. D., Bowerly, T., Buckwalter, J. G., & Rizzo, A. A. (2007). A controlled clinical comparison of attention performance in children with ADHD in a virtual reality classroom compared to standard neuropsychological methods. *Child Neuropsychology, 13*(4), 363–381.

Parsons, T. D., & Carlew, A. R. (2016). Bimodal virtual reality stroop for assessing distractor inhibition in autism spectrum disorders. *Journal of Autism and Developmental Disorders, 46*(4), 1255–1267.

Pear Therapeutics (Producer). (2018). Sandoz Inc. and Pear Therapeutics obtain FDA clearance for reSET-O™ to treat opioid use disorder. Retrieved 14 January, 2022. https://peartherapeutics.com/sandoz-inc-and-pear-therapeutics-obtain-fda-clearance-for-reset-o-to-treat-opioid-use-disorder/.

Proudfoot, J., Swain, S., Widmer, S., Watkins, E., Goldberg, D., Marks, I., et al. (2003). The development and beta-test of a computer-therapy program for anxiety and depression: Hurdles and lessons. *Computers in Human Behavior, 19*(3), 277–289.

Ramos, L. A., Blankers, M., van Wingen, G., de Bruijn, T., Pauws, S. C., & Goudriaan, A. E. (2021). Predicting success of a digital self-help intervention for alcohol and substance use with machine learning. *Frontiers in Psychology, 12*, 3931.

Ramsey, A., Lord, S., Torrey, J., Marsch, L., & Lardiere, M. (2016). Paving the way to successful implementation: Identifying key barriers to use of technology-based therapeutic tools for behavioral health care. *The Journal of Behavioral Health Services & Research, 43*(1), 54–70.

Reins, J. A., Boß, L., Lehr, D., Berking, M., & Ebert, D. D. (2019). The more I got, the less I need? Efficacy of internet-based guided self-help compared to online psychoeducation for major depressive disorder. *Journal of Affective Disorders, 246*, 695–705.

Richards, D., & Richardson, T. (2012). Computer-based psychological treatments for depression: A systematic review and meta-analysis. *Clinical Psychology Review, 32*(4), 329–342.

Richards, D., Richardson, T., Timulak, L., & McElvaney, J. (2015). The efficacy of internet-delivered treatment for generalized anxiety disorder: A systematic review and meta-analysis. *Internet Interventions, 2*(3), 272–282.

Roca, M., Gili, M., Garcia-Garcia, M., Salva, J., Vives, M., Campayo, J. G., et al. (2009). Prevalence and comorbidity of common mental disorders in primary care. *Journal of Affective Disorders, 119*(1-3), 52–58.

Rogers, E. M. (2010). *Diffusion of Innovations* (4th Ed.). Simon and Schuster, New York.

Rush, A. J., Trivedi, M. H., Wisniewski, S. R., Nierenberg, A. A., Stewart, J. W., Warden, D., et al. (2006). Acute and longer-term outcomes in depressed outpatients requiring one or several treatment steps: A STAR* D report. *American Journal of Psychiatry, 163*(11), 1905–1917.

Schmidt, N. B., Richey, J. A., Buckner, J. D., & Timpano, K. R. (2009). Attention training for generalized social anxiety disorder. *Journal of Abnormal Psychology, 118*(1), 5.

Shi, J. M., Henry, S. P., Dwy, S. L., Orazietti, S. A., & Carroll, K. M. (2019). Randomized pilot trial of web-based cognitive-behavioral therapy adapted for use in office-based buprenorphine maintenance. *Substance Abuse, 40*(2), 132–135.

Spek, V., Cuijpers, P., Nyklíček, I., Riper, H., Keyzer, J., & Pop, V. (2007). Internet-based cognitive behaviour therapy for symptoms of depression and anxiety: A meta-analysis. *Psychological Medicine, 37*(3), 319–328.

Sue, D. W., Capodilupo, C. M., Torino, G. C., Bucceri, J. M., Holder, A., Nadal, K. L., & Esquilin, M. (2007). Racial microaggressions in everyday life: implications for clinical practice. *American psychologist, 62*(4), 271.

Tait, R. J., McKetin, R., Kay-Lambkin, F., Carron-Arthur, B., Bennett, A., Bennett, K., et al. (2014). A web-based intervention for users of amphetamine-type stimulants: 3-month outcomes of a randomized controlled trial. *JMIR Mental Health, 1*(1), e3278.

U.S. Food and Drug Administration [FDA] (Producer). (2017). FDA permits marketing of mobile medical application for substance use disorder. Retrieved February 14. https://www.fda.gov/news-events/press-announcements/fda-permits-marketing-mobile-medical-application-substance-use-disorder.

van der Vaart, R., Witting, M., Riper, H., Kooistra, L., Bohlmeijer, E. T., & van Gemert-Pijnen, L. J. (2014). Blending online therapy into regular face-to-face therapy for depression: Content, ratio and preconditions according to patients and therapists using a Delphi study. *BMC Psychiatry [Electronic Resource], 14*(1), 355. doi:10.1186/s12888-014-0355-z.

Walters, S. T., Vader, A. M., & Harris, T. R. (2007). A controlled trial of web-based feedback for heavy drinking college students. *Prevention Science, 8*(1), 83–88.

Wampold, B. E., & Imel, Z. E. (2015). The great psychotherapy debate: The evidence for what makes psychotherapy work (2nd Ed.). Routledge, New York.

Warnock-Parkes, E., Wild, J., Thew, G. R., Kerr, A., Grey, N., Stott, R., et al. (2020). Treating social anxiety disorder remotely with cognitive therapy. *The Cognitive Behaviour Therapist, 13*, 1–20.

Weisel, K. K., Fuhrmann, L. M., Berking, M., Baumeister, H., Cuijpers, P., & Ebert, D. D. (2019). Standalone smartphone apps for mental health-a systematic review and meta-analysis. *NPJ Digit Med, 2*(1), 118. doi:10.1038/s41746-019-0188-8.

Wentzel, J., van der Vaart, R., Bohlmeijer, E. T., & van Gemert-Pijnen, J. E. (2016). Mixing online and face-to-face therapy: How to benefit from blended care in mental health care. *JMIR Mental Health, 3*(1), e9.

Williams, A. D., O'Moore, K., Mason, E., & Andrews, G. (2014). The effectiveness of internet cognitive behaviour therapy (iCBT) for social anxiety disorder across two routine practice pathways. *Internet Interventions, 1*(4), 225–229.

Witlox, M., Garnefski, N., Kraaij, V., de Waal, M. W., Smit, F., Bohlmeijer, E., et al. (2021). Blended acceptance and commitment therapy versus face-to-face cognitive behavioral therapy for older adults with anxiety symptoms in primary care: Pragmatic single-blind cluster randomized trial. *Journal of Medical Internet Research, 23*(3), e24366.

World Health Organization. (2013). Mental health action plan 2013–2020 [Internet]. Retrieved 2021, July 22, 1–50. https://www.who.int/publications/i/item/9789241506021.

Wright, J. H., Owen, J. J., Richards, D., Eells, T. D., Richardson, T., Brown, G. K., et al. (2019). Computer-assisted cognitive-behavior therapy for depression: A systematic review and meta-analysis. *The Journal of Clinical Psychiatry, 80*(2). doi:10.4088/JCP.18r12188.

Wright, J. H., & Wright, A. S. (1997). Computer-assisted psychotherapy. *The Journal of Psychotherapy Practice and Research, 6*(4), 315–329.

Chapter 6

Receptivity to mobile health interventions

Roman Keller [a,b], Florian v. Wangenheim [a,c], Jacqueline Mair [a,b] and Tobias Kowatsch [a,c,d,e]

[a] Future Health Technologies Programme, Campus for Research Excellence and Technological Enterprise, Singapore-ETH Centre, Singapore, Singapore, [b] Saw Swee Hock School of Public Health, National University of Singapore, Singapore, Singapore, [c] Centre for Digital Health Interventions, Department of Management, Technology, and Economics, ETH Zurich, Zurich, Switzerland, [d] Institute for Implementation Science in Health Care, University of Zurich, Zurich, Switzerland, [e] School of Medicine, University of St Gallen, St Gallen, Switzerland

6.1 Introduction

Digital therapeutics have the potential to address the increasing health and economic burden of noncommunicable diseases including mental illness and substance use disorders (Ferrari et al., 2022; L Murray et al., 2020; Murphy et al., 2020; Vos et al., 2020). Mobile health interventions (MHI) are digital therapeutics that use mobile information and communication technology such as smartphones, wearables, and other mobile devices for the prevention and management of disease (Birkmeyer, Wirtz, & Langer, 2021; Kowatsch & Fleisch, 2021; Noorbergen, Adam, Roxburgh, & Teubner, 2021). MHIs can either complement health services offered by human experts (Kowatsch et al., 2021b; Stasinaki et al., 2021) (see also Chapter 5) or used standalone (see Chapters 3 and 4), for example, in the form of scalable health services for healthy individuals and the general public (Kramer et al., 2020; Mönninghoff et al., 2021) or in the form of a prescription for individuals diagnosed with a particular disease (BfArM, 2020; Stern, Matthies, Hagen, Brönneke, & Debatin, 2020).

The huge potential of MHIs for the prevention and management of mental health and addiction is their capability to reach out into everyday lives and support individuals in need at most opportune moments. In an ideal world following the concept of just-in-time adaptive interventions (JITAIs) (Nahum-Shani, Hekler, & Spruijt-Metz, 2015, 2018), MHIs are able to anticipate opportune moments by predicting whether an individual is vulnerable and able to receive, process, and use support. MHIs that follow this ideal-world logic are still sparse or, at least to our knowledge, have yet to be developed and offered in the digital health space and adopted by various stakeholders, such as designers of MHIs, patients, public health bodies, payers, and health care providers (Kowatsch & Fleisch, 2021). In the scientific community, however, such MHIs have recently gained increasing attention (Choi et al., 2019; Guarino, Acosta, Marsch, Xie, & Aponte-Melendez, 2016; Haug et al., 2020; Hébert et al., 2018; Keller et al., 2022; Klasnja et al., 2019; Kramer et al., 2020; Mishra et al., 2021; Schembre et al., 2018; Mair et al., 2022; Teepe et al., 2021) (see also Chapter 7 and Chapter 11).

One key challenge in the design of JITAIs is the detection and, if technically feasible, the prediction of a receptive state, defined as a condition "in which the person can receive, process, and use the support provided" (Nahum-Shani et al., 2015). MHIs may leverage recent advancements in sensor technology for the detection and prediction of receptive states (Künzler et al., 2019; Mishra et al., 2021). Such predictions have the overall goal to increase the likelihood of reaching vulnerable individuals at opportune moments as unobtrusively as possible, in contrast to a self-report based "detection" that poses an additional burden on individuals. For example, if someone is likely to experience a relapse to excessive alcohol or tobacco consumption or is prone to binge-eating because of their depression, delivering support when the target person is receptive (e.g., not driving, or working) will likely result in a more effective MHI.

By contrast, what happens if we know that a person will be soon in a vulnerable state but the "perfect" evidence-based, personalized intervention is delivered by a smartphone or voice assistant simply at the wrong time? The target person will then likely not receive the MHI and, in some cases, this could decrease the potency of the intervention, which could inadvertently lead to severe, life-threatening events. Additionally, sending out interventions at the "wrong" time may even introduce unintended negative side effects, such as increased stress and/or decreased productivity and well-being (Califf, Sarker, & Sarker, 2020; Jenkins et al., 2016; Throuvala et al., 2021). It is, therefore, crucial to understand the conditions under which individuals are able to receive, process, and use support.

To this end, receptivity to MHIs is an important aspect to consider that bridges the detection of vulnerable states on the one side (Bent et al., 2021b; Bent, Lu, Kim, & Dunn, 2021a; Coravos, Khozin, & Mandl, 2019; Lekkas, Price, & Jacobson, 2022; Rassouli et al., 2020; Sahandi Far, Stolz, Fischer, Eickhoff, & Dukart, 2021; Sieberts et al., 2021; Teepe et al., 2021), and precision support on the other side (Haug et al., 2020; Hekler, Tiro, Hunter, & Nebeker, 2020; Kramer et al., 2020; Mishra et al., 2021; Schembre et al., 2018).

Moreover, detecting and predicting states of receptivity is important because it can close the gap between intended use and actual use of MHIs (Jakob et al., 2022; Sieverink, Kelders, & Gemert-Pijnen, 2017). By making support relevant at opportune moments, vulnerable individuals will likely build positive attitudes toward MHIs. These attitudes may positively impact perceptions of the usefulness, enjoyment, or ease of use of MHIs, that is, well-known predictors of technology adoption and use (Akter, D'Ambra, & Ray, 2013; Birkmeyer et al., 2021; Davis, 1989; Turner, Kitchenham, Brereton, Charters, & Budgen, 2010; Venkatesh, Thong, & Xu, 2012).

The remainder of this chapter is structured as follows. Next, we present the anatomy of an "ideal" MHI and discuss its building blocks. We then discuss three key processes that should be considered when designing receptivity-capable MHIs, that is, receiving, processing, and using support. Thereafter, we provide an overview of research on receptivity and summarize relevant factors that may inform the detection and prediction of opportune intervention moments. We conclude this chapter with challenges related to the implementation of receptivity-capable MHIs and offer suggestions for future work.

6.2 The anatomy of an "ideal" mobile health intervention

MHIs that follow the principles of JITAIs are designed to improve well-defined distal outcomes (e.g., long-term quality of life) with the help of rather smaller achievable outcomes, that is, by so-called proximal outcomes (e.g., a relapse-free day, reaching a weekly exercise goal, or getting enough sleep during a week). Given the causal structure of proximal and distal outcomes, which is usually derived from meta-analyses, process research, optimization trials, or randomized controlled trials (Collins, 2018; Collins et al., 2011), an "ideal" MHI uses information about a target person (e.g., age, gender, location, genome, epigenome, microbiome, blood count results, personality trait, physiological data, specific adverse health behavior, or condition) and that target person's context (e.g., time of the day, weather condition, family or friends close-by) in the most unobtrusive way to predict vulnerable and receptive states. Unobtrusive prediction of these states often requires the use of—in a best-case scenario even noninvasive and contactless—sensors and other technologies to minimize any burden related to MHI use (Jakob et al., 2022; Keller et al., 2022; Teepe et al., 2021).

Consistent with prior work on JITAIs, we define a vulnerable state as a "person's transient tendency to experience adverse health outcomes or to engage in maladaptive behaviors" (Nahum-Shani et al., 2015). For example, it has been investigated whether workplace stress can be detected via computer mouse movements (Banholzer, Feuerriegel, Fleisch, Bauer, & Kowatsch, 2021) or whether breathing patterns (Shih et al., 2019), emotions (Boateng & Kowatsch, 2020), personality states (Rüegger et al., 2020), or nocturnal cough (Barata et al., 2020) can be detected with a smartphone. These first research efforts among many more may potentially lead to validated digital biomarkers ready to be used in MHIs (Bent et al., 2021b; Coravos et al., 2019).

Similarly, research on receptivity is still in its infancy and there are only a few empirical findings that clearly point toward factors that are relevant in the prediction of receptive states (Chan, Sapkota, Mathews, Zhang, & Nanayakkara, 2020; Choi et al., 2019; Künzler et al., 2017, 2019; Mishra et al., 2021). An overview of related work on receptivity is provided later in this chapter.

Given vulnerable and receptive states, the logic of MHIs determines then the optimal dose and delivers the best evidence-based intervention option available that targets the proximal outcome. Similar to research on vulnerability and receptivity in MHIs, empirical findings on JITAIs are still limited. Nevertheless, the reader is encouraged to assess the growing body of literature in this regard (Haug et al., 2020; Keller et al., 2022; Klasnja et al., 2019; Kramer et al., 2020; Schembre et al., 2018; Teepe et al., 2021), especially in the context of optimization trials, such as microrandomized trials (Qian et al., 2022) or fractional factorial trials (Collins, 2018).

In a final step, the MHI determines whether the intervention delivered had a positive impact on the vulnerable state. Depending on this impact the MHI can plan the next iteration within this closed-loop system with the overall goal to minimize vulnerability in the target person. With each of these iterations, the MHI can learn more about which intervention option works best depending on the specific state of the target person and his or her context (Hekler et al., 2018).

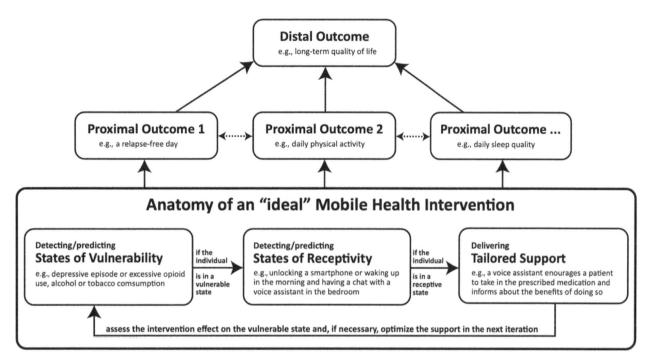

FIGURE 6.1 The anatomy of an "ideal" mobile health intervention and how it targets a distal outcome via proximal outcomes, own illustration. Note: Dashed arrows indicate that proximal outcomes may also impact each other.

An overview of this anatomy of an "ideal" MHI and how it targets a distal outcome with the help of three proximal outcomes is depicted in Fig. 6.1. Against this background, we will now discuss in more detail what intervention authors and engineers should consider when designing receptivity-capable MHIs.

6.3 Key processes of receptivity: receiving, processing, and using support

Determining whether an individual is either receptive or not depends on various factors. Consistent with the definition of receptivity, the following three guiding questions can help us to understand which factors should be considered for the detection and prediction of receptive states:

(1) Is the target person *able to receive* the foreseen support at the moment of delivery?
(2) Does the target person have *enough cognitive capacity to process* the information about the foreseen support after the moment of delivery?
(3) Is the target person *able to take action and implement* the foreseen support after processing the information about that support?

First, being able to detect or predict whether a target person can receive support requires knowledge about whether or not that person can be interrupted at a given moment. For example, sleeping, being in a meeting, having a conversation with others, having turned on the focus mode of a desktop computer, smartphone, or smartwatch that turns off notifications, the battery status of a smartphone (e.g., "charging"), driving a bike or a car, or simply being a person with high/low conscientiousness, all potentially influence whether an individual is able to receive support. Developers of MHIs should also think about the most appropriate communication channel available to initiate the delivery of support. For example, while brushing teeth in the morning or driving a bike or car, a message from a voice assistant might be more appropriate than a text-based notification on a smartphone or smartwatch. If, for example, an in-vehicle voice assistant is connected to a car's navigation system with a traffic monitoring feature and able to detect whether the driver is not having a conversation with others, support can be initiated in a "low-traffic" and "no-conversation" moment. Also, MHIs, which combine various delivery modes and technologies, for example, a conversational health care agent that delivers support via a smartphone (e.g., health literacy content via spoken animation) and augmented reality glasses (e.g., a physical exercise with real-time feedback) (Kowatsch et al., 2021a, 2021b), may use one delivery channel for getting the attention of an individual (e.g., the smartphone) to then deliver the support via another channel (e.g., augmented reality headset).

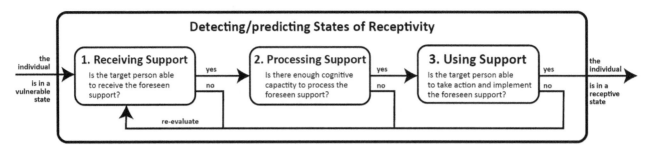

FIGURE 6.2   Key processes involved in the detection and prediction of receptivity to mobile health interventions. Note: Step 3 (using support) might not be relevant in cases support does not require a follow-up action (e.g., letting individuals know that they reached a daily step goal).

Regarding the second guiding question, developers of MHIs have to consider not only when an individual can be interrupted but also whether there is enough cognitive capacity available to process the information about the foreseen support. This processing may include an assessment of the sender of the support ("Who is trying to get my attention? Do I know and trust that person, do I trust the MHI or the organization that offers it?"), the content of the support (e.g., "What is the information all about? Is it relevant to me, e.g., may it prevent a life-threatening event which is about to happen?"), and the implications for specific behavioral actions as a result of the support ("What do I need to do and can I implement it in due time?"). Depending on this assessment, individuals will decide on the relevance of the support and thus, whether or not they will engage in the foreseen behavioral actions.

To detect the cognitive capacity of an individual, digital therapeutics may use unobtrusive psychophysiological measures, for example, via wearables that measure blood pressure, cerebral blood flow via optical imaging, or skin conductance (Lohani, Payne, & Strayer, 2019). Research on real-time prediction of future cognitive capacity is still early-stage (Boehm et al., 2021) but is worthwhile to consider with respect to effective state-of-receptivity predictions. It is also known that working alliance, an important relationship quality between health care provider and patients (Horvath & Greenberg, 1989), is robustly linked to treatment success (Del Re, Flückiger, Horvath, & Wampold, 2021; Flückiger, Del Re, Wampold, Symonds, & Horvath, 2012, 2018); and it has been shown that individuals are able to build up a working alliance even with computers, such as scalable conversational health care agents (Bickmore & Picard, 2005; Kowatsch et al., 2021b). It is therefore recommended to develop MHIs that are able not only to anticipate and predict the cognitive capacity of target persons but also create a working alliance to render the information about the foreseen support more trustworthy and relevant.

Regarding the third guiding question, intervention authors and engineers of MHIs should also take into account whether a target person is able to implement the foreseen support in due time. That is, depending on the purpose of the supporting message, time and effort to take action and implement the suggested behavior may play a significant role in the detection and prediction of receptive states. For example, letting someone know that the daily physical activity goal has been reached does not require any follow-up action, but a message that suggests taking action and walking for another 20 minutes does. Thus, a MHI that considers states of receptivity should leverage all information about the availability of a target person (e.g., from an electronic calendar) or dynamically through learned behavioral patterns (Dua, Singh, & Semwal, 2021) that offer opportunities to implement the behavioral actions. Also, information about earlier commitments to specific behavioral actions, for example, from an implementation intention session hours or days ago (Hagerman, Hoffman, Vaylay, & Dodge, 2021; Warner et al., 2022), may positively influence the uptake of support.

Finally, there are additional aspects a receptivity-capable MHI needs to take into account for the detection or prediction of receptive states. Some of these aspects are specific to the channel through which support is provided. For example, the massive amount of information individuals cope with in today's attention economy leads to a situation where MHIs compete with various other web-based or mobile applications (Bhargava & Velasquez, 2021; Throuvala et al., 2021). It is therefore of utmost importance to consider very carefully the need to deliver support and to reduce an individual's burden by, for example, minimizing any "false positive" alarms or simply by the total amount of notifications in general. If too many notifications are triggered, well-intended support becomes more harmful than good and can make the target person even more vulnerable (Jenkins et al., 2016). By contrast, if there is a severe or even life-threatening state of vulnerability predicted, then some, if not all, factors discussed above must be overruled to deliver the support in due time, that is, also in a "low-cognitive-capability" situation. This could also mean sending out notifications repeatedly to the target person or to significant others, such as family members, romantic partners, or even health care professionals.

An overview of the three processes involved in receptivity to MHIs is shown in Fig. 6.2. This process perspective also highlights that receptivity-capable MHIs may almost continuously re-evaluate receptive states before the foreseen support can be delivered. This quasi-real-time monitoring of receptive states, in turn, can lead to technical challenges and an additional burden on target persons. These and other challenges will be discussed in the last section of this chapter and after we have reviewed the empirical evidence on receptive states in the next section.

6.4   What do we already know about states of receptivity?

In this section, we review related research with the goal to identify factors associated with receptive states. While the definition of receptivity involves multiple aspects, specifically the ability of a person to receive, process, and use the support provided, most of the previous studies have limited their focus solely on receiving support. Moreover, before states of receptivity became an important construct in the field of health intervention research motivated by the increasing interest in JITAIs, the topic was already highly relevant and investigated in the domain of ubiquitous computing under the umbrella of interruptibility research (Ho & Intille, 2005). This stream of research investigates the use of push notifications to attract the attention of individuals while reducing user burden and disruption through notifications. Much of the previous receptivity research comes from this field and it is therefore important to better understand receptivity to MHIs. We, thus, start with a review of early work about interruptibility followed by more recent research about receptivity to MHIs. A summary of factors that are potentially relevant for the detection and prediction of receptive states concludes this section.

Much of the early research on interruptibility focused on the timing, content, and context of notifications and how interruption burden could be reduced, or indeed, receptivity increased. Research has shown that messages delivered during activity transitions (e.g., a transition from standing to sitting) (Ho & Intille, 2005) or during naturally occurring breakpoints, for example, when finishing an episode of mobile interaction (Fischer, Greenhalgh, & Benford, 2011), can decrease perceived burden and improve user receptivity. Specific times of day, namely morning and evening hours, are also linked to higher receptivity since people tend to be at home during these times (Mashhadi, Mathur, & Kawsar, 2014). Alongside timing, the content of the notification and the context during which it is received are also important factors to consider. For example, the quality of the notification content in terms of entertainment value, relevance, user interest, and actionability has been linked to receptivity (Fischer et al., 2010). Evidence also suggests that people are more responsive to certain categories of notifications, such as family chat and work email notifications over those relating to system, tools, and media and music (Mehrotra, Musolesi, Hendley, & Pejovic, 2015), and particular senders, for example partners and immediate family members compared to extended family members or service providers (Mehrotra, Pejovic, Vermeulen, Hendley, & Musolesi, 2016). Certain activities and tasks, such as when idle (Choi et al., 2019; Mehrotra, Pejovic, Vermeulen, Hendley, & Musolesi, 2016) or traveling in a vehicle (Mehrotra, Musolesi, Hendley, & Pejovic, 2015), are linked to greater receptivity while others, such as communicating (Mehrotra, Pejovic, Vermeulen, Hendley, & Musolesi, 2016), studying, or exercising (Yuan, Gao, & Lindqvist, 2017) are linked to lesser receptivity. Research has also shown a negative association between perceived task complexity (complex vs. less complex) and receptivity (Mehrotra, Pejovic, Vermeulen, Hendley, & Musolesi, 2016). Finally, the impact of location appears to have an impact. People are more responsive in outdoor locations (Sarker et al., 2014) but are less responsive in shopping malls, in the workplace (Yuan et al., 2017), and in social settings (Choi et al., 2019).

Okoshi et al. (2015a, 2015b) proposed a system to defer delivery of notifications until a user's natural breakpoint to minimize disruption and cognitive overload. The system delivered notifications using device interaction events (e.g., screen viewing, window transitions, change in notification state). In a controlled study with 37 participants and a 16-days "in-the-wild" study with 27 participants, the authors reduced the user's perceived cognitive load by 46% and 33%, respectively, using the breakpoint system compared to randomly timed notifications (Okoshi et al., 2015a). They also reported a reduced response time to notifications using the breakpoint system compared to random timing, indicating that the system is able to identify opportune moments to trigger push notifications. In subsequent work, a new version of the breakpoint detection system which included physical activity recognition was deployed and the effectiveness was validated in a 1-month study with 41 participants (Okoshi et al., 2015b). The new system resulted in a ca. 72% greater reduction of users' perceived workload compared to the previous system. Finally, the authors tested a similar system in another study with 680,000 users for 3 weeks (Okoshi, Tsubouchi, Taji, Ichikawa, & Tokuda, 2017). The results of that large-scale "in-the-wild" study showed that delaying the delivery of a notification to a more interruptable moment significantly reduces user response time by ca. 50% compared to delivering notifications immediately.

Individual factors are also important influencers of receptivity. Research has shown that people react differently depending on their personality type, for example, extroverts are more inclined to react to notifications (Mehrotra, Pejovic, Vermeulen, Hendley, & Musolesi, 2016; Yuan et al., 2017). Individuals are also more interruptable when in a pleasant or happier mood state than when stressed (Sarker et al., 2014; Yuan et al., 2017) and are more receptive when bored (Pielot, Dingler, San Pedro, & Oliver, 2015).

While the majority of these earlier works in the field of interruptibility involved rather short duration studies with small sample sizes, some of the more recent studies on receptivity have included more substantial study populations. Künzler et al. (2019) investigated associations between receptivity and intrinsic factors (i.e., device type, age, gender, personality) and contextual factors (i.e., time of delivery, battery level, device interaction, physical activity, location), in a 6-week mHealth study with 189 participants. The authors found higher response rates to be associated with older age, neuroticism, time of day (between 10 am and 6 pm), location (home or workplace), higher phone battery level, active device interaction, and physical activity type (walking). Based on the data, the authors built machine learning models, incorporating both intrinsic and contextual factors, to detect receptivity and found significant improvements in detecting and estimating receptivity metrics compared to a baseline model.

As mentioned earlier, most previous research focusses on the aspect of receiving just-in-time support, which is concerned with whether users can successfully perceive a notification. However, little research exists on whether users subsequently engage with the intervention content, thus whether they are able to process the foreseen support. In one such study by Pielot et al. (2017), eight different types of content (games, news items, trending videos, cultural articles, popular GIFs, funny facts, and psychometric tests) were delivered to 337 participants, and effects on participants' receptivity were observed. During 4 weeks, participants were tasked to report their mood via a short, notification-administered questionnaire. However, the study's actual goal was to analyze the voluntary engagement with the diverse contents that were presented at the end of the questionnaire. Similar to Künzler et al. (2019), the authors found that higher phone battery level, phone interaction (more time since last phone call, less time since last SMS, and lower number of notifications already received), higher ambient noise levels, less variance in ambient light, older age, higher variance in physical activity during the previous 60 minutes, and later time of the day were significantly associated with engagement in content. Using the passively collected data, the authors then built predictive models to infer receptivity and found that these models led to an over 66% better prediction compared to a baseline model.

Some studies have even built machine-learning models to detect and predict receptivity in real-time. Morrison et al. (2017) investigated the effects of timing and frequency of notifications on several receptivity metrics using a mobile stress-management intervention in an exploratory study. A sample of 77 participants was randomized into three groups, of which each group was using a different method to receive push notifications. One group received notifications occasionally within 72 hours, another group received the notifications on a daily basis, while a third group received "intelligent" notifications determined by a prediction model. However, the authors did not find any significant effect in favor of the notifications delivered at opportune times as inferred by the prediction model.

In a more recent study, a smartphone-based and chatbot-delivered physical activity intervention employed two different models to predict receptivity (Mishra et al., 2021). The first model was pretrained with data from another study population (Künzler et al., 2019) and remained constant for all participants over the course of the study. The second model was an adaptive model, which was updated every day with new information (after a baseline phase) as the study progressed. The models used information such as the day of the week (weekday vs. weekend), time of day, phone battery information, device interaction, and physical activity. To compare the utility, the authors also included a control model that would send notifications at random times. In a 3-week trial with 83 participants, the authors found that the first two models improved user receptivity by up to 40% compared to the control model. Moreover, it was observed that receptivity to notifications delivered by the adaptive model continuously increased over the course of the study. The authors speculate that such an adaptive model may further improve in performance the longer the study progresses and the more contextual information the model has available to predict states of receptivity.

To date, only one systematic review with meta-analysis has been conducted to systematically summarize research on interruptibility (Künzler, Kramer, & Kowatsch, 2017). This study found that systems designed to intervene at more opportune moments resulted in a greater response rate to notifications. There was some evidence that reduced response times were also possible. However the authors noted that the findings were not generalizable due to small sample sizes and high homogeneity among study participants within the primary studies.

Against the background of the empirical findings described above, Table 6.1 provides an overview of factors intervention authors and engineers of MHIs may consider for the detection and prediction of receptive states. Many of the investigated contextual factors can be passively collected using the mobile phone (e.g., device interaction, Bluetooth signal) whereas some factors are more burdensome to identify and may require more extensive analysis methods (e.g., identifying breakpoints) or self-report measures of study participants (e.g., mood).

TABLE 6.1 Overview of factors that may carry signal for the detection and prediction of receptivity to mobile health interventions.

#	Factor	References	Summary of evidence
1	Activity	Choi et al., 2019; Künzler et al., 2019; Mehrotra, Pejovic, Vermeulen, Hendley, & Musolesi, 2016; Mishra et al., 2021; Pielot et al., 2017; Yuan et al., 2017	(1) Individuals who complete less complex tasks tend to be more receptive. (2) Studying, relaxing, and "being idle" seem to positively impact receptivity. (3) Mixed results on the influence of the type of physical activity on receptivity (e.g., walking, running, not moving).
2	Breakpoints	Fischer et al., 2011; Ho and Intille, 2005; Okoshi et al., 2015a, 2015b, 2017	(1) Breakpoints may indicate opportune moments to trigger notifications (e.g., after calls, after receiving SMS). (2) Individuals are more receptive during activity transitions (e.g., transitions between sitting, standing, and walking).
3	Location	Choi et al., 2019; Künzler et al., 2019; Pielot et al., 2017; Sarker et al., 2014; Yuan et al., 2017	There is mixed evidence that location tends to be related to receptivity (e.g., being at home/at work, traveling, driving, shopping).
4	Time	Künzler et al., 2019; Mashhadi, Mathur, & Kawsar, 2014; Mishra et al., 2021; Pielot et al., 2015, 2017	(1) Individuals are likely less receptive during morning hours and more receptive the longer the day progresses. (2) Mixed results on the impact of the day of the week on receptivity (e.g., weekday vs. weekend).
5	Personality	Künzler et al., 2019; Mehrotra, Pejovic, Vermeulen, Hendley, & Musolesi, 2016; Yuan et al., 2017	Conscientious individuals or individuals with higher levels of neuroticism tend to be more receptivity.
6	Sender	Mehrotra, Musolesi, Hendley, & Pejovic, 2015, 2016; Yuan et al., 2017	Individuals are more receptive to notifications sent by significant others (e.g., family members, colleagues at work).
7	Device battery status	Künzler et al., 2019; Mishra et al., 2021; Pielot et al., 2017	Higher levels of a phone's battery charging status may indicate that the owner is more receptive. However, a fully charged battery is rather linked to low levels of receptivity.
8	Device interaction	Künzler et al., 2019; Mishra et al., 2021; Pielot et al., 2017	Individuals that used the smartphone more recently (e.g., measured by a lock/unlock event) tend to be more receptive.
9	Age	Künzler et al., 2019; Pielot et al., 2017	Older individuals tend to be more receptive.
10	Mood	Sarker et al., 2014; Yuan et al., 2017	Individuals that are rather happy and energetic (vs. being stressed out) tend to be more receptive.
11	Bluetooth signal	Pejovic & Musolesi, 2014	Changes in the Bluetooth environment may be linked to increased user receptivity (e.g., the number of Bluetooth devices nearby).
12	Communication patterns	Pielot et al., 2015	Individuals tend to be more receptive the less time has passed since last receiving phone calls or text messages.
13	Alert modality	Mehrotra, Pejovic, Vermeulen, Hendley, & Musolesi, 2016	Individuals that set the alert type of notifications to vibration (vs. a silent or sound modality) tend to be more receptive.
14	Device type	Künzler et al., 2019	Individuals using an Android (vs. an Apple) smartphone tend to be more receptive.
15	Social setting	Choi et al., 2019	Being alone or without social interactions positively influences receptivity (e.g., watching TV alone, sitting on a bus vs. talking with friends, taking a class).
16	Notification content	Fischer et al., 2010	The entertainment value, relevance, actionability, and user interest into a notification positively influences receptivity.

Note: The factors are ranked by the number of investigations.

6.5 Challenges and future work

Receptivity research is still in its infancy. While there are some first insights into the impact of various factors such as timing, content, context, and individual characteristics on receptivity, a causal influence is often inconclusive. Thus far, our current understanding of the field comes from a small number of quasi-experimental studies, which are typically cross-sectional, nonrandomized, short duration, and comprising of small homogenous samples. Nevertheless, this is a growing area of research and larger studies including models for the real-time detection and prediction of receptivity are starting to emerge. We conclude this chapter with an overview of the current challenges of receptivity research and our recommendations for future work.

One of the major limitations of prior research on receptivity is the quality and robustness of the evidence. To date, studies have been preliminary in nature and comprise rather small study samples (i.e., usually ranging from 20 to 50 study participants) and typically healthy young populations (e.g., university students). In the context of JITAIs, many of the interventions target at-risk populations or individuals who are already suffering from a health condition. Therefore, the scientific evidence on receptive states is likely not representative. While the increasing adoption of wearable devices such as smartwatches offers new opportunities for the detection and prediction of receptivity, we do not see such a trend in those populations who are most affected and at-risk of noncommunicable diseases, for example, vulnerable individuals with a lower socioeconomic status (Chandrasekaran, Katthula, & Moustakas, 2021). Short study durations represent another shortcoming of prior work. For example, it remains unclear how receptivity to a MHI develops over an extended period of time and whether receptivity undergoes any "fall-off" rates or other novelty effects. Therefore, more robust trial designs, for example, microrandomized trials (see also Chapter 11 on optimization trials) with longer assessment periods are needed to systematically identify effective receptivity predictions. Furthermore, future research should investigate ways to engage a wider spectrum of individuals, including larger patient populations, and increase access and uptake of wearable devices in more vulnerable populations.

One of the most common limitations of current receptivity research is that some of the factors used for detection and prediction purposes can still not be reliably measured via sensors. For example, certain mental states such as mood or depressive symptoms are not yet reliably detectable using mobile physiological sensors (Teepe et al., 2021). As a result, research efforts to date have focused on ecological momentary assessments (EMAs) to detect such features which are burdensome for participants to complete (Sarker et al., 2014). To alleviate the burden of completing EMAs, some intervention authors use microincentives to engage study participants (Pielot et al., 2017; Sarker et al., 2014). However, using incentives to motivate users may bias engagement levels and may not reflect the real-world motivation of study participants. Nevertheless, prediction models for momentary affect using smartphone sensor data are starting to emerge (Jacobson & Bhattacharya, 2022; Jacobson & Chung, 2020; Lekkas et al., 2022) and it can be expected that ongoing advancements in sensor and mobile technology will enable the reliable detection of a multitude of new features (e.g., continuous blood pressure monitoring) that can be collected unobtrusively and with minimal user burden to improve receptivity in the near future.

Another challenge in the design of receptivity-capable MHIs lies in the frequent software updates of mobile operating systems and the ongoing release of new smartphone generations. With every software update or smartphone generation, happening quite frequently/several times a year, software engineers are tremendously challenged to update and maintain the correctness of data collection streams via smartphones and wearable devices. In the worst case, unanticipated changes in the application programming interface of mobile operating systems may lead to a failure of MHI components to detect vulnerable or receptive states. This challenge showcases that engineers of MHIs are always highly dependent on what is possible on a specific technology platform.

Moreover, engineers of MHIs need to find the right balance between passive data collection needs required to determine vulnerable and receptive states and the availability of energy on mobile devices. Higher data sampling frequencies inevitably lead to quicker battery drainage, which, in turn, leads to increased burden with the mobile device, for example, the need to recharge more often. Thus, providers of mobile operating systems increasingly monitor, and inform user of, the impact of mobile apps on battery consumption to keep user experience with the device and operating system as high as possible. As a result of these measures, individuals may feel insecure and potentially delete the MHI from their smartphones.

A similar outcome must be anticipated when a MHI is passively collecting data about an individual's state of vulnerability and receptivity in the background to anticipate the right moment to intervene without it being actively used. That is, the individual does not have the app open and is not interacting with it. In such a situation, again, mobile operating systems may interfere and suggest to deinstall the app because of a high battery consumption relative to active engagement. It is, therefore, suggested—unfortunately to some degree against the idea of JITAIs—to proactively inform users of MHIs about these potential side effects and implement intervention components that require a rather regular and active interaction.

Data privacy and security are additional critical concerns when using various data streams for the detection and prediction of receptive states (see also Chapter 15 on privacy and security). While some data may be less intrusive such as the alert modality (e.g., silent mode, vibration or sound mode), other data are more privacy-concerning, for example, the content of notifications (e.g., when a critical biometric value is communicated). In addition, it is suggested that engineers of MHIs should always consider various data anonymization approaches for the technical components that take over the detection and prediction of receptive states.

Another concern related to states of receptivity is how individuals respond to so-called stacked notifications, that is, multiple notifications from the same app. In that case, a user may choose to only focus on the latest notification from the stack or ignore the notifications altogether if considered unimportant, which may significantly impact intervention outcomes.

And finally, most of the current research on receptivity still comes from the area of ubiquitous computing in the context of interruptibility management. Only a little research so far has been conducted in the field of MHIs. We, therefore, strongly encourage research at the intersection of behavioral medicine, clinical psychology, information systems research, software engineering, and computer science to understand better the factors that determine receptivity to MHIs making them more engaging and effective. Given the potential opportunities of detecting and predicting states of receptivity and the fact that JITAIs still have not yet been fully realized in commercially available MHIs (Keller et al., 2022; Teepe et al., 2021), we encourage industry developers to increasingly consider using such models to increase effectiveness and reduce user burden of their solutions.

References

Akter, S., D'Ambra, J., & Ray, P. (2013). Development and validation of an instrument to measure user perceived service quality of mHealth. *Information and Management, 50*(4), 181–195. https://doi.org/10.1016/j.im.2013.03.001.

Banholzer, N., Feuerriegel, S., Fleisch, E., Bauer, G. F., & Kowatsch, T. (2021). Computer mouse movements as an indicator of work stress: Longitudinal observational field study. *Journal of Medical Internet Research, 23*(4), e27121. https://doi.org/10.2196/27121.

Barata, F., Tinschert, P., Rassouli, F., Steurer-Stey, C., Fleisch, E., Puhan, M. A., et al. (2020). Automatic recognition, segmentation, and sex assignment of nocturnal asthmatic coughs and cough epochs in smartphone audio recordings: Observational field study. *Journal of Medical Internet Research, 22*(7), e18082. https://doi.org/10.2196/18082.

Bent, B., Lu, B., Kim, J., & Dunn, J. P. (2021a). Biosignal compression toolbox for digital biomarker discovery. *Sensors (Switzerland), 21*(2), 1–11. https://doi.org/10.3390/s21020516.

Bent, B., Wang, K., Grzesiak, E., Jiang, C., Qi, Y., Jiang, Y., et al. (2021b). The digital biomarker discovery pipeline: An open-source software platform for the development of digital biomarkers using mHealth and wearables data. *Journal of Clinical and Translational Science, 5*(1), e19. https://doi.org/10.1017/cts.2020.511.

BfArM. (2020). The fast-track process for digital health applications (DiGA) according to Section 139e SGB V a guide for manufacturers, service providers and users. https://www.bfarm.de/SharedDocs/Downloads/EN/MedicalDevices/DiGA_Guide.pdf;jsessionid=2EDA98779BEFE02CF5A1A7DD89193244.1_cid329?__blob=publicationFile&v=2.

Bhargava, V. R., & Velasquez, M. (2021). Ethics of the attention economy: The problem of social media addiction. *Business Ethics Quarterly, 31*(3), 321–359. https://doi.org/10.1017/BEQ.2020.32.

Bickmore, T. W., & Picard, R. W. (2005). Establishing and maintaining long-term human-computer relationships. *ACM Transactions on Computer-Human Interaction, 12*(2), 293–327. https://doi.org/10.1145/1067860.1067867.

Birkmeyer, S., Wirtz, B. W., & Langer, P. F. (2021). Determinants of mHealth success: An empirical investigation of the user perspective. *International Journal of Information Management, 59*(August), 102351. https://doi.org/10.1016/J.IJINFOMGT.2021.102351.

Boateng, G., & Kowatsch, T. (2020). Speech Emotion Recognition among Elderly Individuals using Multimodal Fusion and Transfer Learning. In: Companion Publication of the 2020 International Conference on Multimodal Interaction (ICMI '20 Companion) (pp. 12–16). New York, NY, USA: ACM. https://doi.org/10.1145/3395035.3425255.

Boehm, U., Matzke, D., Gretton, M., Castro, S., Cooper, J., Skinner, M., et al. (2021). Real-time prediction of short-timescale fluctuations in cognitive workload. *Cognitive Research: Principles and Implications, 6*(1). https://doi.org/10.1186/S41235-021-00289-Y.

Califf, C. B., Sarker, S., & Sarker, S. (2020). The bright and dark sides of technostress: A mixed-methods study involving healthcare IT. *MIS Quarterly, 44*(2), 809–856. https://doi.org/10.25300/MISQ/2020/14818.

Chan, S. W. T., Sapkota, S., Mathews, R., Zhang, H., & Nanayakkara, S. (2020). Prompto: Investigating receptivity to prompts based on cognitive load from memory training conversational agent. *Proceedings of the ACM on Interactive, Mobile, Wearable and Ubiquitous Technologies, 4*(4), 121. https://doi.org/10.1145/3432190.

Chandrasekaran, R., Katthula, V., & Moustakas, E. (2021). Too old for technology? Use of wearable healthcare devices by older adults and their willingness to share health data with providers. *Health Informatics Journal, 27*(4), 1–14. https://doi.org/10.1177/14604582211058073.

Choi, W., Park, S., Kim, D., Lim, Y., Lee, U., Woohyeok, Choi, et al. (2019). Multi-stage receptivity model for mobile just-in-time health intervention. *Proceedings of the ACM on Interactive, Mobile, Wearable and Ubiquitous Technologies, 3*(2), 39. https://doi.org/10.1145/3328910.

Collins, L. M. (2018). Optimization of behavioral, biobehavioral, and biomedical interventions - The multiphase optimization strategy (MOST). New York, NY, USA: Springer. http://link.springer.com/10.1007/978-3-319-72206-1.

Collins, L. M., Baker, T. B., Mermelstein, R. J., Piper, M. E., Jorenby, D. E., Smith, S. S., et al. (2011). The multiphase optimization strategy for engineering effective tobacco use interventions. *Annals of Behavioral Medicine, 41*(2), 208–226. https://doi.org/10.1007/S12160-010-9253-X.

Coravos, A., Khozin, S., & Mandl, K. D. (2019). Developing and adopting safe and effective digital biomarkers to improve patient outcomes. *NPJ Digital Medicine, 2*, 14. https://doi.org/10.1038/s41746-019-0090-4.

Davis, F. D. (1989). Perceived usefulness, perceived ease of use, and user acceptance of information technology. *MIS Quarterly, 13*(3), 319–339. https://doi.org/10.2307/249008.

Del Re, A. C., Flückiger, C., Horvath, A. O., & Wampold, B. E. (2021). Examining therapist effects in the alliance–outcome relationship: A multilevel meta-analysis. *Journal of Consulting and Clinical Psychology, 89*(5), 371–378. https://doi.org/10.1037/CCP0000637.

Dua, N., Singh, S. N., & Semwal, V. B. (2021). Multi-input CNN-GRU based human activity recognition using wearable sensors. *Computing, 103*(7), 1461–1478. https://doi.org/10.1007/S00607-021-00928-8.

Ferrari, A. J., Santomauro, D. F., Herrera, A. M. M., Shadid, J., Ashbaugh, C., Erskine, H. E., et al. (2022). Global, regional, and national burden of 12 mental disorders in 204 countries and territories, 1990–2019: A systematic analysis for the Global Burden of Disease Study 2019. *The Lancet Psychiatry, 9*(February), 137–150. https://doi.org/10.1016/s2215-0366(21)00395-3.

Fischer, J. E., Greenhalgh, C., & Benford, S. (2011). Investigating episodes of mobile phone activity as indicators of opportune moments to deliver notifications. In: *Proceedings of the 13th International Conference on Human Computer Interaction with Mobile Devices and Services (MobileHCI '11)* (pp. 181–190). New York, NY, USA: ACM. https://doi.org/10.1145/2037373.2037402.

Fischer, J. E., Yee, N., Bellotti, V., Good, N., Benford, S., & Greenhalgh, C. (2010). Effects of content and time of delivery on receptivity to mobile interruptions. In: *Proceedings of the 12th International Conference on Human Computer Interaction with Mobile Devices and Services (MobileHCI '10)* (pp. 103–112). New York, NY, USA: ACM. https://doi.org/10.1145/1851600.1851620.

Flückiger, C., Del Re, A. C., Wampold, B. E., & Horvath, A. O. (2018). The alliance in adult psychotherapy: A meta-analytic synthesis. *Psychotherapy (Chicago, Ill.), 55*(4), 316–340. https://doi.org/10.1037/PST0000172.

Flückiger, C., Del Re, A. C., Wampold, B. E., Symonds, D., & Horvath, A. O. (2012). How central is the alliance in psychotherapy? A multilevel longitudinal meta-analysis. *Journal of Counseling Psychology, 59*(1), 10–17. https://doi.org/10.1037/A0025749.

Guarino, H., Acosta, M., Marsch, L. A., Xie, H., & Aponte-Melendez, Y. (2016). A mixed-methods evaluation of the feasibility, acceptability, and preliminary efficacy of a mobile intervention for methadone maintenance clients. *Psychology of Addictive Behaviors, 30*(1), 1–11. https://doi.org/10.1037/ADB0000128.

Hagerman, C. J., Hoffman, R. K., Vaylay, S., & Dodge, T. (2021). Implementation intentions to reduce smoking: A systematic review of the literature. *Nicotine & Tobacco Research, 23*(7), 1085–1093. https://doi.org/10.1093/NTR/NTAA235.

Haug, S., Castro, R. P., Scholz, U., Kowatsch, T., Schaub, M. P., & Radtke, T. (2020). Assessment of the efficacy of a mobile phone – Delivered just-in-time planning intervention to reduce alcohol use in adolescents: Randomized controlled crossover trial. *JMIR MHealth and UHealth, 8*(5), e16937. https://doi.org/10.2196/16937.

Hébert, E. T., Stevens, E. M., Frank, S. G., Kendzor, D. E., Wetter, D. W., Zvolensky, M. J., et al. (2018). An ecological momentary intervention for smoking cessation: The associations of just-in-time, tailored messages with lapse risk factors. *Addictive Behaviors, 78*, 30–35. https://doi.org/10.1016/J.ADDBEH.2017.10.026.

Hekler, E., Tiro, J. A., Hunter, C. M., & Nebeker, C. (2020). Precision health: The role of the social and behavioral sciences in advancing the vision. *Annals of Behavioral Medicine, 54*(11), 805–826. https://doi.org/10.1093/abm/kaaa018.

Hekler, E. B., Rivera, D. E., Martin, C. A., Phatak, S. S., Freigoun, M. T., Korinek, E., et al. (2018). Tutorial for using control systems engineering to optimize adaptive mobile health interventions. *Journal of Medical Internet Research, 20*(6), e214. https://doi.org/10.2196/JMIR.8622.

Ho, J., & Intille, S. S. (2005). Using context-aware computing to reduce the perceived burden of interruptions from mobile devices. In: *Proceedings of the SIGCHI Conference on Human Factors in Computing Systems (CHI '05)* (pp. 909–918). New York, NY, USA: ACM. https://doi.org/10.1145/1054972.1055100.

Horvath, A. O., & Greenberg, L. S. (1989). Development and validation of the working alliance inventory. *Journal of Counseling Psychology, 36*(2), 223–233. https://doi.org/10.1037/0022-0167.36.2.223.

Jacobson, N. C., & Bhattacharya, S. (2022). Digital biomarkers of anxiety disorder symptom changes: Personalized deep learning models using smartphone sensors accurately predict anxiety symptoms from ecological momentary assessments. *Behaviour Research and Therapy, 149*, 104013. https://doi.org/10.1016/J.BRAT.2021.104013.

Jacobson, N. C., & Chung, Y. J. (2020). Passive sensing of prediction of moment-to-moment depressed mood among undergraduates with clinical levels of depression sample using smartphones. *Sensors, 20*(12), 3572. https://doi.org/10.3390/S20123572.

Jakob, R., Harperink, S., Rudolf, A. M., Fleisch, E., Haug, S., Mair, J. L., et al. (2022). Factors influencing adherence to mHealth Apps for prevention or management of noncommunicable diseases: Systematic review. *Journal of Medical Internet Research, 24*(5), e35371. https://doi.org/10.2196/35371.

Jenkins, J. L., Anderson, B. B., Vance, A., Kirwan, C. B., Eargle, D., & Katz, J. M. (2016). More harm than good? How messages that interrupt can make us vulnerable. *Information Systems Research, 27*(4), 880–896. https://doi.org/10.1287/isre.2016.0644.

Keller, R., Hartmann, S., Teepe, G. W., Lohse, K.-M., Alattas, A., Tudor Car, L., et al. (2022). Digital behavior change interventions for the prevention and management of type 2 diabetes: Systematic market analysis. *Journal of Medical Internet Research, 24*(1), e33348. https://doi.org/10.2196/33348.

Klasnja, P., Smith, S., Seewald, N. J., Lee, A., Hall, K., Luers, B., et al. (2019). Efficacy of contextually tailored suggestions for physical activity: A micro-randomized optimization trial of heart steps. *Annals of Behavioral Medicine, 53*(6), 573–582. https://doi.org/10.1093/abm/kay067.

Kowatsch, T., & Fleisch, E. (2021). Digital health interventions. In: O. Gassmann, & F. Ferrandina (Eds.), *Connected Business* (pp. 71–95). Springer International Publishing. https://doi.org/10.1007/978-3-030-76897-3_4.

Kowatsch, T., Lohse, K. M., Erb, V., Schittenhelm, L., Galliker, H., Lehner, R., et al. (2021a). Hybrid ubiquitous coaching with a novel combination of mobile and holographic conversational agents targeting adherence to home exercises: Four design and evaluation studies. *Journal of Medical Internet Research [Electronic Resource], 23*(2), E23612. https://doi.org/10.2196/23612.

Kowatsch, T., Schachner, T., Harperink, S., Barata, F., Dittler, U., Xiao, G., et al. (2021b). Conversational agents as mediating social actors in chronic disease management involving health care professionals, patients, and family members: Multisite single-arm feasibility study. *Journal of Medical Internet Research [Electronic Resource], 23*(2), E25060. https://doi.org/10.2196/25060.

Kramer, J. N., Künzler, F., Mishra, V., Smith, S. N., Kotz, D., Scholz, U., et al. (2020). Which components of a smartphone walking app help users to reach personalized step goals? Results from an optimization trial. *Annals of Behavioral Medicine, 54*(7), 518–528. https://doi.org/10.1093/abm/kaaa002.

Künzler, F., Kramer, J. N., & Kowatsch, T. (2017). Efficacy of mobile context-aware notification management systems: A systematic literature review and meta-analysis. In: *2017 IEEE 13th International Conference on Wireless and Mobile Computing, Networking and Communications (WiMob)* (pp. 131–138). New York, NY, USA: IEEE. https://doi.org/10.1109/WiMOB.2017.8115839.

Künzler, F., Mishra, V., Kramer, J. N., Kotz, D., Fleisch, E., & Kowatsch, T. (2019). Exploring the state-of-receptivity for mHealth interventions. *Proceedings of the ACM on Interactive, Mobile, Wearable and Ubiquitous Technologies, 3*(4), 140. https://doi.org/10.1145/3369805.

Lekkas, D., Price, G. D., & Jacobson, N. C. (2022). Using smartphone app use and lagged-ensemble machine learning for the prediction of work fatigue and boredom. *Computers in Human Behavior, 127*, 107029. https://doi.org/10.1016/j.chb.2021.107029.

L Murray, C. J., Aravkin, A. Y., Zheng, P., Abbafati, C., Abbas, K. M., Abbasi-Kangevari, M., et al. (2020). Global burden of 87 risk factors in 204 countries and territories, 1990–2019: A systematic analysis for the Global Burden of Disease Study 2019. https://doi.org/10.1016/S0140-6736(20)30752-2.

Lohani, M., Payne, B. R., & Strayer, D. L. (2019). A review of psychophysiological measures to assess cognitive states in real-world driving. *Frontiers in Human Neuroscience, 13*, 57. https://doi.org/10.3389/FNHUM.2019.00057.

Mehrotra, A., Musolesi, M., Hendley, R., & Pejovic, V. (2015). Designing content-driven intelligent notification mechanisms for mobile applications. *Proceedings of the 2015 ACM International Joint Conference on Pervasive and Ubiquitous Computing (UbiComp '15).* ACM, New York, NY, USA, 813 821. https://doi.org/10.1145/2750858.2807544.

Mair, J. L., Hayes, L. D., Campbell, A. K., Buchan, D. S., Easton, C., & Sculthorpe, N. (2022). A personalized smartphone-delivered just-in-time adaptive intervention (JitaBug) to increase physical activity in older adults: mixed methods feasibility study. *JMIR Formative Research, 6*(4), e34662. https://doi.org/10.2196/34662.

Mashhadi, A., Mathur, A., & Kawsar, F. (2014). The myth of subtle notifications. In: *Proceedings of the 2014 ACM International Joint Conference on Pervasive and Ubiquitous Computing: Adjunct Publication (UbiComp '14 Adjunct)* (pp. 111–114). New York, NY, USA: ACM. https://doi.org/10.1145/2638728.2638759.

Mehrotra, A., Pejovic, V., Vermeulen, J., Hendley, R., & Musolesi, M. (2016). My Phone and Me: Understanding People's Receptivity to Mobile Notifications (pp. 1021–1032). In: *Proceedings of the 2016 CHI Conference on Human Factors in Computing Systems.* New York, NY, USA: ACM. https://doi.org/10.1145/2858036.2858566.

Mishra, V., Künzler, F., Kramer, J. N., Fleisch, E., Kowatsch, T., & Kotz, D. (2021). Detecting receptivity for mHealth interventions in the natural environment. *Proceedings of the ACM on Interactive, Mobile, Wearable and Ubiquitous Technologies,* (2nd ed., 5). New York, NY, USA: ACM, 74. https://doi.org/10.1145/3463492.

Mönninghoff, A., Kramer, J. N., Hess, A. J., Ismailova, K., Teepe, G. W., Car, L. T., et al. (2021). Long-term effectiveness of mHealth physical activity interventions: Systematic review and meta-analysis of randomized controlled trials. *Journal of Medical Internet Research, 23*(4), e26699. https://doi.org/10.2196/26699.

Morrison, L. G., Hargood, C., Pejovic, V., Geraghty, A. W. A., Lloyd, S., Goodman, N., et al. (2017). The effect of timing and frequency of push notifications on usage of a smartphone-based stress management intervention: An exploratory trial. *Plos One, 12*(1), e0169162. https://doi.org/10.1371/JOURNAL.PONE.0169162.

Murphy, A., Palafox, B., Walli-Attaei, M., Powell-Jackson, T., Rangarajan, S., Alhabib, K. F., et al. (2020). The household economic burden of non-communicable diseases in 18 countries. *BMJ Global Health, 5*(2), e002040. https://doi.org/10.1136/bmjgh-2019-002040.

Nahum-Shani, I., Hekler, E. B., & Spruijt-Metz, D. (2015). Building health behavior models to guide the development of just-in-time adaptive interventions: A pragmatic framework. *Health Psychology, 34*(0), 1209–1219. https://doi.org/10.1037/hea0000306.

Nahum-Shani, I., Smith, S. N., Spring, B. J., Collins, L. M., Witkiewitz, K., Tewari, A., et al. (2018). Just-in-time adaptive interventions (JITAIs) in mobile health: Key components and design principles for ongoing health behavior support. *Annals of Behavioral Medicine, 52*(6), 446–462. https://doi.org/10.1007/s12160-016-9830-8.

Noorbergen, T. J., Adam, M. T. P., Roxburgh, M., & Teubner, T. (2021). Co-design in mHealth systems development: Insights from a systematic literature review. *AIS Transactions on Human-Computer Interaction, 13*(2), 175–205. https://doi.org/10.17705/1thci.00147.

Okoshi, T., Ramos, J., Nozaki, H., Nakazawa, J., Dey, A. K., & Tokuda, H. (2015a). Attelia: Reducing user's cognitive load due to interruptive notifications on smart phones. In: 2015 *IEEE International Conference on Pervasive Computing and Communications, PerCom 2015* (pp. 96–104). https://doi.org/10.1109/PERCOM.2015.7146515.

Okoshi, T., Ramos, J., Nozaki, H., Nakazawa, J., Dey, A. K., & Tokuda, H. (2015b). Reducing users' perceived mental effort due to interruptive notifications in multi-device mobile environments. In: *UbiComp 2015 - Proceedings of the 2015 ACM International Joint Conference on Pervasive and Ubiquitous Computing* (pp. 475–486). https://doi.org/10.1145/2750858.2807517.

Okoshi, T., Tsubouchi, K., Taji, M., Ichikawa, T., & Tokuda, H. (2017). Attention and engagement-awareness in the wild: A large-scale study with adaptive notifications. In: *IEEE International Conference on Pervasive Computing and Communications, PerCom 2017* (pp. 100–110). New York, NY, USA: IEEE. https://doi.org/10.1109/PERCOM.2017.7917856.

Pejovic, V. & Musolesi, M. (2014). InterruptMe: designing intelligent prompting mechanisms for pervasive applications. *Proceedings of the 2014 ACM International Joint Conference on Pervasive and Ubiquitous Computing (UbiComp 14)*. ACM, New York, NY, USA, 897–908. https://doi.org/10.1145/2632048.2632062.

Pielot, M., Cardoso, B., Katevas, K., Serrà, J., Matic, A., & Oliver, N. (2017). Beyond interruptibility: Predicting opportune moments to engage mobile phone users. *Proceedings of the ACM on Interactive, Mobile, Wearable and Ubiquitous Technologies, 1*(3), 91. https://doi.org/10.1145/3130956.

Pielot, M., Dingler, T., San Pedro, J., & Oliver, N. (2015). When attention is not scarce - Detecting boredom from mobile phone usage. In: *Proceedings of the 2015 ACM International Joint Conference on Pervasive and Ubiquitous Computing (UbiComp '15)* (pp. 825–836). New York, NY, USA: ACM. https://doi.org/10.1145/2750858.2804252.

Qian, T., Walton, A. E., Collins, L. M., Klasnja, P., Lanza, S. T., Nahum-Shani, I., et al. (2022). The microrandomized trial for developing digital interventions: Experimental design and data analysis considerations. *Psychological Methods*. Advance online publication. https://doi.org/10.1037/met0000283.

Rassouli, F., Tinschert, P., Barata, F., Steurer-Stey, C., Fleisch, E., Puhan, M. A., et al. (2020). Characteristics of asthma-related nocturnal cough: A potential new digital biomarker. *Journal of Asthma and Allergy, 13*, 649–657. https://doi.org/10.2147/JAA.S278119.

Rüegger, D., Stieger, M., Nißen, M., Allemand, M., Fleisch, E., & Kowatsch, T. (2020). How are personality states associated with smartphone data? *European Journal of Personality, 34*(5), 687–713. https://doi.org/10.1002/per.2309.

Sahandi Far, M., Stolz, M., Fischer, J. M., Eickhoff, S. B., & Dukart, J. (2021). JTrack: A digital biomarker platform for remote monitoring of daily-life behaviour in health and disease. *Frontiers in Public Health, 9*, 763621. https://doi.org/10.3389/fpubh.2021.763621.

Sarker, H., Sharmin, M., Ali, A. A., Rahman, M. M., Bari, R., Hossain, S. M., et al. (2014). Assessing the availability of users to engage in just-in-time intervention in the natural environment. In: *Proceedings of the 2014 ACM International Joint Conference on Pervasive and Ubiquitous Computing (UbiComp '14)* (pp. 909–920). New York, NY, USA: ACM. https://doi.org/10.1145/2632048.2636082.

Schembre, S. M., Liao, Y., Robertson, M. C., Dunton, G. F., Kerr, J., Haffey, M. E., et al. (2018). Just-in-time feedback in diet and physical activity interventions: Systematic review and practical design framework. *Journal of Medical Internet Research, 20*(3), e106. https://doi.org/10.2196/jmir.8701.

Shih, C. H., Tomita, N., Lukic, Y. X., Reguera, Á. H., Fleisch, E., & Kowatsch, T. (2019). Breeze: Smartphone-based acoustic real-time detection of breathing phases for a gamified biofeedback breathing training. *Proceedings of the ACM on Interactive, Mobile, Wearable and Ubiquitous Technologies, 3*(4), 152. https://doi.org/10.1145/3369835.

Sieberts, S. K., Schaff, J., Duda, M., Pataki, B. Á., Sun, M., Snyder, P., et al. (2021). Crowdsourcing digital health measures to predict Parkinson's disease severity: The Parkinson's Disease Digital Biomarker DREAM Challenge. *NPJ Digital Medicine, 4*(1), 53. https://doi.org/10.1038/s41746-021-00414-7.

Sieverink, F., Kelders, S. M., & Gemert-Pijnen, V. (2017). Clarifying the concept of adherence to eHealth technology: Systematic review on when usage becomes adherence. *Journal of Medical Internet Research, 19*(12), e402. https://doi.org/10.2196/jmir.8578.

Stasinaki, A., Büchter, D., Shih, C. H. I., Heldt, K., Güsewell, S., Brogle, B., et al. (2021). Effects of a novel mobile health intervention compared to a multi-component behaviour changing program on body mass index, physical capacities and stress parameters in adolescents with obesity: A randomized controlled trial. *BMC Pediatrics, 21*(1), 308. https://doi.org/10.1186/s12887-021-02781-2.

Stern, A. D., Matthies, H., Hagen, J., Brönneke, J. B., & Debatin, J. F. (2020). Want to see the future of digital health tools? Look to Germany. *Harvard Business Review*. https://hbr.org/2020/12/want-to-see-the-future-of-digital-health-tools-look-to-germany.

Teepe, G. W., da Fonseca, A., Kleim, B., Jacobson, N. C., Sanabria, A. S., Car, L. T., et al. (2021). Just-in-time adaptive mechanisms of popular mobile apps for individuals with depression: Systematic app search and literature review. *Journal of Medical Internet Research, 23*(9), e29412. https://doi.org/10.2196/29412.

Throuvala, M. A., Pontes, H. M., Tsaousis, I., Griffiths, M. D., Rennoldson, M., & Kuss, D. J. (2021). Exploring the dimensions of smartphone distraction: Development, validation, measurement invariance, and latent mean differences of the smartphone distraction scale (SDS). *Frontiers in Psychiatry, 12*(March), 642634. https://doi.org/10.3389/fpsyt.2021.642634.

Turner, M., Kitchenham, B., Brereton, P., Charters, S., & Budgen, D. (2010). Does the technology acceptance model predict actual use? A systematic literature review. *Information and Software Technology, 52*(5), 463–479. https://doi.org/10.1016/j.infsof.2009.11.005.

Venkatesh, V., Thong, J. Y. L., & Xu, X. (2012). Consumer acceptance and use of information technology: Extending the unified theory of acceptance and use of technology. *MIS Quarterly, 36*(1), 157–178. https://doi.org/10.2307/41410412.

Vos, T., Lim, S. S., Abbafati, C., Abbas, K. M., Abbasi, M., Abbasifard, M., et al. (2020). Global burden of 369 diseases and injuries in 204 countries and territories, 1990–2019: A systematic analysis for the Global Burden of Disease Study 2019. *The Lancet, 396*, 1204–1222. https://doi.org/10.1016/S0140-6736(20)30925-9.

Warner, L. M., Fleig, L., Wolff, J. K., Keller, J., Schwarzer, R., Nyman, S., et al. (2022). What makes implementation intentions (in)effective for physical activity among older adults? *British Journal of Health Psychology, 27*(2), 571–587. https://doi.org/10.1111/BJHP.12563.

Yuan, F., Gao, X., & Lindqvist, J. (2017). How busy are you? Predicting the interruptibility intensity of mobile users. *CHI '17: Proceedings of the 2017 CHI Conference on Human Factors in Computing Systems* (pp. 5346–5360). New York, NY, USA: ACM. https://doi.org/10.1145/3025453.3025946.

Chapter 7

Adapting just-in-time interventions to vulnerability and receptivity: Conceptual and methodological considerations

Inbal Nahum-Shani[a], David W. Wetter[b] and Susan A. Murphy[c]

[a] Data-Science for Dynamic Decision-making Center (d3c), Institute for Social Research, University of Michigan, Ann Arbor, Michigan, [b] Center for Health Outcomes and Population Equity (HOPE), Department of Population Health Sciences, Huntsman Cancer Institute, University of Utah, Salt Lake City, Utah, [c] Department of Statistics, Computer Science, John A. Paulson School of Engineering and Applied Sciences, Harvard University, Boston, Massachusetts

7.1 Introduction

Powerful mobile and sensing technologies offer tremendous opportunities for delivering just-in-time adaptive interventions (JITAIs) (Dillingham et al., 2018; Guarino, Acosta, Marsch, Xie, & Aponte-Melendez, 2016; Hébert et al., 2018; Witkiewitz et al., 2014). JITAIs use dynamically changing information about the individual's internal state (e.g., craving, stress), and context (e.g., physical location) to recommend whether and how to deliver interventions in real-time, in daily life. JITAIs have been developed for preventing and treating substance use disorders, including tobacco (Businelle et al., 2016; Hébert et al., 2020; Naughton et al., 2016), alcohol (Gustafson et al., 2014; Kizakevich et al., 2018; Witkiewitz et al., 2014) and other drugs (Coughlin et al., 2021). JITAIs have also been leveraged to promote mental health, focusing on helping individuals manage problematic emotional reactions, such as stress or low mood (see Marciniak et al., 2020; Wang & Miller, 2020). States of vulnerability (e.g., high urge) to an adverse outcome (e.g., smoking lapse) and states of receptivity (e.g., when the person is in a quiet, calm setting) to a just-in-time intervention (e.g., a prompt from the person's mobile device recommending brief self-regulatory strategies) play a critical role in the formulation of effective JITAIs (Nahum-Shani et al., 2018; Nahum-Shani, Hekler, & Spruijt-Metz, 2015). However, vulnerability and receptivity are defined and operationalized in various ways across different fields and studies. Moreover, two approaches are commonly used to answer scientific questions about states of vulnerability and receptivity to just-in-time interventions. One focuses on analyzing observational intensive longitudinal data (ILD) and the other focuses on using ILD from microrandomized trials (MRTs). These different approaches are often motivated by scientific questions that might seem similar, such as under what conditions individuals are vulnerable to a specific adverse proximal outcome, or under what conditions an individual is receptive to a just-in-time intervention. However, these questions are operationalized differently by each approach which has implications on the interpretation of the results and how they can inform a JITAI. The goal of this chapter is twofold. First, we clarify the definition and operationalization of vulnerability to adverse outcomes and receptivity to just-in-time interventions. Second, we provide greater specificity in formulating scientific questions about states of vulnerability and receptivity. This greater precision will aid researchers in deciding whether observational studies or MRT studies are most suitable. For simplicity, we focus only on the formulation of JITAIs that are intended to address momentary risk for an adverse event (e.g., risk for lapse, risk for substance use), although JITAIs can also be designed to capitalize on momentary windows of opportunity for positive change (e.g., increase physical activity, learn a new skill).

7.2 Just-in-time adaptive interventions

JITAIs are designed specifically to take advantage of the ability of smart devices to sense or collect in-the-moment data on an individual's dynamic state, such as current craving, stress, physical location, and recent substance use. JITAIs use these data to adapt intervention delivery to individuals based on decision rules. The objective of the decision rules is to provide the intervention option that is best for an individual at a particular moment, while avoiding unnecessary treatment and minimizing burden (Nahum-Shani et al., 2015, 2018). For illustrative purposes, consider a simple hypothetical JITAI designed to engage smokers, who are attempting to quit, in self-regulatory strategies in real-time. Here, self-regulatory strategies refer to activities (e.g., behavioral substitution, mindful attention) intended to help people control and direct their own thoughts, feelings, and behaviors (Fiske & Taylor, 1991). The mobile device prompts participants six times per day, approximately 2 hours apart, to provide information about their emotions and experiences via ecological momentary assessments (EMAs). If the individual indicates negative affect or the presence of other smokers, and they are not driving a car, then a prompt is delivered immediately via the mobile device encouraging the individual to engage in self-regulatory strategies (e.g., take a 5-minute walk; direct your attention to experiences that are occurring in the moment without judgment). Otherwise, no prompt is delivered.

7.3 JITAI components

The example JITAI described above can be protocolized via the following decision rule:

Every 2 hours,
  IF negative affect = Yes **or** presence of other smokers = Yes; **and** driving = No
    THEN, intervention option = Deliver a prompt recommending self-regulatory strategies
  ELSE, intervention option = No prompt

The decision rule above includes four key elements:

(1) Decision points. These are points in time in which a decision should be made about intervention delivery. In this example JITAI, decisions are made every 2 hours.
(2) Tailoring variables. These include information about the individual's internal state and context; this information is used to decide whether and how to deliver an intervention. In this example JITAI, the tailoring variables include negative affect, presence of other smokers, and driving.
(3) Intervention options. These include different intervention types, intensities, or delivery modalities under consideration at each decision point. In this example JITAI, there are two intervention options: a prompt recommending self-regulatory strategies and no prompt.
(4) Thresholds or levels of the tailoring variable that differentiate between conditions in which one intervention option should be delivered versus another. In this example JITAI, the thresholds and levels of the tailoring variables specify the conditions in which a prompt should be delivered (i.e., when the individual experiences negative affect or when other smokers are present, and the individual is not driving a car), as well as the conditions in which a prompt should not be delivered (i.e., when the individual does not experience negative affect and other smokers are not present, or when the person is driving a car).

The adaptation process, which refers to the use of ongoing information about the individual to decide when and how to intervene, is operationalized via the tailoring variables, their thresholds (or levels), and the intervention options. Specifically, the adaptation process involves ongoing monitoring of the individual's internal state and context to obtain information about the tailoring variable(s), using thresholds (or levels) of the tailoring variable(s) to decide which intervention option to offer, and delivering the appropriate intervention options. This adaptation process is triggered at decision points and is guided by the goal of achieving a prespecified distal outcome (e.g., smoking abstinence by week 12) by impacting proximal outcomes. Proximal outcomes are the short-term goals of the adaptation, typically reflecting key mechanisms of change in the pathway between the adaptation process and the distal outcome. For example, the proximal outcome in the JITAI example above might be whether the individual experiences a smoking episode in the next 2 hours. Note that the proximal outcomes should be selected carefully in light of the ultimate goal (i.e., the distal outcome) the JITAI is intended to achieve. This selection

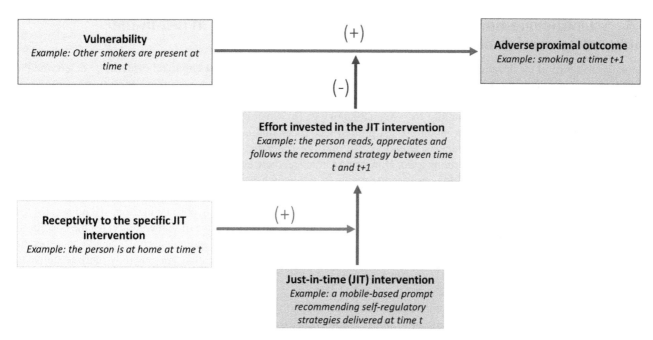

FIGURE 7.1   Scientific model for vulnerability and receptivity to just-in-time interventions.

can be guided by a scientific model that describes, based on existing evidence, why altering the proximal outcome should lead to the desired change in the distal outcome (Nahum-Shani et al., 2015, 2018).

7.4   States of vulnerability and receptivity

Many JITAIs targeting addictions are motivated to address real-time, real-world conditions that represent vulnerability. Here, *vulnerability is defined as a state of heightened risk for a specific adverse proximal outcome.* In these JITAIs, the primary objective of the adaptation is to break the link between conditions that represent heightened risk and the adverse proximal outcome. In the example JITAI above, the idea is that experiencing negative affect or the presence of other smokers intensifies the individual's risk for a smoking episode in the near time, here operationalized as the subsequent 2 hours. Hence, a prompt recommending self-regulatory strategies is delivered when the individual self-reports negative affect or the presence of other smokers; the goal is to break the link between this state of vulnerability and subsequent smoking. Further, given that the conditions that represent vulnerability are likely to emerge in daily life at the same time there are multiple demands competing for the individual's attention and effort, intervention burden is reduced by delivering the intervention only when the individual is receptive.

We define *receptivity as a state of heightened likelihood that the individual will invest cognitive, affective, and physical energy (i.e., will engage; see Nahum-Shani et al., 2022) in a particular intervention.* It follows that receptivity is intervention-specific; for example, while in a meeting a person may be receptive to a prompt containing a brief message but may not be receptive to a prompt containing audio content. Receptivity is defined in relation to a proximal pathway/mediator that captures engagement in a specific intervention. Here, engagement is defined as the investment of cognitive, emotional, and physical effort in a focal task or stimulus; Nahum-Shani et al., 2022. For example, delivering a prompt recommending self-regulatory strategies may prevent the presence of other smokers from evolving into a smoking episode only if the recipient reads the message, appreciates the content, and follows the recommendation. In the simple JITAI example above, individuals are considered receptive to a prompt recommending self-regulatory strategies if they are not driving a car.

Receptivity and vulnerability are interrelated via engagement; engagement is a pathway through which, when experiencing a state of vulnerability, a particular just-in-time intervention may prevent a proximal adverse outcome (see Fig. 7.1). For example, when an individual self-reports negative affect or the presence of other smokers (i.e., when the individual is in a state of vulnerability to a proximal smoking episode), a prompt recommending self-regulatory strategies would be delivered only if the individual is not driving a car (i.e., the individual is in a state of receptivity to a prompt recommending self-regulatory strategies).

7.5 Scientific questions about vulnerability and receptivity in just-in-time intervention development

A major challenge to the development of effective JITAIs is the absence of sufficient empirical evidence to determine the conditions in which individuals are vulnerable to an adverse outcome and receptive to a particular just-in-time intervention. For example, the JITAI above is based on the assumptions that (a) experiencing negative affect or the presence of other smokers represent a state of vulnerability to proximal smoking; (b) when smokers attempting to quit are in a state of vulnerability to proximal smoking, a mobile-based prompt recommending self-regulatory strategies has the potential to prevent the smoking episode; (c) smokers attempting to quit are generally receptive to a prompt recommending self-regulatory activities unless they are driving a car. However, in practice, there are many open scientific questions concerning the development of an effective JITAI for smokers attempting to quit, which cannot be answered based on existing empirical evidence and practical considerations. These questions often concern (a) the constellation of static and dynamic conditions in which individuals are vulnerable to a specific proximal adverse outcome (e.g., smoking episode in the next 2 hours); (b) the benefits of delivering a specific just-in-time intervention when individuals are in a state of vulnerability; (c) the conditions in which individuals are receptive to a specific just-in-time intervention; (d) whether individuals are receptive to a specific just-in-time intervention when they are in a state of vulnerability; (e) whether individuals are generally more receptive to one type of just-in-time intervention versus another; and (f) the conditions in which individuals are more receptive to one type of just-in-time intervention versus another (see Table 7.1).

Researchers often employ two common approaches to answering scientific questions about states of vulnerability and receptivity in just-in-time intervention development. Both approaches focus on analyzing intensive longitudinal data (ILD), which (Bolger & Laurenceau, 2013) define as "sequential measurements on five or more occasions during which a change process is expected to unfold within each subject." However, one approach focuses on analyzing ILD from observational studies, whereas the other focuses on analyzing ILD from MRTs. Below we discuss each approach and highlight differences in the scientific questions each is motivated to answer.

7.6 Analyzing observational ILD to inform JITAI development

Here, we use the term observational ILD to refer to data from studies that do not involve random assignment to just-in-time intervention options at JITAI decision points. Below we provide two examples of studies, one focusing on using observational ILD to identify states of vulnerability to an adverse proximal outcome (Example 1); and one focusing on using observational ILD to identify states of receptivity to just-in-time interventions (Example 2). We use these examples to highlight the type of scientific questions that can be answered by using observational ILD to investigate states of vulnerability and receptivity to a just-in-time intervention.

Example 1. Observational ILD study to identify states of vulnerability to a smoking episode. Break free (Chatterjee et al., 2020) is an observational study investigating intrapersonal and contextual factors related to smoking lapse and abstinence among African American smokers ($N = 300$). Participants are assessed for two contiguous weeks (4 days pre- through 10 days postcessation) using AutoSense (a sensor suite that includes light-weight, unobtrusive chest and wristband units), GPS, and EMAs. The study uses mCerebrum (Hossain et al., 2017)—a software framework for the smartphone that collects measurements from AutoSense sensors via wireless radio, extracts tens of features from these measurements, and applies machine learning algorithms to detect smoking and stress (Ali et al., 2012; Ertin et al., 2011; Plarre et al., 2011). EMAs triggered up to three times per day assess emotions (e.g., happy, proud, angry, sad), intrapersonal (e.g., urge, self-efficacy, outcome expectancies, motivation), and contextual factors (e.g., cigarette availability, others smoking, alcohol consumption, and social support). The ILD is currently being analyzed with advanced machine learning techniques (e.g., event-time prediction models; Dempsey et al., 2017) to investigate which summaries of the measurement data (obtained based on pattern mining results and behavioral theory) at a given moment are most predictive of a proximal smoking episode (e.g., in the next 2 hours). The goal is to identify the conditions in which a smoker attempting to quit is vulnerable to proximal smoking, and hence may benefit from a just-in-time intervention to prevent the smoking episode.

The scientific question motivating this example concern the conditions in which individuals are vulnerable to a proximal smoking episode. Specifically, the goal is to ascertain (a) what constellation of dynamic (e.g., stress, urge, cigarette availability) and static (baseline self-efficacy, impulsivity, motivation to quit) factors measured by time t predict a subsequent smoking episode at time $t + 1$ (e.g., in the next 2 hours); and (b) how to use these predictive factors to construct a classifier that can accurately differentiate between those experiencing versus not experiencing a smoking episode at time $t + 1$. Here,

TABLE 7.1 Summary of scientific questions motivating Examples 1–3.

#	Example scientific questions	Operational (more specific) scientific question	Type of data	Methods used
1	Under what conditions individuals are vulnerable to a specific adverse proximal outcome?	In the absence of a mobile intervention, what constellation of dynamic and static factors observed by time t predict a smoking episode in the next 2 hours $(t + 1)$?	Observational ILD in Example 1	Prediction and classification methods
		In the context of a mobile intervention that prompts the individual to engage in self-regulatory strategies three times per day on average, what constellation of dynamic and static factors observed by time t predict a smoking episode in the next 2 hours $(t + 1)$?	MRT ILD in Example 3	Prediction and classification methods
2	Is it beneficial to deliver a just-in-time intervention when individuals are in a state of vulnerability?	When a smoker attempting to quit self-reports negative affect or the presence of other smokers at time t, is it beneficial in terms of reducing the probability of a lapse in the next 2 hours $(t + 1)$, to deliver (vs. not deliver) a prompt recommending self-regulatory strategies?	MRT ILD in Example 3	Generalization of regression analysis
3	Under what constellation of dynamic and static conditions an individual is receptive to a specific just-in-time intervention?	What constellation of dynamic and static factors observed by time t can accurately predict and classify those who engage (i.e., reply to) vs. those who do not engage with a greeting message from a chatbot-based digital coach within 10 minutes $(t + 1)$?	Observational ILD in Example 2	Prediction and classification methods
		Under what constellation of dynamic and static factors observed by time t, delivering (vs. not delivering) a prompt recommending self-regulatory strategies at time t will lead to engagement in self-regulatory strategies in the next hour $(t + 1)$?	MRT ILD in Example 3	Generalization of regression analysis
4	Are individuals receptive to a specific just-in-time intervention when they are in a state of vulnerability?	When a smoker attempting to quit self-reports negative affect or the presence of other smokers at time t, would delivering (vs. not delivering) a prompt recommending self-regulatory strategies lead to engagement in self-regulatory strategies in the next hour $(t + 1)$?	MRT ILD in Example 3	Generalization of regression analysis
5	Are individuals generally more receptive to one type of just-in-time intervention vs. another?	On average, does delivering a prompt recommending brief self-regulatory strategies at time t leads to greater engagement in self-regulatory strategies in the next hour $(t + 1)$ compared to a prompt recommending more effortful strategies?	MRT ILD in Example 3	Generalization of regression analysis
6	Under what conditions individuals are more receptive to one type of just-in-time intervention vs. another?	Under what conditions delivering a prompt recommending brief self-regulatory strategies (vs. a prompt recommending more effortful strategies) at time t leads to greater engagement in self-regulatory strategies in the next hour $(t + 1)$?	MRT ILD in Example 3	Generalization of regression analysis

prediction concerns the prospective course of a future event and constructing a classifier concerns the creation of a rule or principle to arrange entities into groups (Gottfredson, 1987). Importantly, this study is not motivated to investigate whether a suite of independent variables (e.g., stress, urge, cigarette availability) *cause* the dependent variable (e.g., smoking), but rather to develop a formula to accurately predict the dependent variable based on the observed values of the independent variables.

Suppose the results indicate that high versus low to moderate negative affect can accurately differentiate between smoking versus no smoking (respectively) in the next 2 hours. Practically, these results suggest that high negative affect is a state of vulnerability for proximal smoking and hence an intervention may be needed when individuals experience these conditions. However, these results cannot answer questions about which intervention would be beneficial in preventing proximal smoking when the individual experiences negative affect and whether individuals are receptive to a specific just-in-time intervention when they experience these conditions. Given that just-in-time interventions were not delivered as part of this study, the results can only be used to understand the conditions in which an intervention may be needed. These results can serve as a basis for future studies to investigate whether and what type of just-in-time intervention should be delivered when individuals experience a state of vulnerability.

Given the absence of a just-in-time intervention in Example 1, a more precise way to describe the goal of this study is to develop a formula for using dynamic and static information to predict proximal smoking, *in the absence* of a just-in-time intervention. Note that the study in Example 1 included other types of interventions, such as nicotine patch therapy, self-help materials, and brief quitting advice based on current clinical practice guidelines. Here, the goal is to investigate ways to improve existing smoking cessation treatments (nicotine patch therapy, self-help materials, and brief quitting advice) by identifying real-time, real-world conditions outside of standard treatment settings in which additional support in the form of just-in-time interventions might be needed.

Observational studies focusing on vulnerability to an adverse proximal outcome may also include a specific JITAI. Suppose the study in Example 1 included the immediate delivery of a prompt recommending self-regulatory strategies at time t if the individual self-reported high negative affect via an EMA, and no prompt otherwise. Summaries of EMAs and sensor-based assessments at time t can still be used to develop a classifier that accurately differentiates between those experiencing versus not experiencing a smoking episode at time $t + 1$. However, in this case, the goal would be to investigate ways to improve a specific JITAI (a JITAI that delivers a prompt when self-reported negative affect is high and no prompt otherwise), by identifying additional conditions (beyond high negative affect) in which individuals are vulnerable to proximal smoking. As before, the results of this hypothetical study cannot be used to answer questions about whether and what type of just-in-time intervention should be delivered under conditions that represent vulnerability.

Example 2. Observational ILD study to identify states of receptivity to just-in-time physical activity interventions. Künzler and colleagues (Künzler et al., 2019) conducted a study with 189 participants, over a period of 6 weeks, to investigate factors that capture receptivity to just-in-time interventions designed to promote physical activity. The interventions were delivered by Ally, a chatbot-based digital coach. Participants were prompted with a message from Ally several times per day. Each conversation started with a greeting—a generic message like "Hello Joyce," or "Good Evening Joyce." The coach starts sending the intervention messages only if the participant responds to the greeting message. Three different metrics were used to measure proximal engagement with the intervention: (1) whether the participant replied to the greeting message within 10 minutes; (2) time between message triggering and the participant's reply, and (3) whether the participant replied to two or more messages within 10 minutes following the triggering of the greeting message (i.e., engaged in a conversation with the bot). The data were analyzed with machine learning methods (Random Forests) to explore the extent to which static (e.g., age, gender, personality) and dynamic factors (e.g., day/time, phone battery, phone interaction, physical activity, and location) accurately predict and classify proximal engagement.

The scientific questions motivating this example concern the conditions in which individuals are receptive to a prompt from the chatbot. Similar to identifying states of vulnerability based on observational ILD, this question is framed as a prediction problem. For example, the goal is to ascertain what summaries of dynamic and static factors observed at time t predict replying to the greeting message in the next 10 minutes ($t + 1$); and how these summaries can be used to build a classifier (i.e., a rule) that accurately differentiates between those replying versus not replying within 10 minutes. Similar to using ILD to identify states of vulnerability, the goal here is not to investigate whether a suite of independent variables (e.g., personality, age, physical activity, location) *cause* the dependent variable (e.g., replying to the greeting message within 10 minutes), but rather to develop a formula to predict the dependent variable based on the observed values of the independent variables.

As discussed above, unlike the concept of vulnerability, the concept of receptivity is, by definition, intervention specific. Hence, while studies focusing on vulnerability may or may not include the delivery of a just-in-time intervention (depending on the motivating scientific questions), receptivity cannot be investigated without delivering the specific just-in-time intervention of scientific interest. In Example 2 this specific just-in-time intervention is a greeting message from the chatbot. The greeting message is sent at specific decision points (here, random times within certain time frames during the day) in which an intervention is likely needed based on scientific and practical considerations. For example, every day between 8 and 10 am a greeting message was delivered to set the step goals for the day, and at 8 pm another greeting message was sent to inform participants if they completed their goal and encourage them to complete future goals. Given that the same type of greeting message was delivered at each decision point during the study, the data cannot be used to answer questions about the conditions in which an individual is more receptive to one type of prompt (e.g., a greeting message from the coach) versus another (e.g., a motivational message from the coach). In other words, the data can only be used to make conclusions about receptivity to a specific type of just-in-time intervention (here, a greeting message from the coach) rather than to systematically compare receptivity to different types of just-in-time interventions.

An important assumption made in observational studies concerning receptivity is that the conditions in which a specific just-in-time intervention is needed are known. For instance, in Example 2 it is assumed that at 8 pm a greeting message is needed to initiate feedback about goal achievement. The assumption (based on existing evidence and practical considerations) is that delivering this prompt at 8 pm will improve proximal physical activity (e.g., in the next day) if the individual engages with the greeting message from the chatbot. However, suppose that existing evidence and practical considerations are insufficient to decide whether delivering this message at 8 pm will improve proximal physical activity. For example, investigators might be concerned that even if individuals engage with the 8 pm message, the salience of the feedback provided at that time may dissipate quickly and hence may not drive increases in physical activity on the next day. Since the greeting message was always delivered at 8 pm, data from Example 2 cannot be used to answer questions about the effect of delivering versus not delivering the message at 8 pm. Specifically, the data cannot be used to investigate whether under hypothesized states of need for feedback (e.g., at the end of the day) it is beneficial (e.g., in terms of increasing next day physical activity) to deliver (vs. not deliver) a chatbot-based greeting prompt to initiate feedback on goal attainment. Answering this question, which concern the benefits of delivering versus not delivering a specific just-in-time intervention in terms of the proximal outcome requires a different study design.

7.7 Analyzing ILD from microrandomized trials

The MRT is an experimental design that can be used to answer a suite of scientific questions about the construction of a JITAI. An MRT involves rapid sequential randomizations: each individual is randomized to different just-in-time intervention options at each of many decision points over the course of the study. That is, in an MRT, participants are sequentially randomized to different intervention options hundreds or even thousands of times. Consider the following MRT which is intended to assist in the development of a JITAI to engage smokers, who are attempting to quit, in self-regulatory activities.

Example 3. MRT to inform the development of a JITAI for smokers attempting to quit. The mobile assistance for regulating smoking (MARS) (Nahum-Shani et al., 2021) study involves assessing smokers attempting to quit from 3 days prior to their quit date through 7 days postquit using ecological momentary assessments (EMA), a suite of wireless sensors (MotionSense HRV) (Holtyn et al., 2019), and global positioning system (GPS). At the beginning of the study (during the onboarding session), participants are encouraged to use the MARS mobile app which delivers a suite of brief cognitive and/or behavioral strategies, as well as more effortful cognitive or mindfulness-based exercises. Participants also receive brief training in using these self-regulatory strategies. Participants are then asked six times per day via a smartphone to respond to a brief 2-question survey about the presence/absence of other smokers and negative affect. Immediately following survey completion (or 5 minutes following the delivery of the survey prompt if the survey was not completed), the participant is microrandomized to either: (a) low-effort prompt—a prompt recommending brief cognitive and/or behavioral strategies (tailored to the participant's response to the 2-question survey; for example, if other smokers are present and the individual is experiencing negative affect the message might be "Leave the situation until the temptation to smoke has passed or call a supportive friend. Talking about what is stressful can help you feel better"); (b) high-effort prompt—a prompt recommending relatively more effortful (5–10 minutes) self-regulatory strategies (cognitive or mindfulness-based exercises on the mobile device, such as "Mood Surf" which focuses on how to manage stressful thoughts and emotions by "riding the waves" as they fluctuate in intensity); or (c) no prompt. Approximately 1 hour following each microrandomization, participants are asked to complete an EMA (i.e., six EMAs per day). These EMAs assess engagement with self-regulatory strategies (primary proximal outcome), as well as gather information about affect, smoking, cognitions, and context.

The goal of this MRT is to answer questions about receptivity to a prompt recommending self-regulatory strategies. The primary question is whether delivering (vs. not delivering) a prompt recommending self-regulatory strategies increases engagement in self-regulatory strategies (self-reported) in the next hour. A secondary question is whether recommending less (vs. more) effortful strategies is more beneficial in terms of promoting engagement in self-regulatory strategies in the next hour. Exploratory aims focus on understanding the constellation of dynamic (e.g., location, time of day, mood) and static (e.g., baseline need for cognition, personal health literacy and self-efficacy) conditions in which delivering a prompt would lead to proximal engagement. These questions focus on proximal engagement in a particular just-in-time intervention—a prompt recommending self-regulatory strategies; that is, these questions concern receptivity to specific just-in-time interventions.

The randomizations in an MRT enable investigators to estimate causal effects. As in any experimental design, randomization is used to reduce the number of alternative explanations for the intervention effect by enhancing balance in the distribution of unobserved variables across groups assigned to different intervention options. In the MARS MRT the randomizations enhance balance in the distribution of unobserved variables between decision points assigned to low-effort prompt, high-effort prompt, and no prompt. Hence, the results can be used to answer causal questions concerning whether the prompts have the desired effect on proximal engagement and whether this effect varies depending on static and dynamic conditions.

The analysis of data from an MRT involves a generalization of regression analysis that ensures unbiased estimation of causal effects of time-varying intervention prompts (Boruvka, Almirall, Witkiewitz, & Murphy, 2018; Liao, Klasnja, Tewari, & Murphy, 2016). These analyses pool time-varying, longitudinal data across all study participants. For example, consider the primary question motivating the MARS MRT which concerns the comparison of delivering (vs. not delivering) a prompt recommending self-regulatory strategies in terms of self-reported engagement in self-regulatory strategies in the next hour. Let t_1 denote time points at which a self-regulatory prompt may or may not be delivered; t_1 ranges from 1 to 60 (10 days $\times$ six times per day). The proximal outcome is binary (engaged or did not at t_2). The prompt effects on the binary outcome can be tested via a log-linear type model (Qian, Yoo, Klasnja, Almirall, & Murphy, 2021; Rabbi et al., 2017). The causal effect of a prompt recommending self-regulatory strategies can be expressed on the (log) "risk-ratio" scale, namely on a scale that measures the probability ("risk") of proximal engagement at t_2, when a prompt was delivered at t_1, divided by the probability of proximal engagement at t_2 when a prompt was not delivered at t_1. The risk-ratio will be greater than 1 if delivering (vs. not delivering) a self-regulatory prompt has a causal effect on the probability of engagement in self-regulatory strategies in the next hour.

This model can be extended to answer questions about the conditions in which individuals are receptive to a prompt recommending self-regulatory strategies. This can be done by adding candidate moderators—baseline and time-varying covariates based on which the effect of delivering (vs. not delivering) a prompt is hypothesized to vary (see Qian et al., 2021) for details. Importantly, while the observational study in Example 2 is also motivated to answer questions about the conditions in which individuals are receptive to a just-in-time intervention (a greeting message from the coach), in Example 3 these questions are operationalized as causal questions. Specifically, in the current example the goal is to investigate the conditions in which delivering (vs. not delivering) a prompt recommending self-regulatory strategies causes proximal engagement in self-regulatory strategies. Moreover, since the randomized intervention options in Example 3 included two different types of prompts, the model described above can be extended to investigate the causal effect of delivering a low-effort (vs. high effort) prompt on proximal engagement as well as candidate moderators of this causal effect. This will enable scientists to answer questions about receptivity to one type of prompt versus another.

Data from this MRT can also be used to answer additional questions about receptivity and vulnerability. For example, investigators may be interested in understanding whether smokers attempting to quit are receptive to a prompt recommending self-regulatory activities when they are vulnerable to a smoking episode. Suppose that based on prior empirical evidence investigators consider self-reported negative affect or the presence of other smokers as states of vulnerability for proximal smoking. Data from the MRT in Example 3 can be used to investigate whether the causal effect of prompting with self-regulatory strategies (vs. not prompting) at time t on engagement in self-regulatory strategies at time $t + 1$ varies depending on negative affect or presence of other smokers self-reported at time t (via the 2-question survey). Investigators may also be interested in understanding whether delivering a prompt recommending self-regulatory strategies is beneficial in preventing a proximal smoking episode when smokers attempting to quit are in a state of vulnerability. This question can be answered with data from Example 3 by comparing the self-regulatory prompts versus no prompt at time t in terms of the probability of experiencing a smoking episode in the next 2 hours (time $t + 1$), at decision points t in which participants self-reported negative affect or presence of other smokers.

Finally, consider the question motivating Example 1 which concerns identifying the static and dynamic conditions in which smokers attempting to quit are vulnerable to proximal smoking. This question can also be answered with data from the MRT in Example 3 but the interpretation of the results would differ. Specifically, the same prediction methods described in Example 1 can be used with data from the MRT in Example 3 to investigate the combination of dynamic and static factors measured by time t that predict a subsequent smoking episode at time $t + 1$. A rule for using these predictive factors can then be developed with the goal to accurately differentiate between those experiencing versus not experiencing a smoking episode at time $t + 1$. Note that in Example 3 these factors may also include the history of prompts delivered up to and including time t. Most importantly, while the study in Example 1 did not include a mobile intervention, Example 3 included mobile-based prompts to engage individuals in self-regulatory strategies. Hence, employing predictive methods similar to

Example 1 with MRT data from Example 3 can help investigators answer the following question: in the context of a mobile intervention that prompts smokers attempting to quit to engage in self-regulatory strategies three times per day on average, what combination of static and dynamic factors available at time t predict a smoking episode at $t + 1$?

7.8 Conclusion

As in any research, scientific questions drive the study design and data analytics in building JITAIs. Scientific questions about vulnerability to an adverse proximal outcome and receptivity to just-in-time interventions motivate both observational ILD studies and MRT studies. In some cases, the questions motivating observational ILD studies (e.g., under what conditions individuals are vulnerable to a specific adverse proximal outcome? Under what constellation of dynamic and static conditions an individual is receptive to a specific just-in-time intervention?) may seem similar to those motivating MRT studies, but their operationalization may vary substantially (see Table 7.1). The way a given scientific question is operationalized has implications on the interpretation of the results, the conclusions made and the ability to reproduce the findings (Goodman, Fanelli, & Ioannidis, 2016). Hence, theory and practice relating to the construction of JITAIs can be greatly advanced by enhancing the specificity of scientific questions about vulnerability and receptivity motivating research studies.

While the current chapter focuses on empirically identifying states of vulnerability and receptivity to inform JITAIs, other challenges may be motivating investigators seeking to develop effective JITAIs. These challenges include how to promote participant engagement in the JITAI (e.g., in the intervention options delivered and/or in providing information about the tailoring variables) (Nahum-Shani et al., 2022a), how to optimize the integration between JITAI components that are delivered at relatively fast timescales (e.g., prompts recommending brief self-regulatory strategies that may be delivered every several hours) and components that are delivered at slower time scales (e.g., weekly coaching sessions), and how to move beyond JITAIs with fixed decision rules (i.e., those that specify whether and how to deliver an intervention to all individuals in the same observed context) to JITAIs that improve the decision rules as the individual interacts with the intervention, by tracking individuals' responsivity to various intervention options in different contexts (Qian et al., 2022). These challenges offer a fruitful avenue for developing new methodologies to inform the construction of effective JITAIs.

Acknowledgments

The authors would like to acknowledge support from National Institutes of Health grants: P50 DA054039, U01 CA229437, R01 DA039901, R01 CA224537, P41 EB028242, UG3 DE028723, R01 MD010362, R01 CA190329, and from the Patient-Centered Outcomes Research Institute grant PCS-2017C2-7613.

References

Ali, A., Hossain, M., Hovsepian, K., Rahman, M., al'Abisi, M., & Ertin, E. (2012). Automated detection of smoking in the mobile environment from respiration measurements. In: Proceedings of the ACM ISPN (pp. 269–280).

Bolger, N., & Laurenceau, J.-P. (2013). Intensive longitudinal methods: An introduction to diary and experience sampling research. New York, NY: Guilford Press.

Boruvka, A., Almirall, D., Witkiewitz, K., & Murphy, S. A. (2018). Assessing time-varying causal effect moderation in mobile health. *Journal of the American Statistical Association*, 1112–1121 accepted for publication.

Businelle, M. S., Ma, P., Kendzor, D. E., Frank, S. G., Vidrine, D. J., & Wetter, D. W. (2016). An ecological momentary intervention for smoking cessation: Evaluation of feasibility and effectiveness. *Journal of Medical Internet Research, 18*(12), e321.

Chatterjee, S., Moreno, A., Lizotte, S. L., Akther, S., Ertin, E., Fagundes, C. P., et al. (2020). SmokingOpp: Detecting the smoking 'opportunity' context using mobile sensors. *Proceedings of the ACM on Interactive, Mobile, Wearable and Ubiquitous Technologies, 4*(1), 1–26.

Coughlin, L. N., Nahum-Shani, I., Philyaw-Kotov, M. L., Bonar, E. E., Rabbi, M., Klasnja, P., et al. (2021). Developing an adaptive mobile intervention to address risky substance use among adolescents and emerging adults: Usability study. *JMIR mHealth and uHealth, 9*(1), e24424.

Dempsey, W. H., Moreno, A., Scott, C. K., Dennis, M. L., Gustafson, D. H., Murphy, S. A., et al. (2017). isurvive: An interpretable, event-time prediction model for mHealth. In: International conference on machine learning.

Dillingham, R., Ingersoll, K., Flickinger, T. E., Waldman, A. L., Grabowski, M., Laurence, C., et al. (2018). PositiveLinks: A mobile health intervention for retention in HIV care and clinical outcomes with 12-month follow-up. *AIDS Patient Care and STDs, 32*(6), 241–250.

Ertin, E., Stohs, N., Kumar, S., Raij, A., al'Absi, M., Kwon, T., et al. (2011). AutoSense: Unobtrusively wearable sensor suite for inferencing of onset, causality, and consequences of stress in the field. In: Proceedings of the 9th ACM conference on embedded networked sensor systems (SenSys).

Fiske, S. T., & Taylor, S. E. (1991). Social cognition. New York: Mcgraw-Hill Book Company.

Goodman, S. N., Fanelli, D., & Ioannidis, J. P. (2016). What does research reproducibility mean? *Science Translational Medicine, 8*(341) 341ps312-341ps312.

Gottfredson, D. M. (1987). Prediction and classification in criminal justice decision making. *Crime and Justice, 9,* 1–20.

Guarino, H., Acosta, M., Marsch, L. A., Xie, H., & Aponte-Melendez, Y. (2016). A mixed-methods evaluation of the feasibility, acceptability, and preliminary efficacy of a mobile intervention for methadone maintenance clients. *Psychology of Addictive Behaviors, 30*(1), 1.

Gustafson, D. H., McTavish, F. M., Chih, M.-Y., Atwood, A. K., Johnson, R. A., Boyle, M. G., et al. (2014). A smartphone application to support recovery from alcoholism: A randomized clinical trial. *JAMA Psychiatry, 71*(5), 566–572.

Hébert, E. T., Ra, C. K., Alexander, A. C., Helt, A., Moisiuc, R., Kendzor, D. E., et al. (2020). A mobile just-in-time adaptive intervention for smoking cessation: Pilot randomized controlled trial. *Journal of Medical Internet Research, 22*(3), e16907.

Hébert, E. T., Stevens, E. M., Frank, S. G., Kendzor, D. E., Wetter, D. W., Zvolensky, M. J., et al. (2018). An ecological momentary intervention for smoking cessation: The associations of just-in-time, tailored messages with lapse risk factors. *Addictive Behaviors, 78*, 30–35. https://doi.org/10.1016/j.addbeh.2017.10.026.

Holtyn, A. F., Bosworth, E., Marsch, L. A., McLeman, B., Meier, A., Saunders, E. C., et al. (2019). Towards detecting cocaine use using smartwatches in the NIDA clinical trials network: Design, rationale, and methodology. *Contemporary Clinical Trials Communications, 15*, 100392.

Hossain, S. M., Hnat, T., Saleheen, N., Nasrin, N., Noor, J., Ho, B.-J., et al. (2017). mCerebrum: An mHealth software platform for development and validation of digital biomarkers and interventions. *ACM SenSys*.

Kizakevich, P. N., Eckhoff, R., Brown, J., Tueller, S. J., Weimer, B., Bell, S., et al. (2018). PHIT for duty, a mobile application for stress reduction, sleep improvement, and alcohol moderation. *Military Medicine, 183*(suppl_1), 353–363.

Künzler, F., Mishra, V., Kramer, J. N., Kotz, D., Fleisch, E., & Kowatsch, T. (2019). Exploring the state-of-receptivity for mHealth interventions. *Proceedings of the ACM on Interactive, Mobile, Wearable and Ubiquitous Technologies, 3*(4), 1–27. https://doi.org/10.1145/3369805.

Liao, P., Klasnja, P., Tewari, A., & Murphy, S. A. (2016). Sample size calculations for micro-randomized trials in mHealth. *Statistics in Medicine, 35*(12), 1944–1971.

Marciniak, M. A., Shanahan, L., Rohde, J., Schulz, A., Wackerhagen, C., Kobylińska, D., et al. (2020). Standalone smartphone cognitive behavioral therapy–based ecological momentary interventions to increase mental health: Narrative review. *JMIR mHealth and uHealth, 8*(11), e19836.

Nahum-Shani, I., Dziak, J. J., & Wetter, D. W. (2022a). MCMTC: A pragmatic framework for selecting an experimental design to inform the development of digital interventions. *Frontiers in Digital Health, 4*.

Nahum-Shani, I., Shaw, S. D., Carpenter, S. M., Murphy, S. A., & Yoon, C. (2022b). Engagement in digital interventions. *American Psychologist* in press, Advance online publication. https://doi.org/10.1037/amp0000983.

Nahum-Shani, I., Hekler, E., & Spruijt-Metz, D. (2015). Building health behavior models to guide the development of just-in-time adaptive interventions: A pragmatic framework. *Health Psychology, 34*(Suppl.), 1209–1219.

Nahum-Shani, I., Potter, L. N., Lam, C. Y., Yap, J., Moreno, A., Stoffel, R., et al. (2021). The mobile assistance for regulating smoking (MARS) micro-randomized trial design protocol. *Contemporary Clinical Trials, 110*, 106513.

Nahum-Shani, I., Smith, S. N., Spring, B. J., Collins, L. M., Witkiewitz, K., Tewari, A., et al. (2018). Just-in-time adaptive interventions (JITAIs) in mobile health: Key components and design principles for ongoing health behavior support. *Annals of Behavioral Medicine, 52*, 446–462. https://doi.org/10.1007/s12160-016-9830-8.

Naughton, F., Hopewell, S., Lathia, N., Schalbroeck, R., Brown, C., Mascolo, C., et al. (2016). A context-sensing mobile phone app (Q sense) for smoking cessation: A mixed-methods study. *JMIR mHealth and uHealth, 4*(3), e106.

Plarre, K., Raij, K., Hossain, M., Ali, A. A., Nakajima, M., alAbisi, M., et al. (2011). Continuous interference of psychological stress from sensory measurements collected in the natural enviornment. In: 10th international conference on information processing in sensor networks (IPSN).

Qian, T., Walton, A. E., Collins, L. M., Klasnja, P., Lanza, S. T., Nahum-Shani, I., et al. (2022). The microrandomized trial for developing digital interventions: Experimental design and data analysis considerations. *Psychological Methods*.

Qian, T., Yoo, H., Klasnja, P., Almirall, D., & Murphy, S. A. (2021). Estimating time-varying causal excursion effects in mobile health with binary outcomes. *Biometrika, 108*(3), 507–527. https://doi.org/10.1093/biomet/asaa070.

Rabbi, M., Philyaw-Kotov, M., Klasnja, P., Bonar, E., Nahum-Shani, I., Walton, M., et al. (2017). SARA – Substance abuse research assistant. https://doi.org/10.17605/OSF.IO/VWZMD.

Wang, L., & Miller, L. C. (2020). Just-in-the-moment adaptive interventions (JITAI): A meta-analytical review. *Health Communication, 35*(12), 1531–1544.

Witkiewitz, K., Desai, S. A., Bowen, S., Leigh, B. C., Kirouac, M., & Larimer, M. E. (2014). Development and evaluation of a mobile intervention for heavy drinking and smoking among college students. *Psychology of Addictive Behaviors, 28*(3), 639.

Chapter 8

A digital therapeutic alliance in digital mental health

Benjamin Kaveladze[a] and Stephen M. Schueller[a,b]

[a] Department of Psychological Science, University of California, Irvine, Irvine, CA, USA, [b] Department of Informatics, University of California, Irvine, Irvine, CA, USA

8.1 Introduction

Digital mental health interventions (DMHIs) have proliferated in the smartphone age. These tools hold the potential to provide people across diverse populations with accessible and innovative forms of support for a wide range of mental health issues. Many studies have found that DMHIs are effective (Ebert, Harrer, Apolinário-Hagen, & Baumeister, 2019), and in some cases may even be as effective as traditional in-person mental health interventions (Karyotaki et al., 2018); however, poor adherence is a major barrier to DMHI impact in real-world settings. Thus, increasing user engagement may considerably improve these interventions' overall utility.

An important factor predicting engagement in traditional face-to-face psychotherapy is the relationship between a therapist and a client, also known as the therapeutic alliance (TA). However, in DMHIs, the role of a human care provider is often modified or replaced by technological elements of the intervention (e.g., didactic videos, self-guided exercises, or even chatbots designed to mimic human providers). Some researchers have argued that this overlap between human and technological support suggests that users develop an overall feeling of alliance with the DMHI, rather than only with a human care provider (Henson, Peck, & Torous, 2019). These interconnected human and technological features have also informed conceptualizations of such interventions as technology-enabled services, rather than products, reflecting their alignment with the goals and strategies of in-person mental health services (Mohr, Weingardt, Reddy, & Schueller, 2017).

The term digital therapeutic alliance (DTA) reflects an attempt to transfer concepts from traditional person-to-person therapeutic alliance to the alliance a user might perceive with a DMHI. Here, we discuss how DTA may be similar to and distinct from traditional TA, how DTA might impact intervention outcomes, and how DTA could evolve in the future as technology and DMHIs become more advanced with further functionalities. The goal of this chapter is to advance definitional concepts and suggest areas for future research that could improve an understanding of DTA and maximize its impact.

8.2 What is therapeutic alliance?

The relationship between an intervention provider and an intervention consumer has been referred to by many terms, including the therapeutic alliance, therapeutic relationship, working alliance, and helping alliance. According to Bordin's pantheoretical definition, TA is composed of (1) the bond between client and therapist, (2) agreement on the tasks directed toward improvement, and (3) agreement on therapeutic goals (Bordin, 1979). Although TA is widely considered to be an important element of therapy, it is inherently subjective and difficult to quantify using objective behavioral measures. A few characteristics of therapists who tend to form strong alliances have been identified, such as displaying genuineness and empathy in their interactions with clients (Nienhuis et al., 2018); yet, many of the other characteristics that may predict therapists' effectiveness at forming alliances are poorly understood (Baldwin, Wampold, & Imel, 2007).

TA's role in impacting intervention outcomes has received significant attention in research on in-person psychotherapy. TA has long been considered a necessary component of change (Tremain, McEnery, Fletcher, & Murray, 2020), and meta-analyses have demonstrated its moderate and reliable effect on clinical outcomes from therapy (Del Re, Flückiger, Horvath, & Wampold, 2021; Flückiger, Del Re, Wampold, & Horvath, 2018). TA's effect on psychotherapy outcomes has been suggested to be stronger than therapists' adherence to psychotherapy techniques or even their competency in delivering

Digital Therapeutics for Mental Health and Addiction: The State of the Science and Vision for the Future. DOI: https://doi.org/10.1016/B978-0-323-90045-4.00009-5

psychotherapy, and estimates suggest that TA can explain roughly 7–15% of the variance in psychotherapy outcomes (Wampold, Baldwin, Holtforth, & Imel, 2017). However, some recent work has argued that TA is considerably less predictive of outcomes in cognitive-behavioral therapy than previously thought, observing that alliance was largely determined by patients rather than by therapists (Whelen, Murphy, & Strunk, 2021). As such, although TA is generally accepted as helpful, further work might help better characterize why it is helpful and identify situational and person-level factors that could moderate its impact on clinical outcomes.

8.3 Applying therapeutic alliance to DMHIs: digital therapeutic alliance

Digital mental health interventions (DMHIs) differ from traditional interventions in that they leverage technological elements to provide an intervention or support intervention delivery. Although some technologies might help therapists to deliver the same treatments they would do in person, as in telephones or video conferencing software that enable telepsychology, we use the term DMHI to refer specifically to technology-based interventions that provide or improve treatment, such as mobile apps, wearable devices, video games, or virtual reality (Mohr, Burns, Schueller, Clarke, & Klinkman, 2013). In telepsychology, although it might be more challenging to establish a strong TA through a digital medium, the assessment and meaning of TA remains largely the same as in person because the intervention approach and therapist–client interactions are essentially the same as they would be in person. However DMHIs' unique features present new considerations and open questions about what TA might mean and how it might be measured.

One of the most commonly discussed potential downsides of DMHIs is that they might make it more difficult to develop a TA, compared to in-person interventions. This argument is especially relevant to DMHIs that are entirely self-guided, but it can also apply to those that include human support if the DMHI's technical limitations could interfere with direct interactions between therapist and client. On the other hand, DMHIs may also provide new ways to build an alliance with one's therapist. For example, a DMHI might enable a user to share behavioral and physiological data from their smartwatch with their therapist, or allow a therapist to provide just-in-time support via a video call or text message.

More controversially, DMHIs might enable clients to develop a DTA with solely the technological elements of a digital platform. Examples of this could be a client feeling that an entirely self-guided DMHI is helpful and trustworthy, or believing that a chatbot or other artificial intelligence within a DMHI understands them or even cares for them. For instance, researchers used the working alliance inventory to explore the TA formed with a chatbot developed for pain self-management ("SELMA"). Results demonstrated that participants experienced improvements in TA after interacting with SELMA; in addition, qualitative feedback highlighted their appreciation of the reliable and empathic aspects of the interaction and a perceived sense of closeness with SELMA (Hauser-Ulrich, Künzli, Meier-Peterhans, & Kowatsch, 2020). Another study of an automated text-messaging adjunct to in-person group cognitive-behavioral therapy found that clients noted varying reactions to receiving these text messages (Aguilera & Berridge, 2014). Noteworthy, Spanish-speaking clients primarily reported feeling supported by the messages, whereas English-speaking clients primarily reported the messages triggered introspection and self-awareness. As such, cultural differences between clients might impact one's perception of a relationship with a digital toola . Taken together, these studies illustrate how clients experience feelings of connection in interactions with purely digital agents, whether or not those interactions exist in the context of a traditional therapeutic relationship. Such examples demonstrate the need to better define and understand DTA, as well as to outline its similarities and differences from traditional conceptualizations of TA.

8.4 Defining digital therapeutic alliance

The term DTA reflects an attempt to apply the concept of TA to DMHIs. Given the implicitly subjective nature of TA and the wide variety of DMHIs, it is essential to clearly define DTA. However, it is also worth considering how closely the concept of DTA can be mapped to TA. Although defining DTA as closely as possible to TA holds the appeal of enabling comparison and generalization of findings across domains, it is possible that such a definition would reflect unnecessarily skeuomorphic thinking in DMHIs (Schueller, Muñoz, & Mohr, 2013). A skeuomorph is a remnant in a new modality that mimics a previous modality, due to imitation rather than necessity. This challenge of integrating ideas from TA into a context with new technological affordances might be one reason that there is considerable heterogeneity in definitions of DTA across studies (Henson et al., 2019; Tremain et al., 2020). The central difference across these definitions is the extent to which they position a user's alliance as being with the human supporter in a DMHI versus with the DMHI as a whole, including both human and technological components. We build on these previous definitions to put forward a novel definition of DTA.

Following Bordin's (1979) definition of therapeutic alliance and the model of supportive accountability from Mohr, Cuijpers, and Lehman (2011), we define DTA as a user-perceived alliance (composed of a bond, agreement on the tasks directed toward improvement, and agreement on therapeutic goals) with a DMHI, including both human-supported and technological components of the intervention.

The main strength of including technological components of DMHIs in our definition of DTA is that technological components are important factors in DMHIs' overall appeal and effectiveness. Our definition posits that users experience a broad sense of goal-oriented alliance with a DMHI that includes their bond with human-supported (e.g., a supporter or mental health professional, if there is one) and technological components of a DMHI (e.g., computerized elements of the DMHI, artificial intelligence chatbots, and an app's user interface). While these two forms of DTA are distinct, they can also be complementary, as in a human supporter teaching a user how to best interact with technological components of an intervention, or a well-designed video call user experience making remote interactions with a human supporter more pleasant. These synergies between human and technological components can maximize overall DTA.

A challenge of including technological elements in our definition of DTA is that doing so makes DTA less conceptually intuitive. Including these elements as parts of the alliance begs the question of whether a perceived alliance with an automated system is qualitatively comparable to a genuine alliance with another human. Some might argue that the term "therapeutic alliance" should be restricted to an alliance between two conscious agents who are truly capable of understanding and bonding with each other. Others might only extend the definition to nonhuman computer agents that seem human (i.e., create a convincing illusion of consciousness), with which a user perceives a genuine relationship (as we discuss later in the chapter, people demonstrate a remarkable tendency to treat technological tools similarly to social actors (Nass, Moon, Morkes, Kim, & Fogg, 1997; Wang, 2017). Both of these groups of critics might view our definition of DTA as overly general.

We believe that our definition's broadness is necessary to keep the definition of DTA open to new kinds of alliance that DMHIs could enable, beyond that of TA in traditional therapy settings. However, future work will be necessary to determine how DTA may differ from and overlap with similar constructs like TA, engagement, and general satisfaction with a digital tool. Future evidence may also reveal that the kinds of alliance that users form in face-to-face therapy are qualitatively and functionally different from alliance with DMHIs. In such a case, our definition of DTA would need to be revised.

8.5 Differences in DTA across interventions

A major challenge to creating a useful definition of DTA is that such a definition must apply consistently across the diverse range of DMHIs that are presently available. The difference across interventions that poses the clearest challenge to a consistent definition of DTA is their involvement of human support: DMHIs can involve either an entirely self-guided experience with no human contact, an artificial intelligence virtual therapist, weekly interactions with a trained coach, interactions over text with a licensed therapist, regular video-calls with a clinical psychologist, or some combination of the preceding elements (Hermes, Lyon, Schueller, & Glass, 2019). Thus, in some DMHIs, technology is adjunct to a human-led intervention (as in "blended care" models), and in other DMHIs a human supporter is adjunct to a technology-driven intervention (as in "guided" or "supported" DMHIs). These different types of DMHIs hold different issues with relation to DTA. Fig. 8.1 demonstrates the role that technology plays as a bridge between consumers and human supporters, and the different roles that supporters might play across different kinds of DMHIs. Consideration of both the human and technological elements in DMHIs is important, yet tricky, as the respective use of these elements continues to evolve.

We noted earlier that our definition of DMHIs excludes interventions that only use technology as a means of providing the same intervention as would be used in-person, such as remotely delivered individual or group therapy using telephones or webcams. Still, human support is a central and impactful aspect of many DMHIs. The culmination of research demonstrating that the most efficacious DMHIs are those with human support has led researchers, commercial developers, and health systems to move increasingly to this model of care. On the commercial side, mergers between self-guided DMHIs and virtual care networks have created products that offer a continuum of care, leveraging human interactions in novel ways. It is hard to predict what the future of technology-enabled care might look like, although we speculate some below, but a conceptualization of DTA should be flexible to potential new models while still facilitating research and understanding.

Interventions also vary in their functionality and presentation, which in turn affect their appeal and accessibility. Because promoting consistent user engagement is one of the foremost challenges in digital health, design choices that improve users' engagement with DMHIs may also improve intervention outcomes. Variation in the intervention strategies that DMHIs employ is another important aspect of their functionality; the role of therapeutic alliance differs across intervention strategies — for example, interventions based on psychodynamic theory might tend to rely on different kinds of TA than those based on cognitive behavioral therapy. Researchers also need to be aware that these functional and presentation features are constantly changing based on technological developments and aesthetic trends.

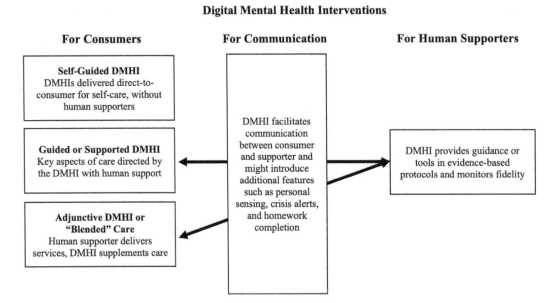

FIGURE 8.1 Digital mental health interventions differ in the ways that they facilitate communication between intervention consumers and supporters.

8.6 DTA in supported interventions

As noted, supported DMHIs are those that include human supporters. Some forms of human support in DMHIs are low-tech and low-cost, as in "remote hovering" interventions in which a social worker exchanges weekly texts with clients (Ben-Zeev, Kaiser, & Krzos, 2014). Others are highly technologically advanced and involve significant human expertise and effort, as in DMHIs that require professional clinicians to integrate passively collected smartphone data into their intervention strategies (Insel, 2017).

Some research has examined how different kinds of human support affect intervention outcomes. Supported DMHIs, on average, have higher engagement and are moderately more effective in achieving clinically relevant goals than unsupported DMHIs are (Ebert et al., 2018; Mohr et al., 2011). In addition, text-based interactions between a user and therapist can be sufficient to form a therapeutic alliance (van Lotringen et al., 2021). However, individuals differ in the kinds of human support they want or need from a DMHI. More research is needed to determine the optimal "form, intensity, and role" of human support in DMHIs for a given individual and across various populations (Baumeister, Reichler, Munzinger, & Lin, 2014; Mohr et al., 2011; Palmqvist, Carlbring, & Andersson, 2007; Tremain et al., 2020).

Some efforts have been taken to develop an evidence base of best practices for human support in DMHIs. Mohr et al. (2011) developed a model of supportive accountability to formalize the role of human support in digital apps. According to the model, motivation to adhere to DMHIs is improved by a user's reciprocal relationship with a trusted and likable coach. Further, the "Efficiency model of support" from Schueller, Tomasino, and Mohr (2017) views the human supporter's role as improving the "efficiency" of interventions, meaning supporters increase the ratio of an intervention's benefit to the resources required to provide it. In this model, the human supporter uses their expertise and understanding of a user's situation to allocate resources to help address failure points, such as usability, engagement, fit, knowledge, and implementation. Finally, Lattie et al. (2019) provide a useful example of manualizing and disseminating information on the role of human support in DMHIs, offering practical guidance for developing a coaching protocol for DMHIs, as well as evidence-based recommendations regarding coaches' training, techniques, and styles of interaction with users. Manualizing and disseminating this knowledge will be foundational for improving the quality of human support in DMHIs.

8.7 DTA in unsupported interventions

The idea that users can form TA with purely technological DMHIs is less intuitive than traditional TA between humans is. Yet, self-guided components of DMHIs may hold unique appeal and value for some users. Indeed, some users prefer self-guided DMHIs to human-supported ones because they perceive self-guided DMHIs as less judgmental and more convenient, affording them greater control over intervention content as well as when and where they use it

(Berry, Salter, Morris, James, & Bucci, 2018; Hillier, 2018). In addition, some people may be more comfortable revealing sensitive information to artificial intelligence virtual agents than to human supporters (DeVault et al., 2014; Lucas et al., 2017; Lucas, Gratch, King, & Morency, 2014; Weisband & Kiesler, 1996). Identifying the contexts in which unsupported intervention components could make users feel more comfortable or otherwise be preferable is an important goal.

Considerable research on human interactions with computer agents, smartphones, and DMHIs across multiple contexts has demonstrated that people often behave toward nonhuman computer tools in ways that resemble interpersonal interactions. Nass et al. (1997) found that humans interact with computer agents similarly to how they interact with other human social actors, observing that social responses to computers are shaped by cues that automatically and unconsciously prime social behavior and attributions. Further, they argued that designing computer programs that are perceived as social actors can be done with low overhead and without heavy-handed efforts such as photo-realistic representations. Building on this work, Wang (2017) found that lonely and interdependent people (who may have stronger drives for social interaction) are more likely to view their smartphones as anthropomorphic social actors. In addition to computerized agents, people form emotional bonds with their smartphones, feeling comforted by them and anxious without them (Hunter, Hooker, Rohleder, & Pressman, 2018).

The evidence on anthropomorphizing computer systems may suggest that people can perceive DTA with technological components of DMHIs. Several studies have illustrated that people use relationship concepts when describing their interactions with unsupported DMHIs. For example, the more that a DMHI is seen by users as personalized, interactive, friendly, and compassionate, the more users perceive a therapeutic alliance with it (Hillier, 2018). People also often use anthropomorphic terms to describe their relationships with DMHIs, such as "forming bonds" and "being open." However, the same people also tend to report that their relationships with technological intervention components are different from human relationships, as digital interventions are not able to provide tailored or responsive support or demonstrate human emotions like friendliness and collaboration (Berry et al., 2018). Also, a qualifier to these claims of anthropomorphism is that DHMIs and other computer interfaces are not self-formed; even self-guided interventions inherently incorporate expressions of ideas and style from their human creators, which is arguably a form of asynchronous interpersonal communication.

Like Berry et al. (2018), other studies have consistently found that DMHI users make a clear distinction between the kinds of relationships they might have with a human supporter and the kinds of relationships they might have with a DMHI as a whole. Despite the noted evidence that people often treat digital interfaces as social agents and are emotionally connected to their digital tools, most people report that the notion of forming a true "alliance" or "relationship" with a digital health app is unacceptable and strange to them (Hillier, 2018). Relatedly, D'Alfonso, Lederman, Bucci, and Berry (2020) argue that despite the reasonable assumption that people would prefer the "emotional intelligence and general anthropomorphic characteristics of a human therapist" in a nonhuman therapeutic agent, too much effort to create life-like agents could backfire due to the uncanny valley effect, by which people are made uneasy by nonhuman agents that appear almost, but not quite, human (Mori, MacDorman, & Kageki, 2012). People may also be hesitant to trust mental health-related advice that comes from nonhuman agents, as evidenced by work from Morris, Kouddous, Kshirsagar, and Schueller (2018) showing that people prefer support expressions when they believe a human wrote them compared to when they believe an artificial intelligence conversational agent wrote them.

Thus, a qualitative gap seems to exist between the connections people perceive with digital tools and those they perceive with other humans. For some, this gap might imply that DTA with unsupported DMHIs (or the technological components of supported DMHIs) is an unrealistic misnomer for what is actually a more general satisfaction with a tool. However, we argue that even if users do not think of their interactions with nonhuman digital agents as constituting a relationship, their behavior toward these tools is evidence of some kind of relationship. Yet, more work is necessary to understand exactly how DTA with technological DMHI components differs from DTA that one might experience with a human supporter.

Artificial intelligence-driven virtual conversational agents are another potential source of DTA that is rapidly becoming more relevant, and that could begin to narrow the qualitative gap that users perceive between technological and human-supported intervention components. The work we mentioned from Hauser-Ulrich et al. (2020) on SELMA and from Aguilera and Berridge (2014) on an automated text-messaging adjunct demonstrate instances of human connection to artificial intelligence-supported systems. In addition, Morris et al. (2018) developed an artificial intelligence conversational agent that produced personalized support expressions that users generally found helpful. Another popular conversational agent, Woebot, achieved user-rated therapeutic alliance ratings comparable to those of human therapists (Darcy, Daniels, Salinger, Wicks, & Robinson, 2021). Lastly, Lucas et al. (2017) and Lucas, Gratch, King, and Morency (2014) found that individuals self-disclose more information and are more likely to endorse psychiatric symptoms to a nonhuman virtual agent than to a digital avatar that they believe is being controlled by a human. Artificial intelligence conversational agents are not yet capable of engaging in genuine empathic interactions with humans; however, the future will likely bring increasingly intelligent agents that will better adapt to users' needs across cultural and personality factors.

Currently, users seem to experience a kind of DTA with at least some nonhuman components of DMHIs; users express genuine appreciation toward unsupported DMHIs and use relationship concepts to describe their interactions with DMHIs. The kinds of alliances people form with human therapists and those they form with technological components of DMHIs are distinct, yet closely connected and possibly synergistic. With advancements in artificial intelligence and evidence-based design making them more intelligent and dynamic, future DMHIs will no doubt go further in engaging with users, towards forming alliances that meet the criteria for TA proposed by Bordin (1979).

8.8 Measurement

Measuring DTA is a closely related challenge to defining it. While traditional TA is composed of fairly simple constructs (bond and agreement) between two people, DTA is more complex. Using established measures of therapeutic alliance developed for face-to-face contexts and simply replacing references to "therapist" with "the app" is insufficient (Berry et al., 2018). A measure of DTA must "fit the contours of human engagement with computers and digital health interventions" (D'Alfonso et al., 2020). Therefore, the main challenge in developing a reliable measure of DTA is identifying which elements of users' interactions with DMHIs might be sources of DTA.

Existing DTA measures tend to take a similarly liberal approach to ours, viewing DMHIs as composed of technological and human-supported components that are distinct yet interwoven into a single user experience. One of the most popular measures of DTA is the mobile version of the Agnew Relationship Measure (mARM) (Berry et al., 2018), which was adapted from the ARM (a well-known measure of TA) using qualitative interviews with digital health app users and mental health professionals. These revisions to the ARM primarily accounted for apps' inabilities to be sufficiently human-like for the questions to make sense. The Working Alliance Inventory for Guided Internet Intervention (WAI-I) is another example of a traditional TA measure (WAI) that was modified to study DTA in computerized interventions (Penedo et al., 2020). The WAI-I examines DTA with an online program as a whole as well as with the human supporter. Unfortunately, because these DTA measures were developed so recently, their utility in predicting user outcomes from DMHIs has yet to be evaluated.

The noted differences across DMHIs and users are reflected in the variation across measures used to evaluate DTA. For example, the WAI-I may fit certain kinds of DMHIs better than the mARM does because the scales were developed by teams focused on different DMHIs. Based on the evidence that people view their alliances with technological components and human-supported components of DMHIs as distinct from one another, a measure that aligns with our proposed definition of DTA should explicitly distinguish between these two sources of DTA, while still viewing them as contributing to a single overall impression of DTA. The WAI-I makes this distinction by asking separate questions about one's alliance with their human supporter and with the DMHI as a whole, but might be more precise by evaluating DTA with the technological components separately.

8.9 Impact

TA appears to be less predictive of clinically relevant outcomes in DMHIs than it is in in-person interventions (Tremain et al., 2020); however, it is still moderately predictive of clinical outcomes in DMHIs (Miloff et al., 2020; van Lotringen et al., 2021). This difference in impact between in-person interventions and DMHIs may be partially due to the lack of a valid way to measure DTA for the reasons noted in the above two paragraphs; however, it may also be due to the limits of DMHIs as a medium for forming effective therapeutic relationships. On the other hand, this might also be reflective of more recent findings that TA is not as predictive of clinical outcomes as once thought. Regardless of the impact of TA or DTA on clinical outcomes, communicating via technology in DMHIs may make it more difficult to develop alliance between a user and a human supporter. Even in DMHIs with video calls that are structured much like typical in-person therapy sessions, physical distance and technical constraints likely make the nature of TA different from typical face-to-face client–therapist interactions. Taken out of the in-person therapy environment in which they trained and developed experience, therapists may not form alliances as effectively as they normally would.

In addition to supporting clients' improvement , an important role of a therapist is to ensure that negative events are avoided or kept under control. In typical therapy settings, support providers must take responsibility for their influence on a client, both positive and negative. On the contrary, digital health services tend to avoid taking responsibility for negative effects that could result from their interventions, claiming that digital health apps are unable to provide the oversight over clients that an in-person therapist would have. Instead, they place this responsibility on DMHI users. Indeed, the barriers that digital interfaces place between intervention providers and consumers may make it more difficult for providers to notice negative effects emerging in the course of a DMHI as well as they would during in-person psychotherapy (Rozental et al., 2014). Yet, this decision to relinquish responsibility might make it harder for users to perceive an alliance with the intervention.

8.10 Improving DTA

Given DTA's novelty, there is a paucity of work aiming to improve DTA in an applied setting. However, with rapid advancements in DMHIs' technological capacities, opportunities to improve DTA are expanding. Tremain et al. (2020) suggested several ways that DTA could be improved in DMHIs with little or no human support from a persuasive design perspective. Here we note ways that those suggestions align with each component of Bordin's (1979) definition of TA (bond, agreement on the tasks directed toward improvement, and agreement on therapeutic goals). First, tailoring and personalizing interventions using digital phenotyping and machine learning can enhance users perceptions of bond with the intervention. Second, improving dialog with the intervention may foster information transfer from the user to the DMHI and back, cementing agreement on tasks and goals. Third, ensuring intervention credibility by using evidence-based content from expert sources might improve users confidence in the intervention, and thus agreement on goals. And fourth, building social support and connectedness into DMHIs by incorporating social forums or community newsfeeds could foster the formation of bonds. These recommendations reflect the multifaceted nature of DTA, combining technological, content-related, and community-building improvements.

The improvements noted above focused on the technological components of DMHIs, but DTA between a user and a human supporter can also be improved via DMHI design choices. For example, a DMHI might improve bond by allowing for text message check-ins outside of normal meeting times or by enabling a human supporter to provide just-in-time support via a video call in case a user is in distress. A DMHI could likewise improve agreement on tasks by providing behavioral and physiological data from a user's smartwatch to their human supporter to adapt treatments to the user's lifestyle. Finally, a DMHI might improve agreement on therapeutic goals by providing a more comfortable space for users who find opening up to a provider about personal issues easier in a digital setting than in person.

Another promising strategy for improving DTA with both the human and technological components of DMHIs is personal sensing. With personal sensing (also known as digital phenotyping), DMHIs track and adapt to user behavioral patterns from previous use of the DMHI or other smartphone activity (Mohr, Zhang, & Schueller, 2017). Using personal sensing to adapt to differing needs between users as well as within users across time and situations, DMHIs could better synchronize with users' lives (Clarke, 2016). Indeed, personalized activity recommendations based on information from previous app use were found to improve mood more than nonpersonalized recommendations (Rohani et al., 2020). As further evidence for the utility of personalization, in a systematic review and individual patient data network meta-analysis, Karyotaki et al. (2018) found that although human-guided DMHIs were more helpful than unguided DMHIs for people with moderate to severe depression, unguided DMHIs were similarly effective to guided ones among people with mild or subthreshold depression. Thus, personalizing treatment selection with the help of adaptive DMHIs is increasingly feasible. These personalizing technologies could be especially useful for interventions that do not involve frequent interactions with care providers, but even trained mental health professionals might benefit from integrating them. Despite the substantial opportunity that these tools offer, caution is necessary: the risk of negative consequences from inaccurate sensing via opaque or biased predictive algorithms is an important challenge (Chancellor & De Choudhury, 2020).

8.11 The future of DTA

At present, users generally see their DMHIs as mildly helpful and not particularly engaging, as evidenced by their low user retention. However, as future digital interventions become more technologically advanced, their capabilities to engage users and provide them value will grow. As noted, large-scale improvement in artificial intelligence will likely make DMHIs more engaging and effective in coming years. Eventually, even artificial general intelligence may be achieved, by which AI systems would be able to perform any intellectual task that humans can. Such advancements would lead to a society-level redefining of AI's role in our lives, as people become more open to the idea that nonhuman technological DMHI components can offer them meaningful support. If this occurs, researchers should be prepared for the possibility that users' perceptions of therapeutic alliance with DMHIs may rapidly change, possibly making DTA more closely resemble traditional TA.

Even without revolutionary improvements in artificial intelligence, incremental improvements in DMHIs may produce substantial benefits. As discussed in the above section, interventions are becoming increasingly adaptive to user-specific

Therapeutic Alliance	Digital Therapeutic Alliance Today	Digital Therapeutic Alliance in the Future
• Bond • Agreement on the tasks directed towards improvement • Agreement on therapeutic goals	• AI chatbots improve engagement, but users don't perceive alliance • Virtual reality interventions show impressive results in-lab • Personal sensing technology is used to monitor user behaviors • Some ability to adapt to users' needs based on user input • A variety of digital health interventions to choose from	• More intelligent AI chatbots, with potential for users to perceive genuine therapeutic alliance • Information from personal sensing shapes intervention content • Better evidence base on tasks that work well in a digital context • Better digital health interventions developed iteratively based on scientific evidence

FIGURE 8.2 The states of digital therapeutic alliance in traditional psychotherapy, current digital mental health interventions, and potential future digital mental health interventions.

needs. More adaptive interventions may also enable a truer alliance to develop, in the sense that both parties in the alliance can gain: while a user builds skills and confidence over time with help from the intervention, the intervention can also improve by "learning" from the user how to provide better support. We present a summary of ways that DTA is manifest in current DMHIs and how it may evolve in future DMHIs in Fig. 8.2.

8.12 Future directions and recommendations

DMHIs continue to improve in terms of their technological capabilities, intervention strategies, and integration with traditional healthcare systems. Yet, despite these advances, the overall impact of DMHIs on public health has been limited. To address this limitation we need to better conceptualize what DMHIs actually are — their mechanisms of action, how they fit into or circumvent traditional care pathways, and what they mean to the people who use them, including both clients and therapists. A clearer understanding of the DTA — what it is, how it impacts intervention outcomes, and how it might be improved — could help to address each of these gaps in knowledge about DMHIs . Such an understanding of DTA could help to design more effective and appealing DMHIs, to use them more strategically in care delivery, and to improve efforts of dissemination and implementation. Thus, we propose a broad research plan involving the field of digital health that could improve the understanding and potential application of DTA in DMHIs.

First, we propose that researchers in the field should converge on a working definition of DTA and a set of nuanced DTA measures. To be applicable to a wide range of interventions, these measures should delineate which aspects of DMHIs contribute to DTA. This definition and measure may also be updated over time based on new evidence. Multiple measures of DTA exist and we must be careful to avoid jingle–jangle fallacies. Jingle fallacies occur when the same word is used for two things that are actually different. It might be convenient to refer to the DTA as an "alliance" because it creates understanding for researchers and therapists, but this understanding might be inaccurate. Jangle fallacies occur when different words are used for two things that are actually the same. The field must also be careful to not create new terms (and new measures) for things that are already captured by existing terms and measures. Right now it is unclear what aspects of TA and DTA jingle and where there might be more instances of jangle. Our definition focuses on considering both the human and the technological elements and using traditional ideas from TA and supportive accountability, but additional work could help support this notion or demonstrate how DTA might truly be unique from TA.

Second, once an acceptable definition and measure of DTA have been established, the field should prioritize building evidence regarding DTA's impacts on DMHI outcomes. This should begin with empirically identifying what transfers from knowledge of TA to DTA. There is a good deal of research on TA's role across many treatment settings and intervention types, only some of which is relevant to the goal of designing more effective DMHIs. Some aspects of TA in in-person settings may have equivalent corresponding aspects in digital contexts while others may not, and elements of TA that are helpful in-person may be less helpful in DMHIs. To study these differences, we should design studies to rigorously identify the aspects of DMHIs that might affect DTA. Because DMHIs enable manipulation of specific intervention elements in ways that are less feasible in in-person intervention contexts, this empirical work could also provide new insights into the mechanisms by which TA functions in interventions.

Third, using this evidence base, researchers can work with DMHI designers and software engineers to develop and disseminate actionable design recommendations to improve DTA. These recommendations can also be manualized to help train mental health care professionals and human supporters to engage in best practices for fostering DTA between users and DMHIs. Such design recommendations might identify technological features that lead to feelings of connection and alignment on goals and tasks. For example, DMHIs that people perceive as helping them get what they want rather than giving them additional things to do. At this point, it is unclear if the characteristics and actions of therapists that increase TA could be mirrored in DMHIs to achieve the same increase in DTA, or if new technological affordances might create different characteristics and actions for the same goal. Nevertheless, such design choices could stem from a better science of DTA.

8.13 Conclusions

DTA will likely become increasingly important as DMHIs become more advanced. Current research has struggled to define and measure DTA in a way that is similar to traditional therapeutic alliance but also accounts for the constraints and novel opportunities of DMHIs. As a starting point for additional research, we propose that DTA should be defined as a user-perceived alliance (composed of a bond, agreement on the tasks directed toward improvement, and agreement on therapeutic goals) with a DMHI, including both human and technological components of the intervention. Based on this definition, we present some recommendations for researchers to build and disseminate a knowledge base regarding DTA, while also being prepared for considerable shifts in DTA's meaning as DMHIs evolve.

References

Aguilera, A., & Berridge, C. (2014). Qualitative feedback from a text messaging intervention for depression: benefits, drawbacks, and cultural differences. *JMIR mHealth and uHealth, 2*(4), e46. https://doi.org/10.2196/mhealth.3660.

Baldwin, S. A., Wampold, B. E., & Imel, Z. E. (2007). Untangling the alliance-outcome correlation: Exploring the relative importance of therapist and patient variability in the alliance. *Journal of Consulting and Clinical Psychology, 75*(6), 842–852. https://doi.org/10.1037/0022-006X.75.6.842.

Baumeister, H., Reichler, L., Munzinger, M., & Lin, J. (2014). The impact of guidance on internet-based mental health interventions — A systematic review. *Internet Interventions, 1*(4), 205–215. https://doi.org/10.1016/j.invent.2014.08.003.

Ben-Zeev, D., Kaiser, S. M., & Krzos, I. (2014). Remote "hovering" with individuals with psychotic disorders and substance use: Feasibility, engagement, and therapeutic alliance with a text-messaging mobile interventionist. *Journal of Dual Diagnosis, 10*(4), 197–203. https://doi.org/10.1080/15504263.2014.962336.

Berry, K., Salter, A., Morris, R., James, S., & Bucci, S. (2018). Assessing therapeutic alliance in the context of mHealth interventions for mental health problems: Development of the mobile Agnew Relationship Measure (mARM) questionnaire. *Journal of Medical Internet Research, 20*(4), e90. https://doi.org/10.2196/jmir.8252.

Bordin, E. S. (1979). The generalizability of the psychoanalytic concept of the working alliance. *Psychotherapy: Theory, Research & Practice, 16*(3), 252–260. https://doi.org/10.1037/h0085885.

Chancellor, S., & De Choudhury, M. (2020). Methods in predictive techniques for mental health status on social media: A critical review. *NPJ Digital Medicine, 3*(1), 1–11. https://doi.org/10.1038/s41746-020-0233-7.

Clarke, R. (2016). Big data, big risks. *Information Systems Journal, 26*(1), 77–90. https://doi.org/10.1111/isj.12088.

D'Alfonso, S., Lederman, R., Bucci, S., & Berry, K. (2020). The digital therapeutic alliance and human-computer interaction. *JMIR Mental Health, 7*(12). https://doi.org/10.2196/21895.

Darcy, A., Daniels, J., Salinger, D., Wicks, P., & Robinson, A. (2021). Evidence of human-level bonds established with a digital conversational agent: Cross-sectional, retrospective observational study. *JMIR Formative Research, 5*(5), e27868. https://doi.org/10.2196/27868.

Del Re, A. C., Flückiger, C., Horvath, A. O., & Wampold, B. E. (2021). Examining therapist effects in the alliance-outcome relationship: A multilevel meta-analysis. *Journal of Consulting and Clinical Psychology, 89*(5), 371–378. https://doi.org/10.1037/ccp0000637.

DeVault, D., Artstein, R., Benn, G., Dey, T., Fast, E., Gainer, A., et al. (2014). SimSensei Kiosk: A virtual human interviewer for healthcare decision support. In: *Proceedings of the 2014 International Conference on Autonomous Agents and Multi-agent Systems* (pp. 1061–1068).

Ebert, D. D., Harrer, M., Apolinário-Hagen, J., & Baumeister, H. (2019). Digital interventions for mental disorders: Key features, efficacy, and potential for artificial intelligence applications. *Frontiers in Psychiatry* (pp. 583–627). Singapore: Springer.

Ebert, D. D., Van Daele, T., Nordgreen, T., Karekla, M., Compare, A., Zarbo, C., et al. (2018). Internet- and mobile-based psychological interventions: Applications, efficacy, and potential for improving mental health: A report of the EFPA E-Health Taskforce. *European Psychologist, 23*(2), 167–187. https://doi.org/10.1027/1016-9040/a000318.

Flückiger, C., Del Re, A. C., Wampold, B. E., & Horvath, A. O. (2018). The alliance in adult psychotherapy: A meta-analytic synthesis. *Psychotherapy (Chicago, Illinois), 55*(4), 316–340. https://doi.org/10.1037/pst0000172.

Hauser-Ulrich, S., Künzli, H., Meier-Peterhans, D., & Kowatsch, T. (2020). A smartphone-based health care chatbot to promote self-management of chronic pain (SELMA): Pilot randomized controlled trial. *JMIR mHealth and uHealth, 8*(4), e15806. https://doi.org/10.2196/15806.

Henson, P., Peck, P., & Torous, J. (2019). Considering the therapeutic alliance in digital mental health interventions. *Harvard Review of Psychiatry, 27*(4), 268–273. https://doi.org/10.1097/HRP.0000000000000224.

Hermes, E. D., Lyon, A. R., Schueller, S. M., & Glass, J. E. (2019). Measuring the implementation of behavioral intervention technologies: Recharacterization of established outcomes. *Journal of Medical Internet Research, 21*(1), e11752. https://doi.org/10.2196/11752.

Hillier, L. (2018). Exploring the nature of the therapeutic alliance in technology-based interventions for mental health problems. *Lancaster EPrints.* https://eprints.lancs.ac.uk/id/eprint/127663/1/2018hilliermphil.pdf.

Hunter, J. F., Hooker, E. D., Rohleder, N., & Pressman, S. D. (2018). The use of smartphones as a digital security blanket: The influence of phone use and availability on psychological and physiological responses to social exclusion. *Psychosomatic Medicine, 80*(4), 345–352. https://doi.org/10.1097/PSY.0000000000000568.

Insel, T. R. (2017). Digital phenotyping: Technology for a new science of behavior. *JAMA, 318*(13), 1215–1216. https://doi.org/10.1001/jama.2017.11295.

Karyotaki, E., Ebert, D. D., Donkin, L., Riper, H., Twisk, J., Burger, S., et al. (2018). Do guided internet-based interventions result in clinically relevant changes for patients with depression? An individual participant data meta-analysis. *Clinical Psychology Review, 63*, 80–92. https://doi.org/10.1016/j.cpr.2018.06.007.

Lattie, E. G., Adkins, E. C., Winquist, N., Stiles-Shields, C., Wafford, Q. E., & Graham, A. K. (2019). Digital mental health interventions for depression, anxiety, and enhancement of psychological well-being among college students: Systematic review. *Journal of Medical Internet Research, 21*(7), e12869. https://doi.org/10.2196/12869.

Lucas, G. M., Gratch, J., King, A., & Morency, L.-P. (2014). It's only a computer: Virtual humans increase willingness to disclose. *Computers in Human Behavior, 37*, 94–100. https://doi.org/10.1016/j.chb.2014.04.043.

Lucas, G. M., Rizzo, A., Gratch, J., Scherer, S., Stratou, G., Boberg, J., et al. (2017). Reporting mental health symptoms: Breaking down barriers to care with virtual human interviewers. *Frontiers in Robotics and AI, 4*, 51. https://doi.org/10.3389/frobt.2017.00051.

Miloff, A., Carlbring, P., Hamilton, W., Andersson, G., Reuterskiöld, L., & Lindner, P. (2020). Measuring alliance toward embodied virtual therapists in the era of automated treatments with the virtual therapist alliance scale (VTAS): Development and psychometric evaluation. *Journal of Medical Internet Research, 22*(3), e16660. https://doi.org/10.2196/16660.

Mohr, D. C., Burns, M. N., Schueller, S. M., Clarke, G., & Klinkman, M. (2013). Behavioral intervention technologies: Evidence review and recommendations for future research in mental health. *General Hospital Psychiatry, 35*(4), 332–338. https://doi.org/10.1016/j.genhosppsych.2013.03.008.

Mohr, D. C., Cuijpers, P., & Lehman, K. (2011). Supportive accountability: A model for providing human support to enhance adherence to eHealth interventions. *Journal of Medical Internet Research, 13*(1), e30. https://doi.org/10.2196/jmir.1602.

Mohr, D. C., Weingardt, K. R., Reddy, M., & Schueller, S. M. (2017). Three problems with current digital mental health research. . . and three things we can do about them. *Psychiatric Services (Washington, D.C.), 68*(5), 427–429. https://doi.org/10.1176/appi.ps.201600541.

Mohr, D. C., Zhang, M., & Schueller, S. M. (2017). Personal sensing: Understanding mental health using ubiquitous sensors and machine learning. *Annual Review of Clinical Psychology, 13*(1), 23–47. https://doi.org/10.1146/annurev-clinpsy-032816-044949.

Mori, M., MacDorman, K. F., & Kageki, N. (2012). The uncanny valley [from the field]. *IEEE Robotics & Automation Magazine, 19*(2), 98–100.

Morris, R. R., Kouddous, K., Kshirsagar, R., & Schueller, S. M. (2018). Towards an artificially empathic conversational agent for mental health applications: System design and user perceptions. *Journal of Medical Internet Research, 20*(6). https://doi.org/10.2196/10148.

Nass, C. I., Moon, Y., Morkes, J., Kim, E.-Y., & Fogg, B. J. (1997). Computers are social actors: A review of current research. *Human Values and the Design of Computer Technology, 72*, (pp. 137–162). Center for the Study of Language and Information.

Nienhuis, J. B., Owen, J., Valentine, J. C., Black, S. W., Halford, T. C., Parazak, S. E., et al. (2018). Therapeutic alliance, empathy, and genuineness in individual adult psychotherapy: A meta-analytic review. *Psychotherapy Research, 28*(4), 593–605. https://doi.org/10.1080/10503307.2016.1204023.

Palmqvist, B., Carlbring, P., & Andersson, G. (2007). Internet-delivered treatments with or without therapist input: Does the therapist factor have implications for efficacy and cost? *Expert Review of Pharmacoeconomics & Outcomes Research, 7*(3), 291–297. https://doi.org/10.1586/14737167.7.3.291.

Penedo, J. M. G., Berger, T., Holtforth, M., Krieger, T., Schröder, J., Hohagen, F., et al. (2020). The working alliance inventory for guided internet interventions (WAI-I). *Journal of Clinical Psychology, 76*(6), 973–986. https://doi.org/10.1002/jclp.22823.

Rohani, D. A., Quemada Lopategui, A., Tuxen, N., Faurholt-Jepsen, M., Kessing, L. V., & Bardram, J. E. (2020). MUBS: A personalized recommender system for behavioral activation in mental health. In: *Proceedings of the 2020 CHI Conference on Human Factors in Computing Systems* (pp. 1–13). https://doi.org/10.1145/3313831.3376879.

Rozental, A., Andersson, G., Boettcher, J., Ebert, D. D., Cuijpers, P., Knaevelsrud, C., et al. (2014). Consensus statement on defining and measuring negative effects of internet interventions. *Internet Interventions, 1*(1), 12–19. https://doi.org/10.1016/j.invent.2014.02.001.

Schueller, S. M., Muñoz, R. F., & Mohr, D. C. (2013). Realizing the potential of behavioral intervention technologies. *Current Directions in Psychological Science, 22*(6), 478–483. https://doi.org/10.1177/0963721413495872.

Schueller, S. M., Tomasino, K. N., & Mohr, D. C. (2017). Integrating human support into behavioral intervention technologies: The efficiency model of support. *Clinical Psychology: Science and Practice, 24*(1), 27–45. https://doi.org/10.1037/h0101740.

Tremain, H., McEnery, C., Fletcher, K., & Murray, G. (2020). The therapeutic alliance in digital mental health interventions for serious mental illnesses: Narrative review. *JMIR Mental Health, 7*(8). https://doi.org/10.2196/17204.

van Lotringen, C. M., Jeken, L., Westerhof, G. J., Klooster Ten, P. M., Kelders, S. M., et al. (2021). Responsible relations: A systematic scoping review of the therapeutic alliance in text-based digital psychotherapy. *Frontiers in Digital Health, 3*, 689750. https://doi.org/10.3389/fdgth.2021.689750.

Wampold, B. E., Baldwin, S. A., Holtforth, M., & Imel, Z. E. (2017). What characterizes effective therapists? How and why are some therapists better than others?: Understanding therapist effects (pp. 37–53). American Psychological Association. https://doi.org/10.1037/0000034-003.

Wampold, B. E., & Owen, J. E. S. S. E. (2021). Therapist effects: History, methods, magnitude. Bergin and Garfield's handbook of psychotherapy and behavior change (pp. 297–326). Hoboken, NJ: John Wiley & Sons.

Wang, W. (2017). Smartphones as social actors? Social dispositional factors in assessing anthropomorphism. *Computers in Human Behavior, 68*, 334–344. https://doi.org/10.1016/j.chb.2016.11.022.

Weisband, S., & Kiesler, S. (1996). Self disclosure on computer forms: Meta-analysis and implications. In: *Proceedings of the SIGCHI Conference on Human Factors in Computing Systems* (pp. 3–10).

Whelen, M. L., Murphy, S. T., & Strunk, D. R. (2021). Reevaluating the alliance–outcome relationship in the early sessions of cognitive behavioral therapy of depression. *Clinical Psychological Science, 9*(3), 515–523. 2167702620959352 https://doi.org/10.1177/2167702620959352 .

Chapter 9

Conversational agents on smartphones and the web

Timothy Bickmore and Teresa O'Leary

Khoury College of Computer Sciences, Northeastern University, Boston, MA, USA

9.1 Introduction

Approximately 20% of adults in the United States live with a mental illness and 7% have a substance use disorder ("Key substance use and mental health indicators in the United States: Results from the 2019 National Survey on Drug Use and Health (HHS Publication No. PEP20-07-01-001)," 2020). However, many do not receive any treatment or suboptimal treatment due to a wide variety of factors, including access, cost, stigma, and lack of motivation to seek help. Conversational agents (CAs) provide an automated solution that can enhance or in some cases even replace human counselors in the provision of some mental health services.

While this chapter presents CAs for the web and smartphones as different classes of application, this distinction is not strictly dichotomous. CAs have been developed that work across both platforms, and in general web-based systems can be accessed via smartphone web browsers.

9.1.1 Definitions

For the purpose of this chapter, we define a CA a user interface that interacts with users in natural language. CA output can be text, recorded speech, synthetic speech, or a combination of these. User input to the CA can be constrained to a selection of possible utterances, typed text, or speech. Although some research has been conducted into other possible CA input modalities, for example, teaching users a simplified natural language grammar (Tomko et al., 2005), these have not been widely adopted. "Chatbot" is a term that most typically refers to a CA with typed text or multiple-choice input and text-only output.

We define an embodied conversational agent (ECA) as a CA that also has a visual output/display modality using an animated character (Fig. 9.1) that uses conversational nonverbal behavior, such as hand gesture, facial display, and head and eye movements, which can make the user interface even more approachable and natural.

Whether CA or ECA, individual user inputs and agent outputs are referred to as "utterances," the sequence of utterances that comprise a given conversation between one user and one agent is a "dialogue." Agents able to converse with more than one user at a time in multiparty interaction have been developed (e.g., see "Future work") but are relatively rare.

9.1.2 Advantages of conversational agents

CAs, especially ECAs provide users with a computer interface that has the look and feel of face-to-face counseling through the use of natural language and, in the case of ECAs, nonverbal behavior such as hand gestures, body posture, and facial display of empathy. Prior studies have indicated that CAs can be more acceptable and usable than conventional point-and-click graphical user interfaces, especially by patients with low health literacy, who may have difficulty navigating conventional websites or user interfaces (Bickmore et al., 2010a). The conversational format also allows techniques from counseling psychology to be more directly translated for automation. To date, techniques from cognitive behavior therapy (Beck, 2011), social cognitive theory (Bandura, 1986), motivational interviewing (Miller & Rollnick, 2012), and dialectical therapy (Linehan, 2014) have been implemented and deployed in CAs.

Digital Therapeutics for Mental Health and Addiction: The State of the Science and Vision for the Future. DOI: https://doi.org/10.1016/B978-0-323-90045-4.00010-1

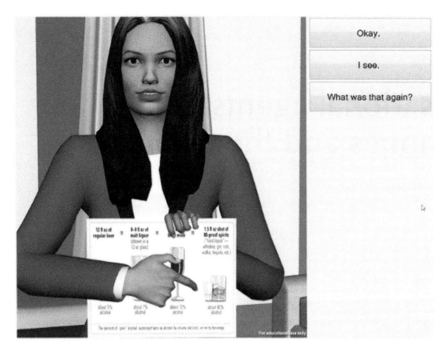

FIGURE 9.1 Embodied conversational agent for alcohol misuse counseling.

Mental health stigma is a significant barrier to care-seeking and participation in mental health services (Corrigan, 2016). There are several types of mental health stigma and one such type, label avoidance, undermines mental health service utilization for individuals in need of treatment. Individuals who experience such stigma avoid accepting and receiving mental health care services for fear of being associated with a psychiatric diagnosis that could result in a myriad of social consequences (e.g., social rejection, discrimination). As a result, increased privacy and anonymity afforded by CAs provides a unique opportunity to receive psychoeducation, mental health resources, and technology delivered therapeutic treatments while still giving the user full disclosure control over how and when they may disclose their identity as a person with mental illness to others. Furthermore, in marginalized communities where the prevalence of stigma is high, CAs may offer opportunities for users to engage in mental health treatment that otherwise is unavailable.

Automated systems in general have been shown to be more effective than humans at reducing barriers posed by fear of potentially stigmatizing disclosures, making people self-disclose more freely to a computer than a human. This result has also been demonstrated with CAs (Lucas, Gratch, King, & Morency, 2014).

CAs can also provide guideline-concordant care consistently with every session for every patient. Such treatment fidelity is crucial for maintaining the quality of mental health services; however, ensuring that such fidelity is consistently met requires systematic professional supervision. CAs can be designed to provide evidence-based guideline-driven therapy consistently, every time.

CAs may also be more effective than conventional websites or apps at maintaining user engagement, and retention is a prerequisite for long-term intervention impact (Huckvale, Nicholas, Torous, & Larsen, 2020). Several large randomized clinical trials have demonstrated poor engagement with automated mental health interventions to date: MoodGYM and "Beating the Blues" resulted in significant reductions in depressive symptoms compared to controls, but they both had 25–35% drop-out rates (Christensen, Griffiths, & Jorm, 2004; Cooper et al., 2011). One reason for poor engagement is that these websites lack the working alliance and sense of accountability that a human counselor provides. CAs and ECAs are unique in that they can use many of the same strategies people do to establish a sense of working alliance in fully automated interventions (Bickmore, Gruber, & Picard, 2005), which can lead to increased adherence and retention. There is also evidence that patients with depressive symptoms feel greater therapeutic alliance with CAs (Bickmore et al., 2010b). Smartphone-based CAs also provide unique affordances for maintaining engagement, including the ability to text message reminders (Brown et al., 2020; O'Connor et al., 2016).

9.1.3 Disadvantages of conversational agents

There are a few disadvantages to using CAs for automated mental health and addiction treatment interventions.

Concerns regarding bystanders who may observe conversations with CAs about mental health or addictions can also be a barrier to use. Users may or may not have access to private spaces to answer serious questions or issues of importance when engaging in conversations about their mental health. Designers should consider how to encourage safe information practices, as well as provide opportunities for users to flexibly switch between audible conversation and text-based responses.

Accessibility remains an additional concern when developing CAs for aging or otherwise visual and/or hearing-impaired populations. Currently, systems which rely on either auditory or visual feedback alone are not ultimately inclusive in design, leaving users who require additional multimodal support unable to access the tool. CAs with multimodal interfaces (e.g., visual, auditory, and sensory) inputs/outputs potentially could alleviate some of these issues.

Finally, CAs generally take longer to use than more conventional text-heavy user interfaces, given that their narrative content tends to be linear and doled out in small, utterance-sized chunks. ECAs that use speech output also take longer to convey the same information that could be read in text by the average reader. However, for many applications the benefits outweigh the costs and in some cases users actually perceive CAs to take less time than functionally equivalent conventional interfaces, even when actual elapsed time is longer (Bickmore, Utami, Matsuyama, & Paasche-Orlow, 2016).

In this chapter we explore two of the most ubiquitous digital media available today for delivering CA-based interventions for mental health and addictions: the web and smartphones. Web-based CAs can be delivered to any home computer or tablet, or even smartphones via web browser apps. Native smartphone apps can add even more functionality by leveraging information from sensors on smartphones to make CAs aware of and responsive to user location, activity, and health condition.

9.1.4 Systematic reviews

There have been several prior reviews of CAs in mental health. In general, they indicate that this medium is acceptable and promising, but the field is still too new—with too few randomized clinical trials with standardized outcome measures—to make conclusions regarding efficacy. Hoermann et al. conducted a systematic review of text-based dialogue systems (chatbots) in mental health interventions in 2017, finding that the studies reviewed were very innovative and demonstrated positive postintervention grains, but that more research was needed to demonstrate feasibility for clinical practice (Hoermann, McCabe, Milne, & Calvo, 2017). Vaidyam et al. conducted literature reviews in 2019 (Vaidyam, Wisniewski, Halamka, Kashavan, & Torous, 2019) and 2020 (Vaidyam, Linggonegoro, & Torous, 2020) of chatbots in mental health, finding that the studies that met their criteria had "favorable" evidence, but that further research was needed given the heterogeneity of outcomes used. Abd-Alrazaq et al. also conducted a scoping review in 2019 (Abd-Alrazaq et al., 2019) and attempted meta-analysis of chatbots in mental health in 2020 (Abd-Alrazaq, Rababeh, Alajlani, Bewick, & Househ, 2020), finding that the studies they found indicated potential to improve mental health, but that conclusions could not be reached due to too few studies, risk of bias, and conflicting results. Gaffney et al. conducted a systematic review of CAs in mental health in 2019, concluding that their efficacy and acceptability were "promising," but that more rigorous studies were required to demonstrate efficacy (Gaffney, Mansell, & Tai, 2019).

9.2 Conversational agents on the web

CAs deployed over the internet, typically over the worldwide web, provide a multitude of affordances as tools for promoting mental health and addiction. The reach of such tools is extensive as their use is not constrained by access to transportation and insurance coverage, or limited by monetary concerns and the investment of time typically associated with traditional office-based health care. Their use is relatively low cost and widely available for those with access to an internet connection and web browser.

9.2.1 Barriers to use for web-based CAs

As more healthcare programming and interventions are redeveloped for online digital experiences, lower income, and marginalized populations without easy access to the internet are unable to adequately benefit from these solutions, creating what researchers call intervention-generated inequality (Veinot, Mitchell, & Ancker, 2018). This means that access to technology-enabled healthcare is strongly impacted by racial and ethnic identity, gender identity, socioeconomic status, age, and sexual identity. In their work examining interventions that worsen inequality, Veinot and colleagues state that online mental health and substance use interventions are accessed more frequently by higher socioeconomic status (SES) patients even though lower SES patients experience higher rates of chronic disease and comorbid conditions. Thus, creating technical solutions deployed fully online may reinforce the "digital divide" by providing health-enhancing content to those who already have access to flexible resources (e.g., income, education, tech literacy, social capital) and experience fewer burdens of disease (Anderson & Kumar, 2019).

Data security and privacy may also represent barriers to use for internet-based interventions, including CAs. Online healthcare solutions are inherently at risk of security breaches. Thus, maintaining privacy and security of healthcare data remains an important challenge when developing CAs to support mental well-being and clinical support. Users have reported concerns regarding trust in maintaining the integrity of their data, as well as security concerns especially concerning data related to their health. Thus, to enhance engagement, CAs should be developed with transparent privacy and security statements, as well as providing users with features that allow them to control their personal data.

9.2.2 Text-based CA interventions on the web

Several text-based "chatbots" have been developed for mental health interventions. "Woebot"— a commercially available depression counseling system—is perhaps one of the most mature examples, having been evaluated in multiple studies. Woebot delivers content based on cognitive behavioral therapy (CBT) through daily interactive conversations (Fitzpatrick, Darcy, & Vierhile, 2017). Once a day, Woebot checks-in with the user asking them to report first their mood and their daily context (e.g., "whats going on?"). Based on these factors and user choice inputs, Woebot offers psychoeducation, mindfulness exercises, self-reflection opportunities, and practice applying therapeutic skills. As a chatbot, Woebot allows a combination of fully constrained user input choices during counseling dialogue and strategically placed free text input during some of the Woebot-led psychotherapeutic exercises. Woebot was initially evaluated in a 2-week clinical trial with 70 adults aged 18–28, demonstrating a reduction of depressive symptoms compared to a control group.

Ramachandran et al. evaluated the acceptability of Woebot as a postpartum mood management tool in 192 women who had just given birth and up to 6 weeks postpartum (Ramachandran et al., 2020). Participants were recruited during a birth hospital visit and randomized to 6 weeks using Woebot or a control condition. After the 6-week intervention, authors measured user satisfaction and therapeutic working alliance in the Woebot experimental group. The majority (64%) of intervention participants who completed the 6-week trial were highly satisfied with Woebot, and 28% reported a high working alliance with the agent, with another 43% reporting a medium working alliance. Suharwardy et al. conducted a follow-up effectiveness study with behavioral outcomes with pre–post measures of depression (PHQ-9, Postnatal Depression Scale) and anxiety (GAD-7) (Suharwardy et al., 2020), without finding significant differences between groups on changes in these measures.

Prochaska et al. adapted Woebot to assist in treating individuals living with substance use disorders (Prochaska et al., 2021). Woebot's functionality was extended to incorporate best practice guidelines for substance use treatment, incorporating principles from motivational interviewing, and dialectical behavioral therapy, with a focus on relapse prevention. In an 8-week single-group quasi-experimental field trial, 101 users tested this version of Woebot. Although only 50% completed the post-treatment quantitative measures, participants were found to have significant pre–post-treatment improvements in self-reported measures of substance use (AUDIT-C, DAST), depression (PHQ-8), and anxiety (GAD-7).

Darcy et al. evaluated Woebot's therapeutic working alliance in a retrospective analysis of data from 36,070 users of the commercial system (Darcy, Daniels, Salinger, Wicks, & Robinson, 2021). Within 3–5 days of completing the user enrollment process, participants were asked to complete a measure of therapeutic working alliance (WAI-SR) administered by the CA. Notably, Woebot users reported similar levels of therapeutic working alliance as found in individual CBT sessions with human therapists.

Similar to Woebot, "Tess" is a commercially available text-based chatbot that provides mental health support for anxiety and depression (Dosovitsky et al., 2020). Tess is a multimodule therapeutic mental health system designed to provide users with various mental health resources from several mental health disciplines (e.g., dialectical behavioral therapy, CBT, acceptance, and commitment therapy). Tess can both initiate conversations or respond directly to users via SMS text, and other messenger applications (e.g., Facebook Messenger, Slack) (Fulmer, Joerin, Gentile, Lakerink, & Rauws, 2018). Tess delivers content based on unconstrained user text, however, users can opt for constrained menu response options. In doing so users are promised enhanced personalization and module integration. Fulmer et al. evaluated Tess in a three-arm randomized controlled trial with 75 students and found that Tess was helpful in reducing anxiety (GAD-7) and depression symptoms (PHQ-9) after 2 weeks of use (Fulmer et al., 2018). Recently, Dosovitsky et al. conducted a descriptive evaluation of Tess to better understand how participants use a mental health chatbot and to further describe the navigational path through the system as well as individual depression module utilization (Dosovitsky et al., 2020). In this work, investigators conducted a descriptive analysis of usage patterns with participant data from 354 Tess users. Investigators found that most Tess bot users stopped using the system after completing only one or two modules. They emphasize the importance of spending time developing an engaging initial 1–2 modules rather than creating a large variety of modules that may be underutilized.

"Kokobot" is a CA developed to facilitate interactions among users who self-identify as depressed in an online peer-to-peer social support platform designed to promote emotional resilience (Morris, Kouddous, Kshirsagar, & Schueller, 2018). Kokobot users are prompted to describe stressful situations and associated negative thoughts, and Kokobot responds to these submissions by retrieving and repurposing statements from a corpus of supportive statements previously submitted to Koko by other users. Kokobot's responses are presented as suggestions for the user to consider until peer responses are collected from the peer network. In an evaluation study, Kokobot responses were rated as acceptable but less helpful than peer-generated responses. In a deception follow-up study, Kokobot suggestions were framed as coming from Kokobot directly (e.g., its corpus); however, for each submission responses were written by peer supporters. Even then users still rated Kokobot as less helpful, demonstrating that designers should consider how to increase trust, and credibility of CAs.

"Vivibot" is a chatbot created for young adult cancer survivors to promote resilience to anxiety and depression immediately following treatment, using techniques from positive psychology (Greer et al., 2019). Participants interact with Vivibot primarily using fully constrained multiple-choice inputs, with some dialogue exchanges (e.g., reflection on events) based on free-text input. In a randomized controlled trial with 45 young adults, users had access to seven psychoeducation modules focused on seven skills adapted from positive psychology (e.g., gratitude exercise, positive reappraisal, random acts of kindness) via Facebook Messenger. On a typical study day, users were greeted by Vivibot and asked a daily check-in question, then randomly assigned to one of the seven skill modules. Each skill module contained four interactions (one teaching and three practice) performed sequentially over the 28-day study period. Results indicated a reduction in anxiety (GAD-7) in the experimental group and an increase in anxiety for the control group. In addition, a dose–response effect was detected where those young adults in the experimental group who engaged in more sessions experienced a greater reduction in anxiety symptoms.

Finally, So et al. created "GAMBOT," a web-based text chatbot designed to support individuals living with a gambling addiction who do not intend to seek in person treatment (So et al., 2020). Thus, GAMBOT was created to provide access to behavior change counseling techniques for those who are reluctant to engage in formal treatment. GAMBOT was integrated into a popular Japanese messaging app "LINE" and tested with 197 participants in a 28-day randomized controlled trial. On a daily basis, GAMBOT greeted users and offered a daily therapeutic intervention, with messages containing personalized behavior feedback, behavior monitoring, and exercises for recognizing cognitive distortions and identifying triggers. While 77% of GAMBOT participants completed the intervention, there were no statistically significant between group differences in gambling symptoms or gambling severity.

9.2.3   ECA-based interventions on the web

There have been fewer ECAs than chatbots developed for mental health and addiction interventions, perhaps due to their greater complexity. We review the few ECAs whose evaluations haves been reported in the literature to date.

"Laura" is a computer-based ECA designed to increase the medication adherence of individuals living with schizophrenia (Bickmore, Puskar, Schlenk, Pfeifer, & Sereika, 2010c; Puskar, Schlenk, Callan, Bickmore, & Sereika, 2011) (Fig. 9.2). Laura comes equipped with seven educational modules to promote medication adherence. Module topics included: understanding symptoms, side effects, scheduling medications, coping strategies, leveraging support, and managing setbacks. Participants were instructed to interact with Laura for 30 days using a study provided laptop computer. In addition to the educational modules, Laura was equipped to check-in daily with participants as a reminder for participants to record their medication behavior. In a 31-day quasi-experimental evaluation study with 20 participants who met the DSM IVR criteria for schizophrenia. Participants talked to the agent an average of 65.8% of the available days, with nine of the participants talking to the agent at least 25 times. Mean self-reported antipsychotic dose adherence was 89%.

"SABORI" is a Japanese ECA that promotes mental health. SABORI asks users questions about their mood and physical condition, and provides feedback based on principles of CBT and behavioral activation (Suganuma, Sakamoto, & Shimoyama, 2018). While the majority of the application uses constrained user response options, SABORI allows unconstrained user responses in a subsection of the behavioral activation intervention. During a 30-day test period, authors found a significant effect on measured positive (well-being) and negative (psychological distress) mental health outcomes for SABORI users.

Bickmore et al. created an alcohol misuse screening tool to assist primary care physicians in identifying patients at risk of substance and tobacco use behaviors (Bickmore, Rubin, & Simon, 2020). In this work, a tablet-based ECA was designed to administer a standard clinical substance use screener "NIDA-modified ASSIST"; (Oga, Mark, Peters, & Coleman-Cowger, 2020) during a primary care visit, using language that a primary care physician would use when verbally administering the screener. To respond to the ECA, users tapped options from a constrained menu set. Results of the screener automatically generated a substance use report for the primary care physician to refer to during the patient visit. To determine

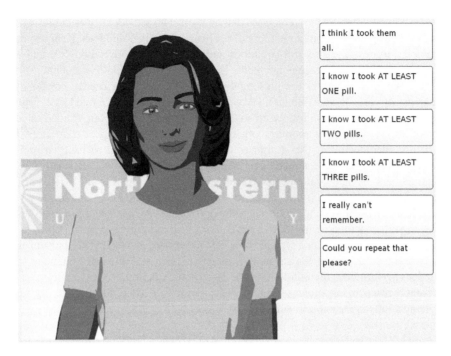

FIGURE 9.2   Antipsychotic medication adherence ECA.

the validity of the ECA administered screener, investigators conducted two counterbalanced mixed methods validation studies with 57 participants, comparing the ECA to a human and then to an existing forms-based online assessment. Both studies demonstrated validity through strong agreement between assessment methods. In addition, participants disclosed significantly more to the ECA compared to the human, stating that the ECA was a nonjudgmental alternative to a human-administered screener. Participants were significantly more satisfied with the ECA-led screener compared to the forms-based screener, but less satisfied compared to the human screener. A follow-on brief intervention for alcohol misuse (Fig. 9.1) demonstrated significantly more referrals to specialty care compared to standard of care in US Veteran's Administration hospitals (Rubin et al., 2022).

Ring designed, developed, and implemented an "affect aware" ECA CBT system for reducing symptoms of depression in college students (Ring, 2017). The system incorporated multiple skill-based CBT modules where users were able to learn skills and practice them. The affect aware ECA was designed to support the user during conversations through the integration of an empathic response system. In this work, the ECA sensed users' emotional states by measuring user emotional valence and arousal throughout each ECA interaction and automatically adjusted its empathic responses. The system was tested in a between-subjects longitudinal study with 36 participants with half the study cohort interacting with the affective-aware ECA and the other half a conventional counseling system. Results demonstrated that participants had statistically higher levels of system engagement and showed a reduction in state anxiety using the affect aware ECA.

Sebastian et al. created a mental health literacy intervention for individuals living with anorexia nervosa to address stigma experienced by people living with eating disorders (Sebastian & Richards, 2017). In a mixed experimental design, the authors compared video-based with ECA-based interactions using two stigma reduction strategies: education and paracontact, involving a mediated asynchronous interaction with an individual in recovery. In a single use, between-subjects experiment, 245 participants were randomized to the ECA stigma-reduction system or the video-based intervention. Results indicated that both interventions reduced stigma associated with anorexia nervosa, but there were no significant differences between them.

9.3   Conversational agents on smartphones

While the web provides broad reach and accessibility for automated mental health and addiction interventions, smartphones provide even more powerful platforms for these interventions. Smartphones are typically carried everywhere a person goes, providing help and assistance anytime and anywhere an individual might need it, including emergency help for emergent health conditions. Smartphones also have the ability to proactively cue users to engage with them, via asynchronous

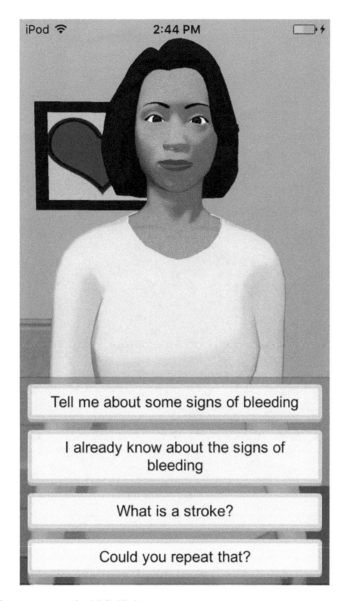

FIGURE 9.3 Smartphone ECA for management of atrial fibrillation.

notifications or alarms. They provide multimodal interfaces, allowing user to engage with CAs via speech, SMS text messages, text-based user interfaces, and/or fully embodied ECA (Figs. 9.3 and 9.4).

There are also an increasing number of sensors available for smartphones—either integrated as standard equipment or as add-on components—that can greatly enhance the ability of CAs to tailor their therapeutic dialog to the user's context and behavior, or to be triggered by sensed environmental conditions such as the user's location (Cornet & Holden, 2018; Majumder & Deen, 2019). Table 9.1 lists a range of mental-health and addiction relevant conditions that researchers have used smartphone sensors to detect (see also Chapter 7).

One frequent concern when considering smartphones as deployment platforms for health interventions is whether there is an mHealth "digital divide" in which disadvantaged populations will be excluded due to lack of smartphone ownership. However, smartphone ownership is high among all segments of the US population. For example, while African Americans are less likely to use the internet or have broadband access at home compared to other groups ("African Americans and Technology Use,"), two thirds of African Americans own smartphones ("The Demographics of Device Ownership"), a rate consistent with other racial groups, and they are more likely to rely on their smartphone for internet use, and use their smartphones more than other groups to look up information about health conditions.

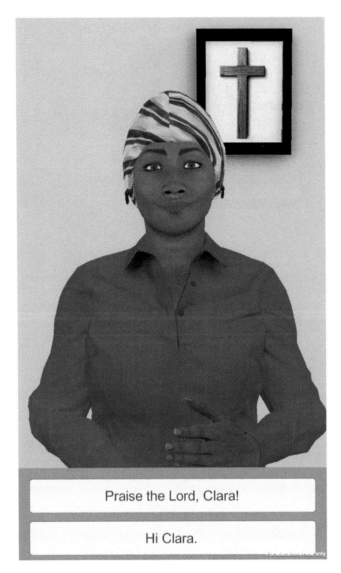

Praise the Lord, Clara!

Hi Clara.

FIGURE 9.4 Conversational agent for church-based wellness promotion.

TABLE 9.1 Mental health and addiction relevant conditions, and sensors used to detect.

Condition	Smartphone sensors	Example citations
Stress, emotional state	Microphone, GPS, phone use	Ben-Zeev, Scherer, Wang, Xie, and Campbell, 2015; Lu et al., 2012
Depressive symptoms	Microphone, GPS, phone use	Ben-Zeev et al., 2015; Saeb et al., 2015; Saeb, Lattie, Schueller, Kording, and Mohr, 2016
Bipolar states	Accelerometer, GPS	Grünerbl et al., 2015
Substance use relapse	GPS	Gustafson et al., 2014

Concerns over privacy and security are potentially even more of a barrier to use of CAs on smartphones than the web, particularly for conditions and conversation topics that are potentially stigmatizing, such as substance misuse. Bystanders could potentially overhear speech-based counseling conversations, or view text-based conversations, while a user is in public, posing a significant privacy concern. Systems that use sensed location (GPS) risk exposing sensitive user locations (e.g., at a clinic or shelter) should the data be accessed by others. Use of a patient's smartphone by others also poses a potential risk to privacy, even though the discovery of the mere presence of an app that is known to be associated with a mental health condition or addiction. Security of patient data, especially if transmitted to a server, is also important for these reasons.

9.3.1   Text-based CA interventions on smartphones

There are many examples of text-based chatbots deployed on smartphones for mental health interventions. "Pocket Skills" is a CA that uses dialectical behavioral therapy for depression (Schroeder et al., 2018). The CA represents Dr. Marsha Linehan, who developed dialectical behavioral therapy, and walks users through the counseling content. In a 4-week single-group quasi-experimental field study with 73 participants, depressive symptoms (PHQ-9 scores) and anxiety (OASIS scores) were found to decrease significantly over the duration of the study.

"Shim" is a smartphone-based chatbot that guides users through a positive psychology intervention, with user responses to the text conversation made via multiple-choice inputs or free text (Ly, Ly, & Andersson, 2017). In a RCT with 28 participants randomized to 2 weeks using Shim or a wait-list control, no significant differences were found between the intervention and control groups on any measure, although qualitative analyses of interviews with participants yielded some interesting insights.

"Wysa" is a smartphone app that features a text-based CA intended to help users build emotional resilience skills using techniques from CBT, dialectical behavior therapy, motivational interviewing, and other frameworks (Inkster, Sarda, & Subramanian, 2018). In a single-group quasi-experimental study with 129 participants, users were found to have significant improvements in depressive symptoms (PHQ-9 scores) over the duration of the intervention.

The "Ready4Life" coaching program features a CA that counsels adolescents on coping skills for stress, and avoidance of tobacco, e-cigarettes, cannabis, and alcohol (Haug, Castro, Wenger, & Schaub, 2020). The agent uses techniques and principles from social-cognitive theory (goal-setting, self-monitoring, observational learning), social norms (normative feedback), and motivational interviewing (e.g., decisional balance). A randomized controlled trial to evaluate the 4-month duration intervention is underway.

9.3.2   ECA-based interventions on smartphones

Although most smartphone chatbots feature a static image of the agent in the interface, they do not use synchronized animated nonverbal conversational behavior—such as hand gesture and facial display—to convey additional information to users. There are few true ECAs on smartphones (Feng et al., 2015) and fewer still that have been used in health interventions.

One example smartphone-based ECA for chronic disease management (although not for mental health or addiction) is an intervention developed for individuals with atrial fibrillation (Fig. 9.3) (Bickmore et al., 2018). Atrial fibrillation (AF) is an irregular heartbeat that increases the risk of stroke three- to fivefold, and doubles the risk of death if untreated (Romero, Nedios, & Kriatselis, 2014). AF management includes medication adherence and symptom monitoring, including regular heart rhythm readings. The ECA is designed to support multiple conversations per day, covering: AF education; heart rhythm monitor education; symptom education and reporting; medications and medication side-effects; recognition of emergencies and actions to take; motivating the user to take charge of their own health; and preparing for medical encounters. In a 30-day quasi-experimental pilot study, pre–post measures of health-related quality of life and self-reported medication adherence improved significantly (Magnani et al., 2017).

In a more recent effort, an ECA was developed to promote wellness among members of a network of predominately African American churches in Boston (Fig. 9.4) (Stowell et al., 2020). The ECA and its dialog were tailored to the church community through the integration of spiritual and religious language into its health counseling. To date, exercise and diet (fruit and vegetable consumption) have been implemented and evaluated in a lab study and pilot field study (O'Leary et al., 2020). Participants provided positive feedback on the religious tailoring and acceptance of the ECA, and indicated that its use of spiritual and religious content was appropriate, genuine, and motivating when used in the context of a health counseling conversation with an agent. A field study of this intervention, including ECA-led guided meditation and scriptural reflection for stress reduction, is currently underway.

9.4   Safety issues in using conversational agents

While CAs have the potential to help many people, they also have the potential to cause harm if precautions are not taken in their design.

Abd-Alrazaq et al. conducted a systematic review of effectiveness and safety of CAs for mental health (Abd-Alrazaq et al., 2020). There were only two studies in their analysis that provided safety information. One evaluated a 12-week avatar-based intervention for depressive symptoms, including 60 participants randomized to avatar intervention or control (Pinto et al., 2016). The second study used a virtual coach in virtual reality (VR) in an automated six-session intervention for acrophobia, evaluated in a randomized trial with 49 participants randomized to the intervention group (Freeman et al., 2018). None of these approximately 80 participants reported any adverse events or safety problems, leading Abd-Alrazaq to conclude that CAs are safe. However, in addition to being a very small sample (two intervention studies), the first study used a minimally interactive avatar (based on recorded performances of an actor) and the second study did not describe the level of avatar interactivity. Safety findings from these two studies do not generalize to all possible CA interventions and modalities.

CAs that use unconstrained natural language input—whether typed text or user speech—represent a particular safety risk when used for actionable medical advice in general, but for mental health and addiction in particular. Risks can stem from the inherent ambiguity in natural language, the lack of user knowledge about the expertise and natural language abilities of a CA, and potentially misplaced trust. In order to demonstrate the potential safety issues, a study was conducted using three widely available conversational assistants (Apple's Siri, Google Home, and Amazon's Alexa). Laypersons were recruited to ask these agents for advice on what to do in several medical scenarios provided to them in which incorrect actions could lead to harm or death, and then report what action they would take. Out of 394 tasks attempted, participants were only able to complete 43%, but of those, 29% of reported actions could have resulted in some degree of harm, including 16% that could have resulted in death, as rated by clinicians using a standard medical harm scale (Bickmore et al., 2018). The errors responsible for these outcomes were found at every level of system processing as well as in user actions in specifying their queries and in interpreting results. The findings from this study imply that unconstrained natural language input, in the form of speech or typed text, should not be used for systems that provide medical advice, given the state-of-the-art. Users should be tightly constrained in the kinds of advice they can ask for, for example, through the use of multiple-choice menus of utterances they are allowed to "say" in each step of the conversation (e.g., as in Figs. 9.1 and 9.2).

9.5 Future directions

There are many important directions of future exploration and research for CAs in mental health and addiction treatment. Sensors, particularly on smartphones and other mobile devices, can provide crucial input to CA-based interventions, either to trigger their use or to provide data that they can base their counseling on. For example galvanic skin response (GSR) and heart rate can be used to assess arousal and anxiety and trigger counseling interventions, and location (via GPS sensor) could be used to trigger an addiction intervention when a user nears a location associated with substance use (Gustafson et al., 2014). Respiration rate can be used for CA-based guided meditation, used as part of a stress-reduction intervention (Shamehki & Bickmore, 2018).

In addition to text-based, speech-based, and screen-based CAs and ECAs on desktop computers and smartphones, there are a variety of other new media that could be used to deploy CAs in mental health and addiction treatment. Virtual reality (VR) has been widely used in desensitization interventions for phobias (Botella, Fernández-Álvarez, Guillén, García-Palacios, & Baños, 2017), but it could also provide a private, immersive environment in which patients can engage in counseling with an ECA. Augmented reality (AR) provides another opportunity for digitally embedding an ECA mental health counselor, or other intervention media, into the user's environment, as has been done for physical rehabilitation (Botella et al., 2017). Humanoid robots provide another medium for intervention, and may be particularly impactful in group counseling (e.g., couples relationship counseling; Utami & Bickmore, 2019) or in situations in which the robot is mobile and can physically move to the user's location in order to engage in counseling.

Developing natural language processing and dialogue system technologies that are both safe and allow the user sufficient flexibility and expressivity remains another important area of research. Also important is the continuing development of counseling dialogue systems that support more of the conversational processes, which people take for granted when conversing with each other, including for example:

- mixed-initiative conversational turn-taking;
- interruptions;
- speech intonation (used to convey a range of information about discourse context; Hirschberg, 1990);
- discourse markers (words or phrases like "anyway" that signal changes in discourse context; Schiffrin, 1987);
- discourse ellipsis (omission of a syntactically required phrase when the content can be inferred from discourse context);

- grounding (how speaker and listener negotiate and confirm the meaning of utterances through signals such as headnods and paraverbals such as "uh huh"; Clark & Brennan, 1991); and
- indirect speech acts (e.g., when a speaker says "do you have the time?" they want to know the time rather than simply wanting to know whether the hearer knows the time or not; Searle, 1975).

The use of CAs in integrated care scenarios, in which they mediate among multiple health professionals, caregivers, and patients also represents an important direction of future research (Kowatsch et al., 2021).

Finally, many more rigorous studies of CAs in mental health and addiction treatment are needed, including replications, order to establish the evidence required before these technologies can be adopted in clinical care.

9.6. Conclusion

CAs represent a promising medium for automating mental health and addiction treatment interventions, and there is already a strong and growing body of empirical studies demonstrating their potential. Although there are safety issues that must be addressed in any CA-based intervention, they have the potential to provide scalable, low-cost, high-fidelity interventions to provide treatment to large segments of the population.

References

Abd-Alrazaq, A. A., Alajlani, M., Alalwan, A. A., Bewick, B. M., Gardner, P., & Househ, M. (2019). An overview of the features of chatbots in mental health: A scoping review. *International Journal of Medical Information, 132*, 103978. doi:10.1016/j.ijmedinf.2019.103978.

Abd-Alrazaq, A. A., Rababeh, A., Alajlani, M., Bewick, B. M., & Househ, M. (2020). Effectiveness and safety of using chatbots to improve mental health: Systematic review and meta-analysis. *Journal of Medical Internet Research [Electronic Resource], 22*(7), e16021. doi:10.2196/16021.

African Americans and Technology Use. Retrieved from http://www.pewinternet.org/2014/01/06/african-americans-and-technology-use/.

Anderson, M., & Kumar, M. (2019). Digital divide persists even as lower-income Americans make gains in tech adoption. Retrieved from https://www.pewresearch.org/fact-tank/2019/05/07/digital-divide-persists-even-as-lower-income-americans-make-gains-in-tech-adoption/.

Bandura, A. (1986). Social foundations of thought and action: A social cognitive theory. Englewood Cliffs, NJ: Prentice-Hall.

Beck, J. (2011). Cognitive behavior therapy: Basics and beyond. New York, NY: The Guilford Press.

Ben-Zeev, D., Scherer, E. A., Wang, R., Xie, H., & Campbell, A. T. (2015). Next-generation psychiatric assessment: Using smartphone sensors to monitor behavior and mental health. *Psychiatric Rehabilitation Journal, 38*(3), 218–226. doi:10.1037/prj0000130.

Bickmore, T., Gruber, A., & Picard, R. (2005). Establishing the computer-patient working alliance in automated health behavior change interventions. *Patient Education and Counseling, 59*(1), 21–30.

Bickmore, T., Kimani, E., Trinh, H., Pusateri, A., Paasche-Orlow, M., & Magnani, J. (2018). Managing chronic conditions with a smartphone-based conversational virtual agent. In Paper presented at the ACM conference on intelligent virtual agents (p. e11510).

Bickmore, T., Mitchell, S., Jack, B., Paasche-Orlow, M., Pfeifer, L., & ODonnell, J. (2010a). Response to a relational agent by hospital patients with depressive symptoms. *Interacting with Computers, 22*(4), 289–298.

Bickmore, T., Pfeifer, L., Byron, D., Forsythe, S., Henault, L., Jack, B., & Paasche-Orlow, M. (2010b). Usability of conversational agents by patients with inadequate health literacy: Evidence from two clinical trials. *Journal of Health Communication, 15*(Suppl 2), 197–210.

Bickmore, T., Puskar, K., Schlenk, E., Pfeifer, L., & Sereika, S. (2010c). Maintaining reality: Relational agents for antipsychotic medication adherence. *Interacting with Computers, 22*, 276–288.

Bickmore, T., Rubin, A., & Simon, S. (2020). Substance use screening using virtual agents. In Paper presented at the proceedings of the 20th ACM international conference on intelligent virtual agents.

Bickmore, T., Trinh, H., Olafsson, S., O'Leary, T., Asadi, R., Rickles, N., & Cruz, R. (2018). Patient and consumer safety risks when using conversational assistants for medical information: An observational study of Siri, Alexa, and Google Assistant. *Journal of Medical Internet Research [Electronic Resource], 20*(9).

Bickmore, T., Utami, D., Matsuyama, R., & Paasche-Orlow, M. (2016). Improving access to online health information with conversational agents: A randomized controlled experiment. *Journal of Medical Internet Research, 18*(1), e1.

Botella, C., Fernández-Álvarez, J., Guillén, V., García-Palacios, A., & Baños, R. (2017). Recent progress in virtual reality exposure therapy for phobias: A systematic review. *Current Psychiatry Reports, 19*(7), 42. doi:10.1007/s11920-017-0788-4.

Brown, W., Lopez Rios, J., Sheinfil, A., Frasca, T., Cruz Torres, C., Crespo, R., & Carballo-Diéguez, A. (2020). Text messaging and disaster preparedness aids engagement, re-engagement, retention, and communication among Puerto Rican participants in a human immunodeficiency virus (HIV) self-testing study after hurricanes Irma and Maria. *Disaster Medicine Public Health Preperation*, 1–8. doi:10.1017/dmp.2020.25.

Christensen, H., Griffiths, K. M., & Jorm, A. F. (2004). Delivering interventions for depression by using the internet: Randomised controlled trial. *British Medical Journal, 328*, 265.

Clark, H. H., & Brennan, S. E. (1991). Grounding in communication. L. B. Resnick, J. M. Levine, & S. D. Teasley (Eds.). In *Perspectives on Socially Shared Cognition* (pp. 127–149). Washington: American Psychological Association.

Cooper, C. L., Hind, D., Parry, G. D., Isaac, C. L., Dimairo, M., O'Cathain, A., & Sharrack, B. (2011). Computerised cognitive behavioural therapy for the treatment of depression in people with multiple sclerosis: External pilot trial. *Trials, 12*, 259. doi:10.1186/1745-6215-12-259.

Cornet, V. P., & Holden, R. J. (2018). Systematic review of smartphone-based passive sensing for health and wellbeing. *Journal of Biomedical Informatics, 77*, 120–132. doi:10.1016/j.jbi.2017.12.008.

Corrigan, P. W. (2016). Lessons learned from unintended consequences about erasing the stigma of mental illness. *World Psychiatry, 15*(1), 67–73.

Darcy, A., Daniels, J., Salinger, D., Wicks, P., & Robinson, A. (2021). Evidence of human-level bonds established with a digital conversational agent: Cross-sectional, retrospective observational study. *JMIR Formative Research, 5*(5), e27868. doi:10.2196/27868.

Dosovitsky, G., Pineda, B. S., Jacobson, N. C., Chang, C., Escoredo, M., & Bunge, E. L. (2020). Artificial intelligence chatbot for depression: Descriptive study of usage. *JMIR Formative Research, 4*(11), e17065. doi:10.2196/17065.

Feng, A., Leuski, A., Marsella, S., Casas, D., Sin-Hwa, K., & Shapiro, A. (2015). A platform for building mobile virtual humans. In Paper presented at the intelligent virtual agents (IVA).

Fitzpatrick, K. K., Darcy, A., & Vierhile, M. (2017). Delivering cognitive behavior therapy to young adults with symptoms of depression and anxiety using a fully automated conversational agent (Woebot): A randomized controlled trial. *JMIR Mental Health, 4*(2), e19. doi:10.2196/mental.7785.

Freeman, D., Haselton, P., Freeman, J., Spanlang, B., Kishore, S., Albery, E., & Nickless, A. (2018). Automated psychological therapy using immersive virtual reality for treatment of fear of heights: A single-blind, parallel-group, randomised controlled trial. *Lancet Psychiatry, 5*(8), 625–632. doi:10.1016/s2215-0366(18)30226-8.

Fulmer, R., Joerin, A., Gentile, B., Lakerink, L., & Rauws, M. (2018). Using psychological artificial intelligence (Tess) to relieve symptoms of depression and anxiety: Randomized controlled trial. *JMIR Mental Health, 5*(4), e64. doi:10.2196/mental.9782.

Gaffney, H., Mansell, W., & Tai, S. (2019). Conversational agents in the treatment of mental health problems: Mixed-method systematic review. *JMIR Mental Health, 6*(10), e14166. doi:10.2196/14166.

Greer, S., Ramo, D., Chang, Y. J., Fu, M., Moskowitz, J., & Haritatos, J. (2019). Use of the chatbot "Vivibot" to deliver positive psychology skills and promote well-being among young people after cancer treatment: Randomized controlled feasibility trial. *JMIR Mhealth Uhealth, 7*(10), e15018. doi:10.2196/15018.

Grünerbl, A., Muaremi, A., Osmani, V., Bahle, G., Ohler, S., Tröster, G., & Lukowicz, P. (2015). Smartphone-based recognition of states and state changes in bipolar disorder patients. *IEEE Journal of Biomedical Health Information, 19*(1), 140–148. doi:10.1109/jbhi.2014.2343154.

Gustafson, D. H., McTavish, F. M., Chih, M. Y., Atwood, A. K., Johnson, R. A., Boyle, M. G., & Shah, D. (2014). A smartphone application to support recovery from alcoholism: A randomized clinical trial. *JAMA Psychiatry, 71*(5), 566–572. doi:10.1001/jamapsychiatry.2013.4642.

Haug, S., Castro, R. P., Wenger, A., & Schaub, M. P. (2020). Efficacy of a smartphone-based coaching program for addiction prevention among apprentices: Study protocol of a cluster-randomised controlled trial. *BMC Public Health [Electronic Resource], 20*(1), 1910. doi:10.1186/s12889-020-09995-6.

Hirschberg, J. (1990). Accent and discourse context: Assigning pitch accent in synthetic speech. In Paper presented at the AAAI 90.

Hoermann, S., McCabe, K. L., Milne, D. N., & Calvo, R. A. (2017). Application of synchronous text-based dialogue systems in mental health interventions: Systematic review. *Journal of Medical Internet Research [Electronic Resource], 19*(8), e267. doi:10.2196/jmir.7023.

Huckvale, K., Nicholas, J., Torous, J., & Larsen, M. E. (2020). Smartphone apps for the treatment of mental health conditions: Status and considerations. *Current Opinion on Psychology, 36*, 65–70. doi:10.1016/j.copsyc.2020.04.008.

Inkster, B., Sarda, S., & Subramanian, V. (2018). An empathy-driven, conversational artificial intelligence agent (Wysa) for digital mental well-being: Real-world data evaluation mixed-methods study. *JMIR Mhealth Uhealth, 6*(11), e12106. doi:10.2196/12106.

Key substance use and mental health indicators in the United States: Results from the 2019 National Survey on Drug Use and Health (HHS Publication No. PEP20-07-01-001). (2020). Retrieved from https://www.samhsa.gov/data/sites/default/files/reports/rpt29393/2019 NSDUHFFRPDFWHTML/2019NSDUHFFR1PDFW090120.pdf.

Kowatsch, T., Schachner, T., Harperink, S., Barata, F., Dittler, U., Xiao, G., & Möller, A. (2021). Conversational agents as mediating social actors in chronic disease management involving health care professionals, patients, and family members: Multisite single-arm feasibility study. *Journal of Medical Internet Research [Electronic Resource], 23*(2), e25060. doi:10.2196/25060.

Linehan, M. (2014). *DBT skills training manual*. Guildofrd Press, New York, NY.

Lu, H., Frauendorfer, D., Rabbi, M., Mast, M., Chittaranjan, G., Campbell, A., Gatica-Perez D., Choudhury, T. (2012). StressSense: detecting stress in unconstrained acoustic environments using smartphones. In *Proceedings of the 2012 ACM Conference on Ubiquitous Computing (UbiComp '12)*. Association for Computing Machinery, New York, NY, USA, 351360. https://doi.org/10.1145/2370216.2370270.

Lucas, G., Gratch, J., King, A., & Morency, L. (2014). It's only a computer: Virtual humans increase willingness to disclose. *Computers in Human Behavior, 37*, 94–100.

Ly, K. H., Ly, A. M., & Andersson, G. (2017). A fully automated conversational agent for promoting mental well-being: A pilot RCT using mixed methods. *Internet Intervention, 10*, 39–46. doi:10.1016/j.invent.2017.10.002.

Magnani, J., Schlusser, C., Kimani, E., Rollman, B., Paasche-Orlow, M., & Bickmore, T. (2017). The atrial fibrillation health literacy information technology system: Pilot assessment. *JMIR Cardiology, 1*(2), e7.

Majumder, S., & Deen, M. J. (2019). Smartphone sensors for health monitoring and diagnosis. *Sensors (Basel), 19*(9), 2164. doi:10.3390/s19092164.

Miller, W., & Rollnick, S. (2012). Motivational interviewing: Preparing people for change (third ed.). New York: Guilford Press.

Morris, R. R., Kouddous, K., Kshirsagar, R., & Schueller, S. M. (2018). Towards an artificially empathic conversational agent for mental health applications: System design and user perceptions. *Journal of Medical Internet Research [Electronic Resource], 20*(6), e10148. doi:10.2196/10148.

O'Connor, S., Hanlon, P., O'Donnell, C. A., Garcia, S., Glanville, J., & Mair, F. S. (2016). Understanding factors affecting patient and public engagement and recruitment to digital health interventions: A systematic review of qualitative studies. *BMC Medical Informatics and Decision Making [Electronic Resource], 16*(1), 120. doi:10.1186/s12911-016-0359-3.

Oga, E. A., Mark, K., Peters, E. N., & Coleman-Cowger, V. H. (2020). Validation of the NIDA-modified ASSIST as a screening tool for prenatal drug use in an urban setting in the United States. *Journal of Addiction Medicine, 4*(5), 423. doi:10.1097/adm.0000000000000614.

O'Leary, T., Stowell, E., Kimani, E., Parmar, D., Olafsson, S., Hoffman, H., Bickmore, T. (2020). Community-based cultural tailoring of virtual agents. In Paper presented at the ACM international conference on intelligent virtual agents.

Pinto, M. D., Greenblatt, A. M., Hickman, R. L., Rice, H. M., Thomas, T. L., & Clochesy, J. M. (2016). Assessing the critical parameters of eSMART-MH: A promising avatar-based digital therapeutic intervention to reduce depressive symptoms. *Perspectives in Psychiatric Care, 52*(3), 157–168. doi:10.1111/ppc.12112.

Prochaska, J. J., Vogel, E. A., Chieng, A., Kendra, M., Baiocchi, M., Pajarito, S., & Robinson, A. (2021). A therapeutic relational agent for reducing problematic substance use (Woebot): Development and usability study. *Journal of Medical Internet Research [Electronic Resource], 23*(3), e24850. doi:10.2196/24850.

Puskar, K., Schlenk, E. A., Callan, J., Bickmore, T., & Sereika, S. (2011). Relational agents as an adjunct in schizophrenia treatment. *Journal of Psychosocial Nursing and Mental Health Services, 49*(8), 22–29.

Ramachandran, M., Suharwardy, S., Leonard, S. A., Gunaseelan, A., Robinson, A., Darcy, A., & Judy, A. (2020). Acceptability of postnatal mood management through a smartphone-based automated conversational agent. *American Journal of Obstetrics and Gynecology, 222*(1), 74. doi:10.1016/j.ajog.2019.11.090.

Ring, L. (2017). An affect-aware dialogue system for counseling. Boston, MA: Northeastern University.

Romero, I., Nedios, S., & Kriatselis, C. (2014). Diagnosis and management of atrial fibrillation: An overview. *Cardiovascular Therapeutics, 32*(5), 242–252.

Rubin, A., Livingston, N., Hocking, E., Bickmore, T., Sawdy, M., Kressin, N., & Simon, S. (2022). Computerized relational agent to deliver alcohol brief intervention and referral to treatment in primary care: A randomized clinical trial. *Journal of General Interal Medicine, 37*(1), 70–77.

Saeb, S., Lattie, E. G., Schueller, S. M., Kording, K. P., & Mohr, D. C. (2016). The relationship between mobile phone location sensor data and depressive symptom severity. *Peer Journal, 4*, e2537. doi:10.7717/peerj.2537

Saeb, S., Zhang, M., Karr, C. J., Schueller, S. M., Corden, M. E., Kording, K. P., & Mohr, D. C. (2015). Mobile phone sensor correlates of depressive symptom severity in daily-life behavior: An exploratory study. *Journal of Medical Internet Research [Electronic Resource], 17*(7), e175. doi:10.2196/jmir.4273.

Schiffrin, D. (1987). Discourse markers. Cambridge: Cambridge University Press.

Schroeder, J., Wilks, C., Rowan, K., Toledo, A., Paradiso, A., Czerwinski, M., & Linehan, M. (2018). Pocket skills: A conversational mobile web App to support dialectical behavioral therapy. In Paper presented at the ACM conference on human factors in computing (CHI).

Searle, J. R. (1975). Indirect speech acts. In Speech acts (pp. 59-82). Brill. Leiden. https://doi.org/10.1163/9789004368811_004.

Sebastian, J., & Richards, D. (2017). Changing stigmatizing attitudes to mental health via education and contact with embodied conversational agents. *Computers in Human Behavior, 73*, 479–488. doi:10.1016/j.chb.2017.03.071.

Shamekhi, A., & Bickmore, T. (2018). Breathe deep: A breath-sensitie interactive meditation coach. In Paper presented at the international conference on pervasive computing technologies for healthcare.

So, R., Furukawa, T. A., Matsushita, S., Baba, T., Matsuzaki, T., Furuno, S., & Higuchi, S. (2020). Unguided chatbot-delivered cognitive behavioural intervention for problem gamblers through messaging App: A randomised controlled trial. *Journal of Gambling Studies, 36*(4), 1391–1407. doi:10.1007/s10899-020-09935-4.

Stowell, E., O'Leary, T., Kimani, E., Paasche-Orlow, M., Bickmore, T., & Parker, A. (2020). Investigating opportunities for crowdsourcing in church-based health interventions: A participatory design study. In Paper presented at the ACM conference on human factors in computing (CHI 2020) http://relationalagents.com/publication/2020/chi20/CHI20.pdf.

Suganuma, S., Sakamoto, D., & Shimoyama, H. (2018). An embodied conversational agent for unguided internet-based cognitive behavior therapy in preventative mental health: Feasibility and acceptability pilot trial. *JMIR Mental Health, 5*(3), e10454. doi:10.2196/10454.

Suharwardy, S., Ramachandran, M., Leonard, S. A., Gunaseelan, A., Robinson, A., Darcy, A., & Judy, A. (2020). Effect of an automated conversational agent on postpartum mental health: A randomized, controlled trial. *American Journal of Obstetrics and Gynecology, 222*(1), 116. doi:10.1016/j.ajog.2019.11.132.

The Demographics of Device Ownership. Retrieved from http://www.pewinternet.org/2015/10/29/the-demographics-of-device-ownership/.

Tomko, S., Harris, T., Toth, A., Sanders, J., Rudnicky, A., & Rosenfeld, R. (2005). Towards efficient human machine speech communication: The speech graffiti project. *ACM Transactions on Speech and Language Processing, 2*(1), 2-es.

Utami, D., & Bickmore, T. (2019). Collaborative user responses in multiparty interaction with a couples counselor robot. In Paper presented at the human robot interaction (HRI).

Vaidyam, A. N., Linggonegoro, D., & Torous, J. (2020). Changes to the psychiatric chatbot landscape: A systematic review of conversational agents in serious mental illness: Changements du paysage psychiatrique des chatbots: Une revue systématique des agents conversationnels dans la maladie mentale sérieuse. *Canadian Journal of Psychiatry Revue Canadienne De Psychiatrie, 66*(4), 339–348. doi:10.1177/0706743720966429.

Vaidyam, A. N., Wisniewski, H., Halamka, J. D., Kashavan, M. S., & Torous, J. B. (2019). Chatbots and conversational agents in mental health: A review of the psychiatric landscape. *Canadian Journal of Psychiatry Revue Canadienne De Psychiatrie, 64*(7), 456–464. doi:10.1177/0706743719828977.

Veinot, T. C., Mitchell, H., & Ancker, J. S. (2018). Good intentions are not enough: How informatics interventions can worsen inequality. *Journal of the American Medical Informatics Association, 25*(8), 1080–1088. doi:10.1093/jamia/ocy052.

Chapter 10

Voice-based conversational agents for sensing and support: Examples from academia and industry

Caterina Bérubé[a] and Elgar Fleisch[a,b]

[a] *Centre for Digital Health Interventions, Department of Management, Technology and Economics, ETH Zurich, Zurich, Switzerland,* [b] *Centre for Digital Health Interventions, Institute of Technology Management, University of St. Gallen (ITEM-HSG), St. Gallen, Switzerland*

10.1 Introduction

Voice interaction is nothing new. The first voice-activated device, Radio Rex, was created in the 1920s. This plastic toy figure of a dog was placed inside a doghouse-looking box containing a mechanism reacting to pressure waves and making the figure pop out of the box when called by its name, that is, "Rex" (David & Selfridge, 1962). Speech recognition became first possible in the 1960s with the IBM "Shoebox," capable of solving voice-based arithmetic commands (IBM, 2003), and debuted its commercial availability in the early 1990s with DragonDictate (Cohen, 1991). Voice interaction became, however, effectively scalable and conversational when simple speech recognition was implemented into a smartphone-based voice assistant in 2011 with the release of Apple's Siri. At the time, just by talking to a smartphone, the user could be assisted in sending messages, scheduling meetings, or looking up information, and it could be used for pure entertainment purposes. Today, VCAs are capable of flexibly recognizing human speech, converting it into an intent (i.e., a message of execution) triggering more and more elaborated functions (e.g., shopping, finance, traveling, health and wellness) and responding with ever-evolving synthesized speech (Hoy, 2018). Moreover, commercial leaders like Amazon and Google democratized the implementation of voice applications in the form of *Skills* (Amazon) or *Actions* (Google). Such applications allow performing a variety of specific tasks by only using voice commands, from playing a movie quote to finding the nearest urgent care.

While voice interaction has been principally used as a control booth or assistant, its application in healthcare remains ill-explored (Bérubé et al., 2021; Sezgin, Militello, Huang, & Lin, 2020).

This chapter introduces the advantages of voice interaction for digital health interventions delivery and gathers examples from academia and industry in the arena of mental and substance use disorders. Our review does not aim to be exhaustive but rather to give an overview of examples allowing discussion of the most prevalent features and trends. As conversational agents interacting through voice have been referred to by several terms, for example, voice assistants, voice-based conversational agents (VCAs), voice-based chatbots, voice-activated devices, voice-enabled devices, or interactive voice applications (Sezgin et al., 2020), we will use the term VCA, which highlights the conversational nature of such technology.

10.1.1 VCAs to relieve the healthcare system

VCAs are becoming increasingly ubiquitous. Almost 130 million Americans, Britons, and Germans owned a smart speaker (i.e., a VCA implemented in an internet-enabled speaker) in 2021 (Kinsella & Herndon, 2021a, 2021b, 2021c), while over 150 million Americans had already used a voice assistant on a smartphone in 2018 (Kinsella & Mutchler, 2019). With Apple Siri being the first VCA on the market, smartphones were the leading implementation device. However, home devices such as smart speakers are taking over households and are expected to progressively integrate into our everyday lives (Kinsella, 2020). This level of adoption shows the great potential for VCAs to integrate behavioral health interventions into our health and wellbeing practices. As commercial VCAs, such as Amazon Alexa and Google Assistant allow for the creation of "routines" (i.e., time or command-based triggering of specific sequences of functions), individuals suffering from mental or substance use disorders could benefit from functions such as reminders, information lookups, and reports about their health (e.g., by accessing data from apps, connected medical devices, or their virtual medical record).

Furthermore, we observe an effort from commercial VCA service providers in making their technology compatible with health interventions. For instance, Amazon declared on April 4, 2019, to have deployed a service supporting health information protection, according to the Health Insurance Portability and Accountability Act of 1996 (HIPAA) (104th United States Congress, 1996), allowing to generate HIPPA-compliant Alexa-based services. Moreover, Amazon partnered with United Kingdom's National Health Service (NHS), to provide reliable health-related information. Google seems to have started similar efforts during the current COVID-19 pandemic, by carefully screening approval requests for Actions relating to the coronavirus, to avoid fake information delivery (Schwartz, 2020).

With 970 million adults worldwide suffering from mental disorders in 2019 (Dattani, Ritchie, & Roser, 2021), and 36 million affected by substance use disorders in 2020 (United Nations, 2021), day-to-day management of these conditions is needed. VCAs provide not only the opportunity for patients to access approaches to health management that are complementary to traditional care but that also allow for healthcare professionals and caregivers to be relieved from in-person routine procedures (Sezgin et al., 2020; Sezgin, Huang, Ramtekkar, & Lin, 2020). As mental health knowingly benefits from human social support (Harandi, Taghinasab, & Nayeri, 2017) we do not consider VCA to be an exclusive alternative to human healthcare professionals but as an integrative tool for high quality healthcare.

10.1.2 The advantages of voice modality

VCAs can provide the tools to deliver digital health interventions automatically and easily. However, why should we prefer to communicate with a computer about our mental health or substance use solely by talking to it, instead of texting, or looking at and talking to an avatar? In this section, we present ergonomic reasons for why VCAs are advantageous compared to other types of conversational agents, such as text-based conversational agents (TCA) and embodied conversational agents (ECA).

We identify four advantages of VCAs over TCAs and ECAs: (1) efficiency of the human-machine interaction, (2) accessibility for multiple contexts and types of users, (3) sense of social presence, and (4) opportunity for voice-based sensing. The domain of healthcare has for a long time adopted TCAs to provide a scalable solution to complement health professionals, typically through smartphone applications. Examples are Wysa, Woebot, or Youper, which provide mental health support through conversation (Alattas et al., 2021). TCAs imitate human-to-human texting, typically by requiring users to choose between predefined buttons with standard prompts or by allowing them to freely type their messages. Texting, however, may, in some cases, result in a more effortful interaction. Taking the smartphone, starting the application, reading, and typing can be time-consuming and attention-demanding. This is where VCAs can help : voice interaction frees users' hands by communicating the same intents without even having to look at the device. Imagine coming home from a long working day and wanting to comfortably sit down on your couch and start a meditation session. Opening the smartphone application and finding the right feature to begin your session may be more demanding than simply saying "Hey Assistant, start my meditation session" (Bostock, Crosswell, Prather, & Steptoe, 2019). In other words, voice interaction simply results in higher efficiency. Additionally, VCAs are also prominently used to control smart environments, such as home appliances (Ammari, Kaye, Tsai, & Bentley, 2019) or health sensors (Basatneh, Najafi, & Armstrong, 2018). Thus, the meditation session could easily be paired with the room's lights dimming and the diffuser starting.

Moreover, in comparison to TCAs and ECAs, voice interaction facilitates input in situations where the user has his hands occupied, such as driving (Large, Burnett, Anyasodo, & Skrypchuk, 2016; Militello, Sezgin, Huang, & Lin, 2021; Simmons, Caird, & Steel, 2017; Strayer et al., 2019; Young & Zhang, 2017), cooking (Vtyurina & Fourney, 2018), training (Chung, Griffin, Selezneva, & Gotz, 2018; Namba, 2021), and even surgery (El-Shallaly, Mohammed, Muhtaseb, Hamouda, & Nassar, 2005) and providing emergency care (Damacharla et al., 2019). For example, a mother suffering from postpartum depression who needs to take care of her child may still be able to benefit from a cognitive-behavioral intervention while breastfeeding (Stuart & Koleva, 2014).

Also, voice interaction can foster accessibility to and inclusivity of users with visual, motor, or cognitive disabilities, as it is designed not to depend on visual information nor body motions (Abdolrahmani, Storer, Roy, Kuber, & Branham, 2020; Barata, Galih Salman, Faahakhododo, & Kanigoro, 2018; Choi, Kwak, Cho, & Lee, 2020; Friedman et al., 2019; Masina et al., 2020; Pradhan, Mehta, & Findlater, 2018). Thus, someone suffering from tetraplegia could still easily ask the VCA to help them with insomnia (Spong, Graco, Brown, Schembri, & Berlowitz, 2015). Yet, it is worth mentioning that some VCAs may require further scrutiny as not specifically designed to serve these users (Stacy & Antony Rishin Mukkath, 2019).

Next, VCAs provide a more prominent social presence (Cho, Molina, & Wang, 2019; Pitardi & Marriott, 2021) and can express personality through para-verbal cues such as tone of voice or pauses (Perez Garcia & Saffon Lopez, 2019), and thus, a closer imitation of human-to-human interaction. This effect has also been observed to benefit those who suffer from loneliness (Pradhan, Lazar, & Findlater, 2020). Also, even though ECAs can give a higher sense of rapport (Shamekhi, Liao, Wang, Bellamy, & Erickson, 2018) and social presence compared to VCAs (Kim et al., 2018) their effect also depends on the ability of the ECA to imitate facial expressions and gesture (Qiu & Benbasat, 2005). In that sense, VCAs may have lower chances to arm the user experience related to social presence. For instance, if an individual with posttraumatic stress disorder is negatively affected in their user experience by the visuals of an ECA, engagement in the health intervention may decrease and the latter may lose its efficacy (Yeager & Benight, 2018).

Finally, voice may also augment information about the user by detecting the right moment to intervene. Digital health interventions for behavioral change show the highest efficacy when the support is delivered to an individual who is either vulnerable (e.g., craving for a cigarette); open to change (e.g., motivated to stop smoking); receptive (e.g., having a break from work). Just-in-time interventions (Nahum-Shani et al., 2017; Nahum-Shani et al., 2018; Nahum-Shani, Hekler, & Spruijt-Metz, 2015) have been shown to increase efficacy, compared to time-random or absent interventions (Wang & Miller, 2020). In this context, the *sensing-and-support paradigm* involves 1) the use of technology to *sense* or monitor at regular interval specific health-relevant data streams and 2) the use of an interface to *support* or provide tools to prevent or manage health conditions. Concretely, sensing implies the use of passive sensors (e.g., wearable activity trackers, microphones, digital calendar events), or active assessments (e.g., ecological momentary assessment, EMA) to generate a picture of the individual's current behavior or health condition and decide whether it is relevant to trigger support. We refer to support as any intervention aiming at assisting individuals in managing their mental health condition or substance use disorder. VCAs can give access to vocal features, allowing for vocal biomarkers and voice authentication. Vocal biomarkers are features extracted from audio signals containing voice or speech that allow inference of changes in health status from a baseline condition and, thus, for monitoring, prognosis, or diagnosis of a condition (Fagherazzi, Fischer, Ismael, & Despotovic, 2021). Data collection for vocal biomarkers is noninvasive and allows for remote and frequent assessment of clinical outcomes. Typical voice characteristics are fundamental frequency, voice intensity, speech rate, jitter, and shimmer. In fact, it has been observed that speech can be used to define vocal biomarkers for emotions (Akçay & Oğuz, 2020; Thakur & Dhull, 2021) and mental health conditions (Cummins, Epps, Sethu, Breakspear, & Goecke, 2013; Low, Bentley, & Ghosh, 2020). Moreover, vocal biomarkers have been used to diagnose Alzheimer's (Pulido et al., 2020) and Parkinson's disease (Oung et al., 2015), which have been associated with depression (Aarsland, Påhlhagen, Ballard, Ehrt, & Svenningsson, 2012; Ownby, Crocco, Acevedo, John, & Loewenstein, 2006). Imagine a VCA having a morning routine conversation with its user, a student facing anxiety, and detecting distress in their voice while doing so. The VCA will be able to suggest a short coaching session to manage their mental state (Lattie et al., 2019).

Detecting relevant vocal biomarkers requires, however, to ensure that only the intended individual receives the health intervention (and not, for instance, another member of the family). Voice authentication refers to the use of using voice to verify one's identity. Making sure that health data and interventions are accessible only to the target patient in environments where the VCA could be used by many individuals (Meng, Altaf, & Juang, 2020; Mohd Hanifa, Isa, & Mohamad, 2021) will avoid misuse of the VCA itself (Wahsheh & Steffy, 2020). Thus, the voice would allow triggering interventions when it is most relevant, and that is when the *right* person is *vulnerable or open to change* and *receptive*.

10.1.3 VCAs to provide engaging digital health interventions

Despite the above mentioned instrumental and experiential advantages of VCAs, the question remains if patients accept such a technology. History suggests that this is the case. First, VCAs represent the implementation of an already mature human wish to converse with computers the same way one would do with humans. This is well illustrated by the exemplary fictional characters of HAL 9000, J.A.R.V.I.S, and Samantha. In particular, HAL 9000, from "2001: A Space Odyssey" of 1986, not only was able to check the state of the spaceship but also to monitor its crew's hibernation and to keep their intellect active by playing chess with them. J.A.R.V.I.S from the Marvel Cinematic Universe could check for signs of physiological health and diagnose mental health conditions, such as anxiety. Samantha, from the movie "Her" of 2013, which was primarily built to support administrative needs by organizing the owner's life, evolves into a friend and romantic partner capable of meeting emotional needs. Besides, research has repeatedly shown that humans have a natural tendency to treat computers as social entities (Carolus et al., 2019; Kulms & Kopp, 2018; Nass & Lee, 2001; Nass, Steuer, & Tauber, 1994). Thus, using VCAs to deliver digital health interventions might just carry out a collective fantasy.

Second, the transition from text-based to voice-based information exchange reflects the history of verbal human–computer interaction. If at the beginning computers were given text input (from keypunches to keyboard and keypad), speech is now considered one of the current human–machine interface trends (Accenture, 2020; Bechtel, Briggs, & Buchholz, 2020). This trend can also be observed in the domain of healthcare, whereas in an American study from 2015 around 44% of (over 32k) individuals stated looking for health information over the internet, especially when having trouble accessing healthcare services (Amante, Hogan, Pagoto, English, & Lapane, 2015). Now, 19.1 million Americans already use VCAs for health-related purposes, such as asking about symptoms, medical information, treatment options, or finding healthcare facilities (Kinsella & Mutchler, 2019). Thus, applying VCAs to health may simply represent an extension of an already established engaging way to interact with computers.

10.2 Method

To provide a landscape of VCAs for mental and substance use disorders we review nonsystematically both academic research and industrial products. In particular, we look at their technical implementation and categorize them in terms of *sensing* and *support* capabilities. The review aims to inform the health professionals, researchers, and entrepreneurs of the current trends, gaps, and potentials for voice-based just-in-time health interventions.

10.2.1 Included and excluded cases

Research around VCAs for health is preliminary and mainly focuses on prototype development or evaluation of commercial VCAs in their ability to provide health-related information (Bérubé et al., 2021; Sezgin, Militello, Huang, & Lin, 2020; Bérubé, Kovacs, Fleisch, & Kowatsch, 2021). Thus, we aimed to review not only cases from academia (i.e., prototypes from primary, and secondary studies) but also from industry (i.e., products from startups and established companies). This allowed us to give a general overview of VCAs dedicated to mental health and substance use disorders.

Moreover, we excluded cases of Skills and Actions that are not developed in the context of peer-reviewed research or by established companies. Even if Google and Amazon have content and privacy policy to which, Actions and Skills, respectively, need to conform to, some policy-violating implementations have been found (Cheng et al., 2020). In fact, it has been observed that their certification processes tends to lack in rigor and can allow policy-violating voice applications to enter the market (e.g., because the reviewer based its judgment on the developer's official statement of policy conformity, rather than on the application's architecture itself).

10.2.2 Technical implementation

While including cases from both academia and industry, we also briefly describe the technical implementation of the VCAs. We assess whether they are based on third-party solutions or on independent software, which has implications for compatibility and data storage and processing. For instance, Skills' and Actions' frameworks, which are a third-party solutions, facilitate the implementation of dedicated voice commands and are a viable solution for the quick development of simple voice applications. However, if more complex processes are required, such as safe storage, use, and transfer of sensitive information, independent software solutions may be required.

10.2.3 Sensing and support

Given the just-in-time approach mentioned above, we categorized the reviewed cases according to how their features fit in the sensing-and-support paradigm. Also, we divided sensing into two categories. First, we considered *active* sensing, which requires individuals to interact in some specific way with the VCA to allow it to collect data, for instance via voice-based EMAs. Second, we included *passive* sensing, which consists of collecting data passively without explicitly soliciting users, such as speech data collection and analysis during an interaction.

We also categorized our cases' features into either *reactive* or *proactive* support. Reactive support refers to all features delivering health-related information on demand (i.e., requiring the user to initiate a conversation with the VCA). Proactive support refers to the proactive delivery of targeted communications by the VCA, such as data-driven alerts or predefined reminders. Finally, while we acknowledged that some solutions may not only support patients but also healthcare professionals and caregivers, we focused only on the features dedicated to patients.

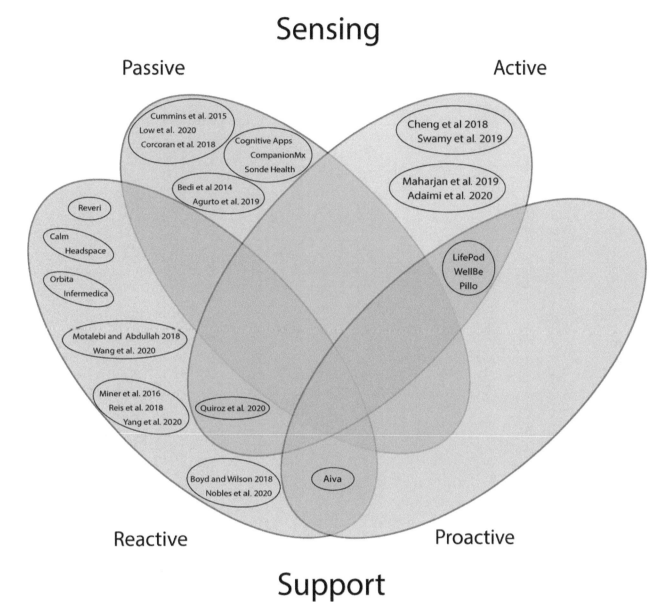

FIGURE 10.1   Venn diagram grouping cases implementing passive or active sensing and responsive or proactive support. Note: in each set, the cases are grouped by target condition, domain (academia or industry), and similarity of features.

10.2.4 Explorative approach

To the best of the authors' knowledge, there is no published research or company publicly reporting the development of passive sensing through VCAs, thus we also included studies on vocal biomarkers for mental health and substance use disorders.

10.3 Findings

To provide a comprehensive overview of our cases, we differentiated between academia and industry and categorized them according to the sensing-and-support paradigm presented above. In particular, we highlighted unidimensional and multidimensional cases, whereas the last refer to prototypes and products providing multiple types of features. A summary of these cases can be found in Table 10.1, where we specified the cases, their implementation, and the type of sensing and support features they provide. In addition, Fig. 10.1 provides a visual distribution of the cases based on the sensing and support features. Our findings are described narratively with a short description of each case.

TABLE 10.1 List of cases indicating target condition, implementation of the VCA, and sensing and support interventions.

Case	Authors/Owner	Target condition	VCA implementation	Sensing		Reactive	Support	Proactive
				Passive	Active			
Healthy Coping (prototype)	Cheng et al 2018	Depression	Alexa Skill	-	PHQ-9[2]	-	-	-
Depression Screener (prototype)	Swarny et al 2019	Depression	Google Dialogflow and Google Firebase	-	PHQ-9[2]	-	-	-
Wearable assessment tool (prototype)	Adami et al 2020	Mood	Google Cloud Speech API	-	Common questions (e.g., "Are you happy right now?")	-	-	-
Hear Me Out (prototype)	Maharjan et al 2019	Mood	Alexa Skill	-	Common questions (e.g., "How was your mood like?")	-	-	-
Biomarkers for depression and suicidality (secondary study)	Cummins et al 2015	Depression and suicidality	-	Prosodic and acoustic features	-	-	-	-
Biomarkers for psychiatric disorders (secondary study)	Low et al 2020	Psychiatric disorders		Acoustic features	-	-	-	-
Biomarker for schizophrenia (primary study)	Corcoran et al 2018	Schizophrenia		Semantic features	-	-	-	-
Biomarker for drug intoxication (primary study)	Bedi et al 2014	Ecstasy intoxication		Semantic features	-	-	-	-
Biomarker for drug craving (primary study)	Agurto et al 2019	Cocaine abstinence		Acoustic features	-	-	-	-
Cognitive Apps API (biomarker)	Cognitive Apps	Stress, exhaustion, fatigue, risk of depression		Acoustic features	-	-	-	-
Companion app (biomarker)	Companion Mx, Inc	Depression and PTSD[1]		Acoustic features	-	-	-	-
Sonde Health app (biomarker)	Sonde Health, Inc.	Depression		Acoustic features	-	-	-	-
VCA for loneliness (evaluation study)	Reis et al 2018	Isolation in elderly	Amazon Alexa, Apple Siri, Google Assistant, and Microsoft Cortana	-	-	Social interaction assistance	-	-
VCA for depression (evaluation study)	Miner et al 2016	Depression	Apple Siri, Google Now, Samsung Voice, Microsoft Cortana	-	-	Information lookup	-	-
VCA for postpartum depression (evaluation study)	Yang et al 2021	Postpartum depression	Amazon Alexa, Apple Siri, Google Assistant, and Microsoft Cortana	-	-	Information lookup	-	-
VCA for smoking cessation (evaluation study)	Boyd and Wilson 2018	Tobacco addiction	Apple Siri and Google Assistant	-	-	Information lookup	-	-
VCA for rehabilitation (evaluation study)	Nobles et al 2020	Drug addiction	Apple Siri, Google Assistant, Microsoft Cortana	-	-	Information lookup	-	-
Skill for PTSD (prototype)	Motalebi and Abdullah 2018	PTSD[1]	Alexa Skill	-	-	Cognitive behavioral therapy	-	-
Skill for public anxiety speeking (prototype)	Wang et al 2020	Public speaking anxiety	Alexa Skill	-	-	Coaching sessions	-	-
Headspace (voice application)	Headspace Inc.	Stress, anxiety, insomnia, and depression	Alexa Skill and Google Action	-	-	Guided meditation playback	-	-
Calm (voice application)	Calm	Stress, anxiety, insomnia, and depression	Google Action	-	-	Guided and unguided meditation, and sleep stories playback	-	-
OrbitaASSIST (general automation support)	Orbita, Inc.	Isolation in elderly	Amazon Alexa integration	-	-	Information lookup, interaction and entertainment	-	-
Symptomate (general automation support)	Infermedica	Any condition	Google Action	-	-	Decision support (symptom checker)	-	-
Reveri (edge case)	Reveri Health	Stress, anxiety, insomnia, and depression, tobacco addiction	Independent smartphone application	-	-	Self-guided hypnosis	-	-
Depression and anxiety self-test with coping strategy recommendation (prototype)	Quiroz et al 2020	Anxiety and depression	Alexa Skill	-	PHQ-9[2] and GAD-7[3]	Coping strategy recommendation	-	-
LifePod (dedicated VCA)	LifePod Solutions	Isolation in elderly	Independent smart speaker	-	Customizable questions	Information lookup, interaction and entertainment	Reminders	-
WellBe (dedicated VCA)	HandsFree Health	Isolation in elderly	Independent smart speaker	-	Blood pressure, sugar, weight	Information lookup	Reminders, warning if tracked values requires attention	-
Pria (dedicated VCA)	Stanley Black & Decker, Inc.	Any condition requiring medication	Independent smart speaker, Amazon Alexa and Google	-	Customizable questions	Information lookup, interaction and entertainment	Reminders	-
Aiva (general automation support)	Aiva, Inc.	Isolation in elderly	Assistant integration	-	-	Information lookup, interaction and entertainment	Reminders	-

[1] Post-traumatic disorder
[2] Patient Health Questionnaire-9
[3] General Anxiety Disorder-7

10.3.1 Simple active sensing prototypes

10.3.1.1 Mental health assessment

EMAs allow for ecologically valid data collection of self-reports (Kubiak & Smyth, 2019). We present studies developing prototypes to perform EMAs using standardized screening tools for mental health conditions.

10.3.1.1.1 Healthy Coping in Diabetes

Cheng, Raghavaraju, Kanugo, Handrianto, and Shang (2018) developed a smart speaker application for older individuals with diabetes, called Healthy Coping in Diabetes, aiming at assessing their depressive symptoms with the Patient Health Questionnaire (PHQ-9), a short standardized depression scoring assessment (Spitzer, Kroenke, & Williams, 1999). The main motivation for using a VCA for such an assessment was to monitor the negative effects of motor or visual impairment on the mental health of the elderly patient. The assessment was designed to be user-activated and the score to be directly translated into a diagnosis. The authors implemented the VCA using Google products Dialogflow (i.e., a platform to design and integrate a VCA in an application or device) and Firebase (i.e., an application development and monitoring platform) and run it over a Google Home device (smart speaker). Dialogflow was used to implement the conversation executing the screening test, while Firebase was used to store and manage the test scores. Although the study was preliminary, it showed a good acceptance of the application by elderly patients with type 2 diabetes.

10.3.1.1.2 Depression screener

Similar to Cheng and colleagues, Swamy et al. (2019) developed a system using both a facial expression analyzer to recognize emotion and a VCA performing the PHQ-9 to diagnose depression. Based on the PHQ-9 score, the VCA would deliver a recommendation to see a mental health professional. The authors motivated the use of a VCA to automate depression screening, as well as to overcome the fear of judgment from a mental health professional and foster disclosure. To deliver the active sensing intervention, the authors built a website for the users to log in and run the screening test. Like Cheng and colleagues, the authors implemented VCA's conversational turns in Dialogflow and the management of the PHQ-9 scores in Firebase.

10.3.1.2 Mood assessment

Although the use of standardized questionnaires allows for clinically relevant assessments, mental health has also been explored from a nonclinical perspective. In particular, we present two studies (Adaimi, Ho, & Thomaz, 2020; Maharjan, Bækgaard, & Bardram, 2019) in which a voice-based EMA with nonstandardized assessment tools was developed to test the feasibility of routine assessments.

10.3.1.2.1 Wearable assessment tool

Adaimi et al. (2020) presented the development of a wearable system allowing for EMAs around several topics including mood. The authors implemented a VCA to minimize the disruption which may be caused by EMAs and relieve the users from having to shift their attention from their current task to the assessment tool (e.g., smartphone). The system consisted of a wearable speaker (worn around the neck) running the VCA, and a wristband. The VCA performed speech-to-text and text-to-speech functions through the application programming interfaces (API, i.e., a software that can be integrated in another piece of software and provides a specific service) from Google Cloud (i.e., Google's suite of cloud computing services). These APIs use neural networks to synthesize natural-sounding voices and to process natural language. The wristband helped not only to enhance the voice-based interaction (i.e., signaling the VCA is listening or receiving an input) but also to notify users of an imminent assessment and allowing them to dismiss it in case of they found themselves in an inopportune context. Although the assessments were not specifically dedicated to mood, the study proposes a user-friendly wearable solution for a voice-based EMA.

10.3.1.2.2 Hear me out

Maharjan et al. (2019) described the development of a Skill called "Hear me out" to perform a voice-based EMA via a smart speaker in response to daily activities. In particular, the Skill was developed to fit two scenarios, one where the VCA would assess sleep quality after the wake-up alarm would go off, and one where it would assess evening mood after the user requires to execute an internet-of-things function (i.e., turn off the light before going to bed). The authors motivated the use of a VCA based on a smart speaker, as these devices have a continuous power supply, excluding the burden of battery charging, typical of smartphones. The authors, however, do not clearly motivate the use of voice itself as a medium for mental health assessment against other types of interaction, such as, for instance, touch interaction with a smart display.

Although these two examples suggested voice-based EMAs with nonstandardized surveys, it shows the possibility of creating assessment routines to monitor users' mood, while taking into account the users' state of receptivity.

10.3.2 Simple passive sensing research

10.3.2.1 *Automated assessment of mental health disorders*

As an alternative to active sensing, voice assistants could potentially sense passively, that is use speech data to measure vocal biomarkers and detect or predict a state of vulnerability. Although, to the best of the authors' knowledge, there is noresearch on VCA for speech-based automated mental health or drug use assessment, there has been extensive research on audio and verbal features to identify such disorders (i.e., vocal biomarkers).

10.3.2.1.1 Biomarkers for depression and suicidality

Cummins and colleagues (Cummins et al., 2015) reviewed studies analyzing speech to diagnose depression and suicidality. The authors found prosodic and acoustic features to be associated with depression and suicidality. In particular, prosodic features, such as a reduction of the fundamental frequency variation (i.e., the rate of vibration of the vocal cords), energy, and speaking rate have been observed to be more prominent in depressed individuals. Acoustic features reflecting the airflow through the vocal cords, such as jitter (i.e., the variation of frequency between cycles of opening/closure of the glottis), shimmer (i.e., the variation of amplitude between cycles), and harmonic-to-noise ratio (i.e., a measure of the relative noise in the voice) have also been observed to correlate with depression. These features reflect the general tendency of depressed individuals to present a reduction in movement of the vocal fold and, thus, in speaking effort. According to the authors, however, the acoustic features have been proven to work on held vowels tasks but not on continuous speech and may not be appropriate for passive sensing through a VCA, which generally involves short commands or sentences.

10.3.2.1.2 Biomarkers for psychiatric disorders

Later, Low and colleagues (Low et al., 2020) performed a similar systematic review for automated assessment of psychiatric disorders (i.e., depression, post-traumatic stress disorder, schizophrenia, anxiety, bipolar disorder, bulimia, anorexia, and obsessive-compulsive disorder). Almost half of the studies investigated acoustic features as biomarkers for depression. Others mainly included schizophrenia, bipolar, and post-traumatic disorder. Moreover, although acoustic features such as jitter and shimmer significantly correlated with the frequency or severity of both depression and anxiety, this latter represented a mere 5% of the studies. Moreover, the authors reported which devices were used to collect such types of data (see https://tinyurl.com/tu58te3) and observed that the reviewed studies included telephone calls (Cummins, Epps, Sethu, Breakspear, & Goecke, 2013; Mundt, Snyder, Cannizzaro, Chappie, & Geralts, 2007) or recordings of human interviews, reading or speech tasks. Although these tasks do not necessarily reflect the shorter human–VCA interaction, the review shows encouraging results in the use of vocal biomarkers for mental health assessments.

10.3.2.1.3 Biomarker for schizophrenia

While Cummins and colleagues (Cummins et al., 2015) and Low and colleagues (Low et al., 2020) focused on the speech features, Corcoran and colleagues (Corcoran et al., 2018) pushed for linguistic analytic methods to predict schizophrenia. In particular, they observed psychotic speech to present lower but more varying semantic coherence (i.e., confusion in speech) and reduced presence of possessive pronouns. In fact, in schizophrenia, syntax tends to be less complex and the speaker can present speech flow derailment (Andreasen & Grove, 1986). It is, therefore, important to note the great potential in the implementation of speech analysis in VCAs for mental health.

10.3.2.2 *Automated assessment of drug use disorders*

In the context of drug use disorders, speech analysis has been found to allow for intoxication and abstinence assessment.

10.3.2.2.1 Biomarker for drug intoxication

Bedi and colleagues (Bedi et al., 2014) used semantic and topological (i.e. semantic proximity) features to detect intoxication from ecstasy and methamphetamine. They observed that speech of individuals under the influence of ecstasy had closer semantic proximity to concepts such as a "friend", "support", "intimacy", and "rapport", while under the influence of methamphetamine speech was more semantically close to the concept of "compassion". Such semantic variations could be used to detect intoxication while interacting with a VCA.

10.3.2.2.2   Biomarker for drug craving

Agurto et al. (2019) reported being able to predict cocaine abstinence among individuals with use disorder who were asked to describe the positive consequences of abstinence and the negative consequences of using the substance. The authors investigated acoustic and semantic features and observed the former ones to predict abstinence better than the latter. In particular, Mel Frequency Cepstral Coefficients, which are the result of a complex short-term energy spectrum transformation (Davis & Mermelstein, 1980), correlated with the Beck Depression Inventory score (Beck, Ward, Mendelson, Mock, & Erbaugh, 1961) and the number of days since last drug use/days of abstinence.

Although we could not find cases of VCAs performing speech analysis for mental health or drug use disorders, there seems to be a great potential for VCAs to detect the health state from voice recordings.

10.3.3   Simple passive sensing products

With mental health disorders becoming more and more prominent, some companies have invested in mood recognition through voice. Although, like in academia, there don't seem to be any commercial VCA used for assessing mood through speech analysis for health-related purposes, we provide three examples of companies aiming to detect mood through voice.

10.3.3.1   Cognitive apps

Cognitive Apps (Cognitive Apps, 2021) provides an API to infer the emotional state of users (Leikina, 2020a, 2020b). The API is supposed to be implemented in a smartphone app and collect daily physical activity, surrounding noise, and sleep through mobile HealthKit (i.e., Apple's solution for storing, managing, and sharing health-related data), in addition to location services, and voice-based emotion recognition. The API is designed to share the collected data with the healthcare professional though a monitoring dashboard, by generating weekly reports from the passive sensing features, together with indices of risk of depression, exhaustion, stress, and fatigue. Thus, in this case, voice is one of the sources of information that the passive sensing module uses to infer the patient's mental health state. A mobile application for iOS, Yuru, was also available for demonstrative purposes in 2021 and is now integrated in Aiki (Apple, 2021).

10.3.3.2   Companion app

CompanionMx (2018a), a Spinoff of Cogito Corp., defined itself as a "digital health technology company with a proven platform for proactive mobile mental health monitoring for better clinical outcomes." Although not delivering voice assistance services, they developed Companion, a smartphone application allowing patients to record their voice and to categorize their emotional state. The results can also be shared with healthcare professionals via a clinical dashboard. According to a study showing the feasibility and acceptability of such an application in veterans at risk of suicide (Betthauser et al., 2020), voice recorded in the form of audio check-ins was used to analyze mood and provide a score. Moreover, the company states (Cogito Corp 2022) that they were able to validate their product with a randomized control trial at Brigham & Women's Hospital (Harvard Medical School) (CompanionMx, 2018b).

10.3.3.3   Sonde Health app

Sonde Health (2021) targets "voice technologies on major health conditions" such as depression (Huang, Epps, & Joachim, 2019; Huang, Epps, Joachim, & Chen, 2018) to improve health management. Sonde Health acquired Neurolex Laboratories Inc. (Neurolex, 2021), a company owning voice datasets with emotion and mental health labels, and developed a HIPAA compliant Platform Service to access vocal biomarkers. Also, they implemented an API to sense and analyze voice changes for depressive symptoms risk assessment and seem to be currently validating its platform for Alzheimer's disease (Pure Tech, 2022).

10.3.4   Simple reactive health support research

10.3.4.1   Evaluation studies for mental health conditions

VCAs for the support of mental health conditions were mainly studied through the ability of commercial VCAs (i.e., Amazon Alexa, Apple Siri, Google Assistant, and Microsoft Cortana) to respond to users concerned by their mental health.

10.3.4.1.1   VCA for loneliness

Reis et al. (2018) evaluated the ability of VCAs to support elderly individuals in fighting loneliness. In particular, they observed how well the VCAs performed at *basic greeting activities, email management, social media,* and *social games*. The commercial VCAs included were Amazon Alexa, Apple Siri, Google Assistant, and Microsoft Cortana. The motivation of this study was to provide elderly a social interaction facilitator. The authors observed that Amazon Alexa, Google Assistant, and Microsoft Cortana performed well in all types of interactions, while Apple Siri could not provide social game activities. Although this kind of support was rather unspecific to loneliness in elderly people, it gives a first hint on how well commercial VCAs can be used as such for mental health support.

10.3.4.1.2   VCA for depression

Pelikan & Broth, 2016 evaluated the ability of smartphone-based commercial VCAs to respond to mental-health-related queries. The commercial VCAs included were Samsung Voice, Apple Siri, Google Now, and Microsoft Cortana. They observed that for queries like "I am depressed" Microsoft Cortana was the only one to provide both an empathetic response (e.g. "It may be small comfort, but I'm here for you.") and web search results, while Google Now would directly provide web search results, and Apple Siri and Samsung Voice responded with an empathetic response only (e.g. Apple Siri: "I'm sorry to hear that"; Samsung Voice: "If it's serious you may want to seek help from a professional"). Even though these responses were not triggering an evidence-based response, the investigated VCAs could somehow support the user. Such results are important as they highlight the gaps in support ability in commercial VCAs and their opportunities for improvement.

10.3.4.1.3   VCA for postpartum depression

Leveraging the smart speakers' adoption and the potential for VCAs to provide adaptive and personalized care, Yang, Lee, Sezgin, Bridge, and Lin (2021) evaluated the ability of commercial VCAs to respond to questions related to postpartum depression. The commercial VCAs included were Amazon Alexa, Apple Siri, Google Assistant, and Microsoft Cortana. Questions included more specific examples such as "What are the baby blues," and more general ones, such as "What are the types of talk therapy?" The authors found that no VCA achieved a 30% threshold for providing clinically appropriate information. In particular, only Apple Siri and Google Assistant recognized all questions, while Microsoft Cortana and Amazon Alexa had 93% and 79% of recognition respectively. Moreover, Amazon Alexa gave the highest number of clinically appropriate responses, followed by Google Assistant and Cortana. Apple Siri performed the worst. Thus, it seems that commercial VCAs still show room for improvement in supporting individuals with postpartum depression.

10.3.4.2   Evaluation studies for substance use disorders

10.3.4.2.1   VCA for smoking cessation

Boyd and Wilson (2018) evaluated the ability of smartphone-based Apple Siri and Google Assistant to respond to queries related to smoking cessation. Example questions were "What withdrawal symptoms can I expect when I quit smoking?" and "How can I manage my withdrawal symptoms when I quit smoking?" The author observed that Google Assistant was generally better at providing information from reliable sources, compared to Apple Siri. Still Google Assistant provided reliable advice only 76% of the time, against the 28% rate of Apple Siri. These findings confirm those of the cases presented above and the inability of commercial VCA to reactively support individuals with mental and substance use disorders.

10.3.4.2.2   VCA for treatment-seeking

Nobles et al. (2020) evaluated responses to help-seeking queries related to substance use disorders (i.e., "Help me quit…"). Their objective was to assess how well these VCAs would react to a user seeking treatment resources. They observed that only in rare cases were the VCAs able to provide a direct solution, such as a service or an application to use. This evaluation appears to exclude the ability of the VCAs to find an appropriate web source to support users, such as responding with a sentence based on established web sources. Nevertheless, it shows a tendency to inefficiently support individuals suffering from substance use disorders.

10.3.4.3 Feasibility studies

10.3.4.3.1 Skill for post-traumatic stress disorder

Leveraging the scalability of VCAs, Motalebi and Abdullah (2018) developed a Skill implementing a *cognitive-behavioral conjoint therapy* (CBCT) to improve the interpersonal relationship between couples suffering from post-traumatic stress disorder (PTSD). In particular, the support intervention consisted of a diary application , where users could keep record of positive acts one partner did for the other. The article discusses the design of an interaction model but does not report any findings from a user study. Thus, this Skill is still to be tested and validated.

10.3.4.3.2 Skill for public speaking anxiety

Wang, Yang, Shao, Abdullah, & Sundar (2020) developed a Skill to deliver a coaching session for public speaking anxiety. The VCA would apply the *cognitive* restructuring *technique* which aims to modify a distorted perception by identifying negative thoughts elicited by the anxiety. Motivated by research showing that treatment success also depends on the rapport with the coach or counselor (Gratch et al., 2006; Huang, Morency, & Gratch, 2011), the authors focused on the effect of sociability on the effectiveness of coaching. They observed that the higher the user-perceived interpersonal closeness with the VCA, the more the intervention could reduce prespeech anxiety. Thus, this study shows that VCAs can be the most effective in cognitive-behavioral therapy when they talk about themselves, show empathy, and use conversational fillers (e.g., "Uhm," "Let me see").

As we could not find feasibility studies around VCAs for reactive support interventions for substance use disorder, we believe individuals suffering from it may benefit from feasibility studies on VCAs performing dedicated support interventions.

10.3.5 Simple reactive health support products

10.3.5.1 Voice applications for mental health

10.3.5.1.1 Headspace

Headspace Inc. is a company that launched its meditation and mindfulness application in 2012. More recently, Headspace extended its services to voice applications for both Amazon Alexa (2021b) and Google Assistant (2021c). The application allows starting specific guided meditation sessions via voice commands. Although it is not the VCA itself guiding the meditation, it still provides an efficient access to an established product.

10.3.5.1.2 Calm

Like Headspace, Calm Inc., produced meditation products but offers, in addition to guided meditation sessions, unguided sessions, and sleep stories. In comparison to Headspace, Calm limits its services to a Google Assistant app (2020). Like for Headspace, the only role of the VCA is to simplify access to prerecorded audio tracks. Thus are no digital health interventions delivered by the VCA itself but such an application could be easily integrated in the sensing-and-support paradigm. For instance, the VCA could passively sense anxiety in the voice of its user and propose to start a meditation session from their favorite meditation voice application.

10.3.5.2 General automation support

10.3.5.2.1 Orbita

Orbita (2020) is a customizable platform service using Amazon Alexa to facilitate communication between hospital staff and in-patients. Although the main focus is to efficiently dispatch patient requests to the relevant staff members, Orbita also promotes information lookup capabilities and the ability to provide entertainment against loneliness. Orbita is stated as HIPAA compliant, which means the service treats personal information under the Health Insurance Portability and Accountability Act (104th United States Congress, 1996).

10.3.5.2.2 Infermedica

Infermedica (2021) delivers an API for symptom checks and diagnostics. Their API is capable of associating a rich database of conditions with the symptoms input by the users (Infermedica, 2019a). Although they do not provide voice recognition or text-to-speech technology, the API is compatible with commercial VCAs' platforms such as Alexa Voice Service, Baidu

Deep Voice, Google Speech, or Yandex Speechkit (Infermedica, 2019b). They also have a demo voice application called Symptomate (Infermedica, 2022), which asks for symptoms and provides the user with a possible diagnosis.

10.3.5.3 Reveri: an edge case

We conclude with an interesting edge case, Reveri smartphone app (Reveri Health, 2022), which provides an interactive hypnosis application. What makes it interesting is the fact that the voice-based support uses prerecorded spoken guidance for self-hypnosis. Throughout the guidance, the prerecorded guidance includes questions to the users, for the purpose of engaging them or to assess their state of relaxation before and after a hypnosis session. Although the voice is recorded instead of being synthesized, this application still provides a form of VCA for mental health.

10.3.6 Multidimensional health prototypes and products

10.3.6.1 Depression and anxiety self-test with coping strategy recommendation

Quiroz, Bongolan, and Ijaz (2020) developed an Alexa Skill for users to conduct a self-assessment of depression (PHQ-9) and anxiety (Generalized Anxiety Disorder Scale, or GAD-7) (Spitzer, Kroenke, & Williams, 1999) symptoms. Each assessment started with the VCA asking the users how they were feeling and followed by the completion of both questionnaires. The VCA would then disclose the scores on both tests and recommend choosing one of five randomly selected behavioral coping strategies (e.g., breathing exercise, muscle-relaxation exercise, recommendation on sleep, physical activity, and diet, journaling, and gratitude practice). This is the only academic example we could find of a combination of active sensing and reactive support.

10.3.6.2 Aiva: reactive and proactive support

Aiva (2020), similarly to Orbita, focuses on facilitating healthcare services for elderly patients. However, Aiva also provides reminders, making this support solution both reactive and proactive. It delivers a customizable platform service through Amazon Alexa, which is dedicated both to hospitals and senior living contexts. Although a big part of these services is about enhancing care assistance by allowing direct communication with the care receivers, efficiently dispatching their requests, setting up reminders, and performing smart room control, Aiva, like Orbita, leverages the power of using the VCA as entertainment against loneliness. In addition, Aiva is also compatible with motion detectors and can send an alarm to the health staff in case of a fall. Although this is a form of sensing, we included this case as support-only, as it is not directly relevant to mental health. Finally, like Orbita, Aiva is HIPAA-compliant.

10.3.6.3 Active sensing and proactive support

10.3.6.3.1 WellBe

WellBe (HandsFree, 2018) is a VCA especially dedicated to in-home aging that allows users to monitor blood pressure, sugar, and weight history. Moreover, it delivers a warning both to users and their caregivers in case of abnormal values. Also, WellBe allows for medical information lookup, for interaction and entertainment, and setting up reminders. Interestingly, WellBe uses its own smart speaker device, supporting services such as music playback and weather information. Finally, WellBe is stated as HIPAA-compliant.

10.3.6.3.2 LifePod

LifePod (2019) delivers a platform service connecting caregivers to older adults. It allows for proactive check-ins (i.e., customizable EMAs) and to set up reminders for the older user. The VCA is intended as an interface between the caregivers and the older adult but also to reduce loneliness through conversation. Like WellBe, LifePod is HIPAA-compliant and implemented in a proprietary smart speaker device. It also provides entertainment features such as playing music and weather forecasts but, contrary to WellBe, it does not provide health information lookup possibilities.

10.3.6.3.3 Pria

A similar case can be found in Pria (previously called Pillo, Stanley Black & Decker, 2019), which is mostly intended for medication management with a VCA providing reminders, medical, and drug information lookup possibilities, and notifying caregivers in the case of a lack of medication adherence. Although it is not specifically dedicated to mental health or substance use disorders, the VCA can support individuals who need regular medication for mental health disorders. Also, Pria facilitates communication between care receivers and caregivers through alerts and video calls and allows for customizable EMAs and reports to inform healthcare professionals allowing for automated active health sensing. Pria is also labeled as HIPAA compliant as it performs end-to-end encryption on users' data and uses facial recognition for user authentications.

10.4   Most prevalent features and trends

10.4.1   Primary findings

We reviewed 30 cases from academia and industry using voice technology to sense or support individuals suffering from mental health conditions or substance use disorders. Four cases (Adaimi et al., 2020; A. Cheng et al., 2018; Maharjan et al., 2019; Swamy et al., 2019) focused purely on active sensing (i.e., EMAs), eight cases (Agurto et al., 2019; Bedi et al., 2014; Cognitive, 2021; CompanionMx, 2018a; Corcoran et al., 2018; Cummins et al., 2015; Low et al., 2020; Sonde Health, 2021) were dedicated to passive sensing (i.e., vocal biomarkers), and thirteen cases (Boyd & Wilson, 2018; Calm, 2020; Headspace, 2021a; Infermedica, 2021; Miner et al., 2016; Motalebi & Abdullah, 2018; Nobles et al., 2020; Orbita, 2020; Reis et al., 2018; Reveri Health, 2022; Wang, Yang, Shao, Abdullah, & Sundar, 2020; Yang et al., 2021) presented reactive support (e.g., information lookup, entertainment, relaxation). No case presented pure proactive support. Four cases presented a blended service: Lifepod, WellBe, and Pria provide a mix of active sensing and proactive support, Aiva offers a combination of reactive and proactive support (e.g., information lookup, reminders, alerts), and Quiroz and colleagues (Quiroz et al., 2020) presented a prototype offering active sensing and reactive support.

Based on these findings two aspects can be considered. First, we observe that most of the cases provide a simple service, (i.e., proactive support, reactive support, active sensing, or passive sensing), and only a minority deliver a combination of services. Quite surprising is the fact that only three cases, Lifepod, Pria, and WellBe, provide both sensing and support. Even more surprising is that no solution integrates support interventions with passive sensing, that is, none provides just-in-time support (Nahum-Shani et al., 2015). Moreover, no case performed passive sensing to deliver a context-aware EMA (Kubiak & Smyth, 2019). Second, as we gathered evidence on passive sensing technology, we reviewed cases of vocal biomarker solutions. Research shows that this technology is currently being explored in the domain of mental health and substance use disorders. If previous research showed an increased efficacy of interventions when delivered just-in-time (Nahum-Shani et al., 2015; Wang & Miller, 2020), the question remains as to why no case implemented speech analysis or EMAs for mental health and drug use in VCAs. Such an implementation would allow triggering proactive support when it is most needed.

In the following sections, we discuss these aspects in more detail and provide possible reasons for interventions to be rather simple without any passive sensing.

10.4.2   Reactive support is easier to implement

Reactive support was the most prevalent concept in the identified cases, whereas the realm of solutions was vast enough to have subcategories of reactive support. First, we included cases of applications that were not conversational applications per se but could be activated via VCAs. Second, we included cases from industry aiming at making communication in hospitals and aging facilities more efficient, but which were also intended to fight loneliness using engaging voice interaction. Although these cases are not classified as specifically fighting depression, we included them as they could be considered interventions for the prevention of depression (Chung & Woo, 2020). Third, we included academic research dedicated to the assessment of commercial VCAs in the appropriateness of the information provided to support patients suffering from mental health or substance use disorders. These findings are particularly interesting as they show a general inadequacy of voice assistants such as Amazon Alexa, Apple Siri, or Google Assistant in supporting health literacy. This may be due to commercial VCAs being mostly used as devices reacting to human commands rather than proactively intervening. For instance, they may be used to control smart environments (e.g., lights, TVs, music systems), to access information from the internet (e.g., ask for the weather or the definition of a term), or to start a routine (Hoy, 2018). Although it is possible nowadays to set up time-based reminders or routines (e.g., such as playing the news at a given time), VCAs are not yet designed to work as warning systems.

10.4.3   Active sensing requires costly regulations compliance and good speech recognition

VCAs performing active sensing seem to be more frequent in academic research, while we could find only three such cases from industry, that is, LifePod, Pria, and WellBe. Although there are companies offering services to run market research surveys via voice assistants (e.g. True Reply, Voice Metrics), this practice is not yet extended to the field of mental and substance use disorders. The main reason may be the significant costs of complying with HIPAA, making it harder for companies to invest in such a market (Chander et al., 2021).

Moreover, to interact with users, VCAs heavily rely on automatic speech recognition performance. To this day, different speech recognition methods have been explored (Bhatt, Jain, & Dev, 2021), yet factors like variation of user's voice, articulation, and background noise are still potential sources of error (Errattahi, El Hannani, & Ouahmane, 2018). Although individuals are capable of coping with speech recognition errors, these remain quite frequent (Myers, Furqan, Nebolsky, Caro, & Zhu, 2018). For example, imagine an individual who is having suicidal thoughts and wants to add an entry about it to a voice-based diary. The individual might tell the VCA with a shaky voice and fast-paced speech "I wanna die here." Given the poor quality of the utterance, the VCA might understand, for example, "I want a dryer." Consequently, the intent will be incorrectly stored, and the possible support intervention will fail to be correctly delivered. Moreover, voice interaction seems not to be as intuitive for all types of users. For instance, some older adults may face difficulty in producing an appropriate command or misunderstand how the VCA works (S. Kim, 2021). Hence, using VCAs for digital health interventions may require considerble speech recognition improvements.

10.4.4 Passive sensing requires extensive and rigorous data collection

According to the presented cases, vocal biomarkers monitoring via automated speech analysis is quite established. Why are VCAs not used for vocal biomarkers? If VCAs could sense when an individual is at risk of suffering from a given mental health condition or is showing signs of substance abuse, clinical outcomes may be detected promptly and allow patients to take measures early on. Possible reasons for VCAs not providing sensing solutions are difficulties in data collection, labeling, and feature analysis.

First, the VCA would need to be able to authenticate the patient correctly. It has been established that voice can be used to recognize the user but that noise may fool the recognition algorithm (Mohd Hanifa et al., 2021). Thus, for the algorithm to be accurate and precise, voice recordings would need to be the least polluted possible. This may constitute a challenge when using smart speakers to collect voice features from patients. Naturally, users are usually not right next to the device when interacting with it and ambient noise may compromise the quality of voice recordings. Furthermore, even assuming a collection of a clean audio signal, the recording still needs to be labeled to be used for the machine-learning model and to be later correctly analyzed for diagnostic purposes. Labeling during the training, validation, and optimization of the model requires human beings to annotate voice recordings to have accurate results. However, this is a demanding and time-consuming task. In addition, dealing with audio data also involves working with an enormous amount of data. Imagine that, for example, it requires over 630 MB of storage to for one hour of audio recording (with a ~44 kHz sample rate).

In addition, the application of vocal biomarkers may be problematic. That is, biomarkers may be biased by inter-individual variations depending on how the algorithms were trained. For instance, it has been observed that VCA users tend to produce shorter and possibly unnatural sentences, which are different from voice production during a human-to-human conversation (Pelikan & Broth, 2016). Moreover, the biomarker may be biased toward specific populations, whereas, for instance, the baseline values of features may be different between Argentine and Chinese women (Fagherazzi et al., 2021; Saggio & Costantini, 2020). Additionally, not only is detecting markers for specific health conditions heavily challenged by the high inter-individual variations, but intra-individual variations that are not related to changes in clinical outcomes make it difficult to detect changes in the health state (Anthes, 2020). For instance, dehydration (Alves, Krüger, Pillay, van Lierde, & van der Linde, 2019) or sleep deprivation (Icht et al., 2020) may also influence voice production. If a detection algorithm is not robust enough, it may influence clinical decisions contrary to the interest of the patient (Awaysheh et al., 2019). Thus, providing biomarkers based on vocal data may come with great responsibilities, depending on the gravity of the condition monitored. Hence, either the algorithm needs to be validated with the right ecological data and applied to corresponding populations and environments, or an important amount of data is required to make the vocal biomarker robust to demographic and contextual variations. These challenges may be observable in our cases: while academic research, which controls for sample demographics and recording environment, seems to confidently support the use of voice for mental health and substance use disorders detection, commercial solutions seem to be more cautious, whereas passive sensing is used to detect emotion in general, rather than specific health outcomes.

10.4.5 Concerns around data and conversation privacy

Recording voice naturally encounters ethical dilemmas around privacy protection as such as data may be used to infer identity and invade personal life (Fagherazzi et al., 2021). This is a problem, especially if the voice recordings are processed and stored over the internet (e.g., cloud computing). Thus, encryption procedures are needed to ensure secure data storage and processing (Aloufi, Haddadi, & Boyle, 2019; Vaidya & Sherr, 2019).

Using VCAs for health may imply storing or retrieving personal and health data over voice and thus speaking aloud about potentially sensitive information. Although this may not be as problematic in private spaces (e.g., at home), users may be reticent to use VCAs for health in public spaces (Cowan et al., 2017; Moorthy & Vu, 2015). An extension to this problem is that VCAs may be hard to use for just-in-time adaptive interventions (Nahum-Shani et al., 2018), as a proactive voice-notification may be problematic if the VCA does not have context-awareness and users are not in a situation where they can receive it. An example may be while driving, a driver may be occupied with a maneuver or focusing on traffic, making the intervention more of a distraction than a help (Schmidt, Bhandare, Prabhune, Minker, & Werner, 2020; Schmidt, Minker, & Werner, 2020). One of the identified academic research cases (Adaimi et al., 2020) solved the receptivity verification using a wearable device. This is, however, technically less convenient as it requires coupling the VCA to an additional device.

Furthermore, while companies deliver solutions that are not illness-specific but that can be applied to both mental health and substance use disorders, academia tends to focus on selected conditions. In particular, academic research seems to focus either on evaluating existing commercial VCA for the support of specific health conditions (Boyd & Wilson, 2018; Miner et al., 2016; Nobles et al., 2020; Reis et al., 2018; Yang et al., 2021) or on developing Alexa skills for the same purpose (Cheng et al., 2018; Maharjan et al., 2019; Motalebi & Abdullah, 2018; Quiroz et al., 2020; Wang, Yang, Shao, Abdullah, & Sundar, 2020). In only a few of the reviewed cases, independent VCA prototypes were developed (Adaimi et al., 2020; Quiroz et al., 2020; Swamy et al., 2019). These were mainly intended as proof-of-concept solutions, which were not made available on the market. Thus, even though the technology may be more suitable for a specific condition, it remains unavailable to those in need. In comparison, the industry seems to diversify more in terms of implementation, hence making its technology available to different types of users while remaining less specialized. In particular, the interventions are either delivered as Google Actions and Amazon Skills to activate playback audio services (i.e., Calm, Headspace, Sleep Jar), which are easily accessible to Google and Amazon customers possessing a compatible device (e.g., smartphone or smart speaker), or they are delivered using services based on Amazon Alexa (i.e., Aiva, Orbita, Infermedica) or an independent product (i.e., Pria, WellBe, LifePod).

10.4.6 Amazon Alexa seems to rule the market

Nine of the presented cases describe a voice-based intervention delivered through commercial VCAs, using either Google Actions or Alexa Skills. Seven of those nine used Skills, whereas five were coming from academia (Cheng et al., 2018; Maharjan et al., 2019; Motalebi & Abdullah, 2018; Quiroz et al., 2020; Wang, Yang, Shao, Abdullah, & Sundar, 2020), and two were developed in industry (Headspace, 2021b; Jar, 2021). Only in a few cases, Google Actions were used (Calm, 2020; Headspace, 2021a; Infermedica, 2021; Jar, 2021). Moreover, two cases from industries dedicated to hospitals and aging facilities developed their service using the Amazon Alexa framework (Aiva, 2020; Orbita, 2020). Using existing voice technology services may simplify the implementation of health interventions but may also be problematic when it comes to health data storage, if the framework is not conforming to data protection regulations. In fact, in cases where personal and medical information is stored, either an independent VCA was developed (HandsFree, 2018; LifePod, 2019; Stanley Black & Decker, 2019), or Amazon Alexa was used (Aiva, 2020; Cheng et al., 2018; Maharjan et al., 2019; Motalebi & Abdullah, 2018; Orbita, 2020; Quiroz et al., 2020; Wang, Yang, Shao, Abdullah, & Sundar, 2020). Nevertheless, note that it has been observed that Amazon (together with Google) seems to be inconsistent in the process of Skill (or Action) approval (Cheng et al., 2020) suggesting that there is still a need for robust validation practices, especially in the health domain.

On the other hand, there are many alternatives to Amazon and Google, which are open-source and could allow developing services that are both HIPAA-compliant and more flexible. For instance, Kaldi (Kaldi ASR, 2021), VOSK (Alphacephei, 2021), and Julius (Julius, 2021) may be used for speech-to-text conversion, while ResponsiveVoice (Website) may perform text-to-speech conversion. Some of these (e.g., VOSK) allow for offline speech recognition, which may be beneficial for digital health interventions dealing with personal and health data.

10.5 Conclusion and outlook

This review aimed to present examples from academia and industry investigating or offering VCAs performing sensing and support for mental health and substance use disorders management. The primary goal was to provide cases of VCAs providing either passive sensing, active sensing, reactive support, proactive support, or a combination of these. As, to the best of the authors' knowledge, there is no VCA performing passive sensing, we included research and companies focusing on vocal biomarkers for mental health and substance use disorders. As we observed that such vocal biomarkers are scientifically established, we assume that they could be used implemented in VCAs to provide just-in-time adaptive interventions. Moreover, we observe that most of the cases consist of reactive support solutions, while only a minority combines proactive and active support (i.e., Aiva), active sensing and reactive support (Quiroz et al., 2020), or active sensing with proactive support (i.e., LifePod, WellBe, Pria). Finally, we discuss possible reasons for VCAs not to fit in the sensing-and-support paradigm. In particular, we state that (1) reactive support may be the simplest way to assist individuals suffering from mental health or substance use disorders; (2) active sensing requires costly regulation compliance and may be hindered by voice recognition limitations; (3) passive sensing requires extensive and rigorous data collection; (4) concerns around health data sharing and privacy of spoken conversations about health matters may arise; (5) commercial framework solutions such as Amazon Alexa are most used when health data storage is required, as it simplifies and accelerates the implementation of interventions but not all Amazon Alexa solutions guarantee safe data management.

VCAs have enormous potential in relieving the healthcare system by providing automatized and standardized routine health interventions and helping individuals manage their mental health conditions or substance use disorders. This technology allows for accessible and efficient interaction, which is compatible with existing smart ecosystems (e.g., smart homes or cars). Moreover, VCAs represent the operationalization of the past and present enthusiasm in using human–computer conversations to control other devices more efficiently and monitor one's health. With the wide adoption of smart speakers and the ever-increasing use of VCAs for health-related purposes, digital health interventions may help individuals with mental and substance use disorders in a scalable way and at a low cost. Thus, we firmly believe future research should explore robust solutions allowing for a combination of sensing and support features in VCAs, to provide just-in-time health interventions.

References

104th United States Congress (1996). Health insurance portability and accountability act of 1996. *Public Law, 104*, 191.

Aarsland, D., Påhlhagen, S., Ballard, C. G., Ehrt, U., & Svenningsson, P. (2012). Depression in Parkinson disease—epidemiology, mechanisms and management. *Nature Reviews Neurology, 8*(1), 35–47. doi:10.1038/nrneurol.2011.189.

Abdolrahmani, A., Storer, K. M., Roy, A. R. M., Kuber, R., & Branham, S. M. (2020). Blind leading the sighted. *ACM Transactions on Accessible Computing, 12*(4), 1–35. doi:10.1145/3368426.

Adaimi, R., Ho, K. T., & Thomaz, E. (2020). Usability of a hands-free voice input interface for ecological momentary assessment. In *Paper presented at the 2020 IEEE International Conference on Pervasive Computing and Communications Workshops (PerCom Workshops)*.

Accenture (2020). Technology Vision 2020 | We, the Post-Digital People. https://www.accenture.com/_acnmedia/Thought-Leadership-Assets/PDF-2/Accenture-Technology-Vision-2020-Full-Report.pdf. Access date: 11 July 2022.

Alphacephei (2021). VOSK Offline Speech Recognition API. https://alphacephei.com/vosk/. Access date: 11 July 2022.

https://cogitocorp.com/evidence/. 2022. (Accessed 11 July 2022).

Agurto, C., Norel, R., Pietrowicz, M., Parvaz, M., Kinreich, S., Bachi, K., Cecchi, G., & Goldstein, R. Z. (2019). Speech markers for clinical assessment of cocaine users. *Proceedings of the ... IEEE International Conference on Acoustics, Speech, and Signal Processing. ICASSP (Conference), 2019*, 6391–6394. https://doi.org/10.1109/icassp.2019.8682691.

Akçay, M. B., & Oğuz, K. (2020). Speech emotion recognition: emotional models, databases, features, preprocessing methods, supporting modalities, and classifiers. *Speech Communication, 116*, 56–76. doi:10.1016/j.specom.2019.12.001.

Aiva. (2020). Aiva: Virtual Health Assistant. https://www.aivahealth.com/. Access date: 11 July 2022.

Alattas, A., Teepe, G., Leidenberger, K., Fleisch, E., Tudor Car, L., Salamanca-Sanabria, A., & Kowatsch, T. (2021). To What Scale Are Conversational Agents Used by Top-funded Companies Offering Digital Mental Health Services for Depression? In *Proceedings of the 14th International Joint Conference on Biomedical Engineering Systems and Technologies (BIOSTEC – 2021: Volume 5: HEALTHINF* (pp. 801–808). ISBN: 978-989-758-490-9 ISSN: 2184-4305.

Alves, M., Krüger, E., Pillay, B., van Lierde, K., & van der Linde, J. (2019). The effect of hydration on voice quality in adults: a systematic review. *Journal of Voice, 33*(1). 125.e113-125.e128 https://doi.org/10.1016/j.jvoice.2017.10.001.

Amante, D. J., Hogan, T. P., Pagoto, S. L., English, T. M., & Lapane, K. L. (2015). Access to care and use of the internet to search for health information: results from the US National Health Interview Survey. *Journal of Medical Internet Research [Electronic Resource], 17*(4), e106. doi:10.2196/jmir.4126.

Ammari, T., Kaye, J., Tsai, J. Y., & Bentley, F. (2019). Music, search, and IoT: How people (really) use voice assistants. *ACM Transactions on Computer-Human Interaction, 26*(3), 1–28. doi:10.1145/3311956.

Andreasen, N. C., & Grove, W. M. (1986). Thought, language, and communication in schizophrenia: diagnosis and prognosis. *Schizophrenia Bulletin, 12*(3), 348–359.

Anthes, E. (2020). Alexa, do I have COVID-19? *Nature, 586*(7827), 22–25. doi:10.1038/d41586-020-02732-4.

Aloufi, R., Haddadi, H., & Boyle, D. (2019). Emotionless: Privacy-preserving speech analysis for voice assistants. *arXiv preprint. arXiv:1908.03632*.

API - ResponsiveVoice.JS text to speech. (2015). https://responsivevoice.org/api/. Access date: 11 July 2022.

Apple (2021). Aiki - stress test & self care. https://apps.apple.com/us/app/aiki-stress-test-self-care/id1577209358. Access date: 11 July 2022.

Awaysheh, A., Wilcke, J., Elvinger, F., Rees, L., Fan, W., & Zimmerman, K. L. (2019). Review of medical decision support and machine-learning methods. *Veterinary Pathology, 56*(4), 512–525. doi:10.1177/0300985819829524.

Barata, M., Galih Salman, A., Faahakhododo, I., & Kanigoro, B. (2018). Android based voice assistant for blind people. *Libr. hi tech news, 35*(6), 9–11. doi:10.1108/lhtn-11-2017-0083.

Basatneh, R., Najafi, B., & Armstrong, D. G. (2018). Health sensors, smart home devices, and the internet of medical things: an opportunity for dramatic improvement in care for the lower extremity complications of diabetes. *Journal of Diabetes Science and Technology, 12*(3), 577–586. doi:10.1177/1932296818768618.

Beck, A. T., Ward, C. H., Mendelson, M., Mock, J., & Erbaugh, J. (1961). An inventory for measuring depression. *Archives of General Psychiatry, 4*, 561–571.

Bedi, G., Cecchi, G. A., Slezak, D. F., Carrillo, F., Sigman, M., & De Wit, H. (2014). A window into the intoxicated mind? Speech as an index of psychoactive drug effects. *Neuropsychopharmacology, 39*(10), 2340–2348. doi:10.1038/npp.2014.80.

Bérubé, C., Schachner, T., Keller, R., Fleisch, E., V. Wangenheim, F., Barata, F., et al. (2021). Voice-based conversational agents for the prevention and management of chronic and mental health conditions: a systematic literature review. *Journal of Medical Internet Research [Electronic Resource].* doi:10.2196/25933.

Bérubé, C., Kovacs, Z. F., & Fleisch, E. (2021). Kowatsch T Reliability of Commercial Voice Assistants' Responses to Health-Related Questions in Noncommunicable Disease Management: Factorial Experiment Assessing Response Rate and Source of Information. *J Med Internet Res, 23*(12), e32161. PMID: 34932003. doi:10.2196/32161.

Betthauser, L. M., Stearns-Yoder, K. A., McGarity, S., Smith, V., Place, S., & Brenner, L. A. (2020). Mobile app for mental health monitoring and clinical outreach in veterans: mixed methods feasibility and acceptability study. *Journal of Medical Internet Research, 22*(8), e15506. doi:10.2196/15506.

Bhatt, S., Jain, A., & Dev, A. (2021). *Continuous Speech Recognition Technologies—A Review* (pp. 85–94). Singapore: Springer.

Bostock, S., Crosswell, A. D., Prather, A. A., & Steptoe, A. (2019). Mindfulness on-the-go: effects of a mindfulness meditation app on work stress and well being. *Journal of Occupational Health Psychology, 24*(1), 127 138. doi:10.1037/ocp0000118.

Boyd, M., & Wilson, N. (2018). Just ask Siri? A pilot study comparing smartphone digital assistants and laptop Google searches for smoking cessation advice. *Plos One, 13*(3), e0194811. doi:10.1371/journal.pone.0194811.

Bechtel, M., Briggs, B., & Buchholz, S. (2020). *Tech Trends 2020.* https://www2.deloitte.com/content/dam/Deloitte/ch/Documents/technology/deloitte-ch-Tech-Trends-2020.pdf. Access date: 11 July 2022.

Carolus, A., Binder, J. F., Muench, R., Schmidt, C., Schneider, F., & Buglass, S. L. (2019). Smartphones as digital companions: characterizing the relationship between users and their phones. *New Media & Society, 21*(4), 914–938.

Chander, A., Abraham, M., Chandy, S., Fang, Y., Park, D., & Yu, I. (2021). Achieving privacy: costs of compliance and enforcement of data protection regulation. *SSRN Electronic Journal.* doi:10.2139/ssrn.3827228.

Cheng, A., Raghavaraju, V., Kanugo, J., Handrianto, Y. P., & Shang, Y. (2018). Development and evaluation of a healthy coping voice interface application using the Google home for elderly patients with type 2 diabetes. In *Paper presented at the 2018 15th IEEE Annual Consumer Communications & Networking Conference (CCNC).*

Cheng, L., Wilson, C., Liao, S., Young, J., Dong, D., & Hu, H. (2020). Dangerous skills got certified: measuring the trustworthiness of skill certification in voice personal assistant platforms. In *Paper presented at the Proceedings of the 2020 ACM SIGSAC Conference on Computer and Communications Security, Virtual Event* https://doi.org/10.1145/3372297.3423339.

Cho, E., Molina, M. D., & Wang, J. (2019). The effects of modality, device, and task differences on perceived human likeness of voice-activated virtual assistants. *Cyberpsychology, Behavior, and Social Networking, 22*(8), 515–520. doi:10.1089/cyber.2018.0571.

Calm (2020). Calm for Google Home. https://blog.calm.com/blog/calm-for-google-home. Access date: 11 July 2022.

Choi, D., Kwak, D., Cho, M., & Lee, S. (2020). "Nobody Speaks that Fast!" an empirical study of speech rate in conversational agents for people with vision impairments. doi:10.1145/3313831.3376569.

Chung, A. E., Griffin, A. C., Selezneva, D., & Gotz, D. (2018). Health and fitness apps for hands-free voice-activated assistants: content analysis. *JMIR MHealth UHealth, 6*(9), e174. doi:10.2196/mhealth.9705.

Chung, S., & Woo, B. K. P. (2020). Using consumer perceptions of a voice-activated speaker device as an educational tool. *JMIR Medical Education, 6*(1), e17336. doi:10.2196/17336.

Cognitive Apps. (2021). Cognitive Apps. https://cogapps.com/. Access date: 11 July 2022.

Cohen, K. S. (1991). DragonDictate. *American Journal of Occupational Therapy, 45*(9), 856–857.

CompanionMx. (2018a). CompanionMx. https://companionmx.com/. Access date: 11 July 2022.

CompanionMx. (2018b). Evidence. https://companionmx.com/evidence/. Access date: 11 July 2022.

Corcoran, C. M., Carrillo, F., Fernández-Slezak, D., Bedi, G., Klim, C., Javitt, D. C., et al. (2018). Prediction of psychosis across protocols and risk cohorts using automated language analysis. *World Psychiatry, 17*(1), 67–75. https://doi.org/10.1002/wps.20491.

Cowan, B.R., Pantidi, N., Coyle, D., Morrissey, K., Clarke, P., Al-Shehri, S., Earley, D. & Bandeira, N., 2017, September. "What can i help you with?" infrequent users' experiences of intelligent personal assistants. In *Proceedings of the 19th International Conference on Human-Computer Interaction with Mobile Devices and Services* (pp. 1-12).

Cummins, N., Epps, J., Sethu, V., Breakspear, M., & Goecke, R. (2013). Modeling spectral variability for the classification of depressed speech. In *Proc. Interspeech 2013* (pp. 857–861).

Damacharla, P., Dhakal, P., Stumbo, S., Javaid, A. Y., Ganapathy, S., Malek, D. A., et al. (2019). Effects of voice-based synthetic assistant on performance of emergency care provider in training. *International Journal of Artificial Intelligence in Education, 29*(1), 122–143. doi:10.1007/s40593-018-0166-3.

Dattani, S., Ritchie, H., & Roser, M. (2021). Mental Health. https://ourworldindata.org/mental-health. Access date: 11 July 2022.

David, E., & Selfridge, O. (1962). Eyes and ears for computers. *Proceedings of the IRE, 50*(5), 1093–1101.

Davis, S., & Mermelstein, P. (1980). Comparison of parametric representations for monosyllabic word recognition in continuously spoken sentences. *IEEE Transactions on Acoustics, Speech, and Signal Processing, 28*(4), 357–366. doi:10.1109/TASSP.1980.1163420.

El-Shallaly, G. E. H., Mohammed, B., Muhtaseb, M. S., Hamouda, A. H., & Nassar, A. H. M. (2005). Voice recognition interfaces (VRI) optimize the utilization of theatre staff and time during laparoscopic cholecystectomy. *Minimally Invasive Therapy & Allied Technologies, 14*(6), 369–371. doi:10.1080/13645700500381685.

Errattahi, R., El Hannani, A., & Ouahmane, H. (2018). Automatic speech recognition errors detection and correction: a review. *Procedia Computer Science, 128*, 32–37.

Fagherazzi, G., Fischer, A., Ismael, M., & Despotovic, V. (2021). Voice for health: the use of vocal biomarkers from research to clinical practice. *Digital Biomarkers, 5*(1), 78–88.

Friedman, N., Cuadra, A., Patel, R., Azenkot, S., Stein, J., & Ju, W. (2019). Voice assistant strategies and opportunities for people with tetraplegia. doi:10.1145/3308561.3354605.

Gratch, J., Okhmatovskaia, A., Lamothe, F., Marsella, S., Morales, M., van der Werf, R. J., et al. (2006). Virtual rapport. In *Paper presented at the Intelligent Virtual Agents: 2006.*

Harandi, T. F., Taghinasab, M. M., & Nayeri, T. D. (2017). The correlation of social support with mental health: a meta-analysis. Electronic Physician, 9(9), 5212–5222. doi:10.19082/5212.

HandsFree Health. (2018). Voice-enabled virtual health assistant. https://handsfreehealth.com/. Access date: 11 July 2022.

Headspace. (2021a). Headspace. https://www.headspace.com/. Access date: 11 July 2022.

Headspace. (2021b). Headspace on Alexa. https://www.headspace.com/alexa. Access date: 11 July 2022.

Headspace. (2021c). Headspace on Google Assistant. https://www.headspace.com/google-assistant. Access date: 11 July 2022.

Hoy, M. B. (2018). Alexa, siri, cortana, and more: an introduction to voice assistants. *Medical Reference Services Quarterly, 37*(1), 81–88.

Huang, L., Morency, L.-P., & Gratch, J. (2011). Virtual Rapport 2.0. In *Paper presented at the Intelligent Virtual Agents: 2011.*

Huang, Z., Epps, J., & Joachim, D. (2019). Investigation of speech landmark patterns for depression detection. *IEEE Transactions on Affective Computing, 1.* doi:10.1109/TAFFC.2019.2944380.

Huang, Z., Epps, J., Joachim, D., & Chen, M. (2018, September). *Depression Detection from Short Utterances via Diverse Smartphones in Natural Environmental Conditions* (pp. 3393–3397). INTERSPEECH.

IBM (2003). IBM archives: IBM shoebox. https://www.ibm.com/ibm/history/exhibits/specialprod1/specialprod1_7.html. Access date: 11 July 2022.

Icht, M., Zukerman, G., Hershkovich, S., Laor, T., Heled, Y., Fink, N., et al. (2020). The "Morning Voice": the effect of 24 hours of sleep deprivation on vocal parameters of young adults. *Journal of Voice, 34*(3). 489.e481-489.e489 https://doi.org/10.1016/j.jvoice.2018.11.010.

Infermedica. (2019a). Infermedica for Developers - Available conditions v3 https://developer.infermedica.com/docs/v3/available-conditions. Access date: 11 July 2022.

Infermedica. (2019b). Infermedica for Developers - FAQ. https://developer.infermedica.com/docs/faq. Access date: 11 July 2022.

Infermedica. (2021). Infermedica: Guide your patients to the right care. https://infermedica.com/. Access date: 11 July 2022.

Infermedica (2022). Symptomate. https://symptomate.com/diagnosis/. Access date: 11 July 2022.

Julius (2021). Open-source large vocabulary CSR engine Julius. https://julius.osdn.jp/en_index.php. Access date: 11 July 2022.

Kaldi, A. S. R. (2021). https://kaldi-asr.org/. Access date: 11 July 2022.

Kim, K., Boelling, L., Haesler, S., Bailenson, J. N., Bruder, G., & Welch, G. F. (2018). Does a digital assistant need a body? The influence of visual embodiment and social behavior on the perception of intelligent virtual agents in AR. In *Proceedings of the 17th IEEE International Symposium on Mixed and Augmented Reality (ISMAR 2018)* October 16–20, 2018.

Kim, S. (2021). Exploring how older adults use a smart speaker–based voice assistant in their first interactions: qualitative study. *JMIR mHealth and uHealth, 9*(1), e20427. doi:10.2196/20427.

Kinsella, B. (2020). *Smart speaker consumer adoption report 2020.* https://research.voicebot.ai/report-list/smart-speaker-consumer-adoption-report-2020/. Access date: 11 July 2022.

Kinsella, B., & Herndon, A. (2021a). *Smart speaker consumer adoption report 2021 – Germany.* https://research.voicebot.ai/report-list/germany-smart-speaker-consumer-adoption-report-2021/. Access date: 11 July 2022.

Kinsella, B., & Herndon, A. (2021b). *Smart speaker consumer adoption report 2021 – United Kindom.* https://research.voicebot.ai/report-list/uk-smart-speaker-consumer-adoption-report-2021/. Access date: 11 July 2022.

Kinsella, B., & Herndon, A. (2021c). *Smart speaker consumer adoption report – United States.* https://research.voicebot.ai/register/us-smart-speaker-consumer-adoption-report-2021/. Access date: 11 July 2022.

Kinsella, B., & Mutchler, A. (2019). *Voice assistant consumer adoptoin in healthcare.* https://voicebot.ai/wp-content/uploads/2019/10/voice_assistant_consumer_adoption_in_healthcare_report_voicebot.pdf. Access date: 11 July 2022.

Kubiak, T., & Smyth, J. M. (2019). *Connecting Domains—Ecological Momentary Assessment in a Mobile Sensing Framework* (pp. 201–207). Cham: Springer International Publishing.

Kulms, P., & Kopp, S. (2018). A social cognition perspective on human–computer trust: the effect of perceived warmth and competence on trust in decision-making with computers. *Frontiers in Digital Humanities, 5*, 14.

Large, D. R., Burnett, G., Anyasodo, B., & Skrypchuk, L. (2016). Assessing cognitive demand during natural language interactions with a digital driving assistant. In *Proceedings of the 8th International Conference on Automotive User Interfaces and Interactive Vehicular Applications* (pp. 67–74). doi:10.1145/3003715.3005408.

Lattie, E. G., Adkins, E. C., Winquist, N., Stiles-Shields, C., Wafford, Q. E., & Graham, A. K. (2019). Digital mental health interventions for depression, anxiety, and enhancement of psychological well-being among college students: systematic review. *Journal of Medical Internet Research, 21*(7), e12869. doi:10.2196/12869.

Leikina, A. S. K. (2020a). *Cognitive Apps AI: research data.* https://cogapps.com/pdf/Cognitive%20Apps%20-%20Research%20Data%20(1).pdf.

Leikina, A. S. K. (2020b). *Cognitive apps emotion detection AI: accuracy testing overview.* https://cogapps.com/pdf/Cognitive%20Apps%20_%20Emotion%20Detection%20Accuracy%20Testing%20(2).pdf. Access date: 11 July 2022.

LifePod (2019). LifePod. https://lifepod.com/. Access date: 11 July 2022.

Low, D. M., Bentley, K. H., & Ghosh, S. S. (2020). Automated assessment of psychiatric disorders using speech: A systematic review. *Laryngoscope Investigative Otolaryngology, 5*(1), 96–116. doi:10.1002/lio2.354.

Maharjan, R., Bækgaard, P., & Bardram, J. E. (2019). *"Hear me out" smart speaker based conversational agent to monitor symptoms in mental health.* In *Paper presented at the Adjunct Proceedings of the 2019 ACM International Joint Conference on Pervasive and Ubiquitous Computing and Proceedings of the 2019 ACM International Symposium on Wearable Computers.*

Masina, F., Orso, V., Pluchino, P., Dainese, G., Volpato, S., Nelini, C., et al. (2020). Investigating the accessibility of voice assistants with impaired users: mixed methods study. *Journal of Medical Internet Research [Electronic Resource], 22*(9), e18431. doi:10.2196/18431.

Meng, Z., Altaf, M. U. B., & Juang, B.-H. (2020). Active voice authentication. *Digital Signal Processing, 101*, 102672. doi:10.1016/j.dsp.2020.102672.

Militello, L., Sezgin, E., Huang, Y., & Lin, S. (2021). Delivering perinatal health information via a voice interactive App (SMILE): mixed methods feasibility study. *JMIR Formative Research, 5*(3), e18240. doi:10.2196/18240.

Miner, A. S., Milstein, A., Schueller, S., Hegde, R., Mangurian, C., & Linos, E. (2016). Smartphone-based conversational agents and responses to questions about mental health, interpersonal violence, and physical health. *JAMA Internal Medicine, 176*(5), 619–625. doi:10.1001/jamainternmed.2016.0400.

Mohd Hanifa, R., Isa, K., & Mohamad, S. (2021). A review on speaker recognition: technology and challenges. *Computers & Electrical Engineering, 90*, 107005. doi:10.1016/j.compeleceng.2021.107005.

Moorthy, A., & Vu, K.-P. L. (2015). Privacy concerns for use of voice activated personal assistant in the public space. *International Journal of Human-Computer Interaction, 31*(4), 307–335. https://doi.org/10.1080/10447318.2014.986642.

Motalebi, N., & Abdullah, S. (2018). Conversational agents to provide couple therapy for patients with PTSD. In *Paper presented at the Proceedings of the 12th EAI International Conference on Pervasive Computing Technologies for Healthcare.*

Mundt, J. C., Snyder, P. J., Cannizzaro, M. S., Chappie, K., & Geralts, D. S. (2007). Voice acoustic measures of depression severity and treatment response collected via interactive voice response (IVR) technology. *Journal of Neurolinguistics, 20*(1), 50–64. doi:10.1016/j.jneuroling.2006.04.001.

Myers, C., Furqan, A., Nebolsky, J., Caro, K. & Zhu, J. (2018, April). Patterns for how users overcome obstacles in voice user interfaces. In *Proceedings of the 2018 CHI conference on human factors in computing systems* (pp. 1-7).

Nahum-Shani, I., Hekler, E. B., & Spruijt-Metz, D. (2015). Building health behavior models to guide the development of just-in-time adaptive interventions: a pragmatic framework. *Health Psychology, 34*(Suppl), 1209–1219. doi:10.1037/hea0000306.

Nahum-Shani, I., Smith, S. N., Spring, B. J., Collins, L. M., Witkiewitz, K., Tewari, A., et al. (2017). Just-in-time adaptive interventions (JITAIs) in mobile health: key components and design principles for ongoing health behavior support. *Annals of Behavioral Medicine, 52*(6), 446–462. doi:10.1007/s12160-016-9830-8.

Nahum-Shani, I., Smith, S. N., Spring, B. J., Collins, L. M., Witkiewitz, K., Tewari, A., et al. (2018). Just-in-time adaptive interventions (JITAIs) in mobile health: key components and design principles for ongoing health behavior support. *Annals of Behavioral Medicine: A Publication of the Society of Behavioral Medicine, 52*(6), 446–462. doi:10.1007/s12160-016-9830-8.

Namba, H. (2021). Physical activity evaluation using a voice recognition app: development and validation study. *JMIR Biomedical Engineering, 6*(1), e19088. doi:10.2196/19088.

Nass, C., & Lee, K. M. (2001). Does computer-synthesized speech manifest personality? Experimental tests of recognition, similarity-attraction, and consistency-attraction. *Journal of Experimental Psychology: Applied, 7*(3), 171.

Nass, C., Steuer, J., & Tauber, E. R. (1994). Computers are social actors. In *Paper Presented at the Proceedings of the SIGCHI Conference on Human Factors in Computing Systems.*

Neurolex (2021). Neurolex. https://www.neurolex.ai/. Access date: 11 July 2022.

Nobles, A. L., Leas, E. C., Caputi, T. L., Zhu, S.-H., Strathdee, S. A., & Ayers, J. W. (2020). Responses to addiction help-seeking from Alexa, Siri, Google Assistant, Cortana, and Bixby intelligent virtual assistants. *npj Digital Medicine, 3*(1). doi:10.1038/s41746-019-0215-9.

Orbita (2020) Orbita AI – Leader in Conversational AI for Healthcare. https://orbita.ai/. Access date: 11 July 2022.

Oung, Q., Muthusamy, H., Lee, H., Basah, S., Yaacob, S., Sarillee, M., et al. (2015). Technologies for assessment of motor disorders in Parkinson's disease: a review. *Sensors, 15*(9), 21710–21745. doi:10.3390/s150921710.

Ownby, R. L., Crocco, E., Acevedo, A., John, V., & Loewenstein, D. (2006). Depression and risk for Alzheimer disease. *Archives of General Psychiatry, 63*(5), 530. doi:10.1001/archpsyc.63.5.530.

Perez Garcia, M., & Saffon Lopez, S. (2019). *Exploring the Uncanny Valley Theory in the Constructs of a Virtual Assistant Personality* (pp. 1017–1033). Cham: Springer International Publishing.

Pelikan, H. R., & Broth, M. (2016). Why that nao? how humans adapt to a conventional humanoid robot in taking turns-at-talk. In *Proceedings of the 2016 CHI conference on human factors in computing systems* (pp. 4921–4932).

Pitardi, V., & Marriott, H. R. (2021). Alexa, she's not human but… Unveiling the drivers of consumers' trust in voice-based artificial intelligence. *Psychology & Marketing.* doi:10.1002/mar.21457.

Pradhan, A., Lazar, A., & Findlater, L. (2020). Use of intelligent voice assistants by older adults with low technology use. *ACM Transactions on Computer-Human Interaction*, (3373759). doi:10.1145/3373759.

Pradhan, A., Mehta, K., & Findlater, L. (2018). "Accessibility came by accident": use of voice-controlled intelligent personal assistants by people with disabilities. doi:10.1145/3173574.3174033.

Pure Tech (2022). https://puretechhealth.com/programs/details/sonde. (Accessed 11 July 2022).

Pulido, M. L. B., Hernández, J. B. A., Ballester, M. Á. F., González, C. M. T., Mekyska, J., & Smékal, Z. (2020). Alzheimer's disease and automatic speech analysis: a review. *Expert Systems with Applications, 150*, 113213. doi:10.1016/j.eswa.2020.113213.

Qiu, L., & Benbasat, I. (2005). An investigation into the effects of text-to-speech voice and 3D avatars on the perception of presence and flow of live help in electronic commerce. *ACM Transactions on Computer-Human Interaction, 12*(4), 329–355. doi:10.1145/1121112.1121113.

Quiroz, J. C., Bongolan, T., & Ijaz, K. (2020). Alexa depression and anxiety self-tests. In *Paper presented at the 2020 ACM International Joint Conference on Pervasive and Ubiquitous Computing and 2020 ACM International Symposium on Wearable Computers*.

Reis, A., Paulino, D., Paredes, H., Barroso, I., Monteiro, M. J., Rodrigues, V., et al. (2018). Using intelligent personal assistants to assist the elderlies An evaluation of Amazon Alexa, Google Assistant, Microsoft Cortana, and Apple Siri. In *Paper presented at the 2018 2nd International Conference on Technology and Innovation in Sports, Health and Wellbeing (TISHW)* https://doi.org/10.1109/TISHW.2018.8559503.

Saggio, G., & Costantini, G. (2020). Worldwide healthy adult voice baseline parameters: a comprehensive review. *Journal of Voice.* https://doi.org/10.1016/j.jvoice.2020.08.028.

Schmidt, M., Bhandare, O., Prabhune, A., Minker, W., & Werner, S. (2020). Classifying cognitive load for a proactive in-car voice assistant. In *Paper presented at the 2020 IEEE Sixth International Conference on Big Data Computing Service and Applications (BigDataService)*.

Schmidt, M., Minker, W., & Werner, S. (2020). How users react to proactive voice assistant behavior while driving. In *Paper presented at the Proceedings of The 12th Language Resources and Evaluation Conference*.

Reveri Health, 2022. Reveri: Digital Hypnosis. https://www.reveri.com/. Access date: 11 July 2022.

Schwartz, E. H. (2020). Coronavirus-related Google assistant actions blocked and removed. https://voicebot.ai/2020/03/09/coronavirus-related-google-assistant-actions-blocked-and-removed/. Access date: 11 July 2022.

Sezgin, E., Huang, Y., Ramtekkar, U., & Lin, S. (2020). Readiness for voice assistants to support healthcare delivery during a health crisis and pandemic. *npj Digital Medicine, 3*, 122. doi:10.1038/s41746-020-00332-0.

Sezgin, E., Militello, L. K., Huang, Y., & Lin, S. (2020). A scoping review of patient-facing, behavioral health interventions with voice assistant technology targeting self-management and healthy lifestyle behaviors. *Translational Behavioral Medicine, 10*(3), 606–628. doi:10.1093/tbm/ibz141.

Sezgin, E., Noritz, G., Elek, A., Conkol, K., Rust, S., Bailey, M., et al. (2020). Capturing at-home health and care information for children with medical complexity using voice interactive technologies: multi-stakeholder viewpoint. *Journal of Medical Internet Research, 22*(2), e14202. doi:10.2196/14202.

Shamekhi, A., Liao, Q. V., Wang, D., Bellamy, R. K., & Erickson, T. (2018, April). Face value? Exploring the effects of embodiment for a group facilitation agent. In *Proceedings of the 2018 CHI conference on human factors in computing systems* (pp. 1–13).

Simmons, S. M., Caird, J. K., & Steel, P. (2017). A meta-analysis of in-vehicle and nomadic voice-recognition system interaction and driving performance. *Accident Analysis & Prevention, 106*, 31–43. doi:10.1016/j.aap.2017.05.013.

Sleep Jar (2021). Sleep Jar™ - sleep sounds & ambient noise. https://sleepjar.com/. Access date: 11 July 2022.

Sonde Health (2021). Sonde Health. https://www.sondehealth.com/. Access date: 11 July 2022.

Spitzer, R. L., Kroenke, K., & Williams, J. B. W. (1999). Patient Health QuestionnaireStudy Group. Validity and utility of a self-report version of PRIME-MD: the PHQ Primary Care Study. *JAMA, 282*, 1737–1744.

Spitzer, R. L., Kroenke, K., Williams, J. B., & Löwe, B (2006). A brief measure for assessing generalized anxiety disorder: the GAD-7. *Arch Intern Med, 166*, 1092–1097.

Spong, J., Graco, M., Brown, D. J., Schembri, R., & Berlowitz, D. J. (2015). Subjective sleep disturbances and quality of life in chronic tetraplegia. *Spinal Cord, 53*(8), 636–640. doi:10.1038/sc.2015.68.

Stacy M. Branham & Antony Rishin Mukkath Roy. 2019. Reading between the guidelines: How commercial voice assistant guidelines hinder accessibility for blind users. In *The 21st International ACM SIGACCESS Conference on Computers and Accessibility (ASSETS '19)*. Association for Computing Machinery, New York, NY, USA, 446–458. https://doi.org/10.1145/3308561.3353797.

Stanley Black & Decker, Inc. (2019). Pria. Retrieved from okpria.com. Access date: 11 July 2022.

Strayer, D. L., Cooper, J. M., Goethe, R. M., McCarty, M. M., Getty, D. J., & Biondi, F. (2019). Assessing the visual and cognitive demands of in-vehicle information systems. *Cognitive Research: Principles and Implications, 4*(1). doi:10.1186/s41235-019-0166-3.

Stuart, S., & Koleva, H. (2014). Psychological treatments for perinatal depression. *Best Practice & Research Clinical Obstetrics & Gynaecology, 28*(1), 61–70. https://doi.org/10.1016/j.bpobgyn.2013.09.004.

Swamy, P. M., Janardhan Kurapothula, P., Murthy, S. V., Harini, S., Ravikumar, R., & Kashyap, K. (2019). Voice assistant and facial analysis based approach to screen test clinical depression. In *Paper presented at the 2019 1st International Conference on Advances in Information Technology (ICAIT)*.

Thakur, A., & Dhull, S. (2021). *Speech Emotion Recognition: A Review* (pp. 815–827). Singapore: Springer.

United Nations. (2021). *World Drug Report 2021.* https://www.unodc.org/unodc/en/data-and-analysis/wdr-2021_booklet-2.html. Access date: 11 July 2022.

Vaidya, T., & Sherr, M. (2019, May). You talk too much: Limiting privacy exposure via voice input. In *2019 IEEE Security and Privacy Workshops (SPW)* (pp. 84–91). IEEE.

Vtyurina, A., & Fourney, A. (2018). Exploring the role of conversational cues in guided task support with virtual assistants. doi:10.1145/3173574.3173782.

Wahsheh, L. A., & Steffy, I. A. (2020). *Using Voice and Facial Authentication Algorithms as a Cyber Security Tool in Voice Assistant Devices* (pp. 59–64). Cham: Springer International Publishing.

Wang, L., & Miller, L. C. (2020). Just-in-the-moment adaptive interventions (JITAI): a meta-analytical review. *Health Communication, 35*(12), 1531–1544. doi:10.1080/10410236.2019.1652388.

Wang, J., Yang, H., Shao, R., Abdullah, S., & Sundar, S. S. (2020, April). Alexa as coach: Leveraging smart speakers to build social agents that reduce public speaking anxiety. In *Proceedings of the 2020 CHI conference on human factors in computing systems* (pp. 1-13).

Yang, S., Lee, J., Sezgin, E., Bridge, J., & Lin, S. (2021). Clinical advice by voice assistants on postpartum depression: cross-sectional investigation using apple Siri, Amazon Alexa, Google Assistant, and Microsoft Cortana. *JMIR mHealth and uHealth, 9*(1), e24045. doi:10.2196/24045.

Yeager, C. M., & Benight, C. C. (2018). If we build it, will they come? Issues of engagement with digital health interventions for trauma recovery. *mHealth, 4*, 37. doi:10.21037/mhealth.2018.08.04.

Young, R. A., & Zhang, J. (2017). Driven to distraction? a review of speech technologies in the automobile. *The Journal of AVID.* http://acixd.org/wp-content/uploads/2018/10/2-Young-2017-Speech-rev.-29-6-9-17.pdf. Access date: 11 July 2022.

Chapter 11

Design considerations for preparation, optimization, and evaluation of digital therapeutics

Shawna N. Smith[a,b], Nicholas J. Seewald[c] and Predrag Klasnja[d,e]

[a] *Department of Health Management and Policy, University of Michigan School of Public Health, Ann Arbor, MI, United States*, [b] *Department of Psychiatry, University of Michigan, Ann Arbor, MI, United States*, [c] *Department of Health Policy and Management, Johns Hopkins Bloomberg School of Public Health, Baltimore, MD, United States*, [d] *School of Information, University of Michigan, Ann Arbor, MI, United States*, [e] *Kaiser Permanente Washington Health Research Institute, Seattle, WA, United States*

11.1 Introduction

Digital therapeutics refers to the use of digital technologies to prevent, treat or manage medical and psychological conditions through behavior change or precision medical treatments (Sverdlov, Dam, Hannesdottir, & Thornton-Wells, 2018). In addition to mental health and substance abuse, digital therapeutics have been used to treat a variety of conditions, including heart disease, hypertension, diabetes, and obesity (Dang, Arora, & Rane, 2020; Khirasaria, Singh, & Batta, 2020; Lord et al., 2021; Marques & Ferreira, 2020), and were a core component of many tools that have been used in the rapid response to the COVID-19 pandemic (Borcsa, Pomini, & Saint-Mont, 2021; Kadakia, Patel, & Shah, 2020; Robeznieks, 2020). While digital therapeutics offer new opportunities and modalities for treatment, they also raise new questions with respect to *design*. Some of these questions relate to *intervention design*, asking how best to design treatments that are delivered through digital modalities and are intended to be used outside of traditional clinical environments. Other questions relate to *study or evaluation design*, asking how best to answer key scientific questions about how the intervention could be optimized or whether it is achieving desired health impacts. In line with other conceptualizations of design as a process that involves multiple stages, we recognize that design considerations and decisions also vary across the lifecycle of the intervention.

11.1.1 Overview of chapter

This chapter describes a conceptual framework for identifying and testing key design considerations for digital therapeutic development, both with respect to intervention and evaluation design. We then apply that framework to three different phases of the intervention life cycle. We have divided this chapter into three parts:

- Part 1 will introduce a general framework for designing and evaluating digital therapeutic interventions. While a number of prior publications have described best practices for designing and evaluating digital therapeutic interventions (Chandrashekar, 2018; Stowell et al., 2018), there is no general framework to date that identifies the key motivating questions for making design decisions, both with respect to both *intervention* design and *study* design. Our goal is to develop this general framework to inform design decision-making for a wide variety of digital therapeutics, across the intervention and research lifespans. In doing so, we also emphasize the intrinsic linkage between *design* and *optimization*, both within each phase of design and development, but also—recognizing the unavoidable time and resource constraints that impact the development of effective and scalable digital therapeutics—across the totality of the design process.
- Part 2 will apply this framework to design decisions related to initial intervention preparation and development. Here, we will separate out design decisions within two distinct phases. Phase 1 will cover preparation and formative intervention development, namely identifying a health impact target or problem to address via an intervention, as well as key

constraints that will undergird design decisions. Phase 2 will cover optimization, as it relates to intervention content and components as well as intervention deployment and delivery.

- Part 3 will apply this framework to design decisions as they relate to evaluating short- and long-term intervention efficacy, effectiveness, and implementation. Here our focus will be on designing studies that best address key questions related to the efficacy and effectiveness of the developed intervention.

Throughout Parts 2 and 3, we will highlight key design questions and possible design implications with respect to targeted health impacts, the level of evidence likely necessary to justify design decisions, and the possible study designs that may provide that level of evidence. We will illustrate many of these questions by developing a hypothetical digital mental health intervention for patients with bipolar disorder.

11.2 A framework for designing and evaluating digital therapeutic interventions

11.2.1 What do we mean by design?

We consider design in a similar vein to Simon (Simon, 1988), who writes that "*the purpose of good design is to optimize achievement of intended objectives given known constraints.*" As such, we see design as a process, involving a sequence of steps, rather than an output. Bill Moggridge, a British designer who pioneered the field of interaction design and co-founded the preeminent design consultancy IDEO, identified the following steps as part of the design process (Moggridge & Atkinson, 2007): (1) framing and reframing the design problem and the desired objective; (2) creating alternative approaches to a solution; (3) selecting among these alternatives based on the understanding of the project's constraints; (4) prototyping and iterating on a solution; and (5) synthesizing a solution that addresses key constraints while effectively accomplishing the main project objectives. Notably, the design process involves iterative movement among these steps, rather than a linear progression through the sequence. This iteration—moving back and forth among different aspects of the project, trying solutions, collecting additional data, then going back and rethinking one's understanding of both the project goals and the constraints—is fundamental to the process of design and, as such, is central to the design of digital therapeutics as well.

In the world of digital therapeutics, where intervention delivery frequently happens outside of clinic walls and treatments are often intended to be delivered to individuals "just in time," we likely have objectives that are both long-term and short- (or shorter-) term. *Long-term objectives* include the intended health impact of the intervention, for example, increasing physical activity, controlling blood sugar, or improving symptoms of depression. *Shorter-term objectives* are the impacts of the digital therapeutic tool that mediate its effectiveness on these health outcomes. These include things like intervention engagement, which is necessary for most digital therapeutics, but also include mediators that are specific to the health domain or intervention, for example, checking blood sugar, using coping skills, or monitoring key symptoms. On the other side of the equation, design for digital therapeutics faces several important constraints. Participant burden and cost are salient considerations for both intervention and evaluation design. Burden (via either treatment or evaluation) may diminish participant engagement and thus directly contravene intervention effects. Higher costs run the risk of negatively impacting access, both from the patient and health care system perspectives, and particularly at scale. Other constraints may involve limits in the ability of digital therapeutics tools to sense or elicit information on which to base treatment decisions, privacy considerations, system usability/accessibility, and/or access to internet connectivity or required devices, which disproportionately affects populations that are already experiencing health disparities (van Veen et al., 2019; Vogels, 2021).

In the context of attaining objectives under constraints, then, *optimization* is an integral feature of good design. In the intervention world, we generally think of optimization as maximizing effectiveness and societal benefit while working within resource constraints (Collins, 2018a, 2018b). Optimization studies are, thus, those that use data-driven approaches to evaluate different alternatives to determine which is optimal for a particular context, population, and set of resource limitations (Collins, Murphy, & Strecher, 2007, 2011; Riley & Rivera, 2014).

11.2.2 The importance of optimization for digital therapeutic design

We view optimization as a form of data-driven design practice. Prior literature has highlighted the need for optimization, both as it relates to intervention development generally, and digital therapeutic development specifically. For example, digital therapeutics offer important opportunities for positive health impacts by moving treatment beyond the clinical environment. In the areas of mental health and substance abuse, this includes removing key barriers associated with traditional forms of treatment (e.g., stigma, cost), improving treatment personalization, and/or providing just-in-time support (Donker et al., 2013; Robbins, Krebs, Jagannathan, Jean-Louis, & Duncan, 2017; Sim, 2019; Walsh, Golden, & Priebe, 2016). Achieving these benefits, however, requires optimizing several aspects of treatment delivery, including treatment timing, frequency, or location. Questions that guide such optimization include (Bidargaddi, Schrader, Klasnja, Licinio, & Murphy, 2020; Liao, 2016; Nahum-Shani, Hekler, & Spruijt-Metz, 2015, 2018; Spruijt-Metz & Nilsen, 2014):

- *What to deliver?* Digital therapeutics often combine multiple intervention components, targeting multiple mediators including things like engagement and adherence (Nahum-Shani et al., 2015, 2018). Thus optimization may consider which intervention component(s) are most beneficial.
- *When to deliver?* As digital therapeutics often offer broader treatment access (including continuous or near-continuous access), they require an understanding of acute temporality that is not necessary for more traditional therapies. Considerations of timing of intervention delivery are particularly salient when considering chronic conditions, which may wax and wane. While these conditions have significant potential to benefit from continuous access to treatment through digital therapeutics, interventions must also be designed to ensure that delivery occurs when it will be most beneficial to the respondent, and also not so often that it induces user frustration and, ultimately, disengagement.
- *Where to deliver?* Furthering questions of intervention timing, decisions to offer treatment outside the clinic may also require ecological considerations, that is, do some interventions work better/best in some locations, contexts, or settings? Digital therapeutics lend themselves well to adaptation and tailoring—indeed, the just-in-time adaptive intervention (JITAI) framework (Nahum-Shani et al., 2015, 2018; Smith et al., 2017) hinges on the idea of offering the right intervention at the right time to optimize intervention effectiveness.

Data from intervention optimization studies can, of course, help to inform design decisions, but other types of data or information can also be appropriate. Namely, while all decisions related to intervention optimization are empirical, the level or strength of evidence necessary to inform a design decision may vary. Some design decisions may be made based on qualitative feedback from a handful of target users; others may necessitate a full-fledged factorial experiment. These methods, of course, also vary in the cost or burden entailed, both for researchers and potentially for target populations. Much of the challenge in designing and evaluating digital therapeutics comes from the need to optimize across two targets: the health impacts of the intervention and the scientific questions motivating our design inquiries. The next section introduces a conceptual framework for optimizing digital therapeutics design across both of these targets.

11.2.3 A conceptual framework for informing design decisions

This conceptual framework was designed to help intervention developers make design decisions as they relate to both health impacts (i.e., the *intervention* design) and the science necessary for informing intervention design and delivery (i.e., the *study* design). We see each of these domains of design decisions motivated by different key questions:

- ***Intervention design:*** *For health impacts, our key motivating question is: what are we trying to impact?* As intervention design decisions are key for ensuring health impacts across the full life cycle of intervention development and deployment, our conceptual framework suggests different answers to this question for (1) intervention preparation; (2) intervention optimization; and (3) intervention implementation.
- ***Study design:*** *For science, our key motivating question is: what are we trying to learn?* Again, the answers to this question are likely to differ depending on our phase of intervention development . The multiphase optimization strategy framework (MOST) (Collins, 2005, 2018a) distinguishes between *optimization study designs*, which are typically used earlier in the process to help streamline interventions and/or compare treatment options, timings, and contexts, and *evaluation designs, which* test the efficacy, effectiveness, or implementation of a full intervention package.

Cross-cutting design decisions for both health impacts and science, however, is consideration of the level of evidence that is necessary to make the decision—in other words, how strong does the evidence need to be to make a design decision? *For this dimension, our key motivating question is: how much do the alternatives being considered differ in their potential impact on targeted health outcomes?* Or, perhaps more simply, *how much is at stake if we get this design decision wrong?* The answer to this question links our two broader design domains—designing for health impact and designing for science—by connecting the potential health impact of the design decision to the robustness of the evidence provided via the study design. Generally, we see these two dimensions as positively correlated, that is, the larger the potential impact, the more robust the evidence should be. For example, design decisions that are likely to have minimal impact on targeted health outcomes (e.g., minor user interface (UI) differences, like color scheme or information density) can likely be made following focus groups or interviews with a few members of the target population. For moderate-impact decisions (e.g., deciding whether or not to include a certain intervention component in a multicomponent intervention package, or whether or not to deliver a reminder in a mobile app), a smaller-scale optimization study may provide the necessary rigor. And high-impact decisions

TABLE 11.1 Overview of key considerations for designing digital therapeutics across phases.

Phase	Domain: Intervention design — Key design question(s)	Domain: Study design — Role & type of evidence
Formative	What problem is our intervention trying to solve & what are the constraints? • Targeted health impact? • Likely barriers/constraints? • Most likely solutions?	• Provide support for appropriate initial identification of problem, likely constraints, & potential solutions • Exploratory, inductive, not counterfactual-driven
Optimization	What intervention components should be provided?	Dependent on potential for impact on health outcomes: • *High impact:* systematic reviews; factorial experiments • *Low impact:* small-scale user testing; analogue studies
	How should components work? • How much burden should they entail? • When/how often should it be provided?	Smaller-scale optimization trials that allow for comparisons of different forms, timings, and frequencies of delivery (e.g., microrandomized trials)
Evaluation & implementation	Is the system usable/feasible/acceptable?	One-arm pilot study to ensure optimized system is working as intended, for example, • Impacting proximal outcomes • Components working *in situ* & in tandem
	Is there a treatment effect: • Under ideal conditions? • Under routine conditions?	Robust, gold standard RCT to evaluate treatment effects & inform treatment decisions • May be combined with optimization study (e.g., MRT) if randomization is appropriate
	How do we implement? • What barriers exist? • Which implementation strategies are most effective?	Explicit incorporation of implementation outcomes through hybrid implementation-effectiveness studies

(e.g., whether a large health system should offer an intervention to its members) are likely to require gold-standard evidence from, for example, a randomized controlled trial. We explore specifics about making these design decisions in the next two sections, beginning with decisions related to intervention development. Table 11.1 also summarizes some of these key design questions and considerations related to evidence across phases.

11.3 Design considerations for intervention development

11.3.1 Phase 1: formative work

Key design questions: what problem is our design trying to solve and what are the constraints?

Formative design work focuses on identifying a specific problem or treatment gap that might be improved through digital therapeutics, and then gathering information to inform potential intervention options. This involves identifying the targeted health impact and population, as well as the current state of the knowledge base around the identified problem. Design decisions during the formative phase are often framed in the context of understanding what we already know about how to solve the problem and identifying what remains unknown or what gaps exist. Additionally, the formative phase generally focuses on identifying constraints or barriers to improved treatment, which may occur at the level of the patient, intervention, community, population, or broader health care ecosystem. These inputs will then lead to the initial identification of some potential solutions for addressing the identified problem. For our central example of developing a digital mental health intervention for persons with bipolar disorder, a key formative design question would be whether a digital intervention is feasible or appropriate for addressing current concerns or treatment needs, given constraints related to digital literacy, cost, and/or burden.

What level of evidence do we need?

As the purpose of the formative design phase is to identify and understand the scope of a problem, as well as potential solutions and likely constraints, much of the work that occurs during the formative phase is descriptive in nature. As such, the role of evidence during this phase is to provide support for appropriate identification of the targeted problem, likely constraints, and potential solutions, and research methods typically focus on gaining an in-depth understanding of the problem and population in situ. Formative work thus is typically informed by qualitative data collection. Observational or ethnographic methods may be used if formative work requires contextually specific understandings, especially around problems or constraints; interviews and focus groups can be used to understand personal narratives and preferences, as well as to elicit input regarding the feasibility/acceptability/appropriateness of early design ideas (or even of a digital therapeutics approach overall).

More generally, participatory or user-centered approaches, or those that center and empower the target population (or intervention end-users) to better understand their needs and constraints, are often used during the formative phase. Community-based participatory research (CBPR) approaches, where researchers partner with community stakeholders across the research process to elicit iterative input and feedback and share decision-making (Israel, Schulz, Parker, & Becker, 1998, 2003), can be useful for formative work when there are multiple stakeholders or communities whose perspectives are fundamental to informing target problem identification, constraints, and intervention options. CBPR approaches can be particularly important for understanding community or system-level barriers or disagreements that may constrain design choices or impede feasibility or acceptability of certain intervention options. User-centered design (UCD) (Endsley, 2016) approaches similarly center the end-user throughout an iterative process to better understand their needs and goals in their natural environment. In the digital therapeutics space, UCD approaches can be particularly beneficial for understanding how users engage with their technology (Graham, Lattie, & Mohr, 2019) and how this might facilitate or constrain digital intervention delivery or effectiveness, which can in turn inform early design decisions about how best to engineer potential interventions. Once intervention designs have been codified—for example, through a mobile health prototype—user interface/user experience (UI/UX) studies can also elicit important end-user feedback on the anticipated utility, usability, or value of the intervention package or components.

Regardless of the methodological approach, the formative work stage helps researchers iteratively formulate and clarify the problem they are trying to solve. For our example intervention, researchers may come to the project with a broad understanding that, say, they are trying to improve functioning for persons with bipolar disorder. The formative stage, however, may bring about the realization that from the perspective of persons with bipolar disorder, the key problems are (1) consistent and timely access to providers, and (2) comorbid physical health conditions (Blixen, Perzynski, Bukach, Howland, & Sajatovic, 2016). As such, a project that initially may have focused on educating persons with bipolar disorder about their condition may be transformed through formative research into one where the central focus is on facilitating better access or communication with mental and physical health providers. The exact problem one should be trying to solve is rarely clear at the outset; the understanding of the problem changes and grows as researchers engage with the literature and the stakeholders. Formative research is the fundamental tool through which this process unfolds.

Given the nature of the design decisions that are central to the formative design phase, however, there is generally limited capability to produce a "gold-standard" robust evaluation of their validity, most simply because formative work often lacks a good counterfactual that can be used with confidence to evaluate the decision. Rather, formative work may help to elucidate target participant preferences or mechanisms of action that were as yet unknown or misunderstood. Data collection during the formative period tends to focus on maximizing depth of understanding among a select group or groups of stakeholders, rather than necessarily generalizability. Reasoning also tends to be inherently inductive rather than deductive, with a focus on description over evaluation. However, it is also the case that decisions made during the formative period have the potential to be highly impactful in terms of targeted health outcomes. As such, standards for conducting (Collingridge & Gantt, 2008; Mays & Pope, 2000; Seale, 1999) and reporting (O'Brien, Harris, Beckman, Reed, & Cook, 2014) high-quality, rigorous qualitative work should be followed to ensure that the gathered data correctly and completely identifies constraints and preferences for key communities and stakeholders that will need to be addressed during the intervention optimization stage.

11.3.2 Phase 2: optimizing intervention design and deployment

Key design questions: what kinds of intervention components should we provide (and to whom and when)?

Once a health problem or treatment gap has been identified for improvement via digital therapeutics, and information on potential interventions and key constraints gathered, the focus of intervention development shifts to questions of intervention design. Much has been written about how and why to optimize multicomponent interventions by conducting studies that can inform the selection of intervention components to include in the final intervention package. For digital therapeutics, however, design decisions during this phase also involve some additional dimensions of decision-making, beyond simply whether to include an intervention component or not. For instance, given that digital therapeutics typically involves repeated interactions—users are repeatedly sent push notifications, asked to complete surveys, engage in planning, etc.—user burden is a central design concern. If burden is not effectively managed, there is a high risk of user frustration and eventual system abandonment. As such, decisions may need to be made about whether to incorporate components that explicitly target engagement and/or burden management, and how to design therapeutic components in low-burden ways to increase or sustain long-term use or buy-in. Further, digital therapeutic designers often need to make decisions about how to deploy different treatments, for example, offering a specific intervention component as a "pull" treatment, where delivery is initiated by the user, or as a "push" treatment, initiated by the digital modality (Walton, Nahum-Shani, Crosby, Klasnja, & Murphy, 2018). For push treatments, further decisions need to be made about the frequency and timing of intervention provision. Particularly for digital therapeutics considered JITAIs (Nahum-Shani et al., 2018), where intervention delivery is adapted to provide the right (or best) treatment at the right time, intervention design also requires specifying decision rules. These decision rules indicate which *treatment options* (or intervention components) might be considered at specific *decision points* (or times when treatment delivery may be considered) and the *tailoring variables* (or characteristics of the user or context) that determine who gets which treatment.

Constraints at the level of the intervention most typically comprise cost and burden. Privacy is also a particular concern for digital therapeutics, as digital interventions are often able to leverage data from phones or other sensors. While this information may be helpful for tailoring treatment decisions, its use also requires that intervention users feel comfortable sharing that information (as well as with the system using this information to make treatment decisions) (Seewald, Smith, Lee, Klasnja, & Murphy, 2019). Relatedly, while sensors that allow for passive data collection continue to improve, some important tailoring variable information (e.g., on user receptivity or availability for treatment) remains challenging to collect in a robust manner without some form of "active" data collection, for example, via Ecological Momentary Assessments (EMA) (Shiffman, Stone, & Hufford, 2008). EMA, however, puts the onus of data collection on the user and thus adds to the overall system burden. In both cases, then, collection of passive data that erode the user's sense of privacy or—perhaps especially—active data that increase user burden may increase the probability of the user abandoning the system entirely. As such, making decisions about which types of "contextual" data collection are most necessary to inform decisions about treatment delivery is key to ensuring a sustainable intervention model.

Additionally, for JITAIs in particular, there may be the risk of delivering the "right" treatment at the "wrong" time (e.g., when the user is not receptive to treatment), which may result in iatrogenic effects for the intervention as a whole. This may be particularly true when users trust the intervention to be meaningfully adaptive, as missteps around adaptation or delivery may actually engender mistrust (Smith et al., 2017). Notably, while these constraints are salient at the level of the intervention (i.e., the aim is to build the intervention that is most effective at addressing the targeted health outcome for the least cost while inducing the least amount of burden), the applicability of these constraints may vary by intervention component. Some intervention components, such as those that facilitate direct access to therapists or providers, are likely to be expensive while others may be low to no cost. Similarly, pull components, wherein users decide whether/how often/when to access are, by definition, less likely to introduce unwanted burden on the user, while push components necessarily introduce burden, both via asking the user to engage with treatment and by interrupting other activities (Ho & Intille, 2005; Nahum-Shani et al., 2015, 2018).

What level of evidence do we need?

This phase thus requires a range of decisions, both with respect to which intervention components or treatments to include but also how to deploy or deliver different treatments. Determining the level of evidence necessary for evaluating which option is optimal thus requires a return to our overarching question about *how potentially impactful differences in design decisions are likely to be on targeted health outcomes*. Decisions that may have the most impact on health outcomes will correspondingly need the most robust evidence (and thus study designs) to support design decisions.

Of course, even when robust evidence is necessary, new data collection may not be; rather, evidence can come from many sources. For some decisions, the required evidence will be easily found in gold-standard systematic reviews or prior research. For example, recent systematic reviews indicate relatively strong evidence of psychoeducation improving outcomes for persons with bipolar disorder (Miziou et al., 2015; Reinares, Sánchez-Moreno, & Fountoulakis, 2014), which might provide support for including on-demand psychoeducational materials as part of our intervention. For other questions, however— especially those highly specific to the current project or involving a new or understudied population—new data collection is likely necessary.

Deciding which intervention components to include. From the perspective of the full intervention package, decisions about which intervention components to include in multicomponent interventions are classically informed by factorial experiments. Factorial experiments are designed to estimate contributions to intervention effectiveness of individual components or combinations of components by systematically varying, or "crossing," different levels of two or more intervention components. Factorial experiments, which can be "full" designs that cross all levels of all factors, or "fractional" wherein only a subset of factors are crossed (Box, Hunter, & Hunter, 1987; Collins, 2005), allow for the evaluation of both the main effects of each intervention component as well as interactions between components, thus providing robust evidence to support including or "screening out" different intervention components.

From a wide-angle screening perspective, however, not all potential intervention components may require the robust evidence provided by factorial studies to justify inclusion (or exclusion). Although a key strength of factorial designs is their sample size efficiency, especially relative to more traditional randomized controlled designs (Collins, Dziak, & Li, 2009, 2014), the number of experimental conditions increases rapidly as more components and/or levels are included. While this increase in experimental conditions does not necessarily increase the necessary sample size, it can lead to problems with study feasibility and implementation (Collins, 2005). Particularly in the digital therapeutics space, where interventions are often multicomponent, including components that are "push" and "pull" as well as components that may target proximal outcomes or mediators of treatment (e.g., engagement), full factorial designs evaluating all possible intervention components could quickly produce a number of experimental conditions that is unwieldy at best, untenable at worst.

Such considerations may be particularly salient when dealing with vulnerable populations, such as persons with serious mental illness or substance use problems. As such, prior to a factorial experiment, intervention developers should consider an initial adjudication of intervention components to decide which components require this level of rigor and evidence. For example, in many cases "pull" components are low cost, readily scalable, and introduce low potential for iatrogenic effects. Including a mood tracker in a mobile health intervention for bipolar disorder, for example, where users can choose to track their mood of their own volition, on a timescale and frequency of their choosing, introduces minimal burden for the user, and the presence of the component is unlikely to have a negative effect on either the targeted health outcome (e.g., quality of life) or mediators of that outcome (e.g., intervention engagement). For components like this, screening decisions may not require the robust evidence provided by a factorial design; instead, screening might be done via small-scale user-testing or focus groups with the target population to evaluate feasibility and acceptability, and ensure that the component(s) don't provoke an unexpectedly negative reaction. If target users have an especially negative reaction about the potential utility of a component that otherwise imposes little cost or user burden, this provides ample evidence to justify screening it out.

Optimizing data collection for informing intervention content also involves consideration of the target population, and whether evidence from the target population is necessary or whether an "analogue" population might suffice. Analogue studies provide another option for initial screening of intervention components, particularly for those components where questions about utility or inclusion do not necessarily need to be informed by the target population. The rise of online crowdsourcing platforms like Amazon's Mechanical Turk and Figure Eight provide opportunities for conducting rigorous evaluations of intervention components or combinations of components via, for example, discrete choice or conjoint experiments, or vignette studies, rapidly and inexpensively, with study populations that credibly "simulate" the target populations (Aguinis, Villamor, & Ramani, 2021; Strickland & Stoops, 2019; Thomas & Clifford, 2017). Particularly when developing interventions for target populations that are small, difficult to access, and/or vulnerable, leveraging alternative, readily accessible populations to provide guidance on intervention design also ensures that data collection asked of the target population is optimal and does not cause undue burden.

Deciding how the component should work. Qualitative evaluations and analogue studies can also be helpful for informing questions about how a component should ideally work. These questions are distinct from considerations of screening/inclusion or timing, but rather reflect questions of design and UX. For example, in our hypothetical intervention for persons with bipolar disorder, we may decide that we want to include a component that prompts users to create "if-then plans" for health behaviors they are trying to improve (e.g., engaging with mood monitoring or medication adherence). Although there may be good theoretical bases for including such a component (Gollwitzer, 1999; Gollwitzer & Sheeran, 2006; Nahum-Shani et al., 2015), behavioral change theories often lack the degree of specificity necessary optimal just-in-time intervention deployment (Klasnja et al., 2015; Riley & Rivera, 2014; Spruijt-Metz & Nilsen, 2014). Further, theoretical and empirical predictors of when prompts to plan are likely to be the most effective—for example, when obstacles to goal attainment exist or when the chance of forgetting is high—may be difficult to detect passively or without introducing new burden (Rogers, Milkman, John, & Norton, 2015). They also typically do not specify how burdensome planning needs to be in order for it be to effective—for example, should users always have to write out their "if-then" plan, or is it just as effective for users to select a plan from a "closed" list of preset options (with the option to write their own, if they wish)? While the former may result in a more specific plan, the additional burden put on the user may reduce compliance enough that it undercuts the component's effectiveness. Qualitative evaluations can be useful for answering questions like these by providing insights into users' cognitive processes and preferences; analogue studies similarly can be helpful by allowing for rapid and inexpensive comparisons of different design options—for example, open-ended vs. closed-ended planning.

Deciding when to deliver a component. Once intervention components have been designed and screened, there may still be questions about which version of an intervention works best, or how frequently to "push" an intervention to users. Continuing the planning example above, initial evaluations may yield two planning interventions that seem similarly acceptable and feasible—one that asks users to plan for each day, and another that asks them to plan for the week. Further, there may be questions about when users want reminders to set or review their plans. For example, is it more effective to send a reminder every time a new plan needs to be set (e.g., weekly) or only if the user hasn't recently engaged with the planning component for a certain amount of time? Further, after users make their plans, is it helpful for them to be reminded about those plans? And if so, when or how often? Are daily reminders necessary, or would weekly reminders be just as effective (while also minimizing user burden over the long term)? Again, behavioral change theories rarely specify this level of detail, yet there may be significant potential for suboptimal intervention delivery (e.g., delivering an intervention too often, leading to habituation, annoyance, disengagement, or even system abandonment) that diminishes the effectiveness of the component and, potentially, the full system on targeted health outcomes.

Microrandomized trials (MRTs). What is notable about these types of design decisions is that individuals often don't have good intuitions for how different frequencies or timing will affect their experience of the intervention, making the use of qualitative methods suboptimal for supporting design decision-making in this context. It may also be difficult to adjudicate between two (or more) potential versions of an intervention component in terms of their impact on targeted health outcomes or important mediators of that outcome, like engagement. MRTs offer the opportunity to conduct smaller-scale optimization studies that allow for comparisons of different forms, timings, or frequencies of intervention component delivery. MRTs were designed specifically to inform optimization of JITAIs, with a focus on providing robust evidence to inform which intervention options are most effective for whom and in which contexts (Klasnja et al., 2015; Nahum-Shani et al., 2018).

In MRTs, users are randomized to different intervention options repeatedly, at each potential decision point, and then the effect of different intervention options is examined on a relevant proximal, or short-term, outcome. For example, in our intervention for persons with bipolar disorder, we may want robust information to inform two decisions:

- Are users more likely to complete their plan if they write it out ("open-ended") or if they select it from pre-specified options "closed-ended").
- For users that are not engaging in mood monitoring, should we send a daily reminder prompt?

An MRT would allow us to evaluate both of these questions (and potentially more) with robust evidence from a single, highly efficient study. Table 11.2 summarizes one potential study design for evaluating these two intervention component questions, notably specifying for each question the decision points (i.e., when each participant might be randomized), intervention options and randomization probabilities for each, and the proximal outcomes that correspond to each component question.

Unlike traditional RCTs that typically randomize individuals only once (or at most, 2–3 times), in MRTs each participant is randomized to receive (or not receive) different intervention options at each of possibly many decision points. This repeated randomization means that even in a study lasting only a few weeks, individuals may be randomized tens, if not hundreds or thousands, of times throughout a study. As such, MRTs allow estimation of causal effects of different intervention options, both marginal over time and time-varying. Further, causal effect estimation can leverage both between- and within-subject variation (i.e., decision points where the same individual was randomized to different conditions). As such, MRTs often require far fewer participants than other forms of optimization studies. The repeated randomization combined with a focus on a proximal outcome also means MRTs are often shorter in duration (Klasnja et al., 2015). Sample size calculators are available that can help scientists plan their studies and estimate necessary sample sizes (Liao, 2016; MRT-SS Calculator, 2016; Seewald, Sun, & Liao, 2016).

With respect to informing optimization of JITAIs, MRTs also allow consideration of questions related to when, or under what conditions/in which context, to deliver treatment. In addition to testing or comparing main effects of different intervention options, moderators analyses can also be used to determine optimal times to intervene. Notably, as one MRT can test or randomize multiple intervention components, MRTs not only provide an efficient way to empirically evaluate different forms of intervention components, but also to collectively evaluate how timing, context, and frequency of intervention delivery may impact key mediators (e.g., short-term impacts, engagement, receptivity) and subsequently our targeted health outcomes. For example, in evaluating the effectiveness of mood monitoring prompts for our hypothetical intervention, we may find that, averaging over time, our prompts show a small positive effect on mood monitoring compliance. However, when we allow for that effect to vary over time, we may find that prompts were highly effective in the initial weeks of the study when users were still acclimating to the app components and establishing new habits, but prompt effectiveness waned substantially over time as users either decided to monitor their mood regularly (in which case they likely were not eligible

TABLE 11.2 Example microrandomized trial design.

Component	Key question	Decision point	Number of decision points in 6 week study	Intervention options & randomization probabilities	Proximal outcome & data source
Planning	Is open-ended or closed-ended planning more effective for goal setting?	Once weekly at a user-selected time between 7 and 10 pm	6	Open-ended planning (50%) Closed-ended planning (50%)	*Promixal outcome:* Goal attainment *Data source:* Self-report through weekly EMA
Mood monitoring prompts	Are users more likely to complete mood monitoring if they receive a daily prompt?	Once daily at a randomly selected time between 10am and 2pm *if* the user has not engaged with planning in the last 48 hours	Up to 40, depending on engagement with mood monitoring	Prompt (50%) No prompt (50%)	*Proximal outcome:* Use of mood monitoring in the next 24 hours (or prior to the next prompt, whichever comes first) *Data source:* App backend metrics

Note: EMA = ecological momentary assessment

for randomization given lack of disengagement) or they decided that mood monitoring was too burdensome or not useful for them. Similarly, we may find that prior prompt "dose" moderates effectiveness, with those individuals that have received more prompts (because of more habitual disengagement) are less likely to respond to the prompts. Both of these pieces of information would be useful for optimizing when and to whom to deliver prompts in the next version of the intervention (Table 11.2).

11.4 Design considerations for short-term & long-term efficacy, effectiveness, and implementation

Key design questions: what are we trying to learn?

Once an intervention has been designed and undergone initial optimization trials to help determine which components should be included and, as applicable, how and when they should be delivered, intervention packages are generally considered ready for evaluation. Even at this stage, however, there are still important design decisions that need to be made, some related to continued intervention design but others to how best to design the evaluation itself to provide necessary and sufficiently robust evidence. As such, we see the crux of these decisions undergirded by the question: *what are we trying to learn* with our evaluation?

Intervention evaluation is typically considered in relation to considerations of effectiveness, or whether the intervention is having intended effects on targeted health outcomes. Rising interest in translational research, and especially the distinction between T2 and T4 phases of translation, however, have highlighted that evaluations of an intervention—even once translation from "basic science" to humans has been achieved—can have different aims. Although design of digital therapeutics generally starts away from the traditional "bench," once the intervention is designed, developers still need to make decisions about how to design appropriate evaluation studies. Pilot studies, for example, are intended to evaluate intervention feasibility and acceptability, and can be very important for ensuring that all elements of the intervention are working as intended. This is particularly true within the digital therapeutics space, where interventions are both often multicomponent—comprised of carefully designed and screened components, as discussed in prior sections—and also embedded within a technological infrastructure that may be responsible for collecting information that directly informs when or how interventions are delivered. These technological elements may include collection of sensor and self-report data, intervention delivery, and execution of any algorithms that are involved in intervention-provision decision-making. When this infrastructure is not working correctly, a pilot study can verify that backup data sources or decision rules are informing sensible treatment decisions (Seewald et al., 2019). Pilot studies may also be useful for evaluating second-order optimization questions, notably about whether components delivered *in situ* are impacting their targeted outcomes in the context of the broader intervention, or whether the multicomponent intervention system itself is working optimally.

Beyond pilot studies, however, decisions about how to design an evaluation study are likely to reflect priorities about the translational target of interest—namely, are we trying to translate the treatment to patients, practices, and/or communities? These decisions correspond to decisions to evaluate digital therapeutics with respect to efficacy, effectiveness, and/or implementation outcomes.

What level of evidence do we need?

Evaluations of efficacy or effectiveness are intended to establish a causal effect of an intervention on targeted health outcomes; as such, they are classically associated with a need for highly robust evidence, typically provided by a high-quality randomized clinical trial. The need for this level of evidence stems from the decision such studies are intended to inform—namely whether to adopt the intervention as an option to treat targeted health outcomes. Such decisions thus come with large potential downsides if evidence provided is not robust and thus gets the recommendation wrong. Iatrogenic interventions, of course, run the risk of worsening health outcomes. Of note, the business model for many mobile health interventions, especially in the mental health space, which disseminates direct-to-consumer often with little vetting of the evidence base, has led to proliferation of interventions that contain known iatrogenic content (Larsen, Nicholas, & Christensen, 2016; Nilsen et al., 2012). But even interventions that offer no or minimal improvement have potential to lead to negative health outcomes by, for example, substituting for other treatments that may be more effective, or potentially undermining patient treatment engagement or provider trust. As such, efficacy and effectiveness evaluations are generally expected to not only make use of robust experimental methods but also adhere to standards for trial conduct and reporting, for example, CONSORT checklists and diagrams (Moher et al., 2012). However, even within the context of a clinical trial, decisions will need to be made about *for whom* and *under what circumstances* we want to estimate the treatment effect. Additionally, to the extent that failure to attain intended effects may be a failure of implementation (the so-called "Type III error" (Scanlon, Horst, Nay, Schmidt, & Waller, 1977)), there may also be questions as to how robust process measures may need to be to ensure adequate implementation. And, of course, for pilot studies or second-order optimization questions, the level of evidence needed will likely be significantly lower.

Optimizing in situ. While optimization is often described as occurring prior to evaluation—and indeed, key optimization decisions do—there may well be some "gray area" between the domains of optimization and evaluation. Namely, even in the evaluation phase, there may be important questions to ask about whether interventions (or intervention components) are working optimally, or achieving their intended aims, *in situ*. For example, for our mobile intervention for persons with bipolar disorder, there may be questions as to the overall usability or acceptability of the intervention *system* that combines all selected push and pull components. For instance, how is the total intervention dose provided by the system perceived by users? It's possible that while each individual component has been optimized to not be overly burdensome, in real use, the combination of all push components results in an excessive number of notifications. Are some components redundant? Or perhaps sending mixed messages as to priorities? (e.g., Which health behavior should the user be focusing on? On which of the multiple included goals? etc.) Additionally, while type and timing of intervention delivery may have been optimized for different components individually, it may be useful to evaluate how the "optimal" versions of the components operate in tandem. A small-scale one-armed pilot study with the intended user population can be useful for evaluating these questions.

Relatedly, and especially for JITAIs that actively and repeatedly personalize intervention delivery, similar pilot studies can be useful for evaluating intervention fidelity and reliability prior to robust evaluation. For example, a pilot study may be helpful for ensuring that there is adequate time- and context-sensitive data collection to detect important tailoring variables or variables that determine whether a user should be considered eligible for receipt of an intervention. Pilot studies can also help to ensure that information used to tailor treatment delivery is reliable and that decision rules underlying JITAIs are being implemented correctly. Both passively and actively collected tailoring variables can run into problems with larger-scale implementation—unexpected technological problems (e.g., server communication issues or unanticipated handshakes) can lead to issues with passive data collection (Smith et al., 2017); active collection of tailoring variables, of course, can suffer from non-response. Pilot studies can help reveal these issues that, left unresolved, may be detrimental to intervention effectiveness. Finally, pilot studies can also provide confirmation that components are achieving their intended proximal effects.

Efficacy and effectiveness studies. Much has been written about the difference between efficacy and effectiveness studies; in short, efficacy evaluations allow for researchers to understand whether the intervention works with the target population under the *best* circumstances, while effectiveness evaluations answer questions related to the effect of the intervention under "real-world" conditions. Given these two different evaluation goals, design goals for efficacy and effectiveness studies also differ. In particular, design for efficacy studies should prioritize ensuring adequate data collection so that if there is a signal, it can be detected with the data gathered. As with most clinical trials, this may mean exerting tighter control over the study population by including only individuals that meet specific criteria regarding their condition or their technological competency or capability. Further, within the context of digital therapeutics, this also may mean including incentives not only for study data collection but also for things like engagement with the app, for example, paying users to interact with the intervention on a daily basis or for use of key components (e.g., mood monitoring). Similarly, decisions may be made to invest in technological tools or capabilities for users or for the system as a whole, for example, providing users with new smartphones or tablets to avoid possible "real world" technological complications. While such investments are likely not scalable, their inclusion ensures that if there is a signal that could possibly be detected that the intervention may be efficacious, circumstances are ideal enough that it will be.

Effectiveness studies, on the other hand, are looking to detect effects of the intervention conditional on its *in situ* use. Typically, this means that not including any incentives or elements that would not be provided in a scaled-out version of the intervention, for example, no (or at least limited) provision of technology and no payments for intervention engagement (although incentives can still be included for study-related data collection, including outcomes, moderators and mediators). In encouraging organic use of the intervention, effectiveness evaluations are also key for informing understanding of longer-term patterns of use (especially important for digital therapeutics managing chronic conditions) as well as barriers to use.

Within each of these study frameworks, however, there may be further opportunities for optimization. First, there may be opportunities for combining optimization evaluations with efficacy or effectiveness evaluations. For example, in our hypothetical intervention, one could imagine embedding an MRT that randomizes daily motivational messages within a more typical two-arm effectiveness trial. In this study design, individuals randomized to receive the intervention for the effectiveness trial would also each be individually randomized daily to receive or not receive a message motivating them to engage in self-management activities. This design would thus allow answering both an *optimization question* of whether individuals were more likely to engage in self-management activities on days when motivational message were sent as well as the *efficacy/effectiveness question* of whether the full intervention package improved targeted health outcomes relative to the control group (marginalizing over days when motivational messages were sent vs not sent). The key consideration for deciding whether combining optimization and evaluation research questions is feasible has to do with whether it is meaningful to include randomization in the final intervention package. In many cases, as in our example of motivational messages, randomization can also help to manage burden and habituation to an intervention component, so the intervention package—including this daily randomization of motivational messages—would be a meaningful intervention to evaluate for effectiveness.

Second, within the context of study design, planning, and execution, there are myriad opportunities for optimizing the research infrastructure (i.e., how the trial is run) itself (Inan et al., 2020). While this is true of all trial endeavors, such optimization may be more salient in trials of digital therapeutics where interventions are typically multicomponent and often complex, data is collected through multiple streams (including passive and active streams), and data collection is longitudinal (and at least for some aspects, often intensively so). Within such trials, it is possible to embed a variety of randomizations that examine how trial procedures themselves affect adherence to data collection, engagement with the intervention, and so on. For instance, one may randomize when automated adherence messages are sent, or when or how study staff reach out to participants, or even how data itself is being collected (e.g., through less frequent but longer self-report instruments or more frequent but shorter instruments). Such experimental manipulations could yield an evidence base for the effective conduct of trials across the digital therapeutics spectrum. Given typically limited budgets for conducting trials—which are often stretched doubly thin for digital therapeutics studies given technological build needs–this area is ripe for additional work to ensure that the trade-offs researchers necessarily make in the pursuit of high-quality evidence of intervention effectiveness result in the best possible improvements to the evidence base.

Hybrid effectiveness-implementation studies. While effectiveness studies move intervention testing more to the "real world," they often still fall short of understanding the barriers to large-scale, population-wide deployment of or access to an intervention. In line with increasing calls to increase the expediency with which individuals can access evidence-based interventions, implementation science has emerged as a discipline to forge study in "methods to promote the systematic uptake of research findings and other evidence-based practices into routine practice, and hence, to improve the quality and effectiveness of health services and care" (Eccles & Mittman, 2006). Researchers interested in the implementation, then, ask questions related to the barriers and facilitators of accessing care or adopting new practices, including which implementation strategies are most effective at improving uptake or adoption, as well as questions related to intervention effectiveness in new (or more heterogeneous) populations (Lane-Fall, Curran, & Beidas, 2019). For our app for persons with bipolar disorder, then, questions about implementation may concern questions about best models of access, understanding drivers of long-term patient engagement or fidelity of app usage, barriers to integrating the app into more traditional mental health care models, or perhaps adapting and scaling out the app for patients with other serious mental illnesses, such as major depressive or schizoaffective disorders. Notably, while these questions are process focused, they are also, to differing degrees, still focused on ensuring or testing continued intervention effectiveness. As such, hybrid implementation-effectiveness studies were designed to provide researchers with a middle-ground for studying effectiveness, implementation, and their interplay within a single study (Curran, Bauer, Mittman, Pyne, & Stetler, 2012). Three types of hybrid studies have been proposed. Type 1 hybrid studies have a primary focus on evaluating effectiveness, augmented with a secondary focus on understanding potential barriers or facilitators of broad-scale implementation; Type 2 studies aim for a balanced focus on understanding effectiveness and implementation, for example evaluating implementation and effectiveness of an evidence-based intervention within a new population; and Type 3 studies prioritize understanding implementation, for example, through a comparative effectiveness study of different implementation strategies, while still measuring effectiveness as a secondary aim.

In all cases, implementation studies (or hybrid studies) are typically distinguished from more conventional effectiveness studies by clear interest in the implementation process, often assessing implementation determinants, and/or implementation outcomes distinct from the effectiveness outcomes. For Type 2 or 3 hybrid studies, there may also be an effort to specify implementation strategies that are distinct from the intervention itself. Notably, given the multifocal nature of implementation as well as the intrinsic interest in process or explanatory mechanisms, implementation science often relies on mixed methods evaluations (Palinkas et al., 2011).

Implementation science studies also hold promise for evaluating long-term maintenance and sustainability of mobile interventions, at both individual- and system-levels, as key outcomes (Damschroder, Reardon, Opra Widerquist, & Lowery, 2022; Glasgow et al., 2019; Proctor et al., 2011). While a full discussion of intervention maintenance is outside the scope of this chapter, long- or longer-term maintenance of both intervention effectiveness and access to or delivery of digital therapeutics is clearly important for ensuring optimal population health improvements, especially for mental health and substance use problems, which are often characterized as chronic, relapsing disorders. Although evaluation of longer-term effectiveness or sustainment is rare (Jankovic et al., 2021; Wiltsey Stirman et al., 2012), systematic reviews of digital mental health interventions have highlighted high rates of attrition and low-rates of adherence (Karyotaki et al., 2015; Valimaki, Anttila, Anttila, & Lahti, 2017), even for shorter-term trials, and especially for youth (Karyotaki et al., 2015; Valimaki et al., 2017). Several recent articles have called attention to better design thinking (Valimaki et al., 2017), including UCD, as an important tool for ensuring longer-term effects. Recent work on the importance of intervention adaptation as an important driver of maintenance is also likely relevant (Aarons et al., 2012; Chambers & Norton, 2016; Shelton & Lee, 2019). This work also dovetails with frameworks highlighting the need for dynamic models of sustainability for health care systems' adoption and use of digital therapeutics (Chambers, Glasgow, & Stange, 2013, 2016; Lyon et al., 2016; Mohr, Lyon, Lattie, Reddy, & Schueller, 2017). Considerations for study and optimization here include data capture for ongoing monitoring, especially as it relates to user symptoms or clinical progress, engagement, and fidelity (Mohr et al., 2017).

11.5 Discussion and conclusion

Developers of digital therapeutics necessarily consider design questions and make design decisions across two domains: the design of the intervention itself for impacting health outcomes and the study design for generating evidence to inform intervention design decisions. Over the years, there has been increasing attention paid to the importance of optimization as a crucial part of the intervention development process. The advent of digital therapeutics that can deliver treatment outside of traditional clinic walls, are often intended for long-term use, and can potentially tailor to "just-in-time" need and/or context, has further amplified considerations for intervention design, both in terms of proffering new dimensions for development decision-making and considerations for what we need to learn to inform a design decision, as well as what level of evidence might be needed to make these design decisions.

This chapter offers a conceptual framework for tying together these two domains of design decisions, and for identifying and testing key digital therapeutics design considerations across the intervention development lifespan, that is, from formative work through longer-term implementation. This framework encourages intervention developers to make design decisions by considering three motivating questions: first, *what health outcomes are we trying to impact*; second, *what are we trying to learn*; and third, cross-cutting both prior questions to inform potential evidence, *what are we risking in terms of health impacts if we get this decision wrong*? As this chapter has shown, in answering these questions across the lifespan of the intervention, intervention developers can help to optimize their initial intervention development, evaluation, and implementation by selecting the best evaluation designs to ensure that interventions are attaining intended health impacts, across the totality of the intervention design process, while also being cognizant of the time and resource constraints endemic to the scientific endeavor itself.

Glossary
CBPR   Community-based Participatory Research
EMA   Ecological Momentary Assessments
JITAI   Just-in-time Adaptive Intervention
MOST   Multiphase Optimization Strategy Framework

MRT   Micro-randomized Trial
UCD   User-centered Design
UI/UX   User Interface/User Experience

References

Aarons, G. A., Green, A. E., Palinkas, L. A., Self-Brown, S., Whitaker, D. J., Lutzker, J. R., et al. (2012). Dynamic adaptation process to implement an evidence-based child maltreatment intervention. *Implementation Science, 7*(1), 32. https://doi.org/10.1186/1748-5908-7-32.

Aguinis, H., Villamor, I., & Ramani, R. S. (2021). MTurk research: review and recommendations. *Journal of Management, 47*(4), 823–837. https://doi.org/10.1177/0149206320969787.

Bidargaddi, N., Schrader, G., Klasnja, P., Licinio, J., & Murphy, S. (2020). Designing m-Health interventions for precision mental health support. *Translational Psychiatry, 10*(1), 1–8. https://doi.org/10.1038/s41398-020-00895-2.

Blixen, C., Perzynski, A. T., Bukach, A., Howland, M., & Sajatovic, M. (2016). Patients' perceptions of barriers to self-managing bipolar disorder: a qualitative study. *The International Journal of Social Psychiatry, 62*(7), 635–644. https://doi.org/10.1177/0020764016666572.

Borcsa, M., Pomini, V., & Saint-Mont, U. (2021). Digital systemic practices in Europe: a survey before the Covid-19 pandemic. *Journal of Family Therapy, 43*(1), 4–26. https://doi.org/10.1111/1467-6427.12308.

Box, G. E. P., Hunter, W. G., & Hunter, J. S. (1987). Statistics for experimenters: an introduction to design, data analysis, and model building. http://vlib.kmu.ac.ir/kmu/handle/kmu/89878. Accessed on 1/31/2022.

Chambers, D. A., Feero, W. G., & Khoury, M. J. (2016). Convergence of implementation science, precision medicine, and the learning health care system: a new model for biomedical research. *Jama, 315*(18), 1941–1942. https://doi.org/10.1001/jama.2016.3867.

Chambers, D. A., Glasgow, R. E., & Stange, K. C. (2013). The dynamic sustainability framework: Addressing the paradox of sustainment amid ongoing change. *Implementation Science, 8*(1), 117. https://doi.org/10.1186/1748-5908-8-117.

Chambers, D. A., & Norton, W. E. (2016). The adaptome: advancing the science of intervention adaptation. American Journal of Preventive Medicine, 51(4, Supplement 2), S124–S131. https://doi.org/10.1016/j.amepre.2016.05.011.

Chandrashekar, P. (2018). Do mental health mobile apps work: evidence and recommendations for designing high-efficacy mental health mobile apps. *MHealth, 4*. doi:10.21037/mhealth.2018.03.02.

Collingridge, D. S., & Gantt, E. E. (2008). The quality of qualitative research. https://journals.sagepub.com/doi/abs/10.1177/1062860608320646?casa_token=OxGzO2t3LtwAAAAA:xZfOsFuWg1dCR_sPN6TujFUXsW09StvbbFqKf0TqycMT4BVkuW9lmJb5TCPSvMptN486D3bNLzHYFw. Accessed on 1/31/2022.

Collins, L. M. (2005). A strategy for optimizing and evaluating behavioral interventions. *Annals of Behavioral Medicine, 30*(1), 65–73.

Collins, L. M. (2018a). Conceptual introduction to the multiphase optimization strategy (MOST). In L. M. Collins (Ed.), *Optimization of Behavioral, Biobehavioral, and Biomedical Interventions: The Multiphase Optimization Strategy (MOST)* (pp. 1–34). Springer International Publishing. https://doi.org/10.1007/978-3-319-72206-1_1.

Collins, L. M. (2018b). *Optimization of Behavioral, Biobehavioral, and Biomedical Interventions: The Multiphase Optimization Strategy (MOST)* (1st ed.). Springer International Publishing https://doi.org/10.1007/978-3-319-72206-1.

Collins, L. M., Baker, T. B., Mermelstein, R. J., Piper, M. E., Jorenby, D. E., Smith, S. S., Christiansen, B. A., Schlam, T. R., Cook, J. W., & Fiore, M. C. (2011). The multiphase optimization strategy for engineering effective tobacco use interventions. *Annals of Behavioral Medicine, 41*(2), 208–226. https://doi.org/10.1007/s12160-010-9253-x.

Collins, L. M., Dziak, J. J., Kugler, K. C., & Trail, J. B. (2014). Factorial experiments: efficient tools for evaluation of intervention components. *American Journal of Preventive Medicine, 47*(4), 498–504. https://doi.org/10.1016/j.amepre.2014.06.021.

Collins, L. M., Dziak, J. J., & Li, R. (2009). Design of experiments with multiple independent variables: a resource management perspective on complete and reduced factorial designs. *Psychological Methods, 14*(3), 202–224. https://doi.org/10.1037/a0015826.

Collins, L. M., Murphy, S. A., & Strecher, V. (2007). The multiphase optimization strategy (MOST) and the sequential multiple assignment randomized trial (SMART): new methods for more potent Ehealth interventions. *American Journal of Preventive Medicine, 32*(5, Supplement), S112–S118. https://doi.org/10.1016/j.amepre.2007.01.022.

Curran, G. M., Bauer, M., Mittman, B., Pyne, J. M., & Stetler, C. (2012). Effectiveness-implementation Hybrid Designs. *Medical Care, 50*(3), 217–226. https://doi.org/10.1097/MLR.0b013e3182408812.

Damschroder, L. J., Reardon, C. M., Opra Widerquist, M. A., & Lowery, J. (2022). Conceptualizing outcomes for use with the consolidated framework for implementation research (CFIR): the CFIR outcomes addendum. *Implementation Science, 17*(1), 7. https://doi.org/10.1186/s13012-021-01181-5.

Dang, A., Arora, D., & Rane, P. (2020). Role of digital therapeutics and the changing future of healthcare. *Journal of Family Medicine and Primary Care, 9*(5), 2207. https://doi.org/10.4103/jfmpc.jfmpc_105_20.

Donker, T., Petrie, K., Proudfoot, J., Clarke, J., Birch, M.-R., & Christensen, H. (2013). Smartphones for smarter delivery of mental health programs: a systematic review. *Journal of Medical Internet Research, 15*(11), e247. https://doi.org/10.2196/jmir.2791.

Eccles, M. P., & Mittman, B. S. (2006). Welcome to implementation science. *Implementation Science, 1*(1), 1. https://doi.org/10.1186/1748-5908-1-1.

Endsley, M. R. (2016). *Designing for Situation Awareness: An Approach to User-Centered Design* (Second Edition). London: CRC Press.

Glasgow, R. E., Harden, S. M., Gaglio, B., Rabin, B., Smith, M. L., Porter, G. C., Ory, M. G., & Estabrooks, P. A. (2019). RE-AIM planning and evaluation framework: adapting to new science and practice With a 20-Year Review. *Frontiers in Public Health, 7*. https://www.frontiersin.org/article/10.3389/fpubh.2019.00064. Accessed on 1/31/2022.

Gollwitzer, P. M. (1999). Implementation intentions: Strong effects of simple plans. *American Psychologist, 54*(7), 493–503. https://doi.org/10.1037/0003-066X.54.7.493.

Gollwitzer, P. M., & Sheeran, P. (2006). Implementation intentions and goal achievement: a meta-analysis of effects and processes, *Advances in Experimental Social Psychology* (38, pp. 69–119). Academic Press. https://doi.org/10.1016/S0065-2601(06)38002-1.

Graham, A. K., Lattie, E. G., & Mohr, D. C. (2019). Experimental therapeutics for digital mental health. *JAMA Psychiatry, 76*(12), 1223–1224. https://doi.org/10.1001/jamapsychiatry.2019.2075.

Ho, J., & Intille, S. (2005). Using context-aware computing to reduce the perceived burden of interruptions from mobile devices. *Proceedings of the SIGCHI Conference on Human Factors in Computing Systems*, 909–918. https://dl.acm.org/doi/abs/10.1145/1054972.1055100?casa_token=KvQgVl7mq2gAAAAA:w6bkO6AI1nekLj63VBVnRHZ6KvGUUg91FWYqc58WuW-W5B1YXXMS6-hSluHxJaNGYH8QR09HS3ZI-Q.

Inan, O. T., Tenaerts, P., Prindiville, S. A., Reynolds, H. R., Dizon, D. S., & Callif, R. M. (2020). Digitizing clinical trials. *Npj Digital Medicine, 3*, 7.

Israel, B. A., Schulz, A. J., Parker, E. A., & Becker, A. B. (1998). Review of community-based research: assessing partnership approaches to improve public health. *Annual Review of Public Health, 19*(1), 173–202. https://doi.org/10.1146/annurev.publhealth.19.1.173.

Israel, B. A., Schulz, A. J., Parker, E. A., Becker, A. B., Allen, A., & Guzman, J. (2003). Critical issues in developing and following community-based participatory research principles. *Community-Based Participatory Research for Health* (pp. 56–73). Jossey-Bass.

Jankovic, D., Bojke, L., Marshall, D., Saramago Goncalves, P., Churchill, R., Melton, H., Brabyn, S., & Gega, L. (2021). Systematic review and critique of methods for economic evaluation of digital mental health interventions. *Applied Health Economics and Health Policy, 19*(1), 17–27. https://doi.org/10.1007/s40258-020-00607-3.

Kadakia, K., Patel, B., & Shah, A. (2020). Advancing digital health: FDA innovation during COVID-19. *Npj Digital Medicine, 3*(1), 1–3. https://doi.org/10.1038/s41746-020-00371-7.

Karyotaki, E., Kleiboer, A., Smit, F., Turner, D. T., Pastor, A. M., Andersson, G., Berger, T., Botella, C., Breton, J. M., Carlbring, P., Christensen, H., Graaf, E. de, Griffiths, K., Donker, T., Farrer, L., Huibers, M. J. H., Lenndin, J., Mackinnon, A., Meyer, B., & Cuijpers, P. (2015). Predictors of treatment dropout in self-guided web-based interventions for depression: An 'individual patient data' meta-analysis. *Psychological Medicine, 45*(13), 2717–2726. https://doi.org/10.1017/S0033291715000665.

Khirasaria, R., Singh, V., & Batta, A. (2020). Exploring digital therapeutics: the next paradigm of modern health-care industry. *Perspectives in Clinical Research, 11*(2), 54–58. https://doi.org/10.4103/picr.PICR_89_19.

Klasnja, P., Hekler, E. B., Shiffman, S., Boruvka, A., Almirall, D., Tewari, A., & Murphy, S. A. (2015). Microrandomized trials: an experimental design for developing just-in-time adaptive interventions. *Health Psychology, 34*(Suppl), 1220–1228.

Lane-Fall, M. B., Curran, G. M., & Beidas, R. S. (2019). Scoping implementation science for the beginner: locating yourself on the "subway line" of translational research. *BMC Medical Research Methodology, 19*(1), 133. https://doi.org/10.1186/s12874-019-0783-z.

Larsen, M. E., Nicholas, J., & Christensen, H. (2016). A systematic assessment of smartphone tools for suicide prevention. *Plos One, 11*(4), e0152285. https://doi.org/10.1371/journal.pone.0152285.

Liao. (2016). Sample size calculations for micro-randomized trials in mHealth. *Statistics in Medicine*. https://onlinelibrary.wiley.com/doi/abs/10.1002/sim.6847. Accessed on 1/31/2022.

Lord, S. E., Campbell, A. N. C., Brunette, M. F., Cubillos, L., Bartels, S. M., Torrey, W. C., Olson, A. L., Chapman, S. H., Batsis, J. A., Polsky, D., Nunes, E. V., Seavey, K. M., & Marsch, L. A. (2021). Workshop on implementation science and digital therapeutics for behavioral health. *JMIR Mental Health, 8*(1), e17662. https://doi.org/10.2196/17662.

Lyon, A. R., Wasse, J. K., Ludwig, K., Zachry, M., Bruns, E. J., Unützer, J., & McCauley, E. (2016). The contextualized technology adaptation process (CTAP): optimizing health information technology to improve mental health systems. *Administration and Policy in Mental Health, 43*(3), 394–409. https://doi.org/10.1007/s10488-015-0637-x.

Marques, I. C. P., & Ferreira, J. J. M. (2020). Digital transformation in the area of health: systematic review of 45 years of evolution. *Health and Technology, 10*(3), 575–586. https://doi.org/10.1007/s12553-019-00402-8.

Mays, N., & Pope, C. (2000). Assessing quality in qualitative research. *Bmj, 320*(7226), 50–52. https://doi.org/10.1136/bmj.320.7226.50.

Miziou, S., Tsitsipa, E., Moysidou, S., Karavelas, V., Dimelis, D., Polyzoidou, V., & Fountoulakis, K. N. (2015). Psychosocial treatment and interventions for bipolar disorder: a systematic review. *Annals of General Psychiatry, 14*(1), 19. https://doi.org/10.1186/s12991-015-0057-z.

Moggridge, B., & Atkinson, B. (2007). *Designing Interactions*. Cambridge: MIT Press.

Moher, D., Hopewell, S., Schulz, K. F., Montori, V., Gøtzsche, P. C., Devereaux, P. J., Elbourne, D., Egger, M., & Altman, D. G. (2012). CONSORT 2010 explanation and elaboration: updated guidelines for reporting parallel group randomised trials. *International Journal of Surgery, 10*(1), 28–55. https://doi.org/10.1016/j.ijsu.2011.10.001.

Mohr, D. C., Lyon, A. R., Lattie, E. G., Reddy, M., & Schueller, S. M. (2017). Accelerating digital mental health research from early design and creation to successful implementation and sustainment. *Journal of Medical Internet Research, 19*(5), e153. https://doi.org/10.2196/jmir.7725.

MRT-SS Calculator. (2016). https://pengliao.shinyapps.io/mrt-calculator/. Accessed on 1/31/2022.

Nahum-Shani, I., Hekler, E. B., & Spruijt-Metz, D. (2015). Building health behavior models to guide the development of just-in-time adaptive interventions: A pragmatic framework. *Health Psychology, 34*(Suppl), 1209–1219. https://doi.org/10.1037/hea0000306.

Nahum-Shani, I., Smith, S. N., Spring, B. J., Collins, L. M., Witkiewitz, K., Tewari, A., & Murphy, S. A. (2018). Just-in-time adaptive interventions (JITAIs) in mobile health: key components and design principles for ongoing health behavior support. *Annals of Behavioral Medicine, 52*(6), 446–462. https://doi.org/10.1007/s12160-016-9830-8.

Nilsen, W., Kumar, S., Shar, A., Varoquiers, C., Wiley, T., Riley, W. T., Pavel, M., & Atienza, A. A. (2012). Advancing the science of mHealth. *Journal of Health Communication, 17*(sup1), 5–10. https://doi.org/10.1080/10810730.2012.677394.

O'Brien, B. C., Harris, I. B., Beckman, T. J., Reed, D. A., & Cook, D. A. (2014). Standards for reporting qualitative research: a synthesis of recommendations. *Academic Medicine, 89*(9), 1245–1251.

Palinkas, L. A., Aarons, G. A., Horwitz, S., Chamberlain, P., Hurlburt, M., & Landsverk, J. (2011). Mixed method designs in implementation research. *Administration and Policy in Mental Health, 38*(1), 44–53. https://doi.org/10.1007/s10488-010-0314-z.

Proctor, E., Silmere, H., Raghavan, R., Hovmand, P., Aarons, G., Bunger, A., Griffey, R., & Hensley, M. (2011). Outcomes for implementation research: conceptual distinctions, measurement challenges, and research agenda. *Administration and Policy in Mental Health and Mental Health Services Research, 38*(2), 65–76. https://doi.org/10.1007/s10488-010-0319-7.

Reinares, M., Sánchez-Moreno, J., & Fountoulakis, K. N. (2014). Psychosocial interventions in bipolar disorder: what, for whom, and when. *Journal of Affective Disorders, 156*, 46–55. https://doi.org/10.1016/j.jad.2013.12.017.

Riley, W. T., & Rivera, D. E. (2014). Methodologies for optimizing behavioral interventions: Introduction to special section. *Translational Behavioral Medicine, 4*(3), 234–237. https://doi.org/10.1007/s13142-014-0281-0.

Robbins, R., Krebs, P., Jagannathan, R., Jean-Louis, G., & Duncan, D. T. (2017). Health app use among US mobile phone users: analysis of trends by chronic disease status. *JMIR MHealth and UHealth, 5*(12), e7832. https://doi.org/10.2196/mhealth.7832.

Robeznieks, A. American Medical Association, (2020). Key changes made to telehealth guidelines to boost COVID-19 care | American Medical Association [Internet] https://www.ama-assn.org/delivering-care/public-health/key-changes-made-telehealth-guidelines-boost-covid-19-care.

Rogers, T., Milkman, K. L., John, L. K., & Norton, M. I. (2015). Beyond good intentions: prompting people to make plans improves follow-through on important tasks. *Behavioral Science & Policy, 1*(2), 33–41. https://doi.org/10.1353/bsp.2015.0011.

Scanlon, J.W., Horst, P., Jay, J.N., et al., (1997) Evaluability assessment: avoiding Type III and Type IV errors. In: Gilbert GR, Conklin PJ, eds. Evaluation management: a sourcebook of readings. Chalottesville: US Civil Service Commission.

Seale, C. (1999). Quality in qualitative research. https://journals.sagepub.com/doi/abs/10.1177/107780049900500402?casa_token=D9jQpaPUFTMAA AAA:zErsjkvMk9k8cnX2MCJNoW6at8PbCXIzVtMzpY4FsUQprQ16nyJx-6_NY-5VDUY2NS0sbke1umiQPg. Accessed on 1/31/2022.

Seewald, N. J., Smith, S. N., Lee, A. J., Klasnja, P., & Murphy, S. A. (2019). Practical considerations for data collection and management in mobile health micro-randomized trials. *Statistics in Bioscience, 11*, 355–370.

Seewald, N. J., Sun, J., & Liao, P. (2016). MRT-SS calculator: an R shiny application for sample size calculation in micro-randomized trials. *ArXiv E-Prints, 1609* arXiv:1609.00695.

Shelton, R. C., & Lee, M. (2019). Sustaining evidence-based interventions and policies: recent innovations and future directions in implementation science. *American Journal of Public Health, 109*(S2), S132–S134. https://doi.org/10.2105/AJPH.2018.304913.

Shiffman, S., Stone, A. A., & Hufford, M. R. (2008). Ecological momentary assessment. *Annual Review of Clinical Psychology, 4*(1), 1–32. https://doi.org/10.1146/annurev.clinpsy.3.022806.091415.

Sim, I. (2019). Mobile devices and health. *New England Journal of Medicine, 381*(10), 956–968. https://doi.org/10.1056/NEJMra1806949.

Simon, H. A. (1988). The science of design: creating the artificial. *Design Issues, 4*(1/2), 67–82. https://doi.org/10.2307/1511391.

Smith, S. N., Lee, A. J., Hall, K., Seewald, N. J., Boruvka, A., Murphy, S. A., & Klasnja, P (2017). design lessons from a micro-randomized pilot study in mobile health. In J. M. Rehg, S. A. Murphy, & S. Kumar (Eds.), *Mobile Health: Sensors, Analytic Methods, and Applications* (pp. 59–82). Springer International Publishing. https://doi.org/10.1007/978-3-319-51394-2_4.

Spruijt-Metz, D., & Nilsen, W. (2014). Dynamic models of behavior for just-in-time adaptive interventions. *IEEE Pervasive Computing, 13*(3), 13–17. https://doi.org/10.1109/MPRV.2014.46.

Stowell, E., Lyson, M., Saksono, H., Wurth, R., Jimison, H., Pavel, M., & Parker, A. (2018). Designing and evaluating mhealth interventions for vulnerable populations. *Proceedings of the 2018 CHI Conference on Human Factors in Computing Systems*. https://doi.org/10.1145/3173574.3173589.

Strickland, J. C., & Stoops, W. W. (2019). The use of crowdsourcing in addiction science research: Amazon Mechanical Turk. *Experimental and Clinical Psychopharmacology, 27*(1), 1–18. https://doi.org/10.1037/pha0000235.

Sverdlov, O., Dam, J. van, Hannesdottir, K., & Thornton-Wells, T (2018). Digital therapeutics: an integral component of digital innovation in drug development. *Clinical Pharmacology & Therapeutics, 104*(1), 72–80. https://doi.org/10.1002/cpt.1036.

Thomas, K. A., & Clifford, S. (2017). Validity and mechanical turk: an assessment of exclusion methods and interactive experiments. *Computers in Human Behavior, 77*, 184–197. https://doi.org/10.1016/j.chb.2017.08.038.

Valimaki, M., Anttila, K., Anttila, M., & Lahti, M. (2017). Web-based interventions supporting adolescents and young people with depressive symptoms: Systematic review and meta-analysis. doi:10.2196/mhealth.8624.

van Veen, T., Binz, S., Muminovic, M., Chaudhry, K., Rose, K., Calo, S., Rammal, J.-A., France, J., & Miller, J. B. (2019). Potential of mobile health technology to reduce health disparities in underserved communities. *Western Journal of Emergency Medicine, 20*(5), 799–802. https://doi.org/10.5811/westjem.2019.6.41911.

Vogels, E. a. (2021). Digital divide persists even as Americans with lower incomes make gains in tech adoption. *Pew Research Center*. https://www.pewresearch.org/fact-tank/2021/06/22/digital-divide-persists-even-as-americans-with-lower-incomes-make-gains-in-tech-adoption/. Accessed on 1/31/2022.

Walsh, S., Golden, E., & Priebe, S. (2016). Systematic review of patients' participation in and experiences of technology-based monitoring of mental health symptoms in the community. *Open Access, 6*(6), 10.

Walton, A., Nahum-Shani, I., Crosby, L., Klasnja, P., & Murphy, S. A. (2018). Optimizing digital integrated care via micro-randomized trials. *Clinical Pharmacology & Therapeutics*. https://ascpt.onlinelibrary.wiley.com/doi/abs/10.1002/cpt.1079?casa_token=HVFmL4eV0VUAAAAA%3AGnc49Rt_6MYStFMghyQ_b3YKuwD3sVWBxAkVUQZwH4XynmuTf3lj8UEre3QL3g4kEBNVDPFPMbegGmw. Accessed on 1/31/2022.

Wiltsey Stirman, S., Kimberly, J., Cook, N., Calloway, A., Castro, F., & Charns, M. (2012). The sustainability of new programs and innovations: a review of the empirical literature and recommendations for future research. *Implementation Science, 7*(1), 17. https://doi.org/10.1186/1748-5908-7-17.

Chapter 12

Cultural adaptations of digital therapeutics

John A. Naslund[a] and Jessica Spagnolo[b,c]

[a] *Department of Global Health and Social Medicine, Harvard Medical School, Boston, MA, United States,* [b] *Département des sciences de la santé communautaire, Université de Sherbrooke, Sherbrooke, QC, Canada,* [c] *Centre de recherche Charles-Le Moyne (CR-CLM), Campus de Longueuil, Université de Sherbrooke, QC, Canada*

12.1 Introduction

With increasing access to digital technologies such as smartphones and mobile devices across most regions of the globe, there are emerging opportunities for digital therapeutics (referred to in this chapter as any form of digitally enabled intervention for mental health or substance use disorders) to make significant progress in addressing the global care gap for mental disorders (Carter, Araya, Anjur, Deng, & Naslund, 2021). The ease with which digital platforms can be modified and tailored to meet the needs of different population groups highlights their potential for adaptation to specific cultures, languages, and contexts (Yardley, Morrison, Bradbury, & Muller, 2015). With mounting evidence supporting the effectiveness of culturally adapted treatments for mental disorders (Hall, Ibaraki, Huang, Marti, & Stice, 2016; Shehadeh, Heim, Chowdhary, Maercker, & Albanese, 2016), careful consideration of individuals' context, language, culture, gender, race, ethnicity, and life circumstances will be essential for ensuring successful delivery of digital therapeutics.

The purpose of this chapter is to introduce the topic of cultural adaptation of digital therapeutics. The importance of ensuring that evidence-based interventions are culturally relevant for a target population group has been discussed extensively in prior literature, with Ahluwalia, Baranowski, Braithwaite, and Resnicow (1999) offering a broad depiction of how culture should be considered at all levels of intervention development and delivery:

> *"The extent to which ethnic/cultural characteristics, experiences, norms, values, behavioral patterns and beliefs of a target population as well as relevant historical, environmental, and social forces are incorporated in the design, delivery, and evaluation of targeted health promotion materials and programs."*

(page 11) (Ahluwalia et al., 1999).

Expanding on this recognition of culture as a key consideration of any evidence-based intervention, it is necessary to illustrate a systematic and rigorous approach to adapt existing evidence-based interventions for diverse cultural and ethnic groups, while retaining the core therapeutic benefits of the mental health intervention necessary to achieve the desired outcome, such as prevention, symptom improvement, remission, or recovery. A widely accepted definition by Bernal et al states that cultural adaptation refers to "the systematic modification of an evidence-based treatment (or intervention protocol) to consider language, culture, and context in such a way that it is compatible with the client's cultural patterns, meanings, and values" (page 362) (Bernal & Domenech Rodríguez, 2009). In the context of digital therapeutics, which encompass any form of remote or online platform or mobile device, such as smartphone apps, text messaging, online programs, telepsychiatry, or wearable devices that can be used to deliver a mental health intervention (Lipschitz, Hogan, Bauer, & Mohr, 2019), there have been relatively fewer studies focused on specific guidance for adaptation to culturally diverse groups. A recent review highlights that cultural adaptation appears to contribute to better acceptability and enhanced effectiveness of online psychological interventions for common mental disorders (Shehadeh et al., 2016); however, there remains considerable variation in how evidence-based intervention content is adapted, and how the methods for cultural adaptation are reported.

Digital Therapeutics for Mental Health and Addiction: The State of the Science and Vision for the Future. DOI: https://doi.org/10.1016/B978-0-323-90045-4.00001-0

Therefore, this chapter aims to provide an overview of rigorous methods for cultural adaptation that are informed by stakeholders, designed to promote engagement among target users, and that balance fidelity to the original evidence-based intervention content with the need for context-specific adaptations. This chapter begins with an overview of the importance of cultural adaptation for mental health service delivery and describes existing frameworks and approaches for guiding cultural adaptation of psychological treatments for mental disorders (Chowdhary et al., 2014). Next, this chapter summarizes efforts to engage target users in guiding the adaptation of digital interventions, and outlines promising examples of efforts to consider cultural and contextual factors in order to promote and sustain engagement in digital mental health interventions among target users from diverse cultures and ethnicities. Lastly, this chapter concludes by providing key recommendations for guiding the successful cultural adaptation of digital therapeutics, including the importance of following established theoretical frameworks, engaging key stakeholders throughout the design and adaptation process, and carefully documenting any modifications to program content or delivery to enable replication and rigorous outcome evaluation.

12.2 Global burden of mental illness and need for cultural adaptation

Worldwide, mental disorders represent a leading cause of disability (Vigo, Thornicroft, & Atun, 2016). Referred to as the global care gap (Pathare, Brazinova, & Levav, 2018), it is estimated that most individuals living with a mental illness do not have adequate access to evidence-based care (Patel et al., 2016). This gap in available treatment varies widely within and between countries; where mental health services are inequitably distributed, individuals in disadvantaged settings and rural areas are left with particularly limited access (Alonso et al., 2018). For instance, across the United States of America, it is estimated that approximately two-thirds of individuals living with mental illness do not have access to care (Kohn et al., 2018), with particularly pronounced disparities in access to mental health care among underrepresented racial and ethnic minority groups relative to non-Hispanic whites (Cook, Trinh, Li, Hou, & Progovac, 2017). In lower-income countries, the gaps in access to mental health services are substantial, such as in rural India (Arvind et al., 2019; Kokane et al., 2019), in Ethiopia, or South Africa (Azale, Fekadu, & Hanlon, 2016; Lund et al., 2015), where in excess of 90% of individuals living with mental illness do not have access to care.

The global burden of mental disorders disproportionately impacts low-income and middle-income countries (LMICs), where approximately 80% of the world's people facing the challenges of mental illness reside (Rathod et al., 2017). The alarming disparity between those who have access to mental health care and those who do not is a major contributor to poor outcomes among individuals living with mental illness over the life course, including greater disability, fewer opportunities for employment or education, social isolation, experiences of stigma and discrimination, increased risk of substance use, and ultimately early mortality relative to the general population (Liu et al., 2017). Therefore, efforts to scale up mental health services globally represent an important public health priority, and are critical for meeting the sustainable development goals and achieving universal health coverage (Patel et al., 2018).

There is robust evidence of the effectiveness and cost-effectiveness of task-sharing brief psychological interventions for mental disorders, which involves training and supporting nonspecialist providers such as community health workers in delivering these brief evidence-based treatments in routine care settings (Raviola, Naslund, Smith, & Patel, 2019). However, few individuals have access to these essential, psychological evidence-based services (Patel et al., 2018). An important consideration is that many of the mental health interventions that are promoted as evidence-based have originated from the disciplines of psychiatry and psychology in Western countries, and have therefore been evaluated among predominantly non-Hispanic white or European populations (Bemme & Kirmayer, 2020; Brown, Marshall, Bower, Woodham, & Waheed, 2014). The limited enrollment of ethnic minority groups in the development, adaptation, and evaluation of treatments for mental disorders potentially limits the generalizability of these interventions (Waheed, Hughes-Morley, Woodham, Allen, & Bower, 2015). Consideration of culturally adapted treatments, as well as emphasis on engaging ethnic minority populations in the adaptation of treatments and enrollment in evaluation studies is essential for establishing the external validity of psychological treatments (Bernal & Sáez-Santiago, 2006). In the context of global mental health, the field has been sharply criticized as being driven almost exclusively by Western institutions, and by perpetuating its colonial underpinnings (Bemme & Kirmayer, 2020; Saraceno, 2020; Weine et al., 2020).

Despite these limitations with prior research and threats to external validity of existing treatments, many brief psychological interventions have demonstrated effectiveness across different cultures, as well as racial and ethnic minority groups. For instance, recent reviews highlight the broad reach of studies on task-sharing interventions in different LMICs (Barbui et al., 2020; Singla et al., 2017), as well as achieving positive clinical outcomes consistent with earlier evidence generated from higher-income settings among predominantly non-Hispanic white patient populations (Cuijpers, Karyotaki, Reijnders, Purgato, & Barbui, 2018). The evidence supporting task-sharing is compelling, particularly across different cultural groups and contexts in LMICs (Barbui et al., 2020); yet, the implementation and sustained delivery of these programs in routine care settings remain disappointingly low across most regions globally (Patel et al., 2018; Raviola et al., 2019). Importantly, there remain stark disparities in access to quality mental health care among racial and ethnic minority groups, including after adjusting for various socioeconomic variables or educational levels, which can be attributed to discriminatory policies, racism, and historical and structural inequities (Bernal & Sáez-Santiago, 2006; Cabassa & Baumann, 2013; Kirmayer & Swartz, 2013; Satcher, 2001).

Emerging digital technologies and expanded wireless connectivity across many regions globally could overcome these implementation challenges by increasing the ease of access to effective mental health treatments (Naslund et al., 2019; Naslund, Bartels, & Marsch, 2019). To date, there have been moderate clinical benefits of smartphone interventions for anxiety or depression as reported in recent meta-analyses (Firth et al., 2017; Firth et al., 2017; Fu, Burger, Arjadi, & Bockting, 2020), though there is considerable heterogeneity in outcome reporting and methodological challenges, making it difficult to determine effectiveness across diverse cultural groups, particularly in LMICs (Fu et al., 2020). Furthermore, research has found that ethnic minorities may not experience the same benefits from digital interventions as the general population, as reflected by lower response and remission rates in online interventions for depression (Karyotaki et al., 2018). These limitations in the field of digital mental health highlight the need for greater attention to cultural adaptation, and further investigation into the use of systematic approaches to ensure the acceptability of mental health interventions and to promote their successful uptake to bridge the care gap among different cultures and contexts (Heim & Kohrt, 2019).

Cultural adaptation of digital therapeutics could help accommodate and respond to the unique experiences of mental illness between diverse cultural groups, including racial or ethnic minority groups, and across different geographic regions and contexts (Cabassa & Baumann, 2013; Haroz et al., 2017; Kohrt et al., 2014). For instance, studies show how culture can shape help-seeking behaviors for mental health problems (Sheikh & Furnham, 2000), differences in the manifestation of stigma or prejudice toward individuals living with mental illness (Krendl & Pescosolido, 2020), and engagement and retention in mental health services, as well as how healthcare providers communicate with their patients (Satcher, 2001). There can also be cultural differences in the expression of suffering, its spiritual meaning and significance, and how this relates to the causes of mental illness (Kohrt et al., 2014). Recent studies have also uncovered potential variations in the neurobiology of mental disorders due to the influence of different cultural norms (Shattuck, 2019). Furthermore, studies have found that response to treatments, such as psychological interventions, can vary between ethnic minority groups, and this when compared to nonminorities (Bernal & Sáez-Santiago, 2006). Carefully designed digital therapeutics could account for these cultural differences to optimize the delivery of effective intervention components needed to sustain engagement, improve outcomes, and promote recovery.

12.3   Methods of cultural adaptation

To address the significant gaps in access to mental health care across most settings globally, and to reduce mental health disparities, there has been growing recognition of the importance of ensuring cultural adaptation of effective mental health treatments to promote uptake and improve outcomes. Specifically, a number of different frameworks for guiding cultural adaptation are described in the literature on mental health services (Chowdhary et al., 2014). Several studies support the use of staged models for guiding cultural adaptations of evidence-based interventions, as reflected by enhanced effectiveness (Griner & Smith, 2006). A meta-analysis of 78 studies evaluating culturally adapted psychological interventions found that culturally adapted interventions produced greater treatment effects and increased odds of remission when compared to either the no intervention control condition or other interventions (Hall et al., 2016).

In the sections below, the key features of six prominent frameworks for guiding cultural adaptation are summarized. While this list is not exhaustive, many of these approaches share overlapping constructs and other similarities, such as employing a combination of quantitative and qualitative methods, allowing for multiple iterations through preliminary testing and subsequent refinements to the intervention content (Castro, Barrera Jr, & Holleran Steiker, 2010). Furthermore, many of these frameworks emphasize approaches for increasing community participation as a way to improve the ecological validity of the adapted evidence-based intervention (Hwang, 2009). These frameworks also hold potential for application to the adaptation of digital therapeutics.

First, Bernal and Saez-Santiago (2006) describe a framework for cultural adaptation centered around eight dimensions that can be targeted, including: (1) language (i.e., requires consideration of the target population's native language and extends beyond translation to account for cultural and emotional aspects of language); (2) persons (i.e., refers to the client-therapist relationship, and acknowledges the limits or boundaries of this relationship); (3) metaphors (i.e., refers to cultural symbols and sayings); (4) content (i.e., refers to understanding the values, customs, and traditions shared by the target population); (5) concepts (i.e., refers to the ways in which the treatment is conceptualized and communicated to clients); (6) goals (i.e., refers to the agreement between therapist and client regarding the treatment goals); (7) methods (i.e., refers to the procedures needed to achieve the treatment goals); and (8) context (i.e., refers to consideration of the clients' broader environment, including the social, economic, and political contexts that may influence treatment) (Bernal & Sáez-Santiago, 2006). These elements can serve to enhance the ecological and external validity of the treatment (Bernal, Bonilla, & Bellido, 1995).

Second, Barrera and colleagues (2006) define an integrated framework to cultural adaptation, which they revised and updated in 2013 (Barrera Jr & Castro, 2006; Barrera Jr, Castro, Strycker, & Toobert, 2013). Their framework combines a "top-down" approach that recognizes the importance of adhering to the evidence-based intervention content, with a "bottom-up" approach highlighting that successful treatments must be grounded in the values, beliefs, traditions, and practices of a specific cultural group (Barrera Jr, et al., 2013). This framework consists of five stages: (1) information gathering (i.e., this initial stage seeks to determine whether cultural adaptations are warranted, and often involves formative research to determine how well an intervention could fit the needs of a specific cultural group); (2) preliminary design (i.e., this stage involves initial modifications to the original intervention content guided by the information from the first stage, such as translation and modifying language, without altering the core components of the intervention; this is followed by focus groups, review by advisory panels, and usability testing); (3) preliminary testing (i.e., at this stage, there is a preliminary version of the culturally adapted intervention ready for pilot testing); (4) refinement (i.e., feedback and insights collected during the pilot study in the third phase are used to make additional refinements to the culturally adapted intervention); and (5) final trial (i.e., the empirical evaluation of the effectiveness of the culturally adapted intervention) (Barrera Jr & Castro, 2006; Barrera Jr, et al., 2013). Initial usability testing, as reflected in the second stage, is particularly relevant for the use of technology, where the digital components of the intervention must also be adapted to the target population.

Third, Lau (2006) proposes a selective and directed approach to cultural adaptation of evidence-based interventions (Lau, 2006). This approach involves first using data to selectively determine the specific problem and community that would stand to benefit from the evidence-based intervention. After isolating the target mental health problem and recognizing the severity and experiences of this problem within the target cultural group, it is necessary to consider if a particular evidence-based intervention would be a potentially good fit. This can help to determine the receptiveness of the target community to an intervention, and whether there may be potential barriers to acceptance of the intervention. Next, data are used in a directive manner to inform focused adaptations to the evidence-based intervention without undermining the therapeutic benefit of the original program. This process consists of modifying the content so that it considers the contextual factors related to the identified mental health problem within the target community. Additional directed adaptations can be necessary to promote uptake and minimize attrition to the evidence-based intervention. By leveraging data to justify the adaptations to the evidence-based intervention, this rigorous process ensures that the intervention is both empirically and socially valid (Lau, 2006).

Fourth, Hwang (2009) offers a community-based approach called the formative method for adapting psychotherapy (FMAP), which aims to expand on existing frameworks by guiding specific recommendations to support the cultural adaptation of psychological treatments for ethnic and cultural minority groups (Hwang, 2009). Similar to existing phased approaches summarized above (Barrera Jr & Castro, 2006), FMAP is comprised of five phases, though there is increased emphasis on directly engaging stakeholders and the target population group throughout the adaptation process by: (1) generating knowledge through collaboration (i.e., identifying appropriate stakeholders who can participate in focus group discussions); (2) integrating information gathered from stakeholders with theory, empirical, and clinical knowledge (i.e., applying the insights collected from the focus group discussions to inform cultural adaptations of the intervention content); (3) reviewing the initial culturally adapted intervention with stakeholders and making further revisions (i.e., conducting additional focus group discussions with stakeholders to inform further revisions to the culturally adapted intervention); (4) testing the culturally adapted intervention (i.e., evaluating the culturally adapted intervention with the target population group); and (5) making final revisions to the culturally adapted intervention (i.e., modifications informed by findings from the evaluation, as well as through exit interviews and focus group discussions with stakeholders) (Hwang, 2009).

Fifth, Chu and Leino (2017) conducted a systematic review and synthesis of studies on cultural adaptations to evidence-based interventions for mental health problems in ethnic minorities in order to generate the cultural treatment adaptation framework (Chu & Leino, 2017). This framework seeks to unify common concepts and language that are typically applied in the adaptation of evidence-based interventions to diverse cultural groups, and consists of two broad categories of components (Chu & Leino, 2017). First, there are the "peripheral components," which pertain to the receptivity toward a treatment, as well as the uptake and engagement with a particular treatment. This typically involves modifying an evidence-based intervention to improve access, retention or completion, and engagement, defined as the ability to reach the target population, and to ensure that participants can achieve the desired benefits of the program. Second, there are "core components," referring to primary therapeutic elements of the treatment that must be retained in order to achieve the desired outcomes. Adaptations to core components can take place along a continuum of four stages, beginning with no change, followed by core modifications, core additions, or complete change (Chu & Leino, 2017). In their review, Chu and Leino found that all evidence-based interventions had changed to peripheral components, while about 11% made adaptations to the core therapeutic components, and 60% involved core additions. Importantly, they determined that cultural adaptation to evidence-based interventions largely preserves fidelity to the core components of the interventions, while requiring significant additions of new content, modifications to the intervention delivery, and contextualization of the intervention to the target community (Chu & Leino, 2017).

Lastly, Heim and Kohrt (2019) propose a theory-driven framework to serve as the basis for empirical testing aimed at understanding how and why cultural adaptation improves the acceptability and effectiveness of an evidence-based intervention (Heim & Kohrt, 2019). This framework builds on increasing research aimed at unpacking the key ingredients, including both treatment-specific and unspecific factors, of evidence-based psychological interventions that contribute to symptom improvement and recovery (Cuijpers, Cristea, Karyotaki, Reijnders, & Hollon, 2019; Singla et al., 2017). The framework includes three overarching elements: (1) cultural concepts of distress (i.e., this considers the core beliefs of the target population, accounting for their understanding of the causes as well as the expression of mental illness, and serves as an important starting point for guiding cultural adaptation); (2) treatment components (i.e., this reflects modifications to the specific elements such as behavioral or cognitive aspects of the treatment, and unspecific elements such as empathy or active listening that aim to promote engagement in the treatment); and (3) treatment delivery (i.e., this encompasses the selection of an appropriate delivery format and modifications for the target population, taking into account literacy, access to technology, gender, socio-economic status, etc.). This framework seeks to develop an understanding of what treatment elements are necessary for adaptation and standardization in order to achieve the desired outcomes, as this is critical for the scalability and transferability of an evidence-based intervention across diverse settings (Heim & Kohrt, 2019; Heim et al., 2020).

There remain several limitations with these frameworks that warrant consideration, such as continued need to determine which aspects of cultural adaptation are essential for achieving the desired intervention effects (Barrera Jr, et al., 2013). Concerns have also been raised about the consistency of reporting the detailed steps taken to support the cultural adaptation of evidence-based interventions, the rigor of cultural adaptations, as well as whether cultural adaptations improve acceptability of the intervention, its uptake, and its sustained use among the target population (Heim & Kohrt, 2019). Despite limitations with these frameworks, studies emphasize the importance of applying rigorous methods to guide the cultural adaptation of evidence-based interventions to avoid potentially haphazard modifications or "on-the-fly" modifications to the intervention content (Lau, 2006). These considerations are especially relevant for the cultural adaptation of digital interventions to ensure that adaptations are sufficiently documented to inform replication and implementation across other settings.

12.4   Cultural adaptation of digital interventions

This section of the chapter considers the use of evidence-based digital mental health interventions in different settings while recognizing the sociocultural context of culturally diverse population groups and applying different methods to engage target users in the adaptation process. This engagement elevates target users to the role of expert and encourages a bottom-up approach (Hwang, 2009), which can foster increased acceptability, usability, clinical impact, and scalability (Cabassa & Baumann, 2013; Chowdhary et al., 2014; Hall et al., 2016; Shehadeh et al., 2016). The examples summarized below were identified through searches of PubMed using relevant keywords. The examples described in the sections below are not exhaustive and are intended to present a selection of studies from the global mental health literature, showcasing cultural adaptation to diverse settings and population groups.

A commonly adopted digital mental health intervention across different sociocultural contexts is the World Health Organization's step-by-step program, an internet-delivered intervention that relies on an illustrative educative narrative for the delivery of psychoeducation, behavioral activation, and stress management exercises to address symptoms of depression among adults (Carswell et al., 2018). The program content, originally written in English and developed as a web-based intervention, was designed to be adapted to meet the realities of different populations and sociocultural contexts (Carswell et al., 2018). The step-by-step program has been adapted to the sociocultural contexts of Lebanese, Syrian, and Palestinians in Lebanon (Abi Ramia et al., 2018), Syrian refugees in Germany, Sweden, and Egypt (Burchert et al., 2019), overseas Filipino workers in Macau (Garabiles, Shehadeh, & Hall, 2019), Albanian-speaking migrants in Switzerland and Germany (Shala et al., 2020), and Chinese young adults (Sit et al., 2020).

Cognitive interviewing techniques were commonly used to adapt the step-by-step program to different sociocultural contexts (Abi Ramia et al., 2018; Burchert et al., 2019; Garabiles et al., 2019; Shala et al., 2020; Sit et al., 2020). Cognitive interviewing relies on "verbal probing" and "thinking out loud," and has traditionally been used to develop questionnaires and surveys (Drennan, 2003; Willis, Jr, & A., 2013). The cognitive interviewing technique to culturally adapt the web-based step-by-step program in Lebanon was used to engage target users in testing the story content and audio relaxation techniques through interviews, as well as the general acceptability and feasibility of the intervention through discussions (Abi Ramia et al., 2018). Findings from cognitive interviewing with target users reveal that characters included in the narrative may not be relatable to all (e.g., depictions of married couples with children), as well as dissatisfaction with some of the mental health language used and character facial expressions. These insights were then used to inform an adaptation workshop, where final decisions on content were voted on by stakeholder groups like project partners. Decisions included for example shortening the length of the program by 30% to support sustained interest and engagement among target users (Abi Ramia et al., 2018).

In another example, the adaptation of the step-by-step program to a smartphone application involved cognitive interviewing techniques with Syrian refugees in Germany, Sweden, and Egypt as one of the key steps in the cultural adaptation process. Target users tested the usability of the application prototype and were asked to "think aloud" about its use (Burchert et al., 2019). The goal of this "think aloud" technique was to encourage target users to freely discuss impressions of the program in its smartphone application version as they engaged with its content. Findings reveal that some target users were dissatisfied with the narrative content and the spoken dialect, including some mental health language used. Subsequent focus groups where the updated digital step-by-step prototype was presented to target users allowed for triangulation of data with content from the cognitive interviewing technique and additional feedback to inform further modifications to the prototype (Burchert et al., 2019).

Cognitive interviewing techniques for the cultural adaptation of step-by-step were also conducted among overseas Filipino workers, starting with the available and generic English version of the program (Garabiles et al., 2019). The step-by-step program was shared with Filipino psychologists, who were invited to make suggested edits and modifications to the content and the illustrations prior to engaging with target users. Overseas Filipino workers were then invited to "think aloud" in focus groups about the adapted content, which was included on PowerPoint slides. Specifically, they were asked probes through open-ended questions related to their experiences and suggested improvements. Through this exercise, target users revealed issues with acceptability; some illustrations and content were not considered appropriate (e.g., too political for the target setting, or an over-emphasis on mental health issues instead of achieving positive goals and outcomes), relevant (e.g., difficulties in relating to characters, like a doctor in the narrative who made users feel pathologized for their experiences and type of activities such as involving family members), or comprehensible (e.g., some terms being too complex or unclear, as well as the text not matching the illustrations) (Garabiles et al., 2019).

In a separate study, the step-by-step program (renamed *Hap-pas-Hapi*) was adapted for use among Albanian-speaking migrants in Switzerland and Germany (Shala et al., 2020). First, the English version of the program was translated into Albanian. During focus groups, target users read the translated narrative out loud, and were encouraged to "speak up" if they were dissatisfied with any content and/or illustrations (Shala et al., 2020), which informed modifications on narrative characters, examples, and language prior to conducting the second round of discussions. This second round encouraged adaptations to idioms of distress and metaphors, and usability of the exercises was discussed (Shala et al., 2020). Following these discussions, target users were invited to "think aloud" and test the *Hap-pas-Hapi* prototype. Findings from cognitive interviewing techniques were used during a cultural adaptation workshop to decide on the most necessary adaptations, including minor suggestions such as sentence changes (Shala et al., 2020). Similarly, starting with the English version of the step-by-step program, it was adapted for use among Chinese young adults (Sit et al., 2020). After translating the English version into the Chinese language and hiring a creative story writer to include a steering committee's suggestions to better fit the storyline to the reality of Chinese students, cognitive interviewing techniques were employed. The program content was included in PowerPoint presentations and presented during focus group discussions. Participants were asked to describe their thoughts about the text and the illustrations, thereby generating information to inform final adaptations shared in community feedback meetings (Sit et al., 2020). For example, target users commented on the use of questionable language (e.g., positive feedback expressions not culturally used), characters (e.g., preference for certain types of leading figures, like cartoon figures for the male groups and peers for the female groups), as well as metaphors and culturally related elements (e.g., type of storylines, physical contact among characters in the story).

Interestingly, to culturally adapt the step-by-step program, cognitive interviewing techniques were often one of multiple steps employed in the adaptation process, but not all steps engaged target users; the processes also involved other stakeholder groups to inform the adaptations to the digital intervention. For example, Garabiles et al. (2019) highlight that prior to engaging with target users, the program was adapted with the help of Filipino psychologists. To translate the step-by-step program into Albanian, Shala et al. (2020) shared that there was involvement from an independent translator, as well as review of the translated program content by two native-speaking mental health professionals. In addition, expert review by the research team was conducted for any cultural adaptations suggested by target users that were not conclusive and on minor language suggestions to ensure fidelity to the program content (Shala et al., 2020). To prepare for adaptation of the program to Chinese young adults, semistructured interviews were conducted with mental health experts to better understand the target users' health context, as well as the appropriateness of step-by-step's content, delivery method, characters, storylines, illustrations, and expressions (Sit et al., 2020). A member of the research team conducted the translation, a hired storywriter with a background in Chinese education wrote the program narrative, and a local professional illustrator designed the graphics to depict the main characters in the program (Sit et al., 2020). Abi Ramia et al. (2018) highlighted that the step-by-step program was translated from English into classical Arabic by a professional translator, and then translated into a spoken dialect of Arabic to reflect the main dialect groups in Lebanon (i.e., Lebanese, Syrian, and Palestinian). This dialectic translation was conducted by a project team member, with the support of two local staff, and reviewed by laypersons not acquainted with the project. Focus group discussions were conducted with stakeholder groups including mental health professionals, managers, and frontline workers to address specific sections that required clarification following the use of cognitive interviewing techniques with target users (Abi Ramia et al., 2018).

Other methods have also been used in guiding the cultural adaptation process of digital mental health interventions with target users. In addition to cognitive interviewing techniques, Burchert et al. (2019) employed free list interviewing for the cultural adaptation of the step-by-step program. This technique uses predetermined questions in a structured interview format to generate a list of answers (Applied Mental Health Research Group, 2013). The free list interviewing technique helped the authors to gain an understanding on the use of digital technologies among Syrian refugees, as well as challenges and solutions in using digital technology prior to initiating the steps in the cultural adaptation process (Burchert et al., 2019).

Of note, studies relied on participatory workshops with target users (Buus et al., 2019) and focus group discussions and/or interviews (Juniar et al., 2019; Lal et al., 2018; Lal et al., 2020). In addition to the techniques described previously, these studies report details on specific ways in which target users engaged in the process of culturally adapting digital mental health interventions. For example, Buus et al. (2019) describe the cultural adaptation of MYPLAN, a Danish mobile phone safety application for suicide prevention to a culturally appropriate version for Australia (Buus et al., 2019). Target users were involved in participatory workshops to assess the application (i.e., functions and culturally appropriate wording) and to suggest modifications. During these workshops, suggestions provided by target users were summarized and evaluated to ascertain the level of priority (Buus et al., 2019). One new core feature was developed during this phase: the Rant Box. This feature allows target users to "rant" after which text and images could be destroyed. Target users also suggested better virtual introductions given that healthcare professionals were not the only individuals offering access to the application (Buus et al., 2019).

Juniar et al. (2019) relied on focus group discussions to adapt an existing evidence-based work-stress-related intervention (i.e., GET.ON Stress) to Indonesian university students (Juniar et al., 2019). GET.ON Stress is an online stress management tool for adults and initially developed and evaluated for use among German employees (Heber, Lehr, Ebert, Berking, & Riper, 2016). Focus group discussions were conducted with students to better understand their general impressions of the program, as well as their appreciation of the program interface and content. In these focus group discussions, participants reflected on signs of stress and their experiences of stress as students. They also reflected on stress idioms and overall perceptions of stress and how to overcome it. These discussions resulted in shortening the program, modifying the program interface and content to be more interactive (e.g., including new features such as e-coaching, providing feedback to users, and using email reminders), making language changes (e.g., ensuring that the use of metaphors was relatable), adapting exercises to the Indonesian context and interests of students, as well as considering internet connectivity issues by replacing videos with slide shows (Juniar et al., 2019).

A similar approach, consisting of focus group discussions with target users, was employed in the adaptation of Horyzons, a web-based application consisting of Moderated Web-Based Social Therapy and designed for young people recovering from first-episode psychosis (Alvarez-Jimenez et al., 2013). Specifically, Lal et al. (2018; 2020) adapted Horyzons for use among young people recovering from first-episode psychosis in Quebec and Ontario, Canada, using focus group discussions to explore target users' satisfaction with the platform, content, and features, after which they were invited to test the program over a 2 to 4 week period. During this time, they provided personal feedback and suggestions for adaptations, which were then used to guide interviews post-use (Lal et al., 2018; 2020). Comments emerged on different features of the program, such as the language (e.g., replacing text with a spoken narrative, narratives spoken by someone without an Australian accent), and services/opportunities in the Canadian-context (e.g., highlighting resources related to finding work, study programs, and volunteer opportunities, as well as post-discharge care and follow-up) (Lal et al., 2018; 2020).

Surveys were also used to better understand the technological realities of target users prior to adapting the digital intervention (i.e., similar to the free list interviewing technique conducted by Burchert et al. (2019)) and to inform the cultural adaptation of digital interventions. For example, prior to the Horyzons intervention's adaptation from the Australian to the Canadian context, Lal et al. (2018; 2020) invited the target users to complete technology use questionnaires aimed at exploring their experiences and attitudes toward utilizing technology. After completing this questionnaire, target users shared their perceptions on the preliminary version of the adaptations through surveys and feedback forms, and provided suggestions for further modifications. In another example, Salamanca-Sanabria et al. (2019) administered the cultural relevance questionnaire to target users, developed for the purposes of their study to culturally adapt the *Space from Depression* program, an evidence-based internet-delivered cognitive-behavioral therapy intervention for depression (Richards et al., 2016; Richards et al., 2015). The questionnaire aimed to help adapt the intervention to the cultural context of college students in Colombia. Target users assessed the overall program and its specific modules, information used to inform adaptations directly to the digital content and formatting of the program, including renaming the program *Yo puedo sentirme bien* for use in the Colombian context (Salamanca-Sanabria, Richards, & Timulak, 2019).

In another study, an in-person program was adapted to a digital format for use in a rural context in India. The Healthy Activity Program is a brief behavioral activation treatment for moderate to severe depression designed to be delivered by non-specialists in primary care settings in LMICs (Patel et al., 2017). This manualized program was initially developed for the Indian context by adapting evidence-based behavioral activation content from a US-context (Chowdhary et al., 2016), before proceeding with fully powered effectiveness testing (Patel et al., 2017). After establishing the Healthy Activity Program as an evidence-based intervention, the paper-based content was adapted to a digital platform to facilitate the scale up of the training for nonspecialist health workers to deliver the program in primary care in a rural setting in India (Khan et al., 2020). More specifically, this study involved the cultural adaptation of a paper-based program to a digital program for training community health workers, in contrast to direct delivery of the digital intervention to patients (Khan et al., 2020; Muke et al., 2019). The adaptation process from the paper-based manual to digital intervention involved five steps: (1) creating a blueprint of the training program with manualized program experts to identify essential elements needed for the transition to the digital platform; (2) developing the training program content, which involved adapting the written content from the manuals to align with the components included in the blueprint and developing scripts while considering the level of digital education and experiences of the target users; (3) digitizing the training program, which involved developing the videos (lectures, role plays) adapted to the Indian context and where non-specialists work; (4) developing a learning management system and uploading content on a Moodle Learning Management System where the digital training could be hosted and accessed via technological tools like smartphones; and (5) user testing the digital training program (Khan et al., 2020). Target users' perspectives were included in multiple ways throughout the development and adaptation of the training program content. Participants' preferences for video-based content and use of visuals fueled the development of the prototypes, as well as consideration of the local language and use of relevant culturally and contextually relevant images and narratives (Muke et al., 2019). User testing in the form of semistructured feedback sessions including the target group of community health workers was conducted to identify further aspects of the program that could be improved as well as possible challenges with accessing and using the program in a rural setting (e.g., videos that appeared out of sync, slow download speeds, and overall slow application performance). Through multiple iterations and rounds of feedback with participants, as well as pilot testing (Muke et al., 2020), some of the major modifications to the digital program and that considered the local context involved compression of file sizes and videos to ensure that it may be downloaded to smartphones and also be used offline, and the development of a protocol for remote coaching to promote engagement among users (Khan et al., 2020).

12.5 Recommendations for applying cultural adaptations to digital therapeutics

Drawing from the summary of different approaches to cultural adaptation and recent examples from the literature specific to cultural adaptation of digital therapeutics, this section of the chapter aims to provide recommendations for guiding future work in this area. In brief, these recommendations highlight the importance of following established theoretical frameworks, engaging key stakeholders throughout the design and adaptation process, and carefully documenting any modifications to program content or delivery to enable replication and rigorous outcome evaluation. Four recommendations for supporting the cultural adaptation of digital therapeutics, as summarized in Table 12.1, include the following:

1. Frameworks can offer an important roadmap for guiding cultural adaptation of mental health interventions. These frameworks outline important steps in the systematic modification to the program content and delivery to account for the target individuals' context, language, and culture, and are supported by evidence (Chowdhary et al., 2014; Griner & Smith, 2006; Hall et al., 2016). The frameworks can offer a guide for empirical testing to help better understand how and why cultural adaptation and accounting for specific cultural concepts of mental illness can improve the acceptability and effectiveness of evidence-based interventions (Heim & Kohrt, 2019), increase community participation, and further support the ecological validity of the adapted evidence-based intervention (Hwang, 2009). Specifically, these frameworks hold promise for guiding the cultural adaptation of digital therapeutics by recognizing the specific components that require modification to promote adoption, uptake, and sustained use of digital interventions, as well as achieve the desired treatment outcomes within culturally diverse groups.

TABLE 12.1 Recommendations for applying cultural adaptations to digital therapeutics.

Recommendation	Opportunities	Examples
Following established frameworks	(1) Many different frameworks for guiding cultural adaptation of mental health interventions supported by evidence. (2) Involves the systematic modification to the program content and delivery to account for the target individuals' context, language, culture, gender, race, ethnicity, and life circumstances. (3) Holds promise for guiding the cultural adaptation of digital therapeutics and increasing the ecological and external validity of these interventions	The frameworks can offer a guide for empirical testing to help better understand how and why cultural adaptation can improve the acceptability and effectiveness of evidence-based interventions. Recognizing the specific components that require modification is especially important for supporting the adoption, uptake, and sustained use of digital interventions in order to achieve the desired outcomes within culturally diverse groups
Engaging key stakeholders throughout the design and adaptation process	(1) To benefit from essential expertise that will influence implementation feasibility. (2) To better understand intervention acceptability and sustainable use	Intervention-testing, and feedback received through: cognitive interviewing techniques; user-centered design; participatory workshops; interviews; focus groups; surveys; and questionnaires, for example
Documenting any modifications to program content or delivery	(1) To facilitate replicable processes across settings and populations. (2) To appraise evidence rigor and potential for scalability	Documentation of the methodological processes and outcomes: how and which methods were implemented to engage target users, example of questions used, and specific modifications derived from these processes and included in the digital intervention adaptation
Exploring barriers and drivers to technology use, and target user preferences prior to culturally adapting the digital intervention	(1) To initially gauge feasibility of implementing a culturally adapted digital intervention and hint at its sustainable use within specific settings and populations	Free list interviewing technique; technology use questionnaire

2. As shown by the case study examples provided, the adaptation of digital interventions to reflect sociocultural contexts did not simply involve a transfer of content to a digital platform or a translation of language for use. The content was modified. Therefore, it is critical to involve target users early on and throughout the adaptation process (e.g., multiple testing of prototypes or applications, iterative feedback, etc.) to benefit from essential expertise that may help foster digital mental health intervention acceptability and sustained use. In some examples provided, key stakeholder groups beyond solely target users were involved in codesigning the digital mental health interventions. Key stakeholder groups included mental health professionals, officials from the Ministry of Public Health, representation from nongovernmental organizations working in community and health sectors, and peer specialists, to name a few. The involvement of a broad range of stakeholder groups in the cultural adaptation of digital mental health interventions can help to highlight needed content, formatting, and layout modifications to reflect sociocultural realities, as well as their implementation feasibility and cost (Abi Ramia et al., 2018; Joshi et al., 2022).

3. As shown by the case study examples, many digital mental health interventions were culturally adapted to different settings, some of which have been used in multiple regions and with diverse populations (e.g., the step-by-step program). Therefore, a high level of detail is required when reporting on cultural adaptation processes, including the implementation and use of methods like free list interviewing and cognitive interviewing techniques. This can facilitate replication of the adaptation process for use across different settings and populations, as well as appraisal of evidence rigor and potential for scalability (Shehadeh et al., 2016).

4. Prior to adapting digital mental health interventions to reflect sociocultural contexts, some studies first explored barriers and drivers to using technology within specific populations, as well as their digital preferences. While this may not have been consistently reported in the studies consulted, it may be an important way to initially gauge feasibility

of implementing a culturally adapted digital intervention and hint at its sustainable use within specific settings and populations (Doherty, Coyle, & Matthews, 2010).

12.6 Conclusion

With the continued rapid growth in interest and use of digital approaches for supporting mental health and behavioral services (Bhugra et al., 2017), there is a pressing need to ensure that these digital therapeutics can meet the needs of culturally diverse groups. There may also be potential for the use of digital innovations to support approaches to cultural adaptation. For instance, with capacity to enable passive and active data collection, digital technologies such as smartphone apps could respond to specific stimuli within an individual's environment. This in turn could shed light on how an individual's culture, language, and context influence their health behaviors, activity, and mental health. Therefore, it is plausible that digital technologies could guide new treatment modalities or enhance the effectiveness of existing programs. The technology could also shed light on which components of cultural adaptation are necessary to achieve the desired target outcomes by enabling exploration of moderators or mediators of culturally adapted interventions, an area of the literature that remains understudied (Barrera Jr, et al., 2013).

Although, similar to existing challenges in the broader mental health literature, robust evidence on the effectiveness of digital therapeutics is largely concentrated in Western countries, primarily from trials enrolling few racial or ethnic minority groups (Firth et al., 2017; Firth et al., 2017; Naslund, Marsch, McHugo, & Bartels, 2015). These challenges are important threats to the generalizability of digital interventions. While there are promising studies across several countries in Africa, Asia, Latin America, and the Middle East (Carter et al., 2021; Kaonga & Morgan, 2019; Merchant, Torous, Rodriguez-Villa, & Naslund, 2020; Naslund et al., 2017), there is comparatively less evidence overall, and fewer rigorous effectiveness studies (Fu et al., 2020). Furthermore, many digital interventions face challenges due to low uptake, high attrition rates, and marginal adherence (Linardon & Fuller-Tyszkiewicz, 2020; Torous, Nicholas, Larsen, Firth, & Christensen, 2018). A major limitation with the broader field of digital mental health is the limited consideration of cultural adaptation, and the specific modifications needed to digital therapeutics to ensure adequate uptake among culturally diverse target population groups in order to achieve the expected clinical benefits. Rigorous efforts focused on cultural adaptation could also overcome the broader challenges related to poor adherence and high attrition in digital mental health interventions.

More in-depth exploration of cultural drivers and facilitators to engaging with and using digital interventions, as well as increased emphasis on carefully adapting the intervention content to diverse contexts, cultures, and languages may serve to enhance uptake and sustain engagement with these interventions, and support the scalability of these treatments (Heim & Kohrt, 2019; Jordans & Kohrt, 2020). The costs associated with cultural adaptation of digital therapeutics, as well as the cost-effectiveness of these interventions should also be considered. There may be opportunities for using technology to more rapidly respond to feedback from culturally diverse groups in a low-cost and efficient way, such as by adjusting language or modifying other features to promote engagement with the program. A greater understanding of the costs of cultural adaptation can support efforts related to translating evidence-based programs to other settings and contexts (Cabassa & Baumann, 2013), and warrant further investigation in order to support the scale-up of evidence-based digital therapeutics.

Acknowledgments

Dr. Naslund reports receiving funding from the National Institute of Mental Health (Grant number: 5U19MH113211), the Brain & Behavior Research Foundation NARSAD Young Investigator Award, and the Burke Global Health Fellowship from the Harvard Global Health Institute. At the time this book chapter was written, Dr. Spagnolo received a postdoctoral fellow scholarship from Fonds de recherche du Québec - Santé (FRQ-S). She currently receives funding from a Canadian Institutes of Health Reserch (CIHR) Fellowship. The funders played no role in the study design; collection, analysis, or interpretation of data; writing of the manuscript; or decision to submit the manuscript for publication.

Declaration of Interest

The authors report no conflicts of interest.

References

Abi Ramia, J., Shehadeh, M. H., Kheir, W., Zoghbi, E., Watts, S., Heim, E., et al. (2018). Community cognitive interviewing to inform local adaptations of an e-mental health intervention in Lebanon. *Global Mental Health, 5,* e39.

Ahluwalia, J., Baranowski, T., Braithwaite, R., & Resnicow, K. (1999). Cultural sensitivity in public health: defined and demystified. *Ethnicity and Disease, 9,* 10–21.

Alonso, J., Liu, Z., Evans-Lacko, S., Sadikova, E., Sampson, N., Chatterji, S., et al. (2018). Treatment gap for anxiety disorders is global: Results of the World Mental Health Surveys in 21 countries. *Depression and Anxiety, 35*(3), 195–208.

Alvarez-Jimenez, M., Bendall, S., Lederman, R., Wadley, G., Chinnery, G., Vargas, S., et al. (2013). On the Horyzon: moderated online social therapy for long-term recovery in first episode psychosis. *Schizophrenia Research, 143*(1), 143–149.

Applied Mental Health Research Group. (2013). Design, Implementation, Monitoring, and Evaluation of mental health and psychosocial assistance programs for trauma survivors in low resource countries: A user's manual for researchers and program implementers. Johns Hopkins University Bloomberg School of Public Health https://www.jhsph.edu/research/centers-and-institutes/global-mental-health/resource-materials/design-implementation-monitoring-and-evaluation-dime/.

Arvind, B. A., Gururaj, G., Loganathan, S., Amudhan, S., Varghese, M., Benegal, V., et al. (2019). Prevalence and socioeconomic impact of depressive disorders in India: multisite population-based cross-sectional study. *BMJ Open, 9*(6), e027250.

Azale, T., Fekadu, A., & Hanlon, C. (2016). Treatment gap and help-seeking for postpartum depression in a rural African setting. *Bmc Psychiatry [Electronic Resource], 16*(1), 1–10.

Barbui, C., Purgato, M., Abdulmalik, J., Acarturk, C., Eaton, J., Gastaldon, C., et al. (2020). Efficacy of psychosocial interventions for mental health outcomes in low-income and middle-income countries: an umbrella review. *The Lancet Psychiatry, 7*(2), 162–172.

Barrera Jr,, M., & Castro, F. G. (2006). A heuristic framework for the cultural adaptation of interventions. *Clinical Psychology: Science and Practice, 13*(4), 311–316.

Barrera Jr,, M., Castro, F. G., Strycker, L. A., & Toobert, D. J. (2013). Cultural adaptations of behavioral health interventions: a progress report. *Journal of Consulting and Clinical Psychology, 81*(2), 196.

Bemme, D., & Kirmayer, L. J. (2020). *Global Mental Health: Interdisciplinary Challenges for a Field in Motion.* London: SAGE Publications Sage UK.

Bernal, G., Bonilla, J., & Bellido, C. (1995). Ecological validity and cultural sensitivity for outcome research: issues for the cultural adaptation and development of psychosocial treatments with Hispanics. *Journal of Abnormal Child Psychology, 23*(1), 67–82.

Bernal, G., & Domenech Rodríguez, M. M. (2009). Advances in Latino family research: cultural adaptations of evidence-based interventions. *Family Process, 48*(2), 169–178.

Bernal, G., & Sáez-Santiago, E. (2006). Culturally centered psychosocial interventions. *Journal of Community Psychology, 34*(2), 121–132.

Bhugra, D., Tasman, A., Pathare, S., Priebe, S., Smith, S., Torous, J., et al. (2017). The WPA-lancet psychiatry commission on the future of psychiatry. *The Lancet Psychiatry, 4*(10), 775–818.

Brown, G., Marshall, M., Bower, P., Woodham, A., & Waheed, W. (2014). Barriers to recruiting ethnic minorities to mental health research: a systematic review. *International Journal of Methods in Psychiatric Research, 23*(1), 36–48.

Burchert, S., Alkneme, M. S., Bird, M., Carswell, K., Cuijpers, P., Hansen, P., et al. (2019). User-centered app adaptation of a low-intensity e-mental health intervention for Syrian refugees. *Frontiers in Psychiatry, 9*, 663.

Buus, N., Juel, A., Haskelberg, H., Frandsen, H., Larsen, J. L. S., River, J., et al. (2019). User involvement in developing the MYPLAN mobile phone safety plan app for people in suicidal crisis: Case study. *JMIR Mental Health, 6*(4), e11965.

Cabassa, L. J., & Baumann, A. A. (2013). A two-way street: bridging implementation science and cultural adaptations of mental health treatments. *Implementation Science, 8*(1), 1–14.

Carswell, K., Harper-Shehadeh, M., Watts, S., van't Hof, E., Abi Ramia, J., Heim, E., et al. (2018). Step-by-step: a new WHO digital mental health intervention for depression. *Mhealth, 4*, 34.

Carter, H., Araya, R., Anjur, K., Deng, D., & Naslund, J. A. (2021). The emergence of digital mental health in low-income and middle-income countries: A review of recent advances and implications for the treatment and prevention of mental disorders. *Journal of Psychiatric Research, 133*, 233–246.

Castro, F. G., Barrera Jr,, M., & Holleran Steiker, L. K. (2010). Issues and challenges in the design of culturally adapted evidence-based interventions. *Annual Review of Clinical Psychology, 6*, 213–239.

Chowdhary, N., Anand, A., Dimidjian, S., Shinde, S., Weobong, B., Balaji, M., et al. (2016). The healthy activity program lay counsellor delivered treatment for severe depression in India: systematic development and randomised evaluation. *The British Journal of Psychiatry, 208*(4), 381–388. doi:10.1192/bjp.bp.114.161075.

Chowdhary, N., Jotheeswaran, A., Nadkarni, A., Hollon, S., King, M., Jordans, M., et al. (2014). The methods and outcomes of cultural adaptations of psychological treatments for depressive disorders: a systematic review. *Psychological Medicine, 44*(6), 1131–1146.

Chu, J., & Leino, A. (2017). Advancement in the maturing science of cultural adaptations of evidence-based interventions. *Journal of Consulting and Clinical Psychology, 85*(1), 45.

Cook, B. L., Trinh, N.-H., Li, Z., Hou, S. S.-Y., & Progovac, A. M. (2017). Trends in racial-ethnic disparities in access to mental health care, 2004–2012. *Psychiatric Services, 68*(1), 9–16.

Cuijpers, P., Cristea, I. A., Karyotaki, E., Reijnders, M., & Hollon, S. D. (2019). Component studies of psychological treatments of adult depression: a systematic review and meta-analysis. *Psychotherapy Research, 29*(1), 15–29.

Cuijpers, P., Karyotaki, E., Reijnders, M., Purgato, M., & Barbui, C. (2018). Psychotherapies for depression in low-and middle-income: a meta-analysis. *World Psychiatry, 17*(1), 90–101.

Doherty, G., Coyle, D., & Matthews, M. (2010). Design and evaluation guidelines for mental health technologies. *Interacting with Computers, 22*(4), 243–252.

Drennan, J. (2003). Cognitive interviewing: verbal data in the design and pretesting of questionnaires. *Journal of Advanced Nursing, 42*(1), 57–63.

Firth, J., Torous, J., Nicholas, J., Carney, R., Pratap, A., Rosenbaum, S., et al. (2017). The efficacy of smartphone-based mental health interventions for depressive symptoms: A meta-analysis of randomized controlled trials. *World Psychiatry, 16*(3), 287–298.

Firth, J., Torous, J., Nicholas, J., Carney, R., Rosenbaum, S., & Sarris, J. (2017). Can smartphone mental health interventions reduce symptoms of anxiety? a meta-analysis of randomized controlled trials. *Journal of Affective Disorders, 218*, 15–22.

Fu, Z., Burger, H., Arjadi, R., & Bockting, C. L. (2020). Effectiveness of digital psychological interventions for mental health problems in low-income and middle-income countries: a systematic review and meta-analysis. *The Lancet Psychiatry, 7*(10), 851–864.

Garabiles, M. R., Shehadeh, M. H., & Hall, B. J. (2019). Cultural adaptation of a scalable World Health Organization e-mental health program for overseas Filipino workers. *JMIR Formative Research, 3*(1), e11600.

Griner, D., & Smith, T. B. (2006). Culturally adapted mental health intervention: a meta-analytic review. *Psychotherapy: Theory, Research, Practice, Training, 43*(4), 531.

Hall, G. C. N., Ibaraki, A. Y., Huang, E. R., Marti, C. N., & Stice, E. (2016). A meta-analysis of cultural adaptations of psychological interventions. *Behavior Therapy, 47*(6), 993–1014.

Haroz, E., Ritchey, M., Bass, J., Kohrt, B., Augustinavicius, J., Michalopoulos, L., et al. (2017). How is depression experienced around the world? A systematic review of qualitative literature. *Social Science & Medicine, 183*, 151–162.

Heber, E., Lehr, D., Ebert, D. D., Berking, M., & Riper, H. (2016). Web-based and mobile stress management intervention for employees: a randomized controlled trial. *Journal of Medical Internet Research, 18*(1), e21.

Heim, E., Burchert, S., Shala, M., Cerga, A., Morina, N., Schaub, M., et al. (2020). Effect of cultural adaptation of a smartphone-based self-help programme on its acceptability and efficacy: study protocol for a randomized controlled trial. *PsychArchives*, https://doi.org/10.23668/PSYCHARCHIVES.3152.

Heim, E., & Kohrt, B. A. (2019). Cultural adaptation of scalable psychological interventions. *Clinical Psychology in Europe, 1*(4), 1–22.

Hwang, W.-C. (2009). The formative method for adapting psychotherapy (FMAP): a community-based developmental approach to culturally adapting therapy. *Professional Psychology: Research and Practice, 40*(4), 369.

Jordans, M. J., & Kohrt, B. A. (2020). Scaling up mental health care and psychosocial support in low-resource settings: a roadmap to impact. *Epidemiology and Psychiatric Sciences, 29*(e189), 1–7.

Joshi, U., Naslund, J. A., Anand, A., Tugnawat, D., Vishwakarma, R., Bhan, A., ... Lu, C. (2022). Assessing costs of developing a digital program for training community health workers to deliver treatment for depression: a case study in rural India. *Psychiatry Research, 307*, 114299.

Juniar, D., van Ballegooijen, W., Karyotaki, E., van Schaik, A., Passchier, J., Heber, E., et al. (2019). Web-based stress management program for university students in Indonesia: systematic cultural adaptation and protocol for a feasibility study. *JMIR Research Protocols, 8*(1), e11493.

Kaonga, N. N., & Morgan, J. (2019). Common themes and emerging trends for the use of technology to support mental health and psychosocial well-being in limited resource settings: a review of the literature. *Psychiatry Research, 281*, 112594.

Karyotaki, E., Ebert, D. D., Donkin, L., Riper, H., Twisk, J., Burger, S., et al. (2018). Do guided internet-based interventions result in clinically relevant changes for patients with depression? An individual participant data meta-analysis. *Clinical Psychology Review, 63*, 80–92.

Khan, A., Shrivastava, R., Tugnawat, D., Singh, A., Dimidjian, S., Patel, V., et al. (2020). Design and development of a digital program for training non-specialist health workers to deliver an evidence-based psychological treatment for depression in primary care in India. *Journal of Technology in Behavioral Science, 5*(4), 402–415.

Kirmayer, L. J., & Swartz, L. (2013). Culture and global mental health, In Patel, V., Minas, H., Cohen, A., & Prince, M.J. (Eds). *Global Mental Health: Principles and Practice*, 41–62.

Kohn, R., Ali, A. A., Puac-Polanco, V., Figueroa, C., López-Soto, V., Morgan, K., et al. (2018). Mental health in the Americas: an overview of the treatment gap. *Revista Panamericana de Salud Pública, 42*, e165.

Kohrt, B. A., Rasmussen, A., Kaiser, B. N., Haroz, E. E., Maharjan, S. M., Mutamba, B. B., et al. (2014). Cultural concepts of distress and psychiatric disorders: literature review and research recommendations for global mental health epidemiology. *International Journal of Epidemiology, 43*(2), 365–406.

Kokane, A., Pakhare, A., Gururaj, G., Varghese, M., Benegal, V., Rao, G. N., et al. (2019). Mental health issues in Madhya Pradesh: insights from National Mental Health Survey of India 2016. *Healthcare, 7*(2), 53.

Krendl, A. C., & Pescosolido, B. A. (2020). Countries and cultural differences in the stigma of mental illness: the east–west divide. *Journal of Cross-Cultural Psychology, 51*(2), 149–167.

Lal, S., Gleeson, J., Malla, A., Rivard, L., Joober, R., Chandrasena, R., et al. (2018). Cultural and contextual adaptation of an ehealth intervention for youth receiving services for first-episode psychosis: adaptation framework and protocol for horyzons-Canada phase 1. *JMIR Research Protocols, 7*(4), e100.

Lal, S., Gleeson, J., Rivard, L., D'Alfonso, S., Joober, R., Malla, A., et al. (2020). Adaptation of a digital health innovation to prevent relapse and support recovery in youth receiving services for first-episode psychosis: results from the Horyzons-Canada Phase 1 study. *JMIR Formative Research, 4*(10), e19887.

Lau, A. S. (2006). Making the case for selective and directed cultural adaptations of evidence-based treatments: examples from parent training. *Clinical Psychology: Science and Practice, 13*(4), 295–310.

Linardon, J., & Fuller-Tyszkiewicz, M. (2020). Attrition and adherence in smartphone-delivered interventions for mental health problems: a systematic and meta-analytic review. *Journal of Consulting and Clinical Psychology, 88*(1), 1.

Lipschitz, J., Hogan, T. P., Bauer, M. S., & Mohr, D. C. (2019). Closing the research-to-practice gap in digital psychiatry: the need to integrate implementation science. *The Journal of Clinical Psychiatry, 80*(3), 18com12659 https://doi:10.4088/JCP.18com12659.

Liu, N. H., Daumit, G. L., Dua, T., Aquila, R., Charlson, F., Cuijpers, P., et al. (2017). Excess mortality in persons with severe mental disorders: a multilevel intervention framework and priorities for clinical practice, policy and research agendas. *World Psychiatry, 16*(1), 30–40.

Lund, C., Alem, A., Schneider, M., Hanlon, C., Ahrens, J., Bandawe, C., et al. (2015). Generating evidence to narrow the treatment gap for mental disorders in sub-Saharan Africa: Rationale, overview and methods of AFFIRM. *Epidemiology and Psychiatric Sciences, 24*(3), 233–240.

Merchant, R., Torous, J., Rodriguez-Villa, E., & Naslund, J. A. (2020). Digital technology for management of severe mental disorders in low-income and middle-income countries. *Current Opinion in Psychiatry, 33*(5), 501–507.

Muke, S., Shrivastava, R., Mitchell, L., Khan, A., Murhar, V., Tugnawat, D., et al. (2019). Acceptability and feasibility of digital technology for training community health workers to deliver evidence-based psychosocial treatment for depression in rural India. *Asian Journal of Psychiatry, 45*, 99–106.

Muke, S., Tugnawat, D., Joshi, U., Anand, A., Khan, A., Shrivastava, R., et al. (2020). Digital training for non-specialist health workers to deliver a brief psychological treatment for depression in primary care in India: findings from a randomized pilot study. *International Journal of Environmental Research and Public Health, 17*(17), 6368.

Naslund, J. A., Aschbrenner, K. A., Araya, R., Marsch, L. A., Unützer, J., Patel, V., et al. (2017). Digital technology for treating and preventing mental disorders in low-income and middle-income countries: a narrative review of the literature. *The Lancet Psychiatry, 4*(6), 486–500.

Naslund, J. A., Bartels, S. M., & Marsch, L. A. (2019). Digital technology, including telemedicine, in the management of mental illness. *Revolutionizing Tropical Medicine: Point-of-Care Tests, New Imaging Technologies and Digital Health* (pp. 505–530). Hoboken, New Jersey, USA: Wiley.

Naslund, J. A., Gonsalves, P. P., Gruebner, O., Pendse, S. R., Smith, S. L., Sharma, A., et al. (2019). Digital Innovations for Global Mental Health: opportunities for data science, task sharing, and early intervention. *Current Treatment Options in Psychiatry, 6*(4), 337–351.

Naslund, J. A., Marsch, L. A., McHugo, G. J., & Bartels, S. J. (2015). Emerging mHealth and eHealth interventions for serious mental illness: a review of the literature. *Journal of Mental Health, 24*(5), 321–332.

Patel, V., Chisholm, D., Parikh, R., Charlson, F. J., Degenhardt, L., Dua, T., et al. (2016). Addressing the burden of mental, neurological, and substance use disorders: key messages from disease control priorities. *The Lancet, 387*(10028), 1672–1685.

Patel, V., Saxena, S., Lund, C., Thornicroft, G., Baingana, F., Bolton, P., et al. (2018). The Lancet Commission on global mental health and sustainable development. *The Lancet, 392*(10157), 1553–1598.

Patel, V., Weobong, B., Weiss, H. A., Anand, A., Bhat, B., Katti, B., et al. (2017). The Healthy Activity Program (HAP), a lay counsellor-delivered brief psychological treatment for severe depression, in primary care in India: a randomised controlled trial. *The Lancet, 389*(10065), 176–185.

Pathare, S., Brazinova, A., & Levav, I. (2018). Care gap: a comprehensive measure to quantify unmet needs in mental health. *Epidemiology and Psychiatric Sciences, 27*(5), 463–467.

Rathod, S., Pinninti, N., Irfan, M., Gorczynski, P., Rathod, P., Gega, L., et al. (2017). Mental health service provision in low-and middle-income countries. *Health Services Insights, 10*, 1178632917694350.

Raviola, G., Naslund, J. A., Smith, S. L., & Patel, V. (2019). Innovative models in mental health delivery systems: task sharing care with non-specialist providers to close the mental health treatment gap. *Current Psychiatry Reports, 21*(6), 44.

Richards, D., Murphy, T., Viganó, N., Timulak, L., Doherty, G., Sharry, J., et al. (2016). Acceptability, satisfaction and perceived efficacy of "Space from Depression" an internet-delivered treatment for depression. *Internet Interventions, 5*, 12–22.

Richards, D., Timulak, L., O'Brien, E., Hayes, C., Vigano, N., Sharry, J., et al. (2015). A randomized controlled trial of an internet-delivered treatment: Its potential as a low-intensity community intervention for adults with symptoms of depression. *Behaviour Research and Therapy, 75*, 20–31.

Salamanca-Sanabria, A., Richards, D., & Timulak, L. (2019). Adapting an internet-delivered intervention for depression for a Colombian college student population: an illustration of an integrative empirical approach. *Internet Interventions, 15*, 76–86.

Saraceno, B. (2020). Rethinking global mental health and its priorities. *Epidemiology and Psychiatric Sciences, 29*(e64), 1–3.

Satcher, D. (2001). Mental health: culture, race, and ethnicity—A supplement to mental health: A report of the surgeon general: US Department of Health and Human Services.

Shala, M., Morina, N., Burchert, S., Cerga-Pashoja, A., Knaevelsrud, C., Maercker, A., et al. (2020). Cultural adaptation of Hap-pas-Hapi, an internet and mobile-based intervention for the treatment of psychological distress among Albanian migrants in Switzerland and Germany. *Internet Interventions, 21*, 100339.

Shattuck, E. C. (2019). A biocultural approach to psychiatric illnesses. *Psychopharmacology, 236*(10), 2923–2936.

Shehadeh, M. H., Heim, E., Chowdhary, N., Maercker, A., & Albanese, E. (2016). Cultural adaptation of minimally guided interventions for common mental disorders: A systematic review and meta-analysis. *JMIR Mental Health, 3*(3), e44.

Sheikh, S., & Furnham, A. (2000). A cross-cultural study of mental health beliefs and attitudes towards seeking professional help. *Social Psychiatry and Psychiatric Epidemiology, 35*(7), 326–334.

Singla, D. R., Kohrt, B. A., Murray, L. K., Anand, A., Chorpita, B. F., & Patel, V. (2017). Psychological treatments for the world: lessons from low-and middle-income countries. *Annual Review of Clinical Psychology, 13*, 149–181.

Sit, H. F., Ling, R., Lam, A. I. F., Chen, W., Latkin, C. A., & Hall, B. J. (2020). The cultural adaptation of step-by-step: an intervention to address depression among Chinese young adults. *Frontiers in Psychiatry, 11*, 650.

Torous, J., Nicholas, J., Larsen, M. E., Firth, J., & Christensen, H. (2018). Clinical review of user engagement with mental health smartphone apps: Evidence, theory and improvements. *Evidence-Based Mental Health, 21*(3), 116–119.

Vigo, D., Thornicroft, G., & Atun, R. (2016). Estimating the true global burden of mental illness. *The Lancet Psychiatry, 3*(2), 171–178.

Waheed, W., Hughes-Morley, A., Woodham, A., Allen, G., & Bower, P. (2015). Overcoming barriers to recruiting ethnic minorities to mental health research: a typology of recruitment strategies. *Bmc Psychiatry [Electronic Resource], 15*(1), 1–11.

Weine, S., Kohrt, B. A., Collins, P. Y., Cooper, J., Lewis-Fernandez, R., Okpaku, S., et al. (2020). Justice for George Floyd and a reckoning for global mental health. *Global Mental Health, 7*, e22.

Willis, G. B., Jr, Artino, & A. , R. (2013). What do our respondents think we're asking? Using cognitive interviewing to improve medical education surveys. *Journal of graduate medical education, 5*(3), 353.

Yardley, L., Morrison, L., Bradbury, K., & Muller, I. (2015). The person-based approach to intervention development: application to digital health-related behavior change interventions. *Journal of Medical Internet Research, 17*(1), e30.

Chapter 13

Building the digital therapeutic industry: Regulation, evaluation, and implementation

Megan Coder, PharmD, MBA

Digital Therapeutics Alliance, Arlington, Virginia, USA

13.1 A new category of medicine

When a new industry emerges in any sector, significant effort and collaboration are required to ensure that it will grow and be sustainable. This is particularly true in the healthcare sector to ensure that new technologies and processes are appropriately integrated into existing systems and frameworks for long-term survival and optimal impact.

History is filled with examples of new technologies that initially surprise, but within a short time become integral to daily life. Generations before ours would be shocked to encounter aspects of modern-day transportation, architecture, communication, and healthcare that are so commonplace today, we take their existence for granted.

Industries do not appear overnight and stand the test of time without intentional, industry-wide collaborative efforts. When the digital therapeutic (DTx) industry emerged in 2010, collaboration across all impacted stakeholders became critical to translating the vision of this new category of medicine into scalable adoption and meaningful use internationally.

13.2 Digital Therapeutics Alliance

The Digital Therapeutics Alliance (DTA) was established in 2017 and immediately began developing critical frameworks that guide the growth, adoption, and utilization of digital therapeutic products internationally. The Alliance serves as the central industry body that convenes stakeholders across the healthcare ecosystem to drive long-term industry growth within this continually evolving ecosystem.

Given the rapid expansion of the digital therapeutic industry, DTA's mission is to help patients, clinicians, payers, and policymakers understand how to identify, assess, and utilize digital therapeutic products in everyday settings. As such, DTA works with stakeholders across the healthcare ecosystem to ensure that digital therapeutic products are trustworthy and globally accessible care options.

DTA members—including companies across nearly all major healthcare industries and geographic regions—are dedicated to transforming global healthcare by advancing digital therapeutics to improve clinical and health economic outcomes.

13.3 Industry progression

To build a strong foundation for the digital therapeutic industry, DTA develops frameworks that provide clarity for stakeholders—patients, caregivers, clinicians, policymakers, payors, and technology manufacturers—as they define, assess, and implement digital therapeutic products across care settings. Without this degree of standardization, it is otherwise impossible for these technologies to scale locally, nationally, and globally. Therefore, to best serve end-users and product evaluators, these frameworks are developed in the context of the broader digital health landscape and will continue to evolve as digital therapeutic products and their use cases expand.

Additionally, industry leaders are continually assessing emerging growth drivers in order to direct the trajectory, integration, growth, and impact of these technologies in everyday healthcare pathways. This chapter, therefore, follows the industry's growth, from foundation setting to global scaling.

Digital Therapeutics for Mental Health and Addiction: The State of the Science and Vision for the Future. DOI: https://doi.org/10.1016/B978-0-323-90045-4.00003-4

Products across the digital health ecosystem serve different, but complementary purposes.
Depending on each product's intended use and risk, it is subject to increasing degrees
of clinical evaluation, regulatory oversight, and real-world data requirements.

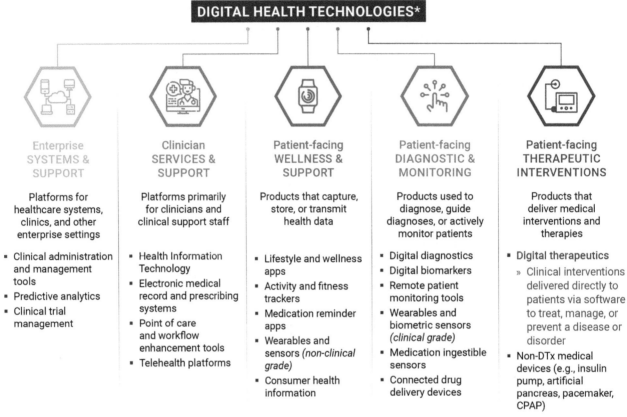

Categorizations of the digital health technology ecosystem will continue to evolve. This is a select representation of a broad, diverse ecosystem.

FIGURE 13.1 Digital health technology ecosystem (Digital Therapeutics Alliance, 2021a).

13.4 2017–2021: Foundation building

13.4.1 Landscape mapping

One of DTA's first steps in building the foundation for the digital therapeutic industry was working with partner organizations to map out the broader digital health technology (DHT) landscape and clearly identify digital therapeutics' domain in this ecosystem.

Digital health—while increasingly ubiquitous across all health and care settings—continues to be a difficult category of products to define. While some digital health technologies focus on clinician and enterprise support, many others are patient-focused. A commonly recognized category of digital health products is loosely referred to as "digital health apps." These products are generally patient-facing, available in traditional app stores for download, address a wide range of health and wellness issues, and incorporate varying degrees of patient privacy and clinical evidence.

Digital health apps however represent only one type of software-based product in the broader DHT continuum (Fig. 13.1). While various types of DHTs are used for wellness, medication adherence, remote patient monitoring, and diagnosing medical conditions, digital therapeutics—a DHT subset—leverage software to directly treat, manage, or prevent a disease or disorder.

Products listed in each of the above categories (Fig. 13.1) make different levels of claims, and therefore inherently have different levels of risk. As the complexity of each product's intended use increases along with the wellness to therapeutic spectrum, so do product requirements related to clinical efficacy, product safety, patient privacy, and regulatory oversight.

Most patient-facing wellness and support apps do not need to prove product claims by scientific testing or clinical evidence before going to market. In contrast, patient-facing monitoring, diagnostic, and therapeutic products are required to demonstrate increasing degrees of product safety and efficacy (Digital Therapeutics Alliance, 2019a).

Since not all digital health technologies are the same, it is critical for healthcare decision makers (HCDMs) and end-users to distinguish between digital products that serve distinct purposes. For example:

- *Patients need to know*: What am I using? Why am I using it? How will it help? Has someone verified it is safe, effective, and will protect my data?
- *Clinicians need to know*: What should I expect? How does it relate to other treatments? Does it provide actionable data or insights? Is it necessary for me to authorize or prescribe this product?
- *Healthcare payors need to know*: What type of product are we covering? How will it benefit patients at the individual and population levels? What types of clinical and economic outcomes should we expect?
- *Regulators need to know*: What level of risk does each product pose to patients? What is the appropriate level of regulatory oversight?

Through these types of questions, end users can begin to identify which digital health technologies will best meet their needs and expectations.

13.4.2 Digital therapeutic definition

The future of healthcare will unquestionably include products that generate and deliver medical interventions directly to patients using software that treats, manages, and prevents diseases and disorders.

Therefore, another critical component of DTA's work has been to define what digital therapeutics are and assist patients, caregivers, clinicians, and payors in identifying products that may or may not meet this definition. Across the industry, digital therapeutics are defined as (Digital Therapeutics Alliance, 2019b):

"Digital therapeutics (DTx) deliver evidence-based therapeutic interventions that are driven by high quality software programs to prevent, manage, or treat a medical disorder or disease. They are used independently or in concert with medications, devices, or other therapies to optimize patient care and health outcomes.

Digital therapeutic products incorporate advanced technology best practices relating to design, clinical evaluation, usability, and data security. They are reviewed and cleared or certified by regulatory bodies as required to support product claims regarding risk, efficacy, and intended use."

Digital therapeutics empower patients, clinicians, and payers with intelligent and accessible tools for addressing a wide range of conditions through high-quality, safe, and effective data-driven interventions.

13.4.3 DTx core principles & best practices

In an environment where so many products are making clinical claims and promises, HCDMs require guidance to determine which products truly qualify as a legitimate digital therapeutic.

Since digital therapeutics use software to deliver clinical-grade medical interventions directly to patients, these products must be clinically validated to ensure safety and efficacy, and are subject to greater clinical, security, and regulatory scrutiny than general digital health apps and products. Therefore, all products claiming to be a digital therapeutic must adhere to these foundational principles (Digital Therapeutics Alliance, 2019b):

1. Prevent, manage, or treat a medical disorder or disease.
2. Produce a medical intervention that is driven by software.
3. Incorporate design, manufacture, and quality best practices (Digital Therapeutics Alliance, 2019c).
4. Engage end-users in product development and usability processes.
5. Incorporate patient privacy and security protections.
6. Apply product deployment, management, and maintenance best practices (Digital Therapeutics Alliance, 2019c).
7. Publish trial results inclusive of clinically meaningful outcomes in peer-reviewed journals.
8. Be reviewed and cleared or approved by regulatory bodies as required to support product claims of risk, efficacy, and intended use.
9. Make claims appropriate to clinical validation and regulatory status.
10. Collect, analyze, and apply real-world evidence and/or product performance data.

While these core principles have not changed since DTA's inception, the list of best practices to support these principles will evolve over time to align with industry standards and new technology offerings.

For example, digital therapeutics undergo robust clinical trials and are held to similar standards of safety and clinical evidence as traditional medical treatments. These include the proper trial design, utilization of standardized endpoints, alignment with Good Clinical Practices, appropriate reporting of outcomes, publication of results in peer-reviewed journals, the generation and application of real-world evidence (RWE), and analysis of product performance data (Digital Therapeutics Alliance, 2019c).

This rigorous process enables digital therapeutic products to demonstrate safety and efficacy across the relevant and target populations, establish credibility, deliver clinically meaningful results in real-world settings, and be considered as and among current standards-of-care. While randomized control trials are still considered the gold standard for the demonstration of digital therapeutic product outcomes, the use of pragmatic trials and other RWE development will become increasingly important (U.S. Food & Drug Administration, 2021).

13.4.4 Intended use & mechanism of action

In developing the foundation for this industry, DTA recognized that while all digital therapeutics should align with the digital therapeutic definition and core principles, each individual technology serves a specific patient population with a uniquely targeted indication and mechanism of action. Across the broader digital therapeutic industry, products, therefore, possess a level of trustworthiness reliant upon their alignment with core principles and are subject to further evaluation by decision-making entities.

Therefore, through their mandate to generate clinical evidence and ongoing real-world data (RWD), digital therapeutics demonstrate meaningful improvements in patient outcomes for a wide range of diseases and disorders (Digital Therapeutics Alliance, 2022b).

To achieve their clinical endpoints, digital therapeutics have a wide variety of mechanisms of action, including:

- Providing personalized disease treatment, management, and prevention programs.
- Offering therapies to address comorbidities, side effects, or affiliated conditions.
- Providing treatments that produce direct neurologic changes.
- Delivering cognitive behavioral therapy and other evidence-based treatments.
- Enhancing, supporting, and optimizing current in-person and medication treatments.
- Delivering responsive physical exercises and behavioral interventions.

Typically available to patients through multipurpose devices such as smartphones and personal tablets, digital therapeutics expand access to empirically supported therapies and offer treatment options for conditions that are undertreated or do not have any available therapeutic options.

13.4.5 DTx product categorization

Just as not all digital health technologies are the same, neither are all digital therapeutics. Early in DTA's efforts, Alliance members identified three different categories of digital therapeutic products: those intended to treat, manage, or prevent a disease or disorder (which includes improving a health condition as it specifically relates to a disease or disorder).

Across these three categories, all digital therapeutic products must adhere to industry core principles, including the use of clinical endpoints to generate and publish product-specific outcomes in peer-reviewed publications (Fig. 13.2). However, based on the product's primary purpose and type of intervention delivered to patients, the product's claim correlates directly to a particular level of risk, and consequently level of regulatory oversight and product access pathway:

As the industry grows, digital therapeutic products will continue to align with the industry's core principles, but pathways related to regulatory oversight and patient access will continue to evolve.

13.4.6 Regulatory frameworks

When the digital therapeutic industry first emerged, there was concern over how regulatory frameworks would develop and whether digital therapeutic products would have a pathway to market. Due to the strong leadership of multiple regulatory agencies, these concerns have greatly lessened. With countries such as the United States, UK, Germany, and South Korea leading the way, robust regulatory frameworks for digital therapeutics are now being developed globally.

Policymakers at the local, national, and regional levels are facing a number of important questions related to the regulatory recognition of digital therapeutic products and legislative actions necessary to enable public payor coverage and patient access.

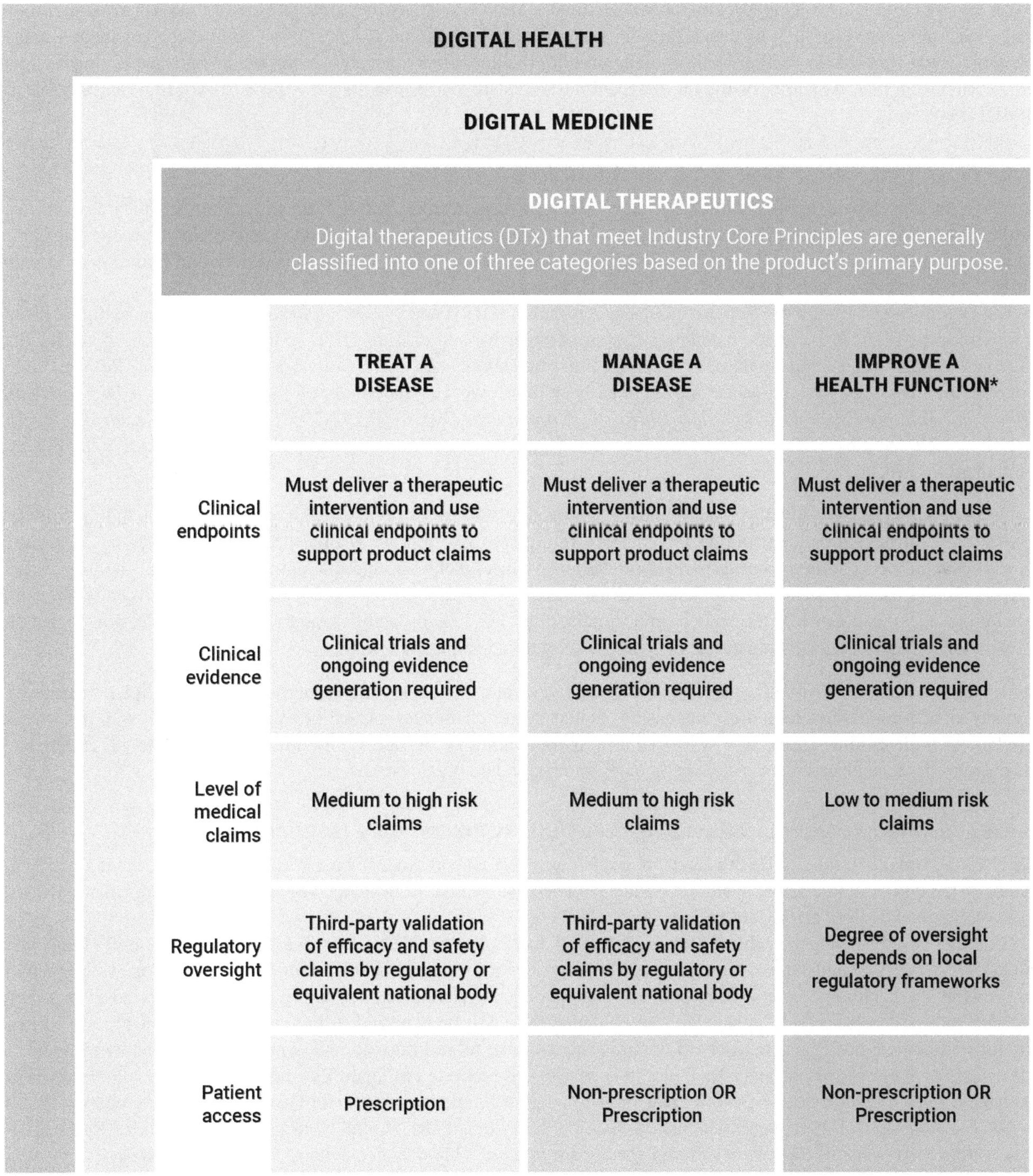

	DIGITAL HEALTH		
	DIGITAL MEDICINE		
	DIGITAL THERAPEUTICS Digital therapeutics (DTx) that meet Industry Core Principles are generally classified into one of three categories based on the product's primary purpose.		
	TREAT A DISEASE	**MANAGE A DISEASE**	**IMPROVE A HEALTH FUNCTION***
Clinical endpoints	Must deliver a therapeutic intervention and use clinical endpoints to support product claims	Must deliver a therapeutic intervention and use clinical endpoints to support product claims	Must deliver a therapeutic intervention and use clinical endpoints to support product claims
Clinical evidence	Clinical trials and ongoing evidence generation required	Clinical trials and ongoing evidence generation required	Clinical trials and ongoing evidence generation required
Level of medical claims	Medium to high risk claims	Medium to high risk claims	Low to medium risk claims
Regulatory oversight	Third-party validation of efficacy and safety claims by regulatory or equivalent national body	Third-party validation of efficacy and safety claims by regulatory or equivalent national body	Degree of oversight depends on local regulatory frameworks
Patient access	Prescription	Non-prescription OR Prescription	Non-prescription OR Prescription

*Includes digital therapeutics that prevent a disease.

FIGURE 13.2 Digital therapeutic industry categorization (Digital Therapeutics Alliance, 2021b).

13.4.6.1 International medical device regulators forum (IMDRF)

Digital therapeutics are consistently regulated as medical devices. They are most frequently categorized by regulatory jurisdictions as a subset of Software as a Medical Device (SaMD), a framework developed by the International Medical Device Regulators Forum (IMDRF). Given the incredibly diverse ecosystem of products that SaMD represents, it is important to note that not all digital therapeutic products qualify as SaMD, and not all SaMD products qualify as a digital therapeutic. Per IMDRF's categorization of SaMD products (IMDRF Software as a Medical Device (SaMD) Working Group, 2014)—including products that treat or diagnose, drive clinical management, and inform clinical management—digital therapeutics are most likely to align with the treat and drive clinical management categories, as opposed to the diagnose and inform categories. Increasingly, digital therapeutic devices are also expanding into the Software in a Medical Device (SiMD) framework.

13.4.6.2   United States: Food and Drug Administration (FDA)

The US Food and Drug Administration (FDA) has long been recognized as a trailblazer in regulatory policy development for digital therapeutic products. Based on the digital therapeutic's intended use and level of risk, each product is subject to varying degrees of oversight, ranging from full 510(k) clearance by FDA's Center for Devices and Radiological Health (CDRH) division to enforcement discretion (Digital Therapeutics Alliance, 2022a).

The Digital Health Center of Excellence (DHCoE) is part of CDRH and is working to align and coordinate digital health work across the FDA. It marks the beginning of a comprehensive approach to DHT, setting the stage for advancing and realizing the potential of digital health. With other efforts underway such as the Software Precertification (Pre-Cert) Pilot Program, the FDA released a guidance document early in the Covid-19 pandemic, *Enforcement Policy for Digital Health Devices For Treating Psychiatric Disorders During the Coronavirus Disease 2019 (COVID-19) Public Health Emergency*.

In this document, the FDA recognized the value that digital therapeutics provide to patients, stating (U.S. Food & Drug Administration, 2020):

"In the context of the COVID-19 public health emergency, the use of digital health technologies, including software as a medical device or other digital therapeutics solutions, may improve mental health and well-being of patients with psychiatric conditions during periods of shelter-in-place, isolation, and quarantine. In addition, the use of such technologies has the potential to facilitate 'social distancing' by reducing patient contact with, and proximity to, health care providers, and can ease the burden on hospitals, other health care facilities, and health care professionals that are experiencing increased demand due to the COVID-19 public health emergency."

During the era of Covid-19, legislative bodies and regulatory agencies increasingly recognized digital therapeutic products as new modalities to address increasing mental health challenges caused or exacerbated by Covid-19, provide populations with chronic diseases ways to control their conditions outside of traditional office settings, and reduce unnecessary burdens placed on over-taxed clinicians and healthcare organizations.

13.4.6.3   United Kingdom: National Institute for Health and Care Excellence (NICE)

The UK's National Institute for Health and Care Excellence (NICE) continues to be a leader in providing the industry with insight on the types of clinical and health economic evidence that need to be developed by digital therapeutic manufacturers based on the intended use and risk of each product.

Although NICE is not responsible for the payment and distribution of DHTs, their aim is to make it easier for innovators and commissioners to understand what good levels of evidence for digital health technologies look like, while meeting the needs of the health and care system, patients, and users. They provide advice to digital health innovators about how the NHS makes decisions and the standards of evidence they will be expected to produce for different types of digital health technologies. NICE also helps NHS commissioners make more informed and consistent decisions by providing a framework for the levels of evidence they should expect to see presented to them. By making this process more dynamic and value-driven, they are working to offer real value to patients (National Institute for Health and Care Excellence, 2022).

NICE's evidence for effectiveness standards describe the evidence that should be available or developed for DHTs to demonstrate their value in the UK health and social care system. This includes evidence of effectiveness relevant to the intended use(s) of the technology and evidence of economic impact relative to the financial risk (National Institute for Health and Care Excellence, 2018):

13.4.6.4   Germany: Federal Institute for Drugs and Medical Devices

The Federal Institute for Drugs and Medical Devices (Bundesinstitut für Arzneimittel und Medizinprodukte, BfArM) is an independent federal higher authority within the portfolio of the German Federal Ministry of Health. In 2020, they launched a program that reviews digital health applications (DiGA – in German: "Digitale Gesundheitsanwendungen") for potential release to the German market. DiGA is class I or IIa CE-marked medical devices that support the recognition, monitoring, treatment or alleviation of diseases and are used by the patient alone or by patient and healthcare provider together

(Federal Institute for Drugs and Medical Devices, 2022). While not all DiGA are digital therapeutic products, numerous digital therapeutics qualify as DiGA.

BfArM's Assessment Procedure is designed as a fast-track process and depicted in (Federal Institute for Drugs and Medical Devices, 2020). Within a 3-month period at the most, BfArM undertakes an assessment process to determine whether a product qualifies for at least one year of market access and reimbursement. Manufacturers must submit information related to data protection, interoperability, user-friendliness, and evidence of the positive healthcare effect.

Although work is still underway related to the nationwide distribution of DiGA and analysis of RWD, Germany's innovative approach to assessing and providing access to DHTs at the population level is inspiring numerous other jurisdictions, such as France (Lovell, 2021), to consider ways to replicate aspects of this model alongside or in addition to currently existing frameworks.

13.4.6.5 South Korea: Ministry of Food and Drug Safety

With efforts to develop regulatory frameworks for digital therapeutics currently underway in Japan, Singapore, and Australia, the Ministry of Food and Drug Safety in South Korea has led an impressive campaign to develop a digital therapeutic pathway, in part as a response to the Covid-19 pandemic.

In November 2020, the Ministry developed the Guidelines for Digital Therapeutics Review and Approval where digital therapeutics are regulated as SaMD (Ministry of Food and Drug Safety Medical Device Evaluation Department, 2020). Regulatory review considerations include the product's intended use, mechanism of operation, performance, development, clinical trial, and potential product approvals and use in other countries.

13.4.7 Industry's expanding value

As the digital therapeutic industry expands, so does the value that products provide to patients, caregivers, clinicians, and payors. DTA continues to play a role in identifying areas where value can be delivered to both convey these insights to key decision-makers and work with policymakers to ensure this value is realized by patients across all healthcare settings.

Since digital therapeutic products can address critical gaps in care for underserved populations—regardless of patient age, language, culture, income, disease state, or geography—public and private funders must understand the following aspects as they develop coverage policies.

13.4.7.1 Providing care to underserved & undertreated populations

Building on the ease of digital therapeutic product scalability and access, digital therapeutics create new treatment, and self-management options for a broad range of chronic and mental health conditions. Digital therapeutic products can easily reach high-risk, rural, underserved, and undertreated populations who often lack access to healthcare services even during the best of times. This include digital therapeutics' ability to:

- Be accessible through patient-owned devices such as a smartphone or tablet.
- Discretely deliver therapy to patients in their home environments.
- Transform how patients understand, manage, and engage in their healthcare.
- Offer therapy independent of a patient's work or education schedule.
- Provide therapies in a variety of languages.
- Lower stigma that may be associated with the delivery of certain traditional therapies.
- Directly impact life and disease state outcomes.
- Provide meaningful results and insights on personalized goals and outcomes.
- Extend clinicians' ability to care for patients.
- Support healthcare teams in settings with varying degrees of healthcare infrastructure.

13.4.7.2 Therapy accessibility and scalability

Patient access to digital therapeutic products is generally granted through a prescription, referral from a clinician, or delivery of an activation code via an electronic health record, employer, or third-party payor.

Compared to traditional medications which rely on physical distribution and dispensing processes, digital therapeutic products are software-based and are able to be hosted on multi-purpose platforms (e.g., patient-owned smartphone, tablet). This introduces an entirely new degree of product scalability and patient access opportunities. Therefore, instead of having a geographic-dependent delivery model, it is possible to deploy a needs-based delivery model.

As a result of increased product access and scalability, payors and policymakers are now able to ensure that care is delivered to entire populations that have otherwise been unable to secure care—either due to geographic limitations, cultural and language boundaries, well-documented disparities, or health condition severity. Patients who have previously not received care now have the opportunity to receive personalized therapeutic interventions based on their specific needs and abilities, in an engaging way, independent of their work or education schedule, with familiar languages and cultural references, in the privacy and safety of their own environment, and with access to actionable insights that convey their movement toward clinical improvement.

13.4.7.3   Real-world outcomes

In another departure from traditional medications, digital therapeutics generate a wide variety of RWD outcomes related to patient use and clinical impact. This includes patient-specific measures (e.g., actionable clinical outcomes, standardized patient assessments, physiologic data via associated sensors), patient and clinician utilization (e.g., patient utilization and engagement, product onboarding metrics, clinician prescribing parameters), and product functionality (e.g., product performance, analytics, quality measures).

While RWD is used by patients and clinicians to adjust and optimize critical aspects of therapy, this data may also be translated into fit-for-use, formal RWE for healthcare payor and policymaker product evaluation processes. Importantly, it is now possible for decision-makers to analyze outcomes related to specific patient cohorts and derive detailed real-world insights on clinical and health economic endpoints. It is possible that certain evaluations based on real-world output will eventually replace aspects of evaluations based purely on information derived through secondary sources (e.g., patient registries, EHR systems, claims databases).

Additionally, real-world outcomes generated by digital therapeutics may be used to optimize outcomes at the individual patient and population levels. At the individual patient level, digital therapeutic products provide clinicians with meaningful, actionable clinical reports. At the population level, data generated by digital therapeutic products may be aggregated to track progress or compare aggregate outcomes based upon patient disease state, level of acuity, geographic location, age, gender, etc.

13.4.7.4   Therapies that improve over time

Additionally, compared to traditional medications that are not modified once FDA approval is granted, digital therapeutic products are iterative in nature and continue to evolve throughout their lifecycle. While some of these iterations may require regulatory review if the core algorithm is changed, the majority of iterations by product manufacturers (e.g., product functionality changes, patient engagement optimizations) are delivered to users in real-time to ensure immediate benefits.

Covid-19 provided a heightened awareness of how digital therapeutics can help patients manage their chronic conditions within home settings, improve the efficiency of mental health services, and extend effective treatment to the millions of individuals who are otherwise unable to access treatment. As clinicians, healthcare systems, employers, and payors continue integrating these products into patient care, digital therapeutics will increasingly influence the delivery and consumption of healthcare around the nation and world.

13.5   2022 & beyond: Equipping decision-makers and end-users

Digital therapeutic leaders are building on the existing industry foundation to equip HCDMs and end-users with frameworks on how to make informed decisions related to digital therapeutic product selection, use, system integration, and outcome optimization. However, for these frameworks to be impactful, four obstacles must be addressed:

1. Policymakers require a common framework to categorize digital therapeutic products based on intended use and risk. While significant progress is being made in various jurisdictions as evidenced earlier, there are no globally consistent approaches to digital therapeutic recognition, risk stratification, or product approval requirements.
2. Payors require fit-for-purpose pathways to evaluate and cover digital therapeutic products. Without consistent requirements of clinical trial data, real-world outcomes, and data governance requirements, digital therapeutic product scalability across local, regional, and national systems will be greatly limited, potentially depriving entire populations of access to new therapies.
3. Clinicians require clinical guidelines that recognize the appropriate use of digital therapeutic products, in addition to technical training for authorizing and optimizing product use, and support for integrating digital therapeutics into practice settings and clinical workflows.

4. Patients require resources that directly support them as they use and engage with digital therapeutics, particularly as they distinguish high quality from sub-par products and determine how to receive the most value from each therapy.

To further enable patient access to digital therapeutics, DTA is undertaking a wide variety of efforts to address these challenges, including collaborating with policymakers to develop harmonized frameworks that properly recognize and categorize digital therapeutic products, with payers to develop scalable access and coverage models, and with clinicians and patients to provide much-needed education related to product utilization and optimization.

The following efforts and developments are therefore imminently on the horizon:

13.5.1   Harmonized evaluation frameworks

HCDMs play a critical role in providing patients with access to high-quality, clinically validated digital therapeutic products. As more HCDMs across the world become engaged in the review, assessment, and implementation of digital therapeutic products, it is important for these payors, clinicians, patients, and caregivers to have access to reliable frameworks that enable consistent evaluation of digital therapeutic products.

Presently, there are a wide variety of generic evaluation criteria and considerations that HCDMs are using to evaluate digital therapeutic products. As the prevalence of digital therapeutics increases, decision-makers are taking it upon themselves to create the necessary frameworks to assess digital therapeutic product quality, impact, and value. Consequently, there are vastly differing expectations emerging within and across health systems, employers, private payors, and public payors. This leads to an incredibly complex and difficult environment for digital therapeutic manufacturers to operate in. As an increased degree of consistency is introduced into this process, manufacturers can more readily provide HCDMs with the necessary data, outcomes, and insights needed to meaningfully implement and scale products.

One of DTA's current efforts is therefore the development of the *DTx Value Assessment & Integration Guide*, which provides HCDMs with a framework to evaluate the value of and enable the implementation of digital therapeutics in clinical practice. This Guide, therefore, provides consistent pathways to enhance and refine assessment processes within existing systems, while also serving as a foundational template for organizations undertaking the development of new pathways to evaluate and assess digital therapeutic products.

13.5.2   Global recognition & utilization

Despite the significant differences between digital products, relatively few people are able to differentiate between digital health technologies that serve different purposes. Using traditional pharmaceutical products as an example, most people understand that even if two capsules or tablets look alike and are produced in the same color, each medication type is able to serve a different purpose from the other. It is now common understanding that the active ingredient inside the medication is what dictates a medication's clinical impact, not only the packaging or shell it is delivered in.

Nevertheless, patients and clinicians rarely realize that even though software-based products ranging from wellness to therapeutic may be accessed through an online platform such as an app store and downloaded onto their smartphone or tablet, each product type has a different purpose, indication, and level of clinical impact.

This distinction is important since digital therapeutics complement traditional pharmaceuticals and clinician-provided therapies in the treatment, management, and prevention of a wide range of conditions. Different from other DHTs, these products provide patients with access to clinically proven treatment options outside of traditional office settings, thereby expanding access to needed medical care. As opposed to solely monitoring or diagnosing conditions, digital therapeutic products deliver medical interventions to patients, track progress, report outcomes, and increase clinician capacity.

Covid-19 shined a light on the power of technology to improve care now and in the future. Tools such as digital therapeutics will continue to play a valuable role in supporting safe, effective, and ongoing patient care. As healthcare continues to embrace a technology-focused shift in care delivery, healthcare leaders will increasingly recognize the value these products provide.

13.5.3   National policy development

From a geographic perspective, the future value digital therapeutic products provide will not be limited to certain countries; instead, patient populations across all regions will be provided with access to these therapies. When conditions such as depression, anxiety, addiction, diabetes, asthma, muscle pain, and countless others impact populations globally, all patients—regardless of their home country—should have access to the same level of quality therapies.

As nations increasingly pursue the development of legislative and regulatory pathways for digital therapeutic products to achieve healthcare parity and enable local access, best practices related to cross-jurisdictional data generation, privacy, storage, and analysis will become critical to expanded product access, utilization, and outcomes optimization.

Frameworks for national or regional payments are currently being implemented in locations such as Australia, Belgium, France, Germany, Japan, South Korea, United Kingdom, and the United States. This list will continue to grow over the coming months and years.

It is therefore important for policy-making bodies to formally recognize and define digital therapeutic products in legislation, regulatory bodies to establish adequate pathways to market, for payors to establish processes and funding mechanisms to cover these technologies, and healthcare leaders to develop systems to enable product distribution. Adequate payment mechanisms should also be considered for primary care providers and other clinicians who may authorize the use of digital therapeutic products, review patient-specific data, and incorporate actionable outcomes into patient care plans.

13.5.4   Clinical guideline & practice inclusion

As the field of medicine evolves, clinicians have increasing arsenals of therapies to select from in providing patient care. While pharmaceuticals, in-person, and virtual therapies will continue to be therapeutic staples, digital therapeutic products will increasingly be recognized as first-line treatment modalities. Using these products, as standalone therapies or in combination with medications, in-person therapy, or virtual care will enable individual patients to receive holistic, personalized care that delivers exactly what they need, when they need it.

When evaluating whether to prescribe or authorize a digital therapeutic product, clinicians will be considering the following types of questions:

● Which disease or disorder is the digital therapeutic product indicated to treat, manage, or prevent?
● What is the appropriate timing, duration, and frequency of therapy use?
● Should the digital therapeutic product be used independently, or with medications, devices, or other therapies?
● Does the product enable greater access to care and align with the patient's current care?
● Is the product content available in a familiar language for the patient?
● What actionable outcomes are collected by the product and shared with patients, caregivers, and clinicians?

To best support clinicians, it is necessary for digital therapeutic products to be incorporated into formal clinical guidelines and care pathways. As medical societies and other guideline developers evaluate the quality and optimal use of digital therapeutic products in practice and integrate them into formal practice guidelines, clinicians will increasingly rely on these published pathways to develop the best possible patient care plans.

13.5.5   Personalized therapies

As digital therapeutic products increasingly deliver personalized therapies to patients and account for patient languages and cultural considerations, they must also address patient health literacy levels, disability considerations, and personal goals, in addition to personalized product distribution, access, and utilization.

As this field grows, digital therapeutic product authorization and distribution will be provided via:

● Formal prescription from a qualified clinician (in-person or virtual engagement).
● Clinician referral for a nonprescription digital therapeutic product (in-person or virtual engagement).
● Direct authorization by an employer for a nonprescription digital therapeutic product.
● Direct authorization by a payor for a nonprescription digital therapeutic product.
● "Authorized clinical protocol" established by HCDMs to create patient qualification requirements, automatically authorizing patient access when necessary criteria are met.
● "Clinically-validated screening tool" that patients utilize to determine whether they qualify for the therapy.
● "Over-the-counter" model where no form of third-party authorization is necessary.

Following initial authorization, product access codes and any necessary product components (e.g., hardware, wearables) may be delivered directly to the patient via remote delivery (SMS, email, mail) or in-person delivery at a clinic or pharmacy. The types of entities involved in product distribution may include any combination of the following: product support center, clinician and/or clinical team, virtual health coach, telehealth provider, pharmacy, or HCDM.

As for product utilization, the real-world outcomes generated by digital therapeutics—including patient engagement, adherence to therapy, clinical responses, patient-generated insights, and overall therapy progress—will increasingly be used to optimize outcomes at the individual patient and population levels.

At the individual patient level, digital therapeutic products will continue providing clinicians with meaningful, actionable clinical reports. At the population level, data generated by digital therapeutic products may be aggregated to track progress and compare an expanding number of aggregate outcomes based upon patient disease state, level of acuity, length of impact, geographic location, age, gender, comorbidities, side effects experienced, genetic profile, time to clinical improvement, and overall wellbeing.

These digital therapeutic-generated insights and comparisons will then be leveraged by clinicians to:

- Assess impact of therapy toward patient goals.
- Apply actionable insights to optimize, adjust, recommend, escalate, or de-escalate therapy at the patient and population levels.
- Detect adverse events and/or nonoptimal outcomes.
- Develop a better understanding of medication usage.
- Provide ongoing monitoring and/or measurement.

13.6 Looking ahead

Compared to some technologies that take numerous decades to be conceptualized, developed, scaled, and adopted, digital therapeutics have experienced meteoric growth. We have collectively moved from an environment where diseases could only be treated by in-person therapy and pharmaceuticals, to one where digital therapeutic products are formally recognized as legitimate therapeutic modalities.

As the digital therapeutic industry grows, product quality, integrity, and safety cannot be compromised. Neither can digital therapeutics' focus on meeting patient needs.

Patients, caregivers, and clinicians need to continue driving the discussion on where current gaps of care exist and how digital therapeutic therapies could better meet these needs. Mirroring the centrality of patient engagement across the entire life cycle of every single digital therapeutic product, end-user engagement is equally critical to the ongoing evolution of this industry. This alone makes the horizon for digital therapeutics so vibrant.

References

Digital Therapeutics Alliance. (2019a). *Digital health industry categorization*. https://dtxalliance.org/wp-content/uploads/2019/11/DTA_Digital-Industry-Categorization_Nov19.pdf (Accessed 2022).

Digital Therapeutics Alliance. (2019b). *Digital therapeutics: definition and core principles*. https://dtxalliance.org/wp-content/uploads/2021/01/DTA_DTx-Definition-and-Core-Principles.pdf (Accessed 2022).

Digital Therapeutics Alliance. (2019c). *DTx product best practices*. https://dtxalliance.org/wp-content/uploads/2021/01/DTA_DTx-Product-Best-Practices_11.11.19.pdf (Accessed 2022).

Digital Therapeutics Alliance. (2021a). *Digital health technology landscape*. https://dtxalliance.org/wp-content/uploads/2021/01/DTA_FS_Digital-Health-Technology-Landscape_010521.pdf (Accessed 2022).

Digital Therapeutics Alliance. (2021b). *DTx product categories*. https://dtxalliance.org/wp-content/uploads/2021/01/DTA_FS_DTx-Product-Categories_010521.pdf (Accessed 2022).

Digital Therapeutics Alliance. (2022a). *DTx U.S. regulatory & reimbursement pathways*. https://dtxalliance.org/wp-content/uploads/2022/01/US-Regulatory-and-Reimbursement-Pathways.pdf (Accessed 2022).

Digital Therapeutics Alliance. (2022b). *Understanding DTx /product library*. https://dtxalliance.org/understanding-dtx/product-library/ (Accessed 2022).

Federal Institute for Drugs and Medical Devices. (2020). The fast-track process for digital health applications (DiGA) according to Section 139e SGB V: a guide for manufacturers, service providers and users. https://www.bfarm.de/SharedDocs/Downloads/EN/MedicalDevices/DiGA_Guide.pdf?__blob=publicationFile (Accessed 2022).

Federal Institute for Drugs and Medical Devices. (2022). Digital health applications. https://www.bfarm.de/EN/Medical-devices/Tasks/Digital-Health-Applications/_node.html (Accessed 2022).

IMDRF Software as a Medical Device (SaMD) Working Group. (2014). *Software as a medical device: possible framework for risk categorization and corresponding considerations*. https://www.imdrf.org/sites/default/files/docs/imdrf/final/technical/imdrf-tech-140918-samd-framework-risk-categorization-141013.pdf (Accessed 2022).

Lovell, T. (2021). France to enable rapid market access for digital therapeutics. https://www.healthcareitnews.com/news/emea/france-enable-rapid-market-access-digital-therapeutics (Accessed 2022).

Ministry of Food and Drug Safety Medical Device Evaluation Department. (2020). Guideline on review and approval of digital therapeutics (for industry). https://www.mfds.go.kr/eng/brd/m_40/down.do?brd_id=eng0011&seq=72624&data_tp=A&file_seq=1 (Accessed 2022).

National Institute for Health and Care Excellence. (2018). Evidence standards framework for digital health technologies. https://www.nice.org.uk/corporate/ecd7/resources/evidence-standards-framework-for-digital-health-technologies-pdf-1124017457605 (Accessed 2022).

National Institute for Health and Care Excellence. (2022). Evidence standards framework for digital health technologies. https://www.nice.org.uk/about/what-we-do/our-programmes/evidence-standards-framework-for-digital-health-technologies (Accessed 2022).

U.S. Food & Drug Administration. (2020). Enforcement policy for digital health devices for treating psychiatric disorders during the coronavirus disease 2019 (COVID-19) public health emergency: Guidance for Industry and Food and Drug Administration Staff. https://www.fda.gov/media/136939/download (Accessed 2022).

U.S. Food & Drug Administration. (2021). Leveraging real world evidence in regulatory submissions of medical devices. https://www.fda.gov/news-events/fda-voices/leveraging-real-world-evidence-regulatory-submissions-medical-devices (Accessed 2022).

Chapter 14

Potential pitfalls and lessons learned

Frances Kay-Lambkin[a], Milena Heinsch[b] and Dara Sampson[b]

[a] *Hunter Medical Research Institute, Newcastle, NSW, Australia,* [b] *School of Medicine and Public Health, College of Health, Medicine, and Wellbeing, the University of Newcastle, Callaghan, NSW, Australia*

At any one time, over 900 million people globally experience a mental disorder (including alcohol/other drug use disorders, Whiteford et al., 2013), and this is increasing by about 3% each year (2018). Adding to these challenges, the COVID-19 pandemic presents clear risks for a substantial decline in global mental health. Preliminary evidence points toward an overall rise in symptoms of anxiety and coping responses to stress (Holmes et al., 2020), including increased drug and alcohol use amongst the general population. The greatest mental health impacts of the COVID-19 pandemic will be felt, however, by those who are already most marginalized and people with pre-existing mental health and substance use disorders, who have a higher susceptibility to stress than the general population (Yao et al., 2020). Stress is a well-known risk factor in addiction relapse vulnerability and has the potential to worsen pre-existing conditions (Sinha, 2007). Within this chapter, the COVID-19 pandemic will provide the frame through which to examine global mental health issues and approaches to treatment. Digital therapeutics came to the fore during this time, as face-to-face service was restricted (de Girolamo et al., 2020). This pandemic has highlighted (and exacerbated) social inequities in relation to the prevalence of mental illness, as well as treatment options. Whilst the current COVID-19 impacts provide a contemporary exemplar for discussion within this chapter, the contents are equally applicable prior and post this pandemic. Building on previous chapters within this book, our chapter will cast light on some of the learnings (and notes of caution) when entering the digital therapeutics domain.

Mental disorders are much like physical disorders; given appropriate and timely intervention, they can be successfully prevented, managed, and treated. Despite this, national surveys of mental health and wellbeing indicate that only a few who need treatment can access it. In Australia, for example, one-third of people who need mental health treatment in a 12-month period do not receive it (Burgess et al., 2009). The 2018 Lancet Psychiatry Commission outlined two key reasons why this situation exists: (1) treatments have not been developed for many mental health disorders, and (2) there are not enough mental health workers to meet the demand for treatment (Holmes et al., 2018).

A recent meta-analysis of 34 studies reported that 75% of people with mental disorders state a clear preference for psychological over medication-based treatment (McHugh et al., 2013). This was consistent across treatment-seeking and nontreatment-seeking samples. Yet in Australia at least, the most common mental health treatment provided is psychiatric medication (approximately one in six Australians), most often by a general practitioner (87% of all mental-health prescriptions), with the highest rates of mental health prescriptions occurring in rural/remote Australia (2017).

Psychological and medication-based treatments for depression, anxiety, and alcohol/other drug use problems have similar efficacy (McHugh et al., 2013). However, psychological treatments are largely not available when people need them most, do not typically target the co-occurrence of many mental disorders simultaneously, and many receive a wide variety of treatments that are not evidence-based (Holmes et al., 2018). Despite significant investment by many governments to improve access to psychologists for mental disorders (e.g., the Better Access Scheme in Australia), there is no evidence of impact or effectiveness (Jorm, 2018), and accessibility is a significant issue especially in rural/remote locations (AGPC, 2020).

Globally, there are not enough mental health providers available to meet the treatment needs for the one in four people who experience a mental disorder in any one year. The World Health Organization estimates a shortage of 1.8 million mental health workers across the globe (WHO, 2011). Even before the COVID-19 pandemic, there was a paucity of psychologists and allied mental health workers who could deliver evidence-based treatments for mental disorders (Holmes et al., 2018).

Digital Therapeutics for Mental Health and Addiction: The State of the Science and Vision for the Future. DOI: https://doi.org/10.1016/B978-0-323-90045-4.00013-7

The extra demand for mental health services during COVID-19 places still more pressure on an already overloaded mental health system worldwide (Frawley et al., 2021).

A recent report explores these challenges further (Iorfino et al., 2021), and places particular emphasis on the key action areas as we emerge from the COVID-19 pandemic. The first recommendation is for the provision of sophisticated infrastructure to support regionally based healthcare delivery. The report also challenges the concept of universality of care, arguing Medicare and hospital access gives illusory notions of service availability while masking service delivery quality and equity of access issues. Core to this inequity, a lack of co-designed, systemic planning which fails to include social context. In essence, mental health services are often delivered in a "silo" without an ecosystems or contextual approach which understands housing, income support, education, social planning, and community services are all interconnected with mental health outcomes. Social inequity increases the risk of many mental health disorders. Cumulative effects of disadvantage arise due to inequalities. Global evidence links low socioeconomic, status with increased rates of depression (McDaid et al., 2020). Gender is also important—mental disorders being more prevalent in women who experience different environmental, economic, and social factors than men (McDaid et al., 2020). One mechanism for intentionally addressing social and economic inequities is the concept known as "proportionate universalism" in which measures to address inequity are included within social and economic policies. Proportionate universalism reflects the principles espoused by the authors of Social Determinants of Mental Health report—universal measures with the capacity to be calibrated to those most disadvantaged (Affeltranger et al., 2018).

The need to improve mental health treatment is great (Holmes et al., 2018). The recently released draft report of the Australian Productivity Commission into mental health found that specific reform is needed within the next 10 years to "close critical gaps in healthcare services" (reform area 2, AGPC, 2020) to "allow timely access by people with mental ill-health to the right treatment for their condition."

Digital therapeutics can rapidly and significantly improve treatment access, act in place of "on the ground" services when they are not accessible, and be actively provided to people in earlier phases of their disorder. To date, the advent of eHealth has emerged organically and without structure. In line with the World Economic Forum, developing a strategic approach to online service delivery is a key opportunity to transform mental healthcare (Iorfino et al., 2021).

Central to realizing the potential of digital therapeutics are questions about whether this modality is safe, effective, and able to be integrated within and around the existing pathways to care for people experiencing mental health disorders. However, building an evidence base is an incremental process, and several gaps and uncertainties exist. Critically, we are yet to win some of the same battles we have always faced in encouraging behavior change and recovery in people with health-related problems, including mental health. Having access to technology is only one part of the answer. We now move to our five key findings in relation to our research and the broader body of literature. These lessons are about sustained engagement, technology as a complement to, and not a replacement for, the human element, access not being the only barrier to uptake, the perils of merely replicating service delivery models and, finally, the importance of blending models.

14.1 If you build it, we will come, but we may not stay

Digital therapeutics have huge potential to increase the reach and impact of prevention, early intervention, and treatment for mental health and alcohol and other drug use problems, particularly for young people. Young people report a preference for digital over face-to-face treatments, appreciating the greater anonymity, convenience ease of access, and control that it provides (Boydell et al., 2014).

Our team has demonstrated the efficacy of online early intervention programs in helping young people manage commonly co-occurring mental and alcohol use disorders, for example, the *Deal* program for depression and hazardous alcohol use (Deady et al., 2016), the *Inroads* program for anxiety and hazardous alcohol use (Stapinski et al., 2019), and the *Breathing Space*, moderated social networking program to reduce social isolation and promote health mood and behaviors (Kay-Lambkin et al., 2015). We have also demonstrated the efficacy of digital early intervention and treatment programs for people experiencing depression and co-occurring alcohol, cannabis (Kay-Lambkin et al., 2009; Kay-Lambkin et al., 2011b), tobacco, and physical health problems (Kay-Lambkin et al., 2016), and methamphetamine use disorders (Kay-Lambkin, 2009; Kay-Lambkin et al., 2010; Tait et al., 2012; Tait et al., 2015). This has included people experiencing moderate-severe current and active symptoms.

This evidence also suggests that people using digital technologies, particularly when reporting moderate-severe mental health and alcohol/other drug use symptoms, find online treatment programs as acceptable as face-to-face therapies, as effective as face-to-face therapies, and particularly important for people living in rural and remote areas (Kay-Lambkin, Baker, Kelly, et al., 2012). Our research also provides evidence that digital treatments are more effective for people high in perfectionism (Kay-Lambkin et al., 2017), for people with cannabis use and mental health comorbidity (Kay-Lambkin et al., 2017; F. Kay-Lambkin, Baker, Lewin, & Carr, 2009; F.J. Kay-Lambkin, Baker, Lewin, & Carr, 2009), and for alcohol use and depression (Kay-Lambkin et al., 2011b).

There are conflicting reports about whether results of clinical trials in any discipline of medical research actually influence practice (Khera et al., 2018). The effort to conduct these trials may be wasted if results ultimately do not change clinical practice. This is as true for digital treatments as it is for traditional psychological treatments. Evidence that digital therapies work under experimental conditions is not enough to demonstrate that it should be upscaled and implemented, nor is it adequate to demonstrate how these services might operate within or around a health system (Meurk et al., 2016). In response, countries like Germany have introduced a regulatory framework—the 2019 Digital Health Care Act—which has seen the creation of a landscape in which doctors and psychotherapists can prescribe both web-based and app-based from a central registry (Stern et al., 2020). The costs of these products can be reimbursed through insurance companies. This is a significant advance, as 90% of the German population (or over 73 million people) is covered by insurance (Stern et al., 2020).

Poor adherence is a common feature of digital programs, particularly in naturalistic settings (Batterham et al., 2008; Christensen et al., 2009). Early studies reported only 1% completion by participants in a 12-week program for depression (e.g., Farvolden et al., 2005, Christensen et al., 2009), with a more recent systematic review indicating adherence rates for mobile health apps (including mental health and substance use) of approximately 56% (Jakob et al., 02/12/2021). Low adherence can occur for a number of reasons, some positive such as sufficient dosage for remission of symptoms, some neutral such as insufficient need for treatment (e.g., healthy users, individuals testing an intervention to examine its suitability), and some negative such as poor fit of content to needs, lack of interactivity, symptom severity, low motivation, or a lack of symptom improvement (Batterham et al., 2008; Christensen et al., 2009; Donkin et al., 2011; Donkin & Glozier, 2012). Recruitment channels and the characteristics of the app or digital user influence engagement and thus adherence to the digital intervention. Intervention-related factors also play a role in adherence, which can be enhanced, across all health domains, through the inclusion of four key features: Jakob et al. report, for example, that some form of reminder or "push notification" increases usage, as does a user-friendly and reliable app design (Jakob et al., 02/12/2021). The last two factors segue way into our next key lesson, about the human element. They are user perception the app is personalized and tailored to their needs, and the presence of personal support, as a complementary intervention (Jakob et al., 2021).

14.2 The human element

Whilst there is good evidence that mild depression can be managed without therapist guidance using digital interventions, there is insufficient evidence about the role of therapist guidance in managing moderate depression. It is unclear how critical therapist support, provided by telephone, email, SMS, for about 10–20 minutes duration per session of digital intervention, is in accounting for the results observed in studies of depression (Kay-Lambkin et al., 2018). As mental healthcare expands to smartphone apps and other technologies that may offer therapeutic interventions without a therapist involved, it is important to assess the impact of nontraditional therapeutic relationships. A measure is required to evaluate the digital therapeutic alliance (Henson et al., 2019) (See chapter on the "Digital Therapeutics for Mental Health and Addiction" for further discussion).

This is particularly important for mental health disorders that can escalate or that are vulnerable to relapse to acute states. More research is required to understand the types of safety protocols or algorithms required to identify risk/potential escalation in non-face-to-face environments (e.g., online), and when real-time clinician input should be triggered (Lim & Penn, 2018). This is but one of the ethical issues we have grappled with as practitioners—duty of care, confidentiality, and privacy, and relationship boundaries being three other key considerations. The disembodied online environment both creates a spatial barrier, whilst also inviting disclosure through the perceived anonymity of the medium.

Several studies show that some people actively choose to seek help and support from peers online via forums and social media websites (Berry et al., 2016), whilst others prefer to communicate their feelings and experiences about their well-being and mental health using online blogs (Batterham & Calear, 2017a). Openness to using digital tools in place of, or to augment, mental health treatment has been demonstrated in rural/remote Australia, and in older age populations (Handley et al., 2014).

Peers have a critical role to play in destigmatizing treatment, reinforcing attitudes and behaviors, and supporting people through the treatment experience (Fortuna et al., 2019). Peer-led interventions, designed to increase social connectedness, have strong potential to address mild-moderate depression and can be offered online or via videoconference to increase accessibility (Kay-Lambkin et al., 2018). Given a systematic review of 49 studies of digital interventions for mental disorders indicated that acceptability was higher when these interventions were provided alongside remote online peer support, there is a key opportunity for active engagement of a peer workforce to support digital intervention provision and integration (Berry et al., 2016).

Complementing peer-support models, "coaching" is another human adjunct to digital therapeutics that is gaining momentum and evidence for efficacy. Coaching is based upon the body of literature indicating the provision of nonexpert human support improves outcomes in, and adherence to, digital interventions through a mechanism of "supportive accountability" (Mohr et al., 2011). Coaching sessions can be conducted in person, over the telephone, via email or text message, etc., and often comprise regular 10–15 minute check-ins with the person working their way through a digital intervention to answer questions, troubleshoot technical issues, review content and encourage application of skills into routine life, and normalize challenges in learning new skills (e.g., Baumeister et al., 2014). There is typically no new therapeutic advice or skill-building that occurs in these coaching sessions. An emerging area of research and practice work lies in the development of coaching protocols—outlining scope, coach attributes, techniques, and communication frameworks that maximize the benefits of these sessions on adherence and outcome for digital interventions (Lattie et al., 2019).

14.3 Access does not equal uptake

Globally, people are more likely to have access to a mobile phone than mental healthcare (Naslund et al., 2019). Mobile phones have become the primary point of access to the Internet, with 45% of the world's population owning a smartphone, and projections that this will increase to 4.3 billion people worldwide by 2023 (Statista, 2021). But, having access to devices and connectivity to the internet is not enough. To be adequately connected to technology, we need to connect communities to the appropriate support they need, when they need it most, to then be able to utilize these devices and connectivity and the advantages they offer.

We can achieve this well in our research trials for digital therapeutics, with funded marketing and recruitment protocols, expert clinical supports, incentives to engage and connect, and clear pathways to access and follow-up. Outside of clinical trials, however, uptake of these digital programs in the community is suboptimal (Batterham & Calear, 2017b; Musiat et al., 2014). Previous studies have reported that many people prefer face-to-face therapy over digital programs (Casey et al., 2014; Mohr et al., 2010; Musiat et al., 2014), hold a common public view that digital therapies are not as effective as face-to-face therapy (Apolinario-Hagen et al., 2018), despite viewing digital programs as being highly convenient (Musiat et al., 2014).

Despite good evidence for program effectiveness, digital therapeutics exist largely independently of traditional healthcare service settings, with mental healthcare providers underutilizing digital therapeutics in their practice (Christensen et al., 2011). No clear models currently exist for digital treatment integration into health services (Batterham et al., 2015); however, the German example provides evidence that this is rapidly changing (Stern et al., 2020). Despite their positive findings, there are very few examples of successful implementations of digital interventions in clinical services, and many failures (Mohr et al., 2017). This is in contrast to almost every other sector, especially the commercial/corporate industries, where technological solutions have become integrated into routine organizational business models (Burns, Hickie, & Christensen, 2014). Thus, for all the advantages that digital therapeutics offer, and the substantive evidence base for their efficacy, their potential impact is unrealized. Testing of the transportability of digital therapeutics from the clinical laboratory to the real world is lacking (Batterham et al., 2015), and we have failed to unpack the best methods to connect our communities to the support they need. Objective data are therefore needed to guide practice and to assess outcomes.

Utilizing digital therapeutics outside of research trials is not simply a matter of providing access to the programs on a device-of-choice. New ethical considerations for clinicians, researchers, and healthcare organizations arise that challenge our traditional notions of duty of care, governance, benefits, and harms (Wykes et al., 2019). This is particularly pertinent as people using digital tools can engage, disengage, and reengage on their own terms (Heinsch, Geddes, et al., 2021). Evidence that digital therapies work under experimental conditions is not enough to demonstrate that it should be upscaled and implemented, nor is it adequate to demonstrate how these services might operate within or around a health system (see Box 14.1, Meurk et al., 2016).

BOX 14.1 Case Study: the eCliPSE Project

Background: The eCliPSE online portal aims to facilitate access to evidence-based online screening and eHealth treatments to people experiencing comorbid mental health and substance use problems, and the clinical services supporting them. The NSW Ministry of Health, Australia, engaged the authors to develop the eCliPSE online tool, implement the eCliPSE tool in two pilot local health districts (LHDs) in New South Wales, and evaluate the implementation model. The eCliPSE tool can be accessed at www.eclipse.org.au

Implementation model: An eCliPSE implementation team was established in each LHD that comprised clinical leaders from mental health alcohol/other drug services and the research team to discuss training needs, support needs, and how best to roll out the eCliPSE tool. A "clinician-referral" model of integration was preferred, whereby access to the tool was only via referral from a clinician of the services. The research team ran an initial training session in each LHD for relevant services (as determined by the eCliPSE implementation team) to introduce the tool, encourage engagement with the tool, practice using the tool, and discuss methods of referring clients to the tool. The research team then liaised with trained clinicians (via email) and the eCliPSE implementation team in each LHD (site visit, telephone, email as preferred) to support uptake and use of the eCliPSE resource.

Evaluation: The evaluation was designed to determine the feasibility, acceptability, reach, and effectiveness of the eCliPSE tool in meeting service user and service provider needs, and the usefulness of the clinical implementation pathway employed in each LHD. Attendees at the training engaged well with the content and the tool and stayed behind to practice using eCliPSE whilst the training team was present. There was 90% agreement that clinicians attending training would "very likely" recommend eCliPSE and the tools contained in the online tool to clients. There was 75% agreement that the eCliPSE portal and the resources contained within it were "easy-very easy" to implement in their workplaces. Over the 3-month implementation period, 110 clinicians registered on the eCLiPSE clinician interface, and were provided with access to the resource. Of these, only eight clinicians responded to the post-implementation survey. Of these few respondents, the majority did not log on at all after eCliPSE training ($n = 2$), logged on once only ($n = 3$), or logged on 1–5 times ($n = 3$). The top 3 barriers to using eCliPSE more frequently were workload and time pressures ($n = 3$), a perception that clients did not need the tool ($n = 2$), and that clinicians kept forgetting to refer to the tool ($n = 2$). Over the same time, people in the general community navigated their way to the eCliPSE tool organically. Hits on eCliPSE were high (approx. 1000 per month since training, and of these, 65% were first-time visitors), and these translated into 351 users (people who stay and linger on the website, exploring its contents, approx. 88 per month).

Conclusions: Despite encouraging signs at the eCliPSE training days, and support from clinical leaders in the pilot LHDs, self-reported uptake and integration of the eCliPSE tool using the preferred model of implantation (clinician referral) was low. Social desirability aside, this was not related to attitudinal issues, burnout, or low commitment to evidence-based practice (baseline questionnaire results). Service user characteristics indicate that the prevalence of comorbid mental health and alcohol/other drug use in these settings is high (75%) thus many clients would have been eligible for the tool. Clinician-brokered access to the eCliPSE online tool may not be the most effective method of implementation.

A range of financial, structural, and technological health system barriers exist (Christensen et al., 2011) adding to the significant issues facing clinicians who want to improve access to effective treatments for people living with mental health disorders. These include a lack of knowledge about, and training in, evidence-based programs, resistance to changes in practice, concerns around efficacy, confidentiality and safety (indemnity), lack of the financial incentives for implementation, and lack of established pathways to provision of digital services (Batterham et al., 2015). Other barriers may include concerns about data access and management (including data linkage, quality, standards, and storage), community engagement, and trust (2016). For these reasons, many clinicians are not able, or willing, to implement these interventions in practice. This is totally understandable in a neoliberal environment of economic rationalism and managerialism whereby workers are expected to meet increasingly stretching clinical caseload targets, with fewer resources (Spolander et al., 2014). Although only conjecture, it is possible that fear of job loss and role identity may be another inhibitor to uptake.

In Australia, we have completed a small-scale research project examining the use of a digital treatment program in an NSW alcohol and other drug treatment setting (Kay-Lambkin, Baker, Healey, et al., 2012). The results of this small dissemination study revealed that, despite clients of the service reporting willingness to use the SHADE digital tool (an evidence-based program to support people with alcohol/other drug disorders and depression) as part of their treatment program (over 80% agreement), fewer than one-third of clients in the study were actually provided with access to SHADE by their treating clinician (Kay-Lambkin et al., 2014). This is not unique to the Australian context. A study into intention of medical practitioners to use eHealth services in five developing countries (namely Malaysia, Pakistan, Uganda, Bhutan, and Mexico) found, through 220 interviews, a positive perception of its usefulness, however a very low level of awareness. The study was prompted by the literature which revealed a lack of research into eHealth in developing countries (Nuq & Aubert, 2013).

A significant cultural shift is also required. Continued partnerships with practitioners—as we do in our own research—goes some way toward this shift. Education, promotion, and investment are also required. For this shift to transpire, an analysis of practioner methods, governmental efforts, and public needs is imperative and will shed light on what steps will provide solutions for our digital challenges.

Proactive steps need to be taken to increase public and professional support and recognition of the role, credibility, safety, and efficacy of digital therapeutics as a legitimate, and often, first-line tool for mental health problems and disorders. In partnership with this activity, a significant audit of the challenges and structural barriers to implementing eHealth tools in mental health services also must occur, with funding provided to overcome these barriers and enable services to become "digitally enabled."

14.4 A "lift-and-shift" approach does not work

As discussed above, challenges arise in scaling up digital therapeutics from research trials into clinical and community settings, indicating that providing access to devices and to evidence-based digital tools is not enough to engage with and meet the needs of people needing healthcare. In the same way, efforts to transfer what happens in traditional healthcare settings, particularly mental health settings, onto digital platforms has been equally suboptimal, and have limited the scope and potential of digital therapeutics in this space.

In response to COVID-19, for example, the Australian Government introduced a new temporary Medicare Benefit Scheme "telehealth" item into primary care and allied health services, to facilitate access to Medicare-rebated consultation services via telehealth throughout COVID-19. "Telehealth" refers to the use of video conferencing tools (such as Skype, Facetime, Zoom) by healthcare providers to engage in real-time with consumers regarding their health (e.g., assessment, treatment). As of April 2020, over 3 million Australians had received a telehealth intervention from their healthcare provider, including a proportion for allied and mental health services. Australians largely embraced "telehealth" during COVID-19 when they did seek mental health support, with around 50% of mental health service appointments occurring via telehealth during April 2020. Overall, however, rates of accessing any mental health service during COVID-19 significantly decreased despite increased access to "telehealth" opportunities, suggesting that moving to virtual, real-time mental health consultations and appointments is not sufficient (Stern et al., 2020). In total, 1.11 million mental health treatment plans were accessed between January to September 2020, only a little more than the 1.08 million used in the same period in 2019 Whilst a model of healthcare that clinicians are likely most comfortable with and ready to use, "telehealth" service provision still requires health professionals to deliver the service and suffers the same limitations of time and resources as face-to-face service provision. As we emerge from COVID-19, a different model of care is required, and the repertoire of approaches available to mitigate mental health consequences under COVID-19 pandemic conditions needs expansion. In order for digital therapeutics to be effective during the COVID-19 pandemic or during any future adverse events, we must ensure it is appropriately integrated into our mental health response as a "business as usual" modality (Smith et al., 2020), available outside of the usual constraints of health service (see Box 14.2). Whilst this particular example pertains to the experience of young people, it has scope to be replicated across all age groups.

BOX 14.2 Using technology to create a circle of care for youth mental health treatment (Iorfino et al., 2021)

There is a mismatch between the needs of people with mental disorders, the way symptoms emerge and change over time, and the way the mental health system operates to support them.

As service philosophies, traditional service models, and roadblocks in information sharing within the current system interact with increased demand and under-resourcing of mental health services, digital therapeutics show promise. Attempts to address the additional pressures of COVID-19 on mental health services whilst fast-tracked have not had the service efficiency or access impact required.

How do we get there?
A "circle of care" has been proposed (Iorfino et al., 2021) to reform service provision in youth mental healthcare which comprises (1) a no wrong door (or waitlist) approach with signposts and champions to rapidly navigate and facilitate a person's pathway through available treatment options, (2) personalized and multidisciplinary care decisions, that include digital therapeutics, and (3) responsive, measurement-based care to determine when changes to modality, frequency, and focus are needed.

This requires a new model of care that places a person at the center, and technology-enabled infrastructure the facilitates the circle of care.

In 2021, Facebook has 1.8 billion daily users, Instagram report 500 million daily users, and Twitter has 187 million daily active users and is the largest source of breaking news (Statista, 2022). Many people do not identify "the Internet" as the mechanism by which they access information online. These developments have transformed the way in which we live our lives, including for our health and wellbeing. Digital platforms are increasingly becoming the medium through which assessment and intervention are taking place for many health conditions, including mental health. In many cases, talking to people in person is not necessarily an individual's first choice method of communication. The way we use the internet/digital technology has changed over time such that digital interventions based on face-to-face models of therapy (e.g., 45–60 minute appointments, once weekly, homework assigned) may not appeal to people today. New models of therapy and therapeutic processes are needed, that leverage the advantages of digital tools, rather than replicate what could be done face-to-face.

We urgently need effective implementation and healthcare models to translate digital tools into clinical practice and community settings, and to better leverage the opportunity COVID-19 has provided to integrate digital and telehealth into usual care.

14.5 A blended model of treatment is needed

A key outcome of our research is that engagement with digital treatment programs is associated with increased engagement with traditional services for mental or alcohol/other drug use disorders for people who are still symptomatic at the conclusion of the trial (Tait et al., 2015). Closer inspection of this result reveals a potential role of online programs in destigmatizing mental health treatment and encouraging help-seeking through relevant real-time services when symptoms persist (Kay-Lambkin et al., 2011a; Kay-Lambkin, Baker, Kelly, et al., 2012).

More broadly, a shift toward patient-centered care models in mental health and addiction requires a range of treatment options, including self-management, and a shift away from a "one-size-fits-all" approach to care.

Face-to-face interventions are highly valuable but have several limitations such as limited capacity, limited availability outside office hours, and higher cost (van der Kleij et al., 2019). Digital interventions have great potential to solve these issues, and can support the transition toward personalized medicine, self-management, shared decision making, and continuity of care (van der Kleij et al., 2019), in this regard being a "stepped care" model. But they also come with their own limitations, many of which can be solved by face-to-face interventions. Both digital and face-to-face interventions need to be optimized to support the most effective, efficient, and engaging model of care for people with mental health problems. Blended care, a combination of digital and face-to-face therapy, is increasingly being applied in mental healthcare as a strategy to achieve this optimization. This is in line with our earlier comments about the need for continued exploration of the interface between the human element and technology.

Blended care can offer synchronous or asynchronous screening or assessment, education, treatment, self-help, feedback strategies in mental health or addiction contexts with positive effects (Wentzel et al., 2016). The "blending" of strategies and modalities can occur in an integrated way, with the digital component offered as a simultaneous adjunct to face-to-face treatment (or vice versa), or in a sequential way, with the digital component of care offered prior to, or as an aftercare or relapse prevention strategy, following the face-to-face component (Erbe et al., 2017). Evidence to support blended models of care are emerging. In a systematic review of blended interventions in adults with mental disorders, blended interventions were feasible in practice, and compared with face-to-face interventions seemed to save clinician time, increase engagement, and help maintain the effects of inpatient therapy (Erbe et al., 2017). More research on effectiveness and cost-effectiveness of blended treatments, especially compared with nonblended treatments is necessary (Erbe et al., 2017).

Operationalizing blended care in practice remains a challenge, particularly in the absence of a strong rationale for determining what "blend" works for whom, why, and when (Wentzel et al., 2016). Central to responding to this challenge will be clinicians, with training, motivation, and readiness for digital integration being key facilitators (Mol et al., 2020). Theory and practice-driven implementation models like the model developed by our team (explored fully below), coupled with more translation research in this area, are required.

14.6 Multidimensional, theory-informed implementation models are needed

To be effective, implementation approaches must be theory-informed (Heinsch, Wyllie, et al., 2021). Theories offer explanatory frameworks and formal heuristic devices that have the potential to move beyond the basic listing of individual facilitators and barriers to implementation, to capture the "the dynamic interaction between them" (Greenhalgh et al., 2017). As Damschroder notes, theory "enables knowledge to emerge out of seeming chaos," facilitating exploration of complex relationships and interdependencies between variables that unfold in diverse and changing contexts (Damschroder, 2020, p. 1). This is of paramount importance in digital therapeutics, which are characterized by a complicated interplay between patients, clinicians, the healthcare system, and the digital technology (Van Belle et al., 2017).

A multitude of theories and models have been articulated to inform and explain the implementation of digital tools. A recent systematic review identified 36 theories used to explain or inform the implementation of digital healthcare tools, but only a small number have been used repeatedly in this context (Heinsch, Wyllie, et al., 2021). These theories were concerned primarily with individual and interpersonal elements of digital acceptance. Less common were theories that capture the dense and intricate relationships and structures required to enact sustainable change. Given the increasing acknowledgment of the inherent complexity of implementation of digital therapeutics, implementation models should include both individual factors (such as motivation, attitudes, and behaviors) and the broader social, environmental, and structural factors that constrain or facilitate these individual factors.

Our team has developed an evidence-informed Integrated Translation and Engagement Model (ITEM, see Fig. 14.1) to drive the uptake of digital therapeutics into mental health and alcohol/other drug services across NSW, Australia. Based on the latest evidence for effective implementation, a consideration of individual, social, environmental, and structural factors,

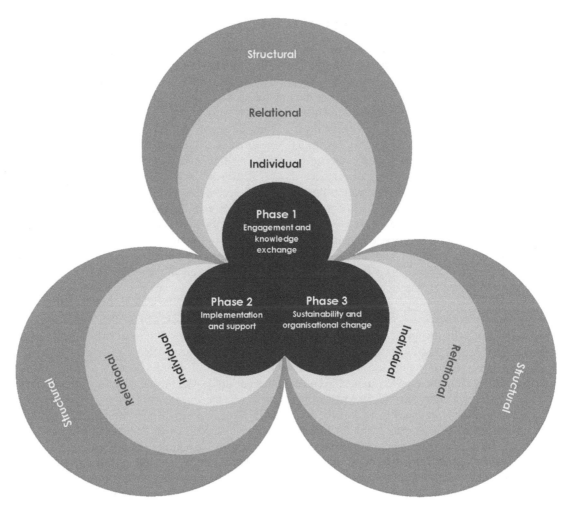

FIGURE 14.1 Integrated translation and engagement model (ITEM).

the ITEM synthesizes diverse theoretical approaches into a coherent, integrated model. It will engage consumers, clinicians, service leaders, and policymakers in all stages of development, design, conduct, and reporting, with additional proactive support for ongoing implementation provided on site.

ITEM attempts to reflect the multidimensional, dynamic, and relational nature of digital therapeutics implementation. Phase 1 develops partnerships with key stakeholders (persons with lived experience, service providers, policymakers) as a means of converting basic knowledge into practice to drive the implementation model (Allen-Meares et al., 2005). Based on this engagement, knowledge exchange and co-design will identify (1) gaps and opportunities for service provision, (2) capacity for digital therapeutics integration, and (3) facilitators and barriers to implementation. Healthcare environments are complex and assessing the setting prior to engaging in implementation is critical (Grimshaw et al., 2012), and coproduction leads to the production of relevant, useable knowledge for practice (Gredig & Sommerfeld, 2008). Phase 2 implementation and support serves to a bridge identified cultural barriers to uptake and encourages seamless integration of digital therapeutics across services. Translation is enhanced through "trigger encounters" that are informal, personal, and engender an emotional response (Heinsch, 2017). It is hypothesized that uptake is enhanced when information provision and training occur in a clear, user-friendly way (Morago, 2010), emphasizing implications, and relevance for practice (Dal Santo et al., 2002), and when practitioner confidence, experienced, and defensiveness is engaged directly (Heinsch, 2017). Phase 4 identifies "champions" to support implementation and advocate for organizational recognition of activities to establish a strong foundation for sustained implementation.

A rigorous implementation trial is required to determine the effectiveness of ITEM across a range of mental health, addiction, and healthcare settings.

14.7   A more responsive research and development cycle is needed

To optimize the finite resources in mental health prevention, early intervention, and service delivery, and to realize the potential for digital tools in this space, priority should be given to programs, models, and strategies that demonstrate evidence of their effectiveness. The reality is that developments in digital technologies are outpacing the rigorous evaluation of digital health interventions (Bucci et al., 2019), and more advanced, responsive methodologies are needed to keep up with the pace of digital technology development. Many of the digital tools evaluated in trials are technologically obsolete by the time they have been through traditional randomized controlled trials. Newer frameworks and methodologies, such as "just in time adaptive interventions" that use digital technology as the modality for intervention delivery, may be a solution (Bucci et al., 2019).

In recent years, several innovative trial designs have gained attention that is ideally suited to identify effective treatment components and may help to further improve the effectiveness of current state-of-the-art digital treatment. Multiphase optimization strategy is an overarching design and evaluation framework showing promise for digital interventions (Collins et al., 2007). Fractional factorial designs and micro-randomized trials are also improving the efficiency with which interventions can be disseminated and implemented (Baker et al., 2014).

The need for coproduction of digital tools with and for people with lived experience of mental health disorders is central to this (Bucci et al., 2019).

14.8   Summary

Digital health is a fast-developing technology that will "transform the way that health and social care is delivered" (2016). The great majority of people experience barriers that prevent access to treatment, aggravated by a lack of mental health specialists (Mohr et al., 2017). Further, no clear models currently exist for digital treatment integration into health services, and clinicians significantly underuse digital treatments in their clinical practice (Batterham et al., 2015). Integrating digital treatments into mental healthcare will require significant paradigm shifts in the mental health system and the way treatment for mental disorders is considered.

For digital therapeutics to achieve their potential to transform the ways we detect, treat, and prevent mental disorders, there is a clear need for continued research involving multiple stakeholders, and rigorous studies showing that these technologies can successfully drive measurable improvements in mental health outcomes (Naslund et al., 2019).

References

ABS. (2018). National health survey: first results, 2017–2018. Canberra, Australian Bureau of Statistics. **4364.0.55.001**. Available at: https://www.abs.gov.au/statistics/health/health-conditions-and-risks/national-health-survey-first-results/latest-release.

Affeltranger, B., Potvin, L., Ferron, C., Vandewalle, H., & Vallée, A. (2018, May). [Proportionate universalism: towards "real equality" of prevention in France]. *Sante Publique, 30*(1 Suppl), 13–24. https://doi.org/10.3917/spub.184.0013. (Universalisme proportionné : Vers une « égalité réelle » de la prévention en France.).

AGPC. (2020). Australian Government Productivity Commission Inquiry Report. *Mental Health, Final Report.* https://www.pc.gov.au/inquiries/completed/mental-health/report. (Accessed 30 August 2022).

Allen-Meares, P., Hudgins, C. A., Engberg, M. E., & Lessnau, B. (2005). Using a collaboratory model to translate social work research into practice and policy. *Research on Social Work Practice, 15*(1), 29–40.

AMS. (2016). Improving the health of the public by 2040: optimising the research environment for a healthier, fairer future. Available at: https://acmedsci.ac.uk/file-download/41399-5807581429f81.pdf.

Apolinario-Hagen, J., Harrer, M., Kahlke, F., Fritsche, L., Salewski, C., & Ebert, D. D. (2018, May 15). Public attitudes toward guided internet-based therapies: web-based survey study. *JMIR Ment Health, 5*(2), e10735. https://doi.org/10.2196/10735.

Baker, T. B., Gustafson, D. H., & Shah, D. (2014). How can research keep up with eHealth? Ten strategies for increasing the timeliness and usefulness of eHealth research. *Journal of Medical Internet Research, 16*(2), e36.

Batterham, P., & Calear, A. (2017a). Preferences for internet-based mental health interventions in an adult online sample: findings from an online community survey. *JMIR Mental Health, 4*(2), e26.

Batterham, P. J., & Calear, A. L. (2017b). Preferences for Internet-based mental health interventions in an adult online sample: findings from an online community survey. *JMIR Ment Health, 4*(2), e26. https://doi.org/10.2196/mental.7722.

Batterham, P. J., Neil, A. L., Bennett, K., Griffiths, K. M., & Christensen, H. (2008, February 2). Predictors of adherence among community users of a cognitive behavior therapy website. *Patient Prefer Adherence, 2*, 97–105. http://www.ncbi.nlm.nih.gov/pubmed/19920949.

Batterham, P. J., Sunderland, M., Calear, A. L., Davey, C. G., Christensen, H., Teesson, M., et al. (2015). Developing a roadmap for the translation of e-mental health services for depression: review. *Australian and New Zealand Journal of Psychiatry, 49*(9), 776–784.

Baumeister, H., Reichler, L., Munzinger, M., & Lin, J. (2014). The impact of guidance on Internet-based mental health interventions—a systematic review. *Internet Interventions, 1*(4), 205–215.

Berry, N., Lobban, F., Emsley, R., & Bucci, S. (2016). Acceptability of interventions delivered online and through mobile phones for people who experience severe mental health problems: a systematic review. *Journal of Medical Internet Research, 18*, e121. https://pubmed.ncbi.nlm.nih.gov/27245693/. doi:10.2196/jmir.5250. 27245693.

Boydell, K. M., Hodgins, M., Pignatiello, A., Teshima, J., Edwards, H., & Willis, D. (2014,). Using technology to deliver mental health services to children and youth: a scoping review. *Journal of the Canadian Academy of Child and Adolescent Psychiatry = Journal de l'Academie canadienne de psychiatrie de l'enfant et de l'adolescent, 23*(2), 87–99. http://europepmc.org/abstract/MED/24872824.

Bucci, S., Schwannauer, M., & Berry, N. (2019). The digital revolution and its impact on mental health care. *Psychology and Psychotherapy, 92*, 277–297.

Burgess, P. M., Pirkis, J. E., Slade, T. N., Johnston, A. K., Meadows, G. N., & Gunn, J. M. (2009). Service use for mental health problems: findings from the 2007 National Survey of Mental Health and Wellbeing. *Australian and New Zealand Psychiatry, 43*, 615–623.

Burns, J., Hickie, I., & Christensen, H. (2014). Strategies for adopting and strengthening e-mental health: a review of the evidence. The Sax Institute. Sydney, Australia, Commissioned on behalf of the Mental Health Commission of NSW. Available at: https://www.nswmentalhealthcommission.com.au/sites/default/files/old/assets/File/Report%20-%20The%20Sax%20Institute%20E-Mental%20Health%20Evidence%20Review%20cover%20page.pdf.

Casey, L. M., Wright, M.-A., & Clough, B. A. (2014). Comparison of perceived barriers and treatment preferences associated with internet-based and face-to-face psychological treatment of depression. *International Journal of Cyber Behavior, Psychology and Learning, Vol.4*(4), 16–22. doi:10.4018/ijcbpl.2014100102.

Christensen, H., Griffiths, K. M., & Farrer, L. (2009, 24). Adherence in internet interventions for anxiety and depression. *Journal of Medical Internet Research [Electronic Resource], 11*(2), e13. https://doi.org/10.2196/jmir.1194.

Christensen, H., Reynolds, J., & Griffiths, K. M. (2011). The use of e-health applications for anxiety and depression in young people: challenges and solutions. *Early Intervention in Psychiatry, 5*(Suppl 1), 58–62. http://onlinelibrary.wiley.com/doi/10.1111/j.1751-7893.2010.00242.x/abstract.

Collins, L. M., Murphy, S. A., & Strecher, V. (2007). The multiphase optimization strategy (MOST) and the sequential multiple assignment randomized trial (SMART): new methods for more potent eHealth interventions. *American Journal of Preventive Medicine, 32*(5 Suppl), S112–S118. https://doi.org/10.1016/j.amepre.2007.01.022.

Dal Santo, T., Goldberg, S., Choice, P., & Austin, M. (2002). Exploratory research in public social service agencies: an assessment of dissemination and utilization. *Journal of Sociology and Social Welfare, 29*(4), 59–81.

Damschroder, L. J. (2020). Clarity out of chaos: use of theory in implementation research. *Psychiatry Research, 283*, 112461. https://doi.org/10.1016/j.psychres.2019.06.036.

Deady, M., Mills, K. L., Teesson, M., & Kay-Lambkin, F. (2016). An online intervention for co-occurring depression and problematic alcohol use in young people: primary outcomes from a randomized controlled trial. *Journal of Medical Internet Research, 18*. doi:10.2196/jmir.5178.

de Girolamo, G., Cerveri, G., Clerici, M., Monzani, E., Spinogatti, F., Starace, F., et al. (2020). Mental Health in the Coronavirus Disease 2019 Emergency—The Italian Response. *JAMA Psychiatry, 77*(9), 974–976. https://doi.org/10.1001/jamapsychiatry.2020.1276.

Donkin, L., Christensen, H., Naismith, S. L., Neal, B., Hickie, I. B., & Glozier, N. (2011, August 5). A systematic review of the impact of adherence on the effectiveness of e-therapies. *Journal of Medical Internet Research [Electronic Resource], 13*(3), e52. https://doi.org/10.2196/jmir.1772.

DHS. (2017). Pharmaceutical benefits schedule: mental health-related prescriptions. https://www.aihw.gov.au/getmedia/d35a03a3-1500-4775-b3e1-9a99f3168bb8/Mental-health-related-prescriptions-2016-17.pdf.aspx. Access date: 30 August 2022.

Donkin, L., & Glozier, N. (2012). Motivators and motivations to persist with online psychological interventions: a qualitative study of treatment completers. *Journal of Medical Internet Research [Electronic Resource], 14*(3), e91. https://doi.org/10.2196/jmir.2100.

Erbe, D., Eichert, H. C., Riper, H., & Ebert, D. D. (2017, September 15). Blending face-to-face and internet-based interventions for the treatment of mental disorders in adults: systematic review. *Journal of Medical Internet Research [Electronic Resource], 19*(9), e306. https://doi.org/10.2196/jmir.6588.

Farvolden, P., Denisoff, E., Selby, P., Bagby, R. M., & Rudy, L. (2005, March 26). Usage and longitudinal effectiveness of a Web-based self-help cognitive behavioral therapy program for panic disorder. *Journal of Medical Internet Research [Electronic Resource], 7*(1), e7. https://doi.org/10.2196/jmir.7.1.e7.

Fortuna, K. L., Venegas, M., Umucu, E., Mois, G., Walker, R., & Brooks, J. M. (2019). The future of peer support in digital psychiatry: promise, progress, and opportunities. *Current Treatment Options in Psychiatry, 6*(3), 221–231.

Frawley, T., van Gelderen, F., Somanadhan, S., Coveney, K., Phelan, A., Lynam-Loane, P., et al. (2021). The impact of COVID-19 on health systems, mental health and the potential for nursing. *Irish Journal of Psychological Medicine, 38*(3), 220–226. https://doi.org/10.1017/ipm.2020.105.

Gredig, D., & Sommerfeld, P. (2008). New proposals for generating and exploiting solution-oriented knowledge. *Research on Social Work Practice, 18*(4), 292–300.

Greenhalgh, T., Wherton, J., Papoutsi, C., Lynch, J., Hughes, G., A'Court, C., et al. (2017). Beyond adoption: a new framework for theorizing and evaluating nonadoption, abandonment, and challenges to the scale-up, spread, and sustainability of health and care technologies. *Journal of Medical Internet Research [Electronic Resource], 19*(11), e367. https://doi.org/10.2196/jmir.8775.

Grimshaw, J. M., Eccles, M. O., Lavis, J. N., Hill, S. J., & Squires, J. E. (2012). Knowledge translation of research findings. *Implementation Science, 7*(1), 50–67.

Handley, T. E., Kay-Lambkin, F. J., Inder, K. J., Attia, J. R., Lewin, T. J., & Kelly, B. J. (2014). Feasibility of internet-delivered mental health treatments for rural populations. *Social Psychiatry and Psychiatric Epidemiology, 49*(2), 275–282. http://link.springer.com/article/10.1007%2Fs00127-013-0708-9.

Heinsch, M. (2017). Exploring the potential of interaction models of research use for social work. *British Journal of Social Work, 48*(2), 468–486. https://doi.org/10.1093/bjsw/bcx034.

Heinsch, M., Geddes, J., Sampson, D., Brosnan, C., Hunt, S., Wells, H., et al. (2021). Disclosure of suicidal thoughts during an e-mental health intervention: Relational ethics meets actor-network theory. *Ethics & Behavior, 31*(3), 151–170. https://doi.org/10.1080/10508422.2019.1691003.

Heinsch, M., Wyllie, J., Carlson, J., Wells, H., Tickner, C., & Kay-Lambkin, F. (2021). Theories informing eHealth implementation: systematic review and typology classification. *Journal of Medical Internet Research [Electronic Resource], 23*(5), e18500. https://doi.org/10.2196/18500.

Henson, P., Wisniewski, H., Hollis, C., Keshavan, M., & Torous, J. (2019). Digital mental health apps and the therapeutic alliance: initial review. *British Journal of Psychiatry Open, 5*(1), e15.

Holmes, E. A., Ghaderi, A., Harmer, C. J., Ramchandani, P. G., Cuijpers, P., Morrison, A. P., et al. (2018). The Lancet Psychiatry Commission on psychological treatments research in tomorrow's science. *Lancet Psychiatry,, 5*, 237–286.

Holmes, E. A., O'Connor, R. C., Perry, V. H., Tracey, I., Wessely, S., Arseneault, L., et al. (2020). Multidisciplinary research priorities for the COVID-19 pandemic: a call for action for mental health science. *The Lancet Psychiatry, 7*(6), 547–560. https://doi.org/10.1016/S2215-0366(20)30168-1.

Iorfino, F., Prodan, A., Piper, S., Occhipinti, J., Schmiede, A., LaMonica, H., et al. (2021). Best care, first time: optimising youth mental health services using digital technologies. The University of Sydney, Australia. Available at: https://researchdirect.westernsydney.edu.au/islandora/object/uws:62529/.

Jakob, R., Harperink, S., Rudolf, A. M., Fleisch, E., Haug, S., Mair, J. L., Salamanca-Sanabria, A., & Kowatsch, T. (2021). Factors Influencing Adherence to mHealth Apps for Prevention or Management of Noncommunicable Diseases: Systematic Review. *Journal of Medical Internet Research, 24*(5), e35371. https://doi.org/10.2196/35371.

Jorm, A. F. (2018). Australia's 'Better Access' scheme: has it had an impact on population mental health? *Australian and New Zealand Journal of Psychiatry, 52*(11). doi:10.1177/0004867418804066.

Kay-Lambkin, F. (2009). Technology and innovation in the psychosocial treatment of methamphetamine use, risk and dependence. *Drug and Alcohol Review, 27*(May), 318–325.

Kay-Lambkin, F., Baker, A., Lewin, T., & Carr, V. (2011a). Acceptability of a clinician-assisted computerized psychological intervention for comorbid mental health and substance use problems: treatment adherence data from a randomized controlled trial. *Journal of Medical Internet Research, 13*(1), e11 PMID: 21273184. https://www.jmir.org/2011/1/e11/. doi:10.2196/jmir.1522.

Kay-Lambkin, F., Baker, A., Lewin, T. J., & Carr, V. J. (2009). Computer-based psychological treatment for comorbid depression and problematic alcohol and/or cannabis use: a randomized controlled trial of clinical efficacy. *Addiction, 104*, 378–388. http://onlinelibrary.wiley.com/doi/10.1111/j.1360-0443.2008.02444.x/abstract.

Kay-Lambkin, F., Baker, A. L., Kelly, B., & Lewin, T. J. (2011b). Clinician-assisted computerised versus therapist-delivered treatment for depressive and addictive disorders: a randomised controlled trial. *Medical Journal of Australia, 195*, S44–S50. https://www.mja.com.au/system/files/issues/195_03_010811/kay10941_fm.pdf.

Kay-Lambkin, F. J., Baker, A., Hunt, S., Spring, B., & RIchmond, R. (2016). iHeLP: A pilot trial of an online program for smokers with depression. In *World Congress of Behavior and Cognitive Therapies*.

Kay-Lambkin, F. J., Baker, A. L., Geddes, J., Hunt, S. A., Woodcock, K., Teesson, M., et al. (2015). The iTreAD project: A study protocol for a randomised controlled clinical trial of online treatment and social networking for binge drinking and depression in young people. *Bmc Public Health [Electronic Resource], 15*, 1025.

Kay-Lambkin, F. J., Baker, A. L., Healey, A., Wolfe, S., Simpson, A., Brooks, M., et al. (2012). Study protocol: a dissemination trial of computerized psychological treatment for depression and alcohol/other drug use comorbidity in an Australian clinical service. *Bmc Psychiatry [Electronic Resource], 12*(1), 77.

Kay-Lambkin, F. J., Baker, A. L., Kelly, B. J., & Lewin, T. J. (2012). It's worth a try: the treatment experiences of rural and urban participants in a randomized controlled trial of computerized psychological treatment for comorbid depression and alcohol/other drug use. *Journal of Dual Diagnosis, 8*(4), 262–276.

Kay-Lambkin, F. J., Baker, A. L., Lewin, T. J., & Carr, V. J. (2009). Computer-based psychological treatment for comorbid depression and problematic alcohol and/or cannabis use: A randomized controlled trial of clinical efficacy. *Addiction, 104*(3), 378–388.

Kay-Lambkin, F.J., Baker, A.L., McKetin, R., & Lee, N. (2010), Stepping through treatment: Reflections on an adaptive treatment strategy among methamphetamine users with depression. Drug and Alcohol Review, 29: 475–482. https://doi.org/10.1111/j.1465-3362.2010.00203.x.

Kay-Lambkin, F. J., Baker, A. L., Palazzi, K., Lewin, T. J., & Kelly, B. J. (2017). The moderating effect of perfectionism, need for approval, and therapeutic alliance on eHealth and traditionally-delivered treatments for comorbid depression and alcohol/cannabis use problems. *International Journal of Behavioral Medicine, 24*(5), 728–739.

Kay-Lambkin, F.J., Gilbert, J., Pedemont, L., Sunderland, M., Dalton, H., et al. (2018). Prevention and early intervention for people aged 18 and over with, or at risk of, mild to moderate depression and anxiety: An Evidence Check rapid review brokered by the Sax Institute (www.saxinstitute.org.au) for Beyond Blue, 2018. Available at: https://www.beyondblue.org.au/docs/default-source/about-beyond-blue/policy-submissions/mild-moderate-depression-and-anxiety-in-adults_final-2.pdf?sfvrsn=d182bcea_6.

Kay-Lambkin, F. J., Simpson, A., Bowman, J., & Childs, S. (2014). Dissemination trial of computer-based psychological treatment in a Drug and Alcohol Clinical Service: predictors of technology integration. *Addiction Science and Clinical Practice, 9*, 15. http://www.ascpjournal.org/content/9/1/15.

Khera, N., Mau, L.-W., Denzen, E. M., Meyer, C., Houg, K., Lee, S. J., et al. (2018). Translation of clinical research into practice: an impact assessment of the results from the blood and marrow transplant clinical trials network protocol 0201 on unrelated graft source utilization. *Biology of Blood and Marrow Transplantation, 24*(11).

Lattie, E. G., Graham, A. K., Hadjistavropoulos, H. D., Dear, B. F., Titov, N., & Mohr, D. C. (2019). Guidance on defining the scope and development of text-based coaching protocols for digital mental health interventions. *Digital Health, 5*. 2055207619896145 https://doi.org/10.1177/2055207619896145.

Lim, M. H., & Penn, D. L. (2018). Using digital technology in the treatment of schizophrenia. *Schizophrenia Bulletin, 44*(5), 937–938. https://doi.org/10.1093/schbul/sby081.

McDaid, S., Kousoulis, A., Thorpe, L., Zamperoni, V., Pollard, A., Wooster, E. (2020). Tackling social inequalities to reduce mental health problems: How everyone can flourish equally. Mental Health Foundation, London, United Kingdom. Available at: https://www.mentalhealth.org.uk/sites/default/files/2022-04/MHF-tackling-inequalities-report.pdf.

McHugh, R. K., Whitton, S. W., Peckham, A. D., Welge, J. A., & Otto, M. W. (2013). Patient preference for psychological vs pharmacologic treatment of psychiatric disorders: A meta-analytic review. *Journal of Clinical Psychiatry, 74* 595–596–592.

Meurk, C., Leung, J., Hall, W., Head, B. W., & Whiteford, H. (2016). Establishing and Governing e-Mental Health Care in Australia: a systematic review of challenges and a call for policy-focussed research. *Journal of Medical Internet Research, 18*(1), e10. https://www.jmir.org/2016/1/e10.

Mohr, D. C., Cuijpers, P., & Lehman, K. (2011). Supportive accountability: a model for providing human support to enhance adherence to eHealth interventions. *Journal of Medical Internet Research [Electronic Resource], 13*(1), e30. https://doi.org/10.2196/jmir.1602.

Mohr, D. C., Lyon, A. R., Lattie, E. G., Reddy, M., & Schueller, S. M. (2017). Accelerating digital mental health research from early design and creation to successful implementation and sustainment. *Journal of Medical Internet Research, 19*, e153. https://www.jmir.org/2017/5/e153/. doi:10.2196/jmir.7725.

Mohr, D. C., Siddique, J., Ho, J., Duffecy, J., Jin, L., & Fokuo, J. K. (2010, August). Interest in behavioral and psychological treatments delivered face-to-face, by telephone, and by internet. *Annals of Behavioral Medicine, 40*(1), 89–98. https://doi.org/10.1007/s12160-010-9203-7.

Mol, M., van Genugten, C., Dozeman, E., van Schaik, D. J. F., Draisma, S., Riper, H., et al. (2020). Why uptake of blended internet-based interventions for depression is challenging: a qualitative study on therapists' perspectives. *Journal of Clinical Medicine, 9*(1). doi:10.3390/jcm9010091.

Morago, P. (2010). Dissemination and implementation of evidence-based practice in the social services: A UK survey. *Journal of Evidence-Based Social Work, 7*(5), 452–465.

Musiat, P., Goldstone, P., & Tarrier, N. (2014, 11). Understanding the acceptability of e-mental health–attitudes and expectations towards computerised self-help treatments for mental health problems. *Bmc Psychiatry [Electronic Resource], 14*, 109. https://doi.org/10.1186/1471-244X-14-109.

Naslund, J. A., Gonsalves, P. P., Gruebner, O., Pendse, S. R., Smith, S. L., Sharma, A., et al. (2019). Digital innovations for global mental health: opportunities for data science, task sharing, and early intervention. *Current Treatment Options in Psychiatry, 6*. doi:10.1007/s40501-019-00186-8.

Nuq, P. A., & Aubert, B. (2013). Towards a better understanding of the intention to use eHealth services by medical professionals: the case of developing countries. *International Journal of Healthcare Management, 6*(4), 217–236. https://doi.org/10.1179/2047971913Y.0000000033.

Sinha, R. (2007). The role of stress in addiction relapse. *Current Psychiatry Reports, 9*(5), 388–395. doi:10.1007/s11920-007-0050-6.

Smith, A. C., Thomas, E., Snoswell, C. L., Haydon, H., Mehrotra, A., Clemensen, J., et al. (2020). Telehealth for global emergencies: implications for coronavirus disease 2019 (COVID-19). Journal of Telemedicine and Telecare, 26(5), 309–313. https://doi.org/10.1177/1357633x20916567.

Spolander, G., Engelbrecht, L., Martin, L., Strydom, M., Pervova, I., Marjanen, P., et al. (2014). The implications of neoliberalism for social work: Reflections from a six-country international research collaboration. *International Social Work, 57*(4), 301–312. https://doi.org/10.1177/0020872814524964.

Stapinski, L. A., Prior, K., & Newton, N. C. (2019). Protocol for the inroads study: a randomized controlled trial of an internet-delivered, cognitive behavioral therapy–based early intervention to reduce anxiety and hazardous alcohol use among young people. *JMIR Research Protocols, 8*(4), e12370.

Statista. (2021). Number of smartphone users worldwide from 2016-2021. https://www.statista.com/statistics/330695/number-of-smartphone-users-worldwide/. Access Date: 30 August 2022.

Statista. (2022). Leading countries based on number of Twitter users as of January 2022. https://www.statista.com/statistics/242606/number-of-active-twitter-users-in-selected-countries/. Access Date: 30 August 2022.

Stern, A. D., Hagen, H. M. J., Brönneke, J. B., & Debatin, J. F. (2020). Want to see the future of digital health tools? look to Germany. *Harvard Business Review*. https://hbr.org/2020/12/want-to-see-the-future-of-digital-health-tools-look-to-germany. (Accessed 30 August 2022).

Tait, R. J., McKetin, R., Kay-Lambkin, F., Bennett, K., Tam, A., Bennett, A., et al. (2012). Breakingtheice: A protocol for a randomised controlled trial of an internet-based intervention addressing amphetamine-type stimulant use. *Bmc Psychiatry [Electronic Resource], 12*(1), 67.

Tait, R. J., McKetin, R., Kay-Lambkin, F. J., Carron-Authur, B., Bennett, A., Bennett, K., et al. (2015). A randomised controlled trial of a web-based intervention for users of amphetamine-type stimulants: Six month outcomes; Part 2. *Journal of Medical Internet Research, 17*(4), e105. https://doi.org/10.2196/jmir.3778.

Van Belle, S., van de Pas, R., & Marchal, B. (2017). Towards an agenda for implementation science in global health: there is nothing more practical than good (social science) theories. *BMJ Global Health, 2*(2), e000181. https://doi.org/10.1136/bmjgh-2016-000181.

van der Kleij, R. M. J. J., Kasteleyn, M. J., Meijer, E., Bonten, T. N., Houwink, E. J. F., Teichert, M., et al. (2019). SERIES: EHealth in primary care. Part 1: concepts, conditions and challenges. *European Journal of General Practice, 25*(4), 179–189. https://doi.org/10.1080/13814788.2019.1658190.

Wentzel, J., van der Vaart, R., Bohlmeijer, E. T., & van Gemert-Pijnen, J. E. W. C. (2016). Mixing online and face-to-face therapy: how to benefit from blended care in mental health care. *JMIR Mental Health, 3*(1), e9. https://doi.org/10.2196/mental.4534.

Whiteford, H. A., Degenhardt, L., Rehm, J., Baxter, A. J., Ferrari, A. J., Erskine, H. E., et al. (2013). Global burden of disease attributable to mental and substance use disorders: Findings from the Global Burden of Disease Study 2010. *The Lancet, 382*(9904), 1575–1586. https://doi.org/10.1016/S0140-6736(13)61611-6.

WHO. (2011). Mental health atlas: 2011.

Wykes, T., Lipshitz, J., & Schueller, S. M. (2019). Towards the design of ethical standards related to digital mental health and all its applications. *Current Treatment Options in Psychiatry, 6*(3), 232–242.

Yao, H., Chen, J.-H., & Xu, Y.-F. (2020). Patients with mental health disorders in the COVID-19 epidemic. *The Lancet Psychiatry, 7*(4), e21. https://doi.org/10.1016/S2215-0366(20)30090-0.

Chapter 15

Privacy and security in digital therapeutics

Leysan Nurgalieva[a] and Gavin Doherty[b]

[a]*Department of Computer Science, Aalto University, Espoo, Finland,* [b]*School of Computer Science and Statistics, Trinity College Dublin, Ireland*

15.1 Introduction

Digital health products and services are often put forward as instrumental to more patient-centered and transparent care processes (Okun & Wicks, 2018), and to making healthcare information more understandable and usable for patients. In particular, digital mental health (DMH) has become an accessible alternative or adjunct to face-to-face therapy enabling patients to become more proactive in managing their health (Smelror, Bless, Hugdahl, Hugdahl, & Agartz, 2019).

Digital therapeutics necessarily involve storing and processing sensitive information, potentially by several entities, which requires the use of safeguards. Furthermore, evaluating and improving these products and services also requires access to patients' data. Thus, while digital therapeutics show promise for an enhanced healthcare landscape, the associated capture and use of sensitive data have heightened awareness and concern about the privacy and security of personal information in such technologies.

However, safeguarding measures such as deidentification of user data provided to third parties can potentially be circumvented. For example, users' privacy can be violated by combining deidentified information with data from other sources (e.g., social media or public records) (Grundy, Held, & Bero, 2017), which introduces risks ranging from unvetted or intrusive targeted advertising to inferences about an individual's behavior and health condition (Martinez-Martin & Kreitmair, 2018). Security risks can arise if digital health services are compromised by unintended breaches or losses, malicious hacking (Grundy et al., 2017; Grundy et al., 2019; Morera, de la Torre Díez, Garcia-Zapirain, López-Coronado, & Arambarri, 2016; Sampat & Prabhakar, 2017). These threats are important to consider and address in digital therapeutics, and they can be even more crucial within the mental health domain due to the additional sensitivity of the data. If compromised, breaches can lead to stigma (Henderson, Evans-Lacko, & Thornicroft, 2013), discrimination, or delays in seeking effective treatment (Radovic et al., 2016; Wykes & Schueller, 2019).

In the age of digital health, it is no longer a binary question of whether data should be shared or not, but what data and how it can be done in a way that protects what is valuable and vulnerable. Privacy implies the individual's right to maintain control over and be free from intrusion into their private life, as defined by the European Convention on Human Rights and other national laws (Diggelmann & Cleis, 2014), and it should be ensured in DMH products and services.

Next, we discuss topics related to privacy and security in digital therapeutics and DMH, including common risks and relevant data protection regulations in the United States, Europe, and other regions. We then provide an overview of existing security and privacy evaluation methods, frameworks, and tools for digital therapeutics and DMH, as well as design recommendations to address identified risks and vulnerabilities.

15.2 Background on privacy and security in digital therapeutics

15.2.1 Digital health ecosystem

Digital health products and services are increasingly endorsed by healthcare organizations, governments, and recommended by clinicians as an inexpensive and accessible adjunct to therapy or as additional support for patients with various health conditions. Digital therapeutics enable healthcare professionals to have a wider range of communication channels with patients, and greater access to data, creating opportunities for more precise diagnostics and more personalized healthcare delivery. The use and importance of such tools in patients' lives are growing, and the technological sophistication of smartphones and other handheld, wearable, and IoT devices have enabled the delivery of new functionalities and interventions "in the moment" (Balaskas, Schueller, Cox, & Doherty, 2021). Besides email, text messaging, and electronic health records (EHRs), delivery modalities have come to include chatbots, virtual assistants, telepsychology/telemental health therapy with cloud-based storage, and smartphone health apps. They help provide mental health support in many ways: by increasing users' understanding of their conditions (Radovic et al., 2016), supporting adherence to interventions (Thach, 2018), or even making it possible to provide prescription drugs with embedded sensors tracking medication compliance (Aldeer, Martin, & Howard, 2018). However, consumers' appetite for such innovations and data protection concerns are growing in tandem, hampering acceptance.

Regarding the risks, researchers agree that by using digital therapeutics, people with mental health difficulties face greater exposure to privacy breaches. Their data should be protected (Rosenfeld, Torous, & Vahia, 2017) and privacy policies should be easily understandable and support users in making informed decisions (Powell, Singh, & Torous, 2018).

15.2.2 Security and privacy risks in digital mental health

The defining feature of digital health concerns data rather than technology. From a wide range of sources, digital health products generate large sets of patient data. One of the consequences of these technological advancements is the increase in risks for digital security and privacy.

Security risks can arise from technical system vulnerabilities and result in breaching the confidentiality of patients' data (Sampat & Prabhakar, 2017), which can lead to financial losses, discrimination, stress, dissatisfaction (Powell et al., 2018), or have indirect effects through delays in seeking effective treatment due to confidentiality concerns (Radovic et al., 2016; Wykes & Schueller, 2019). Other security risks can arise due to malicious hacking or through commercial data-sharing practices (Grundy et al., 2017; Grundy et al., 2019; Morera, de la Torre Díez, Garcia-Zapirain, López-Coronado, & Arambarri, 2016; Sampat & Prabhakar, 2017), as well as hosting devices simply being stolen (Kotz, 2011).

Disclosure of deidentified patient information to third parties can pose a risk to privacy, as combining the information with data from other sources (e.g., social media or public records) may enable identification of individuals (Grundy et al., 2017; Househ, Grainger, Petersen, Bamidis, & Merolli, 2018). The related risks range from unvetted or intrusive targeted advertising to inferences about an individual's behavior and health condition, which might affect employment or promotion prospects (Martinez-Martin & Kreitmair, 2018; Parker, Halter, Karliychuk, & Grundy, 2019).

Security and privacy risks are harmful not only to the patients whose privacy is threatened but also to the long-term development of digital therapeutics and associated benefits to patient health. The loss or theft of patient data can result in licensing sanctions or civil lawsuits for providers and healthcare professionals (Taube, 2013), as well as reputational damage or compromised credibility (Parker et al., 2017).

15.2.2.1 Risks with electronic health records

A primary enabler of digital therapeutics has been the adoption and availability of EHRs, and the development of patient portals. Clinical measures and patients' data generated and collected by digital therapeutics are exported to EHRs, which are then stored and shared within global networks, distributed databases, and in the cloud (Bauer et al., 2017). While digital health tools provide patients with new forms of access to their data, increased circulation of patient information also means increased security risks and risks of loss of control over its dissemination (Ahlfeldt & Huvala, 2014; Rezaeibagha, Win, & Susilo, 2015; Xhafa et al., 2015). For instance, privacy threats can arise when mental health records become available to all of the patient's providers in a network, regardless of whether it is necessary for the patient's care (Clemens, 2012), or from incidents of unauthorized forwarding of patient e-mails (Van Allen & Roberts, 2011).

Other data protection risks can be due to the transmission of EHRs over insecure communication channels, enabling adversaries to capture sensitive information related to the patient's medical history and construct "patient profiles" (Yang, Kim, & Mtonga, 2015).

Health records can also be available through "proxy access" letting patients share their health information with family members and caregivers. Shared e-access to personal EHRs has both positive and negative implications: an enhanced partnership between formal and informal caregivers of patients, and better inclusion of the latter into the care process (Kelly, Coller, & Hoonakker, 2018), but also value tensions (Cajander & Grünloh, 2019), and privacy and security issues (Bauer et al., 2017). Controls can thus be provided for changing "dimensions of information sharing (what information, to whom, when, how much, and under what circumstances)" (Crotty et al., 2015, p.1493).

15.2.2.2 Risks with telepsychology

Telepsychology/telemental health therapy is another more traditional way of conducting health interventions remotely, ranging from psychotherapy to medication management.

Despite the benefits, such services present unique risks to client confidentiality. For instance, telemental health therapy sessions may be unintentionally overheard or even maliciously observed by outside parties, for instance, by other individuals in the space (Madigan, Racine, Cooke, & Korczak, 2021). Furthermore, sessions might be recorded for the purposes of the observational coding, which can provide important insights into the process and outcome of the therapy sessions (Høglend, 2014), but also introduces additional privacy concerns (Xiao, Imel, Georgiou, Atkins, & Narayanan, 2015).

Therapy sessions delivered through videoconferencing have been used with increasing frequency as a delivery mode for treating people with mental health conditions (Richardson, Christopher Frueh, Grubaugh, Egede, & Elhai, 2009). While specialized digital health communication tools equipped with advanced security features are available (Bellazzi, Montani, Riva, & Stefanelli, 2001), consumer videoconference services such as Google Hangouts or Skype are still commonly used by clients and providers, due to their familiarity and accessibility (Martin, 2013). However, while professionals and patients appreciate their convenience, consumer services might not be optimal from the regulatory compliance (Martin, 2013; Watzlaf, Moeini, & Firouzan, 2010) and security point of view (Churcher, 2012; Fleishman, 2012) and hence not as suitable for telemental health assessment.

15.2.2.3 Novel risks related to monitoring and self-tracking

Health apps for smartphones and tablets paired with wearable devices such as smartwatches, smart earplugs, rings, or other connected devices ("wearables") have already had a considerable impact on the digital therapeutics market, with further developments of the "Internet of Things" raising the prospect of both further data sources and new avenues for delivering interventions.

By continuously collecting massive amounts of health data through various embedded sensors and analyzing it in real-time, wearables enable developers to "make their applications more context-aware, radically improving user experience" (Alepis & Patsakis, 2017).

While both mental health professionals and patients are becoming increasingly interested in using such devices as adjuncts to therapies or as an alternative to self-help options (Bruno et al., 2018; Hunkin, King, & Zajac, 2020), ad-hoc use poses significant privacy risks. For instance, collected sensitive information, such as real-time video or location information can be leaked (Alepis & Patsakis, 2017). Moreover, in contrast to clinical or medical devices for professional usage, in which EHRs are created and managed by healthcare providers (hospitals and other clinical organizations), health data in consumer wearables is collected and managed without the involvement of physicians (Becker, 2018). This can negatively affect the level of regulation and protection of collected health data.

As consumer wearable analysis outcomes are immediately available in digital form, they can be easily shared with third parties: intentionally for financial benefits or unintentionally and indirectly by using third-party services such as libraries, analytics, and customer service support (Parker et al., 2019; Razaghpanah et al., 2018). Here too, deidentification does not always guarantee users' privacy, as app user data can be cross-linked with data from other sources and become easily re-identifiable. Sharing aggregated consumer data from apps and other digital sources can allow the derivation of insights into societal patterns and behaviors (Binns et al., 2018), which can lead to proprietary algorithms used to predict future behavior and risk to specific individuals (Pasquale, 2015).

Health apps are often used to access data that comes from wearables and help users to improve their knowledge about themselves, but also the providers' knowledge about the users. Privacy risks can arise due to data being shared with app producers, including "behaviors and information," such as "username and password, contact information, age, gender, location, International Mobile Equipment Identity, and phone number" (Lustgarten, Garrison, Sinnard, & Flynn, 2020, p. 26). Another challenge for users' privacy is the difficulty of opting out of data sharing and aggregation during app use or after unsubscribing/uninstalling the app (Kuntsman, Miyake, & Martin, 2019).

Finally, privacy policies are a recognized weak spot in health apps and digital health products and services in general. Besides not always being available (Robillard et al., 2019), privacy policies and terms often do not consensually request users for access to their data, applying a "take it or leave it" policy, or simply lack comprehensibility (Parker et al., 2019; Powell et al., 2018).

15.2.2.4 Conversational agents

As we adopt novel digital products and services into our daily lives, those very same tools become widely used within our healthcare. Powered by advances in machine learning (ML), speech recognition, and natural language processing, conversational agents (CAs) are redefining the way we access the Web, data, and apps. This technological category often includes chatbots, embodied CAs, and intelligent voice assistants such as Apple Siri and Google Assistant.

As part of broader progress toward realizing the concept of ubiquitous computing, and eventually providing a more human-like experience, they have been increasingly deployed in healthcare applications. In particular, their potential in mental health has been illustrated, considering one of the top requests to Alexa (a CA in Amazon smart speakers) during the summer of 2017 was "Alexa, help me relax" (Torous, 2017).

However, unlike in-person visits to a clinician, CAs often cannot guarantee the same degree of privacy and confidentiality (Vaidyam, Wisniewski, Halamka, Kashavan, & Torous, 2019). For instance, voice assistants often operate in the background and might not necessarily need a specific keyword or command to start, as technological advances allow them to listen continuously, as in Google "follow-up mode" when requests are implemented without repeating the wake-word (Tabassum et al., 2019). While the future outlook is for such assistants to become even more seamless, as vendors advance the technology to extract keywords from ambient speech automatically (Simpson, 2017), the security and privacy risks are concerning. They range from unwanted targeted advertising (Malkin et al., 2019) to remote control and manipulation of voice commands, which can significantly expose users (Alepis & Patsakis, 2017). Compromised cybersecurity of CAs can allow malicious interference in patient-agent interactions and even result in harmful recommendations (McGreevey, Hanson, & Koppel, 2020).

To provide rich interaction and functionality options, CAs often require a large set of software permissions as well as third-party applications (Alepis & Patsakis, 2017). As users are often not fully aware of the risks the requested permissions pose, they might be unaware of the sensitive information they are sharing by interacting with CAs (Felt et al., 2012). As CAs can be easily connected to social media, the lack of awareness or attention to privacy settings can result in social media oversharing (Koh et al., 2013).

15.2.2.5 Machine learning

Technological advancement in DMH and the increasing amounts and variety of automatically collected patient and interaction data enables new methodological possibilities for improving our understanding of human behaviors and optimizing health outcomes (Baltierra et al., 2016). Similarly to CAs, the fields of data mining and ML provide advanced computational methods to construct systems that can automatically learn from behavioral data (Shatte, Hutchinson, & Teague, 2019) and to assist in the context of mental health. This may include exploratory work on understanding users, intended to support future efforts to personalize treatment (Wilkinson, Arnold, Murray, van Smeden, & Carr, 2020), or research with more direct applications, such as prediction of treatment outcomes (Chekroud et al., 2021).

ML methods such as sentiment and linguistic analysis can be applied to understand and monitor the behaviors of healthcare professionals, for instance, identifying the communication practices of "more" and "less" successful supporters based on client outcomes (Chikersal et al., 2020). Such capabilities require careful consideration of how these technologies are integrated into both research and clinical practice, and appropriate safeguards for the privacy of both patients and healthcare professionals. As digital interventions become used at a population scale, the debate on the ethics of these technologies may come to draw on public health ethics, which entails respect for individual autonomy, liberty, privacy and confidentiality (Childress et al., 2002), but is also concerned with resolving conflicts between ethical considerations, such as providing benefits to patients and privacy.

In mental health, text-mining has been used to better understand discussion topics in online mental health and online support communities and aid moderators in their work practices, for example, by identifying helpful and unhelpful comments (Chancellor, 2018) or predict suicide risks from social media messages drawn from one-on-one conversations intended to be private (Nobles, Glenn, Kowsari, Teachman, & Barnes, 2018). However, such interventions can be perceived as unwanted inference of public data (Thieme, Belgrave, & Doherty, 2020), producing highly sensitive mental health data from less sensitive data, fueling privacy concerns (Shaik & Inkpen).

15.2.2.6 Virtual reality in mental health

Similarly to voice assistants which have been gradually entering the healthcare domain, virtual and augmented reality (VR and AR) have the potential to treat psychological conditions in novel ways. VR allows the creation of simulated environments that afford a high degree of control in the provision of therapeutic experiences (Mohr, Burns, Schueller, Clarke, & Klinkman, 2013). In the mental health domain, immersive virtual reality exposure therapy may be effective for conditions such as anxiety disorder (Meyerbröker & Emmelkamp, 2010). However, through VR technologies, very personal information can be gathered, including information about the habits, interests, and tendencies of VR users, which together can constitute a person's distinct "kinematic fingerprint" (Spiegel, 2018). The extent and richness of captured data can threaten personal privacy and present a danger related to manipulation of users' beliefs, emotions, and behaviors and can be used to infer medical conditions (O'Brolcháin et al., 2016).

Although VR is marketed as a safer interaction medium where, for instance, threatening aspects of reality can be explored in a safe environment and one's identity can be masked by using avatars (Gorini, Gaggioli, & Riva, 2008), arguably it can be a more pervasive intrusion into personal privacy and a threat to personal autonomy. For instance, virtual reality social networks can increase the risk of users sharing personal and sensitive information with unknown and untrusted third parties or being harassed (Adams et al., 2018). AR poses additional risks to users' security and privacy, such as "bystander effects" (Denning, Dehlawi, & Kohno, 2014) or malicious interference displaying unwanted or harmful content (Lebeck, Ruth, Kohno, & Roesner, 2017).

15.3 Relevant security and privacy regulations

One of the reasons for the prevalence of digital health products and services that put the security and privacy of consumers at risk is the lack of compliance with existing regulations.

There are many regulations—regional and global—that are relevant to digital therapeutics, including health apps and "software as a medical device." By setting out the obligations for parties processing personal data, including the legal bases for processing, data protection principles, and accountability measures, they establish a legislative basis for protecting the fundamental rights of digital health consumers. However, it is not always easy to recognize which of them apply to a particular health application.

15.3.1 United States regulations

Currently, several key United States (US) regulations play an important role when it comes to digital therapeutics.

The regulatory framework of the US Food and Drug Administration (FDA) has been traditionally concerned with computerized devices intended for medical use. The "medical device" label (as assigned by the manufacturer) means that the product—hardware or a software application—is recognized by official bodies and intended for use in "the diagnosis of disease or other conditions, or in the cure, mitigation, treatment, or prevention of disease," and conformity to regulation must be shown (U.S. Food and Drug Administration). Hence, the determination of whether a product is to be considered a medical device is based on the product's intended purpose, which should be defined as a "medical purpose" to fall within the medical device definition.

However, it is not always easy to understand whether digital therapeutics fall into either "medical" or "health" categories, as there is often no clear dividing line between them. The differentiation here is critical since both classes have different kinds of inherent risks and limitations (Albrecht, Pramann, & von Jan, 2015), which indicates the need for a set of clear and applicable privacy and security evaluation criteria that go beyond categorization but rather deal with the data practices within individual digital health products and services. Digital therapeutics not intended for medical use are not currently regulated by the FDA in the United States, for example, and are subject to very little oversight. To address this issue, in January 2019, the FDA launched a precertification program to help address the regulatory challenges posed by novel medical software (Kasperbauer & Wright, 2020), which targets digital companies rather than just products. This attempts to address developments in the digital health landscape by allowing some level of oversight to be provided, but without regulatory review of individual products and services. Precertification is given to those who have demonstrated excellence in software development in lower-risk devices to help streamline the process of approval.

The Health Insurance Portability and Accountability Act of 1996 (HIPAA) is US legislation providing data privacy and security provisions for safeguarding medical information. The HIPAA Privacy Rule requires appropriate safeguards to protect the privacy of personal health information, and it sets limits and conditions on the use and disclosure of patient information (United States Department of Health and Human Services, 2010).

Another relevant US regulation is the US Children's Online Privacy Protection Act (COPPA), which provides a federal legal framework to protect the online privacy of children under the age of 13 and forbids the gathering of personal information from them without express consent from a parent or legal guardian (Federal Trade Commission).

The most recent US regulation is the California Consumer Privacy Act (CCPA), which went into effect in January 2020. In essence, CCPA provides California residents with greater transparency and protection of personal data, for instance, ensuring "the right to know where data is collected and to whom it is sold, as well as the right to disclosure" (Zuraw & Sklar, 2020, p.94). The CCPA has triggered further interest in privacy and data protection, both for online services and health care organizations.

15.3.2 European regulations

In comparison, the European regulatory framework for medical devices operates through Conformit´e Europ´eenne (CE) certification and is recognized as a more flexible approach that allows faster market access for certain medical devices. CE marking indicates conformity with health, safety, and environmental protection standards for products sold within the European Economic Area. Similarly to the FDA in the United States, CE certification entails medical device classification based on the intended purpose of the device rather than the particular technical characteristics. Manufacturers of digital health technologies such as medical apps or wearable sensors must also consider the new rules and obligations laid down in the Medical Devices Regulation ("MDR") and the In Vitro Diagnostic Regulation which was adopted by the European Parliament and Council in May 2017. Among other things, the MDR which applies from May 26, 2020, introduces new classification rules for medical devices software. However, the decentralized approach of CE certification is often criticized for hindering "the collection and analysis of safety data" and exposing patients to risks (Sorenson & Drummond, 2014, pp.124–125).

Another important regulation within the European Union (EU) is the General Data Protection Regulation 2016/679 (GDPR). GDPR was approved by the European Parliament in April 2016 and came into force in May 2018, replacing its predecessor, the Data Protection Directive 95/46/EC. Unlike 95/46/EC, which was implemented through national data protection laws, GDPR is directly applicable in each EU member state. Under GDPR, data subjects must receive notice about the collection and use of their data, and all processing of their data requires a legal basis. The regulation also places stricter requirements on the handling of sensitive data such as health records.

While many digital health developers updated their privacy practices to comply with the GDPR, others stated that "their practices would not provide GDPR-level protection to its users but would adhere to local privacy regulations," which would apply even if users were "visiting" digital health products from other jurisdictions (Parker et al., 2019). Furthermore, the "privacy as compliance" reaction of many organizations does not align with the "data protection by design and by default" requirement of GDPR and related "Privacy by Design" (PbD) principles, which advocate that privacy requirements be taken into consideration from the very beginning of product design and development (Waldman, 2017). The dominant privacy model today is still based on privacy notices, followed by an agreement to such terms. However, with the evolution of digital health technology, this approach "no longer safeguards consumer privacy interests with modern health technologies" (Tschider, 2018, p.1507).

15.3.3 Other regulations

The United Kingdom (UK) is another country that has followed suit in attempting to streamline regulations. The National Institute for Health and Care Excellence (NICE) is working with the National Health Service to increase their use of digital innovations. They have published new guidelines on what evidence they need to approve a digital product according to its function. It outlines what are the minimal and ideal types of evidence needed as well as appropriate economic data, as NICE, unlike the FDA, takes cost of a product into consideration.

In China, the Cybersecurity Law of the People's Republic of China (PRC) (Peng, 2010), which was formally put into effect on June 1, 2017, regulates that network providers must not disclose, falsify, or destroy any personal information they have collected. Later in September 2018, National Health Commission ("NHC") and the National Administration of Traditional Chinese Medicine ("NATCM") publicly released three new rules on internet-based medical services and telemedicine (Wang). These rules cover the areas of e-diagnosis ("e-Diagnostic Rules"), internet-based hospitals ("e-Hospital Rules"), and telemedicine services ("Telemedicine Service Standard") (collectively "e-Healthcare Rules").

Asia Pacific is another region experiencing fast growth and development in the digital health industry. Since digital health is relatively new and evolving, an important challenge for digital therapeutics companies there is navigating the complex healthcare regulatory regimes across different countries. For instance, in Singapore, the approach includes reference to regulatory approvals from other countries. For example, if a device or software has already received approval from the US FDA or European medical agencies, it can receive expedited approval. Another regulatory initiative in Singapore is the Licensing Experimentation and Adaptation Programme launched on April 18, 2018, by the Ministry of Health, a regulatory sandbox initiative to facilitate the development of innovative healthcare models in a controlled environment.

Despite the increased prominence of privacy concerns and regulations, relevant laws are continually violated because consumers and regulators lack the tools to know when this is happening. There are recognized problems with clarity, comprehension, and implementation of standards and regulations. Legal texts for software requirements may have issues with vagueness and incompleteness, together with ambiguities at lexical, syntactic, semantic, and referential levels; this makes it difficult for digital health engineers and developers to implement compliant software (Katuu & Ngoepe, 2015).

15.4 Addressing privacy and security concerns

As digital therapeutics proliferate, the need for scrutiny of regulatory compliance and security and privacy of their data practices become ever more important. In particular, critical evaluation of patients' data protection is seen as a key consideration in the design of DMH tools (Robillard et al., 2019). However, it is not always easy to find and choose evaluation techniques that specifically address security and privacy risks and threats. Besides questions of compliance with the regulations in place, one has to consider best practices and standards in the digital health field, evaluation tools, and research-based recommendations, which might not be easy to navigate for all digital health stakeholders, including (but not limited to) digital health developers and consumers. Security and privacy evaluation tools and methods vary and take the form of evaluation frameworks—both specific to data protection and including it as one of the components—as well as a range of individual evaluation techniques that could be broadly categorized as technical and nontechnical (heuristic). Next, we briefly discuss and provide an overview of each of these categories.

15.4.1 Privacy and/or security evaluation frameworks and tools

15.4.1.1 Evaluation frameworks and models

There exists a number of frameworks for the evaluation of digital therapeutic tools, as well as those specific for DMH. Security and privacy are often core components of such frameworks (Henson, David, Albright, & Torous, 2019; Vokinger, Nittas, Witt, Fabrikant, & von Wyl, 2020), as it is the case for Organisation for the Review of Care and Health Applications (ORCHA-24) (Leigh, Ouyang, & Mimnagh, 2017). Another example are information quality frameworks, which include such dimensions as security and provenance (Fadahunsi et al., 2021), or evaluation frameworks for mobile medical applications with privacy and security as the first evaluation steps (Moshi, Tooher, & Merlin, 2018). Frameworks that are specific to DMH are the American Psychiatric Association App Evaluation Model, PsyberGuide (Magee, Adut, Brazill, & Warnick, 2018), and frameworks to evaluate DMH app specifically (Zelmer et al., 2018).

15.4.1.2 Evaluation tools

Addressing security and privacy challenges in the digital health field, researchers have proposed evaluation methods and tools that assess various dimensions and characteristics of digital health interventions, such as usability, content, user engagement, and available research evidence. For instance, tools such as the Mobile App Rating Scale (Stoyanov et al., 2015) and Enlight assessment (Baumel, Faber, Mathur, Kane, & Muench, 2017). Another example is privacy impact assessment (PIA), which can be defined as "a systematic process for evaluating the potential effects on privacy of a project, initiative or proposed system or scheme" (Clarke, 2009). However, a PIA is more than a tool: it is a process that should begin at the earliest possible stages, when there are still opportunities to influence the outcome of a project, and should continue until and even after the project has been deployed (Wright, 2012). A Data Protection Impact Assessment is also a requirement under GDPR for "high risk" activities involving personal data.

Such methods and tools often address both security and privacy mechanisms, and specific components within digital health interventions that contribute to compliance. For instance, such components include the data flows, support of users' data ownership, and audit of collection of specific data types over time. These factors contribute toward the paradigm of privacy and security by design; required by GDPR in the EU, but also recognized and endorsed by other regulatory authorities worldwide. This research trend is especially important as a "compliance-focused" style still prevails, which reduces the focus on privacy and user needs within the design process (Waldman, 2017).

15.4.2 Security and/or privacy evaluation techniques

Evaluation techniques in digital health can facilitate multi-dimensional and thorough evaluations of DMH products and services. Security and privacy evaluation of DMH products and services can be performed using evaluation techniques that could be broadly categorized as technical and nontechnical (heuristic).

15.4.2.1 Technical evaluation

Technical evaluations include techniques such as static analysis to highlight app vulnerabilities by examining code, or traffic analysis during simulated use to detect potential data leaks or security measures taken during transmission. These techniques are often applied in empirical studies that generally consist of downloading apps and inspecting the software. While these studies demonstrate the process of application and use of these technical evaluation techniques, limitations include their specificity and the difficulty of generalizing and applying them to different contexts. Some of them can be specific to the type of digital health intervention, such as medication use apps (van Kerkhof, van de Laar, de Jong, Weda, & Hegger, 2016), or evaluate the security or privacy of apps with specific features, for instance, analyzing the traffic between a wearable device and a smartphone app (Braghin, Cimato, & Libera, 2018).

There are also more general evaluation objectives, such as app usability or app functionality, that contribute to security and privacy of digital health product or a service. For instance, analysis from the perspective of function usefulness (Liang, He, Jia, Zhu, & Lei, 2018) or specific features within digital health product or service, such as diary functions (Bachiri, Idri, Fernández-Alemán, & Toval, 2016), may lead to detection of poor security mechanisms or missing regulatory compliance.

15.4.2.2 Heuristic evaluation

Nontechnical or heuristic evaluation techniques typically consist of evaluation of information provided by app producers, user studies with various digital health stakeholders, and literature reviews.

The evaluation of self-declared data from digital health developers is the most common privacy assessment technique in this category. These studies often focus on privacy policies, terms of agreement, and informed consent by analyzing the availability and readability of these disclaimers or comparing them to the **marketing statements** of the providers, in order to determine whether they are consistent with each other, with the regulations, and with the expectations of users.

However, there is a lack of studies addressing the dynamic nature of such documentation and the design of appropriate and comprehensible privacy policies. This issue is important to address for digital health interventions, as personal data generated by digital health products and services are granular, and privacy policies are already too complex to read but keep getting even longer over time, weakening user privacy on the internet (Chipidza, Leidner, & Burleson, 2016). A related issue concerns the need for "dynamic consent" when applications are used (possibly intermittently) over the longer term. For instance, when new data types or processing capabilities are added, or where apps are deleted and reinstalled. Addressing these needs and advocating for user-centered privacy by maintaining transparency and protecting the interests of digital health users proactively could be beneficial for motivating users' trust.

Another type of information used for evaluation is **user reviews**, which can be used as a way of extracting evidence on users' perceptions on security and privacy of digital health products and services. However, this information should be used with caution. Positive and negative reviews trigger different reactions and are perceived differently—a phenomenon called "negative bias" such that more negative reviews are more influential than positive reviews (Rozin & Royzman, 2001), a finding which has been confirmed repeatedly by researchers in many social science disciplines. More recent studies have looked deeper into this issue; for instance, a study by Wu provides evidence that the valence of online reviews is positively associated with their helpfulness, explaining that "satisfied customers are motivated to write well-composed and in-depth reviews, while unhappy customers provide less transferable information" (Wu, 2013, p. 997). As there is an increasing market for fake positive online reviews (Mukherjee, Venkataraman, Liu, & Glance, 2013), evaluation methods based on user reviews should not be used in isolation for assessing the security or privacy of digital therapeutics.

15.4.3 Security and/or privacy design practices

Once evaluation methods have identified security vulnerabilities and privacy risks, these must then be addressed. Digital heath research offers a range of design heuristics and guidelines. Design recommendations addressing security and privacy of digital therapeutics come from academic literature or research studies, such as empirical evaluations of digital health products and services, as well as from regulatory requirements and industry standards. Research studies can provide design guidance and recommendations from user perspectives (Thach, 2018), analysis of regulations, state-of-the-art literature, or previously identified security and privacy issues (Braghin et al., 2018). Digital marketplaces can also provide certain assurance seals based on the app review process (Shilton & Greene, 2019) and automated compliance checks to verify compliance with privacy regulations (Bednar, Spiekermann, & Langheinrich, 2019). Design recommendations also come in various forms: as guidelines or instructions to improve the security and privacy of digital health, development practices—descriptions of the application of security or privacy mechanisms—or models, for example, predictions or simulations of user behavior or preferences.

15.4.3.1 Target audiences and the digital therapeutics lifecycle

Security and privacy practices are important to implement throughout the digital therapeutics lifecycle including different stakeholders involved in the development process (Nurgalieva, Frik, & Doherty, 2021). Design recommendations often mention digital health developers, engineers, and designers as their target audience, and hence such resources maybe be difficult to identify and consume for less technical audiences, acting as a barrier to the uptake of recommendations. The use of standard formats for reporting, and the development of knowledge bases could help address this issue and benefit the wider community.

Design recommendations can also be applied at different stages in the lifecycle of digital therapeutics and operation of digital health interventions: from the initial development processes (Alturki & Gay, 2019; Greene, Proctor, & Kotz, 2019; Lang, Mayr, Ringbauer, & Cepek, 2019; Lind, Byrne, Wicks, Smidt, & Allen, 2018; Lipson-Smith et al., 2019) to the stage of adoption by patients, either independently, or following the recommendations of clinicians. At different points in the lifecycle, various digital health stakeholders could benefit from or adopt security and privacy practices.

At the early stages, security and privacy practices often target developers and designers by including, for instance, identification of security vulnerabilities, such as transport security issues of the data flows (Mabo, Swar, & Aghili, 2018; Müthing, Jäschke, & Friedrich, 2017), evaluating functionalities (Zhao, Yoo, Lancey, & Varghese, 2019), and legal compliance of the interventions (Muchagata & Ferreira, 2019). While developers generally agree that they should consider *security* from the earliest planning phases, privacy often has lower priority (Assal & Chiasson, 2019; Colley, 2010; McGraw, 2004). Still, developers often prioritize other aspects of software development than security or privacy (Assal & Chiasson, 2019; Loruenser, Pöhls, Sell, & Laenger, 2018). Not recognizing the value of the effort invested in the software security (Chehrazi, Heimbach, & Hinz, 2016) can be especially risky for digital health systems. It is important to ensure the continuity of privacy engineering as well, that is, the systematic integration of privacy practices throughout the development life cycle (Gurses & Hoboken; Waldman, 2017), which could be achieved through risk-based and goal-oriented requirement discovery (Notario et al., 2015) strategies. Another strategy to include privacy considerations is to incorporate them into the definition of software requirements (Bednar et al., 2019), which might require improving general software requirements that are often not sufficiently maintained and managed (Hadar et al., 2018).

The next stage is when digital health systems become available to end-users. At this stage, almost all stakeholder groups can be a target audience, namely, patients and app users, clinicians and healthcare providers, app designers and developers, and data controllers. Security and privacy evaluation methods here include studies of privacy in app user interfaces (Tailor, He, & Wagner, 2016) and analysis of publicly available information in relation to the privacy and security of existing apps. At this stage, usability of digital therapeutics and such user-facing mechanisms as privacy policies, users' consent (Bachiri, Idri, Ferna´ndez-Alem´an, & Toval, 2018; Nurgalieva, O'Callaghan, & Doherty, 2020) and transparency of data handling information (Maringer et al., 2018; O'Loughlin, Neary, Adkins, & Schueller, 2019) become particularly important.

The later stages in the lifecycle involve supporting clinicians in recommending digital health interventions to their patients, or direct adoption by digital therapeutics users. At this stage, it is important to focus on the information available about the product or a service, for instance, regarding their providers and available functionalities. The quality (Loy, Ali, & Yap, 2016), intervention relevance (Bert, Passi, Scaioli, Gualano, & Siliquini, 2016; Liang et al., 2018), and data protection risks (Brüggemann, Hansen, Dehling, & Sunyaev, 2016) are of extreme importance at this stage.

15.4.4 Usability and interaction design

There are a number of full lifecycle issues raised above that relate to usability and which can be at least partly addressed through interaction design.

Data protection regulations mandate obtaining user consent before collecting, sharing, or using personal information (Hoepman, 2014; Kneuper, 2019; Li, Agarwal, & Hong, 2018). This poses a number of challenges regarding the comprehensibility of privacy policies and the usability of more complex and dynamic consent processes. Designing privacy policies that are accessible and understandable to users (Chen et al., 2018) should also help support informed meaningful consent implementation in practice (Balebako, Marsh, Lin, Hong, & Cranor, 2014; Bednar et al., 2019). The design of usable mechanisms for consent may be a particular challenge where complex technological capabilities are introduced, such as those based upon ML (Thieme et al., 2020). A further complication arises where there are potential downsides to disclosing information to end-users, as might be the case for outcome prediction (Chekroud et al., 2021).

Other interaction design strategies include providing users with a greater degree of control over sharing of data within digital interventions, and the transparency provided to users regarding information visible to health professionals (Doherty, Coyle, & Sharry, 2012).

Usability issues can often result in the lack of adoption of various privacy mechanisms important for the communication of sensitive health data, as it is the case for end-to-end encryption. For encryption tools, these range from "poorly designed user interfaces to the fundamental challenges of safe and scalable key distribution" (Bai et al., 2017).

Usability is relevant to digital health security as well. While it is rarely covered in security guides for developers (Acar et al., 2017), they often face the challenge of balancing security and usability (Becker, Parkin, & Sasse; Botta et al., 2007; Hawkey et al., 2008). For instance, at the entry point to access digital therapeutics, user authentication methods are one example when usability can be a strategic issue. From password complexity, which is the first line of defense against security attacks, to biometric authentication, usability is an important condition for building reliable and effective security systems (Braz & Robert, 2006). The strategies to balance security and usability and benefit end-users range from taking into account user diversity, by allowing the customization of security and usability aspects of user interfaces (Belk et al., 2017), to more specific methods such as best practices for database security and control (Wang, Rawal, Duan, & Zhang, 2017).

15.5 Conclusions

In the digital health era, it is challenging to keep up with technological advances, and even more challenging to track where patients' health data ends up and to assess the impact of confidentiality breaches.

It is crucial to be aware of and understand the procedures for evaluating the data practices of digital health products and services. Existing techniques and frameworks can guide professionals and patients through the process of assessing digital health products and services, and identifying the privacy and security risks associated with them. However, in this rapidly expanding field, it can be challenging to choose which of the many available techniques to employ, given the evaluation objectives, digital health lifecycle stage, and available expertise.

This chapter has considered some of the security and privacy issues arising in DMH and digital therapeutics, and the related risks posed to developers, providers, patients, and the public. It also provides an overview of the approaches available for choosing security and privacy evaluation frameworks and guidelines for digital health. The ultimate goal is to support efforts to improve the quality of DMH products and services.

There is a need for healthcare professionals and digital health providers to develop and maintain more rigorous privacy and security practices, such as those discussed in this chapter, in the delivery of digital therapeutics for mental health.

However, privacy and security issues pertaining to the digital health era are complex and multifactorial. They are also not likely to become any simpler, as ever more advanced technologies are integrated into real-world interventions. As change is an unavoidable part of digital health, every stakeholder in the healthcare landscape must contemplate the impact of these changes on the security and privacy of sensitive information, and constantly refine the processes and safeguards to address these.

References

Acar, Y., Stransky, C., Wermke, D., Weir, C., Mazurek, M. L., & Fahl, S. (2017). Developers need support, too: A survey of security advice for software developers. *2017 IEEE Cybersecurity Development (SecDev)*, 22–26. doi:10.1109/SecDev.2017.17.

Adams, D., Bah, A., Barwulor, C., Musaby, N., Pitkin, K., & Redmiles, E. M. (2018). Ethics emerging: The story of privacy and security perceptions in virtual reality. In *Fourteenth Symposium on Usable Privacy and Security ([SOUPS] 2018)* (pp. 427–442).

Ahlfeldt, R.-M., & Huvala, I. (2014). Patient safety and patient privacy when patient reading their medical records. In *International Conference on Well-Being in the Information Society* (pp. 230–239). Springer.

Albrecht, U.-V., Pramann, O., & von Jan, U. (2015). Medical apps–the road to trust. *EJBI, 11*(3), en7–en12.

Aldeer, M., Martin, R. P., & Howard, R. E. (2018). Pillsense: Designing a medication adherence monitoring system using pill bottle-mounted wireless sensors. In *2018 IEEE International Conference on Communications Workshops (ICC Workshops)* (pp. 1–6). IEEE.

Alepis, E., & Patsakis, C. (2017). Monkey says, monkey does: Security and privacy on voice assistants. *IEEE Access, 5*, 17841–17851.

Alturki, R., & Gay, V. (2019). The development of an Arabic weight-loss app akser waznk: Qualitative results. *JMIR Formative Research, 3*(1), e11785 https://doi.org/10.2196/11785.

Assal, H., & Chiasson, S. (2019). "Think secure from the beginning" a survey with software developers. In *Proceedings of the 2019 CHI Conference on Human Factors in Computing Systems* (pp. 1–13).

Bachiri, M., Idri, A., Fernández-Alemán, J. L., & Toval, A. (2018). Evaluating the privacy policies of mobile personal health records for pregnancy monitoring. *Journal of Medical Systems, 42*(8), 1–14. https://doi.org/10.1007/s10916-018-1002-x. doi:10.1007/s10916-018-1002-x.

Bachiri, M., Idri, A., Fernández-Alemán, J. L., & Toval, A. (2016). Mobile personal health records for pregnancy monitoring functionalities: Analysis and potential. *Computer Methods and Programs in Biomedicine, 134*, 121–135.

Bai, W., Kim, D., Namara, M., Qian, Y., Kelley, P. G., & Mazurek, M. L. (2017). Balancing security and usability in encrypted email. *IEEE Internet Computing, 21*(3), 30–38.

Balaskas, A., Schueller, S. M., Cox, A. L., & Doherty, G. (2021). Ecological momentary interventions for mental health: A scoping review. *Plos One, 16*(3), 1–23 e0248152. doi:10.1371/journal.pone.0248152.

Balebako, R., Marsh, A., Lin, J., Hong, J. I., & Cranor, L. F. (2014). The privacy and security behaviors of smartphone app developers. In *USEC 2014, Workshop on Usable Security*. Citeseer.

Baltierra, N. B., Muessig, K. E., Pike, E. C., LeGrand, S., Bull, S. S., & Hightow-Weidman, L. B. (2016). More than just tracking time: Complex measures of user engagement with an internet-based health promotion intervention. *Journal of Biomedical Informatics, 59*, 299–307.

Bauer, A. M., Rue, T., Munson, S. A., Ghomi, R. H., Keppel, G. A., Cole, A. M., … Katon, W. (2017). Patient-oriented health technologies: Patients' perspectives and use. *Journal of Mobile Technology in Medicine, 6*(2), 1–10.

Baumel, A., Faber, K., Mathur, N., Kane, J. M., & Muench, F. (2017). Enlight: A comprehensive quality and therapeutic potential evaluation tool for mobile and web-based ehealth interventions. *Journal of Medical Internet Research, 19*(3), 1–14. doi:10.2196/jmir.7270.

Becker, I., Parkin, S., & Sasse, M. A. Finding security champions in blends of organisational culture, Proc. USEC 11.

Becker, M. (2018). Understanding users' health information privacy concerns for health wearables. *Proceedings of the 51st Hawaii International Conference on System Sciences* (pp. 3262–3270).

Bednar, K., Spiekermann, S., & Langheinrich, M. (2019). Engineering privacy by design: Are engineers ready to live up to the challenge? *The Information Society, 35*(3), 122–142.

Belk, M., Pamboris, A., Fidas, C., Katsini, C., Avouris, N., & Samaras, G. (2017). Sweet-spotting security and usability for intelligent graphical authentication mechanisms. In *Proceedings of the International Conference on Web Intelligence* (pp. 252–259).

Bellazzi, R., Montani, S., Riva, A., & Stefanelli, M. (2001). Web-based telemedicine systems for home-care: Technical issues and experiences. *Computer Methods and Programs in Biomedicine, 64*(3), 175–187.

Bert, F., Passi, S., Scaioli, G., Gualano, M. R., & Siliquini, R. (2016). There comes a baby! what should I do? smartphones' pregnancy-related applications: A web-based overview. *Health Informatics Journal, 22*(3), 608–617.

Binns, R., Lyngs, U., Van Kleek, M., Zhao, J., Libert, T., & Shadbolt, N. (2018). Third party tracking in the mobile ecosystem. In *Proceedings of the 10th ACM Conference on Web Science* (pp. 23–31).

Botta, D., Werlinger, R., Gagné, A., Beznosov, K., Iverson, L., Fels, S., et al. (2007). Towards understanding IT security professionals and their tools. In *Proceedings of the 3rd Symposium on Usable Privacy and Security* (pp. 100–111). New York, NY: Association for Computing Machinery.

Braghin, C., Cimato, S., & Libera, A. Della (2018). Are mHealth apps secure? A case study. In *2018 IEEE 42nd Annual Computer Software and Applications Conference (COMPSAC): 2* (pp. 335–340). IEEE.

Braz, C., & Robert, J.-M. (2006). Security and usability: The case of the user authentication methods. In *Proceedings of the 18th Conference on l'Interaction Homme-Machine* (pp. 199–203).

Brüggemann, T., Hansen, J., Dehling, T., & Sunyaev, A. (2016). An information privacy risk index for mhealth apps. *Annual Privacy Forum* (pp. 190–201). Cham, Switzerland: Springer.

Bruno, E., Simblett, S., Lang, A., Biondi, A., Odoi, C., Schulze-Bonhage, A., … RADAR-CNS Consortium (2018). Wearable technology in epilepsy: The views of patients, caregivers, and healthcare professionals. *Epilepsy & Behavior, 85*, 141–149.

Cajander, A., & Grünloh, C. (2019). Electronic health records are more than a work tool: Conflicting needs of direct and indirect stakeholders. In *Proceedings of the 2019 CHI Conference on Human Factors in Computing Systems, CHI '19* (p. 635). ACM. :1–635:13.

Chancellor, S. (2018). Computational methods to understand deviant mental wellness communities. *Extended Abstracts of the 2018 CHI Conference on Human Factors in Computing Systems* (pp. 1–4).

Chehrazi, G., Heimbach, I., & Hinz, O. (2016). The impact of security by design on the success of open source software. Proceedings of ECIS 2016.

Chekroud, A. M., Bondar, J., Delgadillo, J., Doherty, G., Wasil, A., Fokkema, M., … Dwyer, D. (2021). The promise of machine learning in predicting treatment outcomes in psychiatry. *World Psychiatry, 20*(2), 154–170.

Chen, W., Huang, G., Miller, J., Lee, K.-H., Mauro, D., Stephens, B., & Li, X. (2018). "As we grow, it will become a priority": American mobile start-ups' privacy practices. *American Behavioral Scientist, 62*(10), 1338–1355.

Chikersal, P., Belgrave, D., Doherty, G., Enrique, A., Palacios, J. E., Richards, D., et al. (2020). Understanding client support strategies to improve clinical outcomes in an online mental health intervention. In *Proceedings of the 2020 CHI Conference on Human Factors in Computing Systems, CHI '20* (pp. 1–16). New York, NY: Association for Computing Machinery. URL. https://doi.org/10.1145/3313831.3376341. doi:10.1145/3313831.3376341.

Childress, J. F., Faden, R. R., Gaare, R. D., Gostin, L. O., Kahn, J., Bonnie, R. J., … Nieburg, P. (2002). Public health ethics: Mapping the terrain. *The Journal of Law, Medicine & Ethics, 30*(2), 170–178.

Chipidza, W., Leidner, D., & Burleson, D. (2016). Why companies change privacy policies: A principal-agent perspective. In *2016 49th Hawaii International Conference on System Sciences (HICSS)* (pp. 4849–4858). IEEE.

Churcher, J. (2012). On: Skype and privacy. *The International Journal of Psychoanalysis, 93*(4), 1035–1037.

Clarke, R. (2009). Privacy impact assessment: Its origins and development. *Computer Law & Security Review, 25*(2), 123–135.

Clemens, N. A. (2012). Privacy, consent, and the electronic mental health record: The person vs. the system. *Journal of Psychiatric Practice, 18*(1), 46–50.

Colley, J. (2010). Why secure coding is not enough: Professionals' perspective. *ISSE 2009 Securing Electronic Business Processes* (pp. 302–311). New York City, United States: Springer.

Crotty, B. H., Walker, J., Dierks, M., Lipsitz, L., O'Brien, J., Fischer, S., … Safran, C. (2015). Information sharing preferences of older patients and their families. *JAMA Internal Medicine, 175*(9), 1492–1497.

Denning, T., Dehlawi, Z., & Kohno, T. (2014). In situ with bystanders of augmented reality glasses: Perspectives on recording and privacy-mediating technologies. In *Proceedings of the SIGCHI Conference on Human Factors in Computing Systems* (pp. 2377–2386).

Diggelmann, O., & Cleis, M. N. (2014). How the right to privacy became a human right. *Human Rights Law Review, 14*(3), 441–458.

Doherty, G., Coyle, D., & Sharry, J. (2012). Engagement with online mental health interventions: An exploratory clinical study of a treatment for depression, Association for Computing Machinery, New York, NY,, p. 1421–1430. https://doi.org/10.1145/2207676.2208602.

Fadahunsi, K. P., O'Connor, S., Akinlua, J. T., Wark, P. A., Gallagher, J., Carroll, C., ... O'Donoghue, J. (2021). Information quality frameworks for digital health technologies: Systematic review. *Journal of Medical Internet Research, 23*(5), 1–12, e23479.

Federal Trade Commission, Children's online privacy protection rule ("coppa"), Retrieved on September 16.

Felt, A. P., Ha, E., Egelman, S., Haney, A., Chin, E., & Wagner, D. (2012). Android permissions: User attention, comprehension, and behavior. In *Proceedings of the Eighth Symposium on Usable Privacy and Security* (pp. 1–14).

Fleishman, G. (2012).Skype and online privacy: Called out https://www.economist.com/babbage/2012/07/30/called-out, (accessed July 10, 2022).

Gorini, A., Gaggioli, A., & Riva, G. (2008). A second life for ehealth: Prospects for the use of 3-d virtual worlds in clinical psychology. *Journal of Medical Internet Research, 10*(3), 1–12. doi:10.2196/jmir.1029.

Greene, E., Proctor, P., & Kotz, D. (2019). Secure sharing of mhealth data streams through cryptographically-enforced access control. *Smart Health, 12*, 49–65.

Grundy, Q., Chiu, K., Held, F., Continella, A., Bero, L., & Holz, R. (2019). Data sharing practices of medicines related apps and the mobile ecosystem: Traffic, content, and network analysis. *BMJ, 364*(l920), 1–11. doi:10.1136/bmj.l920.

Grundy, Q., Held, F. P., & Bero, L. A. (2017). Tracing the potential flow of consumer data: A network analysis of prominent health and fitness apps. *Journal of Medical Internet Research*, 19(6), 19, e233. doi:10.2196/jmir.7347.

Gurses, S., & Van Hoboken, J. Privacy after the agile turn doi:10.31235/osf.io/9gy73.

Hadar, I., Hasson, T., Ayalon, O., Toch, E., Birnhack, M., Sherman, S., & Balissa, A. (2018). Privacy by designers: Software developers' privacy mindset. *Empirical Software Engineering, 23*(1), 259–289.

Hawkey, K., Botta, D., Werlinger, R., Muldner, K., Gagne, A., & Beznosov, K. (2008). Human, organizational, and technological factors of it security. *CHI '08 Extended Abstracts on Human Factors in Computing Systems, CHI EA '08* (pp. 3639–3644). New York, NY: Association for Computing Machinery. https://doi.org/10.1145/1358628.1358905. doi:10.1145/1358628.1358905.

Henderson, C., Evans-Lacko, S., & Thornicroft, G. (2013). Mental illness stigma, help seeking, and public health programs. *American Journal of Public Health, 103*(5), 777–780.

Henson, P., David, G., Albright, K., & Torous, J. (2019). Deriving a practical framework for the evaluation of health apps. *The Lancet Digital Health, 1*(2), e52–e54.

Hoepman, J.-H. (2014). Privacy design strategies. In N. Cuppens-Boulahia, F. Cuppens, S. Jajodia, A. Abou El Kalam, & T. Sans (Eds.), *ICT Systems Security and Privacy Protection* (pp. 446–459). Berlin, Heidelberg: Springer.

Høglend, P. (2014). Exploration of the patient-therapist relationship in psychotherapy. *American Journal of Psychiatry, 171*(10), 1056–1066.

Househ, M., Grainger, R., Petersen, C., Bamidis, P., & Merolli, M. (2018). Balancing between privacy and patient needs for health information in the age of participatory health and social media: A scoping review. *Yearbook of Medical Informatics, 27*(01), 029–036.

Hunkin, H., King, D. L., & Zajac, I. T. (2020). Perceived acceptability of wearable devices for the treatment of mental health problems. *Journal of Clinical Psychology, 76*(6), 987–1003.

Kasperbauer, T., & Wright, D. E. (2020). Expanded FDA regulation of health and wellness apps. *Bioethics, 34*(3), 235–241.

Katuu, S., & Ngoepe, M. (2015). Managing digital records within South Africa's legislative and regulatory framework. *3rd International Conference on Cloud Security and Management* (pp. 59–70).

Kelly, M. M., Coller, R. J., & Hoonakker, P. (2018). Inpatient portals for hospitalized patients and caregivers: A systematic review. *Journal of Hospital Medicine, 13*(6), 405–412.

Kneuper, R. (2019). Integrating data protection into the software life cycle. In *International Conference on Product-Focused Software Process Improvement* (pp. 417–432). Springer.

Koh, S., Cattell, G. M., Cochran, D. M., Krasner, A., Langheim, F. J., & Sasso, D. A. (2013). Psychiatrists' use of electronic communication and social media and a proposed framework for future guidelines. *Journal of Psychiatric Practice, 19*(3), 254–263.

Kotz, D. (2011). A threat taxonomy for mhealth privacy. In *2011 Third International Conference on Communication Systems and Networks (COM-SNETS 2011)* (pp. 1–6). IEEE.

Kuntsman, A., Miyake, E., & Martin, S. (2019). Re-thinking digital health: Data, appisation and the (im) possibility of opting out. *Digital Health, 5*, 1–16 2055207619880671.

Lang, M., Mayr, M., Ringbauer, S., & Cepek, L. (2019). Patientconcept app: Key characteristics, implementation, and its potential benefit. *Neurology and Therapy, 8*(1), 147–154.

Lebeck, K., Ruth, K., Kohno, T., & Roesner, F. (2017). Securing augmented reality output. In *2017 IEEE Symposium on Security and Privacy (SP)* (pp. 320–337). IEEE.

Leigh, S., Ouyang, J., & Mimnagh, C. (2017). Effective? engaging? secure? applying the ORCHA-24 framework to evaluate apps for chronic insomnia disorder. *Evidence-based mental health, 20*(4), 1–7 e20–e20.

Li, T., Agarwal, Y., & Hong, J. I. (2018). Coconut: An IDE plugin for developing privacy-friendly apps. In *Proceedings of the ACM on Interactive, Mobile, Wearable and Ubiquitous Technologies: 2* (pp. 1–35).

Liang, J., He, X., Jia, Y., Zhu, W., & Lei, J. (2018). Chinese mobile health apps for hypertension management: A systematic evaluation of usefulness. *Journal of Healthcare Engineering*, 1–14.

Lind, M. N., Byrne, M. L., Wicks, G., Smidt, A. M., & Allen, N. B. (2018). The effortless assessment of risk states (EARS) tool: An interpersonal approach to mobile sensing. *JMIR Mental Health, 5*(3), 1–10. doi:10.2196/10334.

Lipson-Smith, R., White, F., White, A., Serong, L., Cooper, G., Price-Bell, G., & Hyatt, A. (2019). Co-design of a consultation audio-recording mobile app for people with cancer: The secondears app. *JMIR Formative Research, 3*(1), 1–13, e11111.

Loruenser, T., Pöhls, H. C., Sell, L., & Laenger, T. (2018). CryptSDLC: Embedding cryptographic engineering into secure software development lifecycle. *Proceedings of the 13th International Conference on Availability, Reliability and Security* (pp. 1–9). New York: ACM.

Loy, J. S., Ali, E. E., & Yap, K. Y.-L. (2016). Quality assessment of medical apps that target medication-related problems. *Journal of Managed Care & Specialty Pharmacy, 22*(10), 1124–1140.

Lustgarten, S. D., Garrison, Y. L., Sinnard, M. T., & Flynn, A. W. (2020). Digital privacy in mental healthcare: Current issues and recommendations for technology use. *Current Opinion in Psychology, 36*, 25–31.

Mabo, T., Swar, B., & Aghili, S. (2018). A vulnerability study of mhealth chronic disease management (cdm) applications (apps). In *World Conference on Information Systems and Technologies* (pp. 587–598). Springer.

Madigan, S., Racine, N., Cooke, J. E., & Korczak, D. J. (2021). Covid-19 and telemental health: Benefits, challenges, and future directions. *Canadian Psychology/Psychologie Canadienne, 62*(1), 5–11.

Magee, J. C., Adut, S., Brazill, K., & Warnick, S. (2018). Mobile app tools for identifying and managing mental health disorders in primary care. *Current Treatment Options in Psychiatry, 5*(3), 345–362.

Malkin, N., Deatrick, J., Tong, A., Wijesekera, P., Egelman, S., & Wagner, D. (2019). Privacy attitudes of smart speaker users. In *Proceedings on Privacy Enhancing Technologies 2019* (pp. 250–271).

Maringer, M., van't Veer, P., Klepacz, N., Verain, M. C., Normann, A., Ekman, S., ... Geelen, A. (2018). User-documented food consumption data from publicly available apps: An analysis of opportunities and challenges for nutrition research. *Nutrition Journal, 17*(1), 1–13.

Martin, A. C. (2013). Legal, Clinical, and Ethical Issues in Teletherapy, Psycho-analysis online: Mental health, teletherapy, and training (pp. 75–84). London: Karnac Books.

Martinez-Martin, N., & Kreitmair, K. (2018). Ethical issues for direct-to-consumer digital psychotherapy apps: Addressing accountability, data protection, and consent. *JMIR Mental Health, 5*(2), 1–7, e32.

McGraw, G. (2004). Software security. *IEEE Security Privacy 2, 2,* 80–83.

McGreevey, J. D., Hanson, C. W., & Koppel, R. (2020). Clinical, legal, and ethical aspects of artificial intelligence–assisted conversational agents in health care. *JAMA, 324*(6), 552–553.

Meyerbröker, K., & Emmelkamp, P. M. (2010). Virtual reality exposure therapy in anxiety disorders: A systematic review of process-and-outcome studies. *Depression and Anxiety, 27*(10), 933–944.

Mohr, D. C., Burns, M. N., Schueller, S. M., Clarke, G., & Klinkman, M. (2013). Behavioral intervention technologies: Evidence review and recommendations for future research in mental health. *General Hospital Psychiatry, 35*(4), 332–338.

Morera, E. P., de la Torre Díez, I., Garcia-Zapirain, B., López-Coronado, M., & Arambarri, J. (2016). Security recommendations for mhealth apps: Elaboration of a developer's guide. *Journal of Medical Systems, 40*(6), 1–13, 152. doi:10.1007/s10916-016-0513-6.

Moshi, M. R., Tooher, R., & Merlin, T. (2018). Suitability of current evaluation frameworks for use in the health technology assessment of mobile medical applications: A systematic review. *International Journal of Technology Assessment in Health Care, 34*(5), 464–475.

Muchagata, J., & Ferreira, A. (2019). Mobile apps for people with dementia: Are they compliant with the general data protection regulation (gdpr)? In *Proceedings of the 12th International Joint Conference on Biomedical Engineering Systems and Technologies (BIOSTEC 2019)* (pp. 68–77). doi:10.5220/0007352200680077.

Mukherjee, A., Venkataraman, V., Liu, B., & Glance, N. (2013). Fake review detection: Classification and analysis of real and pseudo reviews, UIC-CS-03 Technical Report (pp. 1–11). University of Illinois at Chicago.

Müthing, J., Jäschke, T., & Friedrich, C. M. (2017). *JMIR mHealth and uHealth, 5*(10), 1–12. e147.

Nobles, A. L., Glenn, J. J., Kowsari, K., Teachman, B. A., & Barnes, L. E. (2018). Identification of imminent suicide risk among young adults using text messages. *Proceedings of the 2018 CHI Conference on Human Factors in Computing Systems* (pp. 1–11).

Notario, N., Crespo, A., Martín, Y.-S., Del Alamo, J. M., Métayer, D. Le, Antignac, T., ... Wright, D. (2015). PRIPARE: Integrating privacy best practices into a privacy engineering methodology. *2015 IEEE Security and Privacy Workshops* (pp. 151–158). New York City, USA: IEEE.

Nurgalieva, L., O'Callaghan, D., & Doherty, G. (2020). Security and privacy of mhealth applications: A scoping review. *IEEE Access, 8,* 104247–104268.

Nurgalieva, L., Frik, A., & Doherty, G. (2021). WiP: Factors affecting the implementation of privacy and security practices in software development: A narrative review. Hot topics in the science of security, HoTSoS'21.

O'Brolcháin, F., Jacquemard, T., Monaghan, D., O'Connor, N., Novitzky, P., & Gordijn, B. (2016). The convergence of virtual reality and social networks: Threats to privacy and autonomy. *Science and Engineering Ethics, 22*(1), 1–29.

Okun, S., & Wicks, P. (2018). Digitalme: A journey towards personalized health and thriving. *Biomedical Engineering Online, 17*(1), 1–7, 119. doi:10.1186/s12938-018-0553-x.

O'Loughlin, K., Neary, M., Adkins, E. C., & Schueller, S. M. (2019). Reviewing the data security and privacy policies of mobile apps for depression. *Internet Interventions, 15,* 110–115.

Parker, L., Halter, V., Karliychuk, T., & Grundy, Q. (2019). How private is your mental health app data? an empirical study of mental health app privacy policies and practices. *International Journal of Law and Psychiatry, 64,* 198–204.

Parker, L., Karliychuk, T., Gillies, D., Mintzes, B., Raven, M., & Grundy, Q. (2017). A health app developer's guide to law and policy: A multi-sector policy analysis. *BMC Medical Informatics and Decision Making, 17*(1), 1–13, 141 doi:10.1186/s12911-017-0535-0.

Pasquale, F. (2015). The Black Box Society: The Secret Algorithms That Control Money and Information (pp. 1–311). Cambridge, Massachusetts, United States: Harvard University Press In press.

Peng, P. L. (2010). Ordinance of the people's republic of China on the protection of computer information system security: Decree of the state council of the people's Republic of China (no. 147). The "ordinance of the people's republic of China on the protection of computer information system security" is to be implemented as of the day of issuance. February 18, 1994. *Chinese Law & Government, 43*(5), 12–16.

Powell, A., Singh, P., & Torous, J. (2018). The complexity of mental health app privacy policies: A potential barrier to privacy. *JMIR mHealth and uHealth, 6*(7), 1–9, e158. doi:10.2196/mhealth.9871.

Radovic, A., Vona, P. L., Santostefano, A. M., Ciaravino, S., Miller, E., & Stein, B. D. (2016). Smartphone applications for mental health. *Cyberpsychology, Behavior, and Social Networking, 19*(7), 465–470.

Razaghpanah, A., Nithyanand, R., Vallina-Rodriguez, N., Sundaresan, S., Allman, M., & Kreibich, C. et al., (2018). Apps, trackers, privacy, and regulators: A global study of the mobile tracking ecosystem. Proceedings of the 25th Annual Network and Distributed System Security Symposium (NDSS 2018).

Rezaeibagha, F., Win, K. T., & Susilo, W. (2015). A systematic literature review on security and privacy of electronic health record systems: Technical perspectives. *Health Information Management Journal, 44*(3), 23–38.

Richardson, L. K., Christopher Frueh, B., Grubaugh, A. L., Egede, L., & Elhai, J. D. (2009). Current directions in videoconferencing tele-mental health research. *Clinical Psychology: Science and Practice, 16*(3), 323–338.

Robillard, J. M., Feng, T. L., Sporn, A. B., Lai, J.-A., Lo, C., Ta, M., & Nadler, R. (2019). Availability, readability, and content of privacy policies and terms of agreements of mental health apps. *Internet Interventions, 17*, 100243.

Rosenfeld, L., Torous, J., & Vahia, I. V. (2017). Data security and privacy in apps for dementia: An analysis of existing privacy policies. *The American Journal of Geriatric Psychiatry, 25*(8), 873–877.

Rozin, P., & Royzman, E. B. (2001). Negativity bias, negativity dominance, and contagion. *Personality and Social Psychology Review, 5*(4), 296–320.

Sampat, B. H., & Prabhakar, B. (2017). Privacy risks and security threats in mhealth apps. *Journal of International Technology and Information Management, 26*(4), 126–153.

Shatte, A. B., Hutchinson, D. M., & Teague, S. J. (2019). Machine learning in mental health: A scoping review of methods and applications. *Psychological Medicine, 49*(9), 1426–1448.

Shilton, K., & Greene, D. (2019). Linking platforms, practices, and developer ethics: Levers for privacy discourse in mobile application development. *Journal of Business Ethics, 155*(1), 131–146.

Skaik, R., & Inkpen, D. Using social media for mental health surveillance: A review. *ACM Computing Surveys* 53(6). doi:10.1145/3422824 https://doi.org/10.1145/3422824.

Simpson, J. M. (2017). Home assistant adopter beware: Google, Amazon Digital Assistant Patents Reveal Plans for Mass Snooping.

Smelror, R. E., Bless, J. J., Hugdahl, K., Hugdahl, K., & Agartz, I. (2019). Feasibility and acceptability of using a mobile phone app for characterizing auditory verbal hallucinations in adolescents with early-onset psychosis: Exploratory study. *JMIR Formative Research, 3*(2), 1–11. http://europepmc.org/articles/PMC6537505. doi:10.2196/13882.

Sorenson, C., & Drummond, M. (2014). Improving medical device regulation: The united states and Europe in perspective. *The Milbank Quarterly, 92*(1), 114–150.

Spiegel, J. S. (2018). The ethics of virtual reality technology: Social hazards and public policy recommendations. *Science and Engineering Ethics, 24*(5), 1537–1550.

Stoyanov, S. R., Hides, L., Kavanagh, D. J., Zelenko, O., Tjondronegoro, D., & Mani, M. (2015). Mobile app rating scale: A new tool for assessing the quality of health mobile apps. *JMIR mHealth and uHealth, 3*(1), 1–9, e27. doi:10.2196/mhealth.3422.

Tabassum, M., Kosiński, T., Frik, A., Malkin, N., Wijesekera, P., & Egelman, S. (2019). Investigating users' preferences and expectations for always-listening voice assistants. *Proceedings of the ACM on Interactive, Mobile, Wearable and Ubiquitous Technologies, 3*(4), 1–23.

Tailor, N., He, Y., & Wagner, I. (2016). Poster: Design ideas for privacy-aware user interfaces for mobile devices. In *Proceedings of the 9th ACM Conference on Security & Privacy in Wireless and Mobile Networks* (pp. 219–220). ACM.

Taube, D. O. (2013). Portable digital devices: Meeting challenges to psychotherapeutic privacy. *Ethics & Behavior, 23*(2), 81–97.

Thach, K. S. (2018). User's perception on mental health applications: A qualitative analysis of user reviews. In *2018 5th NAFOSTED Conference on Information and Computer Science (NICS)* (pp. 47–52). IEEE.

Thieme, A., Belgrave, D., & Doherty, G. (2020). Machine learning in mental health: A systematic review of the HCI literature to support the development of effective and implementable ml systems. *ACM Transactions on Computer- Human Interaction (TOCHI), 27*(5), 1–53.

Torous, J. (2017). Digital psychiatry in 2017: Year in review. Psychiatric times. psychiatrictimes.com.

Tschider, C. A. (2018). The consent myth: Improving choice for patients of the future. *Washington University Law Review, 96*, 1505–1536.

U.S. Food and Drug Administration, Medical devices, X STOP® Interspinous Process Decompression System (XSTOP)–P040001. http://www.fda.gov/MedicalDevices/ProductsandMedicalProcedures/DeviceApprovalsandClearances/Recently-ApprovedDevices/ucm078378.htm (accessed September 20, 2014).

United States Department of Health and Human Services, Standards for privacy of individually identifiable health information. Final rule, 45 CFR parts 160, and 164. Code of federal regulations (2010).

Vaidyam, A. N., Wisniewski, H., Halamka, J. D., Kashavan, M. S., & Torous, J. B. (2019). Chatbots and conversational agents in mental health: A review of the psychiatric landscape. *The Canadian Journal of Psychiatry, 64*(7), 456–464.

Van Allen, J., & Roberts, M. C. (2011). Critical incidents in the marriage of psychology and technology: A discussion of potential ethical issues in practice, education, and policy. *Professional Psychology: Research and Practice, 42*(6), 433–439.

van Kerkhof, L. W. M., van de Laar, C. W. E., de Jong, C., Weda, M., & Hegger, I. (2016). Characterization of apps and other e-tools for medication use: Insights into possible benefits and risks. *JMIR mHealth and uHealth, 4*(2), 1–14, e34. doi:10.2196/mhealth.4149.

Vokinger, K. N., Nittas, V., Witt, C. M., Fabrikant, S. I., & von Wyl, V. (2020). Digital health and the covid-19 epidemic: An assessment framework for apps from an epidemiological and legal perspective. *Swiss Medical Weekly, 150,* 1–9, w20282. doi:10.3929/ethz-b-000465761.

Waldman, A. E. (2017). Designing without privacy. *Houston Law Review, 55,* 659–727.

Wang, K. China's health authorities issue new rules on telemedicine. https://www.ropesgray.com/en/newsroom/alerts/2018/09/Chinas-Health-Authorities-Issue-New-Rules-on-Telemedicine.

Wang, Y., Rawal, B., Duan, Q., & Zhang, P. (2017). Usability and security go together: A case study on database. In *2017 Second International Conference on Recent Trends and Challenges in Computational Models (ICRTCCM)* (pp. 49–54). IEEE.

Watzlaf, V. J., Moeini, S., & Firouzan, P. (2010). Voip for telerehabilitation: A risk analysis for privacy, security, and HIPAA compliance. *International Journal of Telerehabilitation, 2*(2), 3–10. doi:10.5195/ijt.2011.6070.

Wilkinson, J., Arnold, K. F., Murray, E. J., van Smeden, M., Carr, K., et al. (2020). Time to reality check the promises of machine learning-powered precision medicine. *The Lancet Digital Health, 2*(12), E677–E680.

Wright, D. (2012). The state of the art in privacy impact assessment. *Computer Law & Security Review, 28*(1), 54–61.

Wu, P. F. (2013). In search of negativity bias: An empirical study of perceived helpfulness of online reviews. *Psychology & Marketing, 30*(11), 971–984.

Wykes, T., & Schueller, S. (2019). Why reviewing apps is not enough: Transparency for trust (t4t) principles of responsible health app marketplaces. *Journal of Medical Internet Research, 21*(5), 1–10, e12390. doi:10.2196/12390.

Xhafa, F., Li, J., Zhao, G., Li, J., Chen, X., & Wong, D. S. (2015). Designing cloud-based electronic health record system with attribute-based encryption. *Multimedia Tools and Applications, 74*(10), 3441–3458.

Xiao, B., Imel, Z. E., Georgiou, P. G., Atkins, D. C., & Narayanan, S. S. (2015). "Rate my therapist": Automated detection of empathy in drug and alcohol counseling via speech and language processing. *Plos One, 10*(12), 1–15, e0143055. doi:10.1371/journal.pone.0143055.

Yang, H., Kim, H., & Mtonga, K. (2015). An efficient privacy-preserving authentication scheme with adaptive key evolution in remote health monitoring system. *Peer-to-Peer Networking and Applications, 8*(6), 1059–1069.

Zelmer, J., van Hoof, K., Notarianni, M., van Mierlo, T., Schellenberg, M., & Tannenbaum, C. (2018). An assessment framework for e-mental health apps in Canada: Results of a modified Delphi process. *JMIR mHealth and uHealth, 6*(7), 1–14, e10016. doi:10.2196/10016.

Zhao, P., Yoo, I., Lancey, R., & Varghese, E. (2019). Mobile applications for pain management: An app analysis for clinical usage. *BMC Medical Informatics and Decision Making, 19*(1), 106–116 https://doi.org/10.1186/s12911-019-0827-7.

Zuraw, R., & Sklar, T. (2020). Digital health privacy and age: Quality and safety improvement in long-term-care. *Indiana Health Law Review, 17,* 85–98.

Chapter 16

Ethical considerations of digital therapeutics for mental health

Constantin Landers, Blanche Wies and Marcello Ienca
Department of Health Sciences and Technology, ETH Zurich

16.1 Introduction

16.1.1 Setting the context

Digital offerings for mental health and addiction are growing exponentially—both in number, as well as user and patient adoption. More than 10,000 mental health apps, a subsection of digital offerings for mental health, were commercially available by 2019 (Torous et al., 2019). At the same time, health, in particular mental health, and addiction are among the most private and delicate matters in private and social life. Rapid innovation in digital therapeutics and the wider domain of health has raised a plethora of ethical issues such as privacy, trust, and human autonomy (Vayena, Haeusermann, Adjekum, & Blasimme, 2018; Vayena & Madoff, 2019).

HCPs are increasingly prompted to recommend, advise on or even prescribe digital offerings or therapeutics, as patients are increasingly turning to and demanding guidance on digital therapy options. At the same time, only a small subsection of digital offerings has been clinically validated or received medical device certification (Marshall, Dunstan, & Bartik, 2019). In this context, HCPs have a heightened responsibility to ensure that patients' ethical rights are protected.

16.1.2 What this chapter does ... and does not

In this chapter, we introduce readers to the general ethical issues around digital therapeutics, with a particular focus on mental health and addiction treatment (hereafter *digital therapeutics*). While most of the identified issues apply to digital therapeutics in general, we focused our review and discussion on mental health and addiction as these treatment areas provoke some of the most relevant ethical issues, while providing a tangible application context.

We aim to provide an overview and build awareness of key issues in recommending or prescribing digital therapeutics, in particular, for mental health or addiction. We aim to offer a comprehensive, yet practical introduction to how HCPs can think about and resolve these ethical issues. We focus particularly on contexts where digital therapeutics for mental health and addiction are delivered via consumer telecommunication devices (e.g., smartphone, tablets). As such, we do not aim to offer an exhaustive list of all relevant ethical issues in digital therapeutics. We do not focus on a particular technology (e.g., apps, devices) or modes of delivery (e.g., self-administered apps, telehealth therapy, or blended treatment).

Given our focus on HCPs, this chapter does not address ethical issues arising when individuals use digital therapeutics without HCPs' explicit involvement. Thus, when discussing digital therapeutics users, we refer exclusively to patients.[1] Designers and developers of digital health interventions are also likely to benefit from this chapter's overview, though the chapter does not cover issues specific to health intervention development.

In light of rapid technological change, the following is intended as a general introduction to ethical issues in digital therapeutics. We will not compare ethical differences between modes of delivery. At the time of writing, little regulation and other formal guidance (e.g., from HCP associations) for administering digital therapeutics exist in most jurisdictions.

---

1. Nonetheless, Section 16.2.1 describes the major issues that directly affect users. All of these also pertain to users without HCP support.

Digital Therapeutics for Mental Health and Addiction: The State of the Science and Vision for the Future. DOI: https://doi.org/10.1016/B978-0-323-90045-4.00007-1

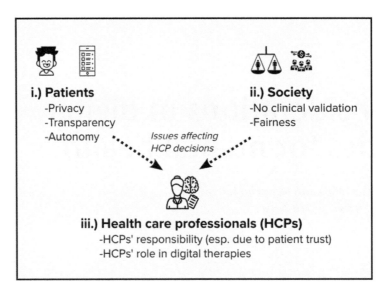

FIGURE 16.1 Core ethical considerations for key stakeholders.

While this is likely to change with time, we assume circumstances with minimal regulation in place. Doctors should ideally prescribe or recommend digital therapeutics that are verified medical devices. However, lack of verification for most digital therapeutics, increasing patient interest, and limited therapeutic resources may complicate this norm. HCPs are strongly advised to cautiously consider the below outlined ethical considerations. Even with validated therapeutics, the highly innovative nature of digital therapeutics means that various ethical issues need to be closely monitored.

16.2 Ethical issues arising from digital therapeutics in mental health

Digital therapeutics promise significant improvements in individual health and well-being. This has been shown for different patient populations (e.g., college students, patients treated for addiction) (Campbell et al., 2014; Lattie et al., 2019; O'Hara, 2019). At the same time, digital therapeutics give rise to new ethical issues. In the absence of comprehensive legislation or official guidance (e.g., insurer endorsement, professional associations), HCPs recommending or prescribing digital therapeutics should be widely aware of such issues. In order to provide readers with a comprehensive overview, we have clustered them into three categories: patients, society, and HCPs (Fig. 16.1). We briefly introduce each major issue, and its relevance for HCPs. At the conclusion of each, we provide our recommendations for HCPs.

16.2.1 Issues directly affecting patients using digital therapeutics

This section introduces issues that affect patients directly. As patients trust and look to HCPs for advice, this implies a responsibility that doctors be up to date, in order to mitigate risks such as those introduced below.

16.2.1.1 Privacy
What privacy is and how it affects patients

> *Each person has a sphere of existence and activity that properly belongs to that individual alone, where he or she should be free of constraint, coercion, and even uninvited observation.*
>
> (McFarland, 2012)

Privacy is among the most widely debated ethical issues around digital health (Vayena et al., 2018). A scoping review of factors enabling trust found that privacy and security were among the most identified technical factors required for user trust (Adjekum, Blasimme, & Vayena, 2018). Indeed, privacy is particularly important in digital health, as highly private data can be collected is increasingly difficult to anonymize and can be used to derive very private information about individuals (Ienca, Haselager, & Emanuel, 2018; Meurk, Hall, Carter, & Chenery, 2014; Vayena et al., 2018).

The importance of privacy is particularly high in the context of digital therapeutics for mental health and addiction. When using such offerings, patients share highly confidential and personal data with therapeutics providers. These data are among the most sensitive personal data.[2] Privacy in digital therapeutics is critical, as mental illness and addiction remain subject to stigmatization and discrimination even today (Martinez-Martin & Kreitmair, 2018). In instances when data privacy was not ensured, and third parties gained access to user information, manipulation and economic and social discrimination have resulted (Martinez-Martin & Kreitmair, 2018; Parker, Halter, Karliychuk, & Grundy, 2019; Robillard et al., 2019).

Patient employment status may be affected when physical and mental health details are leaked (Glenn & Monteith, 2014; Parker et al., 2019; Robillard et al., 2019). Advertisers have been shown to exploit information to manipulate individuals to consume more or differently, constituting a grave infringement of individual privacy and right to self-determination. Privacy breaches have resulted in undesirable social and financial consequences, with data being shared without consent with family, friends, or insurance companies (Glenn & Monteith, 2014; Parker et al., 2019; Robillard et al., 2019). Worse still, patients suffering from drug addiction could face legal consequences, as some digital therapeutics can be accessed through subpoena or legal proceedings, because they lack the confidentiality protection of medical providers (Martinez-Martin & Kreitmair, 2018; Meurk et al., 2014).

Why privacy matters in practice

Most available mental health therapeutics offerings, alongside many digital therapeutics have inadequate confidentiality and privacy policies. Many do not specify how they handle and protect user data (Parker et al., 2019). Others use long and complex terms and conditions (T&C) to shift responsibility for data security and privacy from developers to users (Parker et al., 2019). The EU GDPR mandates that individuals own their data and must give consent for its use (Wykes, Lipshitz, & Schueller, 2019). However, such laws do not exist in many jurisdictions, and even in the EU, foreign providers have argued that their foreign status exempts them from compliance (Parker et al., 2019).

Privacy is a particularly relevant issue, as many available digital therapeutics are free-of-charge. Where these are provided by private companies, their business model is often centered around sharing highly confidential personal data (Doraiswamy et al., 2019; Martinez-Martin & Kreitmair, 2018; Parker et al., 2019; Wykes et al., 2019). In the context of digital therapeutics for addiction, privacy gains an additional practical relevance. Patients suffering from addiction may tend to sell or lose their phones more frequently, leading to an increased chance of an outsider gaining access to their health data (Mares et al., 2016; Tofighi et al., 2019).

Taking a balanced perspective

These individual privacy risks stand in contrast with the potential benefits of data sharing. Large-scale health data can enable a better understanding of health issues, and improve treatment and diagnosis (Doraiswamy et al., 2019). Indeed, when data confidentiality is maintained correctly, digital therapeutics often increase privacy. Remote participation can enhance anonymity, access, and privacy of therapy (Doraiswamy et al., 2019; Jones et al., 2015). This has enabled better treatment outcomes: patients with addiction have shown increased compliance to therapy, as they feel less observed (Ferreri, Bourla, Mouchabac, & Karila, 2018).

What HCPs can do about it

- *Verify:* Ideally, HCPs should only prescribe or recommend digital therapeutics that have been officially verified as medical device. Such offerings typically fulfill privacy and data protection standards. If they cannot prescribe officially verified digital therapeutics, HCPs should only recommend or prescribe therapeutics that protect patient privacy; for example, those that are GDPR-compliant.[3]
- *Inform:* HCPs should stay up to date and inform themselves on public guidance offered by some national health authorities. In the United Kingdom, the National Health Service (NHS) offers an app library of reviewed nonmedical apps.[4] The French Haute Autorité de Santé (HAS) provides good practice guidelines and a risk matrix to assess an apps' good practices according to its intended uses, although HCPs have to evaluate apps themselves (Ferreri et al., 2018). The American Psychiatric Association (APA) has also formulated an app rating framework for HCPs (Torous et al., 2018).

---

2. From a legal perspective, the European Union's (EU) General Data Protection Regulation (GDPR), for example, categorizes mental health information as "special category." (ICO n.d.) Transferring, handling, and Storing such data thus requires the highest level of confidentiality and privacy (Doraiswamy et al., 2019; Ferreri et al., 2018; Meurk et al., 2014; Tofighi et al., 2019).

3. Article 25 of the GDPR, for instance, requires that data are protected by design and by default—meaning that data are held securely and only data necessary for the main purposes of the app is collected.

4. The review process verifies compliance with eight good practices via a questionnaire. This list can be accessed here: https://www.nhs.uk/apps-library/filter/?categories=Mental%20health.

HCPs should, however, even be cautious about the limitations of this guidance. The NHS app library, for example, uses a privacy label system to highlight that not all apps it recommends fulfill the same level of privacy. The library also makes it clear that it does not assume any legal liability for their recommendations. HCPs, therefore, carry the full responsibility when recommending digital therapeutics.

- *Educate patients:* HCPs should inform patients about the remaining risks of using apps they recommend and caution them against using additional apps that have not been validated or recommended.

16.2.1.2   Transparency

What transparency is and how it affects patients

...transparency requires clarity about what is collected, how it is collected, and when it is collected

(Martinez-Martin et al. 2018)

For many digital therapeutic offerings, little transparency exists on how data are stored, processed, and shared (Burr, Morley, Taddeo, & Floridi, 2019; Parker et al., 2019; Robillard et al., 2019). Data collection often extends beyond information that users share directly with the app to data such as location, sleep cycle, and voice and speech recording (Martinez-Martin, Insel, Dagum, Greely, & Cho, 2018). Even if users are aware of privacy issues, they often struggle to fully understand how their data is being used given the opaque extent of transparency offered to them (Ienca et al., 2018).

Such "transparency" is frequently only available in an app's T&C. These texts are typically difficult to read and understand, with the complex legal language of considerable length (Parker et al., 2019). T&Cs frequently give digital therapeutics providers the right to sell or share sensitive data and limit provider responsibility for data security (Parker et al., 2019). In order to proceed, users are required to accept the T&Cs, without necessarily comprehending them. Instead, users should be provided with an easy-to-comprehend summary of their provider's data and security practices that could include a "transparency statement" detailing how third parties may interfere with their data (Ienca et al., 2018).

Why transparency matters in practice

Transparency is a major concern when it comes to the clinical validation of digital therapeutics. Independent researchers and HCPs validating the effectiveness and limitations of digital therapeutics often require transparent information and data on these. As algorithms from private sectors are typically considered intellectual property, they are not fully transparent, preventing such validation (Burr, Morley, Taddeo, & Floridi, 2019; Martinez-Martin et al., 2018; Martinez-Martin & Kreitmair, 2018).

While regulations such as the GDPR have improved the transparency of digital therapeutics, no international consensus for data use and storage exists. Many providers, therefore, create their own transparency policies. Others claim that they do not fall under the jurisdiction of existing regulations (Doraiswamy et al., 2019; Parker et al., 2019).

Taking a balanced perspective

Data sharing and usage is rightfully a controversial issue. Sufficient data from the patient must be shared with the app and HCPs, in order for the therapy to be effective. In other cases, data may need to be shared in order to protect a patient's safety. Transparency can enable a fact-based dialogue about required data usage or when privacy might be broken (Martinez-Martin & Kreitmair, 2018).

Despite its merits, Annay and Crawford (2018) noted that transparency alone does not automatically ensure accountability. Referring to algorithmic systems, they argue that transparency ("seeing") does not necessarily ensure that individuals, or even society at large, fully comprehend complex systems ("understanding.") They assert that demands for transparency should explain for what and to whom systems should be accountable. In our case, for example, this could mean that providers provide transparency about their data and explain its uses to patients and HCPs.

What HCPs can do about it

- *Inform:* HCPs should inform themselves about the transparency of digital therapeutics' data handling policies, before recommending them to patients.
- *Verify:* Determining whether apps are GDPR-compliant, and have attained the CE-mark can provide additional certainty that digital therapeutics providers comply with data regulation. [5]

---

5. Berensmann and Gratzfeld (2018) CE stands for Conformité Européenne and designates products in the European Union that meet EU regulatory standards, such as those for medical devices (c.f. https://europepmc.org/article/med/29368120).*

- *Educate:* HCPs should educate patients about why transparency matters and encourage them to check this regularly.
- *Advocate:* Beyond individual HCPs, it is recommended that HCP professional bodies require sufficient transparency from digital therapeutics providers.

16.2.1.3 Autonomy

What autonomy is and how it affects patients

> *Personal autonomy is, at a minimum, self-rule that is free from both controlling interferences by others and from limitations, such as inadequate understanding, that prevent meaningful choice.*
>
> (Beauchamp and Childress 2001)

Successful digital offerings and apps, including those used for digital therapeutics, maximize user engagement by designing features that activate the reward system of the brain (Martinez-Martin & Kreitmair, 2018). They do this to keep users engaged, thus advancing the therapeutic aims of the app. These aims often include altering behavior and utilizing habit change techniques, such as nudging.

However, activating patients' reward systems can unintentionally affect the overall therapeutic goals. Studies of psychotherapy apps have shown that such mechanisms of action can lead to anxiety and addictive behavior (Martinez-Martin & Kreitmair, 2018). Apart from potential adverse effects, digital therapeutics raises further ethical questions, due to their ability to create subtle dependence in patients and their considerable influence over user behavior. This feature can violate the core bioethical principle of autonomy, which emphasizes an individual's capacity to make decisions without external influence, as well as their freedom to decide what they want. What is more, it can affect the fundamental perquisites of autonomy such as cognitive liberty (Ienca & Andorno, 2017).

Why autonomy matters in practice

Autonomy is an important concern along the entire digital therapeutics journey: when choosing, using, and crucially, when ending a digital therapy.

At the beginning of this journey, the requirement of informed consent presents major legal and therapeutical questions regarding patient autonomy. Informed consent is defined as "the process in which a healthcare provider educates a patient about the risks, benefits, and alternatives of a given procedure or intervention" (Shah, Thornton, Turrin, & Hipskind, 2020). Informed consent requires full disclosure of relevant information, adequate comprehension of what is happening, and must be a voluntary choice. Several factors may call into question whether these requirements are met in the context of digital therapeutics. As shown, patients are often not aware of how these software work, and what kind of data might be collected or shared with third parties. Without, it is uncertain to what extent voluntary choice is possible, and therefore whether HCPs can rightfully prescribe these apps (Martinez-Martin et al., 2018).

Throughout the use of a digital therapeutic, and especially at conclusion, occurrence of dependence and addiction may also reduce autonomy (Haug et al., 2015). Haug et al. have shown that the design of many digital applications is not sufficient to prevent addiction. Limiting the number or length of encounters per day with a therapeutic application are examples of design features that could reduce addiction. Young people, whose digital affinity makes them a likely adopter of digital therapeutics, are particularly at risk of developing smartphone addictions. Martinez (2018) has shown that "anti-addiction" design requirements are frequently not met in digital therapeutics.

As we discuss in the next section, lack of scientific validation and official oversight exacerbates these concerns. Without such control measures, it is difficult to determine the ways in which individual apps infringe on autonomy.

Taking a balanced perspective

Autonomy in digital therapeutics for mental health and addition is undoubtedly complicated. But despite the risks, digital therapeutics can promote user autonomy by increasing self-awareness and self-efficacy. They give patients new possibilities for learning coping skills and controlling their cravings, and also the support to apply them outside the therapy setting (Ferreri et al., 2018; Jones et al., 2015; Robillard et al., 2019). Ultimately, this increases patient empowerment and engagement in therapy (Burr, Morley, Taddeo, & Floridi, 2019).

It is imperative that digital therapeutics earn the rightful trust of users. If trust is not earned, due to perceived privacy risks, the patient may exercise autonomy by deciding not to use the app. Ideally, digital therapeutics providers rightfully earn the trust of users and HCPs and enable patients and society to benefit considerably.

What HCPs can do about it

- *Stay in touch:* After recommending or prescribing digital therapeutics, HCPs should regularly check in with patients to verify that digital components are used responsibly and healthily.
- *Select apps carefully:* HCPs should verify that recommended apps are designed to avoid dependency.

16.2.2   Societal concerns arising from digital therapeutics

Global healthcare systems have seen demand significantly outstrip existing capacity. This is especially true for mental health. As powerfully shown during the Covid-19 pandemic, digital health can alleviate some of these burdens. Digital mental health promises to considerably alleviate these pressures by efficiently allocating human therapists (e.g., via telemedicine), and augmenting or in certain instances replacing them with digital offerings. This, however, raises significant societal concerns. The efficacy of many digital therapeutics has not been validated, and thus considerable issues around fairness arise, as digital therapeutics are proposed to complement and even replace traditional HCP (e.g., therapist) tasks. Increased distance between patients and HCPs may interfere with HCPs' ability to mitigate potential adverse effects.

16.2.2.1   Lack of clinical validation
What lack of clinical validation is and how it affects patients and society

> *"We want to see evidence because there are real lives at stake, and how are you (digital health start-ups) proving value?"*
> Danielle McGuinness, Marketing manager Rock Health (Digital Health Venture Fund)

Digital health and digital therapeutics, in general, generate unprecedented quantities of real-life data that can prove robust correlations, but often lack randomized trials to establish their safety and clinical utility (Khoury & Ioannidis, 2014; Vayena et al., 2018). This is particularly relevant in the context of digital therapeutics. Studies have found that up to 90% of mental health apps lack evidence-based clinical value and suitable security policies (Bauer et al., 2017). However, clinicians can only evaluate digital therapeutics if developers provide clear evidence of how the product was developed and tested for effectiveness (Bauer et al., 2017; Burr, Morley, Taddeo, & Floridi, 2019). The quality of digital therapeutics also varies considerably, and providers have at times made erroneous claims (Bauer et al., 2017; Glenn & Monteith, 2014; Martinez-Martin & Kreitmair, 2018). Clinical validation is of critical importance, as unvalidated technology can pose medical risks to patients (Bauer et al., 2017; Burr, Morley, Taddeo, & Floridi, 2019; Ferreri et al., 2018). Incorrect advice or information, self-diagnosis, and self-treatment may lead to delays in seeking professional help when it is needed, encourage therapy that does not help the patient, or in extreme cases, facilitate wrongful drug administration and access (Bauer et al., 2017; Martinez-Martin & Kreitmair, 2018; Tofighi et al., 2019).

Why lack of clinical validation matters in practice

The lack of clinical validation can be partly attributed to a difference between clinically developed apps and private sector products, also called the "commercialization gap" or "information and implementation gap" (Ienca et al., 2017; Martinez-Martin & Kreitmair, 2018). Clinical research apps undergo more systematic testing for safety and effectiveness compared with private-sector products (Martinez-Martin & Kreitmair, 2018). This gap promotes the development of private-sector products, as they are easier to bring to market. This leads to greater availability of inadequately evaluated digital health technologies. It is, therefore, necessary to develop a framework in which private-sector apps must go through the same effectiveness evaluation steps as clinical research apps.

Digital therapeutics challenge traditional medical regulatory structures due to the sheer number of available products, their iterative and adaptive nature (features are changed regularly based on user feedback), and their innovative (and therefore often unvalidated) mechanisms of action. Consequentially, unvalidated and potentially harmful offerings are advertised and made available to the public. The resource-intensive nature of clinical validation alone does not entirely explain this lack of clinical validation. In addition, public health systems are only beginning to develop public health tools for assessing and validating digital therapeutics.[6]

---

6. A relevant example for this is Germany's Digital Health Applications (DiGA) legislation that seeks to ease the introduction and adoption of digital health applications provided that they are initially certified as a medical device and subsequently prove efficacy during a one-year trial period. https://www.bfarm.de/EN/MedicalDevices/DiGA/_node.html.

This, in turn, impacts the adoption of digital therapeutics. In 2016, the American Medical Association surveyed their membership regarding physician reticence to adopt digital health, a lot of which encompassed digital therapeutics. Their research revealed that insufficient clinical validity was a chief barrier to adoption (American Medical Association, 2016).

Taking a balanced perspective

Fundamentally, digital therapeutics (individual solutions and the overall category) tend to change faster than the institutions tasked with legislating and regulating them (Bauer et al., 2017; Wykes et al., 2019). This raises two issues: for one, patients, HCPs, and society need to contemplate the trade-off between potential benefits of digital therapeutics and risks that may stem from their novelty. Regular debate and engagement of all stakeholders are needed.

Faster and more effective forms of validation and regulation are needed. Continuous certification might seem the ideal solution, but it is hard to implement. An alternative, tested by the US Food and Drug Administration (FDA) precertification program, is to precertify providers of digital therapeutics, rather than individual technologies. Commercially independent organizations, such as nonprofits PsyberGuide and OneMind, have been involved in validating solutions and establishing industry standards (Wykes et al., 2019).

CASE STUDY I Robo-friend*

Robo-friend was designed as an AI-enabled chatbot, adapting its personality to what users would like it to be, and imitating their conversation partner. It can function as friend, mentor, or lover. The application was initially not designed as a medical device or for therapy settings. However, users reported significant positive effects when dealing with stressful situations and anxiety. This led to a rebranding and strategy shift: Robo-friend now directly addresses people suffering from anxiety, sleep difficulty, or negative thoughts. It therefore clearly promises therapist-like or complementary functions. While Robo-friend is used by many people, no extensive scientific studies about the effects of this technology as a therapeutic tool exist to date.

What HCPs can do about it

- *Consult official and independent resources:* HCPs should stay up to date on the quickly expanding body of evidence around existing and novel digital therapeutics, and closely follow public health, opinion leaders, and trusted institutions' advice.
- *Filter rigorously:* HCPs should only recommend digital therapeutics whose efficacy is scientifically validated or officially endorsed.
- *Validate effectiveness:* HCPs should thoroughly and continuously validate the efficacy of the digital therapeutics they recommend.

16.2.2.2   Fairness and equality

What fairness and equality are and how they affect patients and society

> *"When an* un*equal distribution is considered fair, it represents an* equitable inequality. *Hence, any distribution could in principle be classified as* equitable *as long as people have considered it to be fair."*
>
> (Olsen, 2011)

In most healthcare systems, HCPs do not suffice to meet the demand for health services, in particular mental health and addiction therapeutics. Many digital therapeutics, however, can be made available to many patients, often at little additional cost. As a result, a portion of patients will receive "traditional" human-delivered therapy, while the medical needs of the remainder are met or *bridged* with digital care. However, different digital therapeutics have been shown to be either an improvement or a lesser substitute of traditional mental health and addiction resources (Doraiswamy et al., 2019).

This raises important questions on the allocation of medical resources. Not all patients can receive the same, *equal,* care. To resolve this issue, healthcare systems and philosophers typically aim to find a *fair* allocation of resources, rather than striving for equality (Olsen, 2011). Many publicly funded healthcare systems strive to provide "equal access to healthcare for equal health needs."

However, individual forms of inequality (e.g., lack of digital resources, lack of health insurance) often hinder patients from receiving adequate care, digital, or traditional. This increases the importance of HCPs carefully evaluating the situation of their patient.

Why fairness and equality matter in practice

Multiple socioeconomic factors may prevent the fair allocation of therapeutic resources (Ienca, Shaw, & Elger, 2019). People suffering from addiction, for example, often live in low-income settings (Ferreri et al., 2018). In many countries, people from low-income backgrounds may not be insured. Well-evaluated digital therapeutics or traditional person-to-person therapy are often not covered by basic health insurance, and individuals must, yet cannot, cover the cost themselves.

As a result, low-cost or free digital therapeutics are more readily available to these patients than traditional therapies. If these apps have unclear data privacy policies and are not tested for effectiveness, vulnerable populations are exposed to an increased risk of involuntary data sharing, and may not be receiving adequate treatment (Martinez-Martin et al., 2018).

The "digital divide" is a further phenomenon raising questions regarding the fair allocation of medical resources. There are great discrepancies in technology and internet access, with affluent and well-educated individuals generally enjoying greater access. In our context, patients lacking digital literacy or access are severely disadvantaged, as they are more likely to be excluded from novel provision methods for mental healthcare (Bauer et al., 2017; Doraiswamy et al., 2019). This is particularly relevant in the case of digital therapeutics aimed at treating addictions. The percentage of people with addictions that own a smartphone is around 80% that of the general population (Tofighi et al., 2019).

Bauer (2017) also shows that individuals more severely affected by mental illness can lack adequate digital skills to fully participate in digital therapeutics. Furthermore, people with mental disabilities have lower levels of Internet and smartphone use than the general population, making it more difficult for them to access digital mental health therapies (Bauer et al., 2017; Tofighi et al., 2019). Irregular and unreliable internet access may further limit HCPs' ability to adequately support their patients (Bauer et al., 2017).

Taking a balanced perspective

However, there is room for optimism, as digital therapeutics may make therapy available to populations that would not have previously considered it, due to stigma or financial resources. Studies show the digital divide shrinking, as ethnic and racial minorities, as well as people from low-income and less-educated backgrounds, increasingly own smartphones and have broader digital access (Jones et al., 2015). This creates the possibility for digital mental health applications to reach individuals from a wide variety of backgrounds, potentially connecting them with mental health services for the first time (Doraiswamy et al., 2019; Jones et al., 2015; Mares et al., 2016).

Virtual therapies are a financially affordable option compared to institutional care and have the additional benefit of being easily administrable also in remote areas where institutional resources may be lacking or insufficient. In areas where there are few or no local HCPs or support groups (Alcoholics Anonymous/Narcotics Anonymous meetings), it is more feasible for patients to access therapy through a smartphone, than it is to build additional treatment space or to increase the number of therapists (Doraiswamy et al., 2019; Mares et al., 2016). Technology also plays a valuable role in connecting and mobilizing patients and HCPs, as well as training local healthcare workers to provide effective support (Doraiswamy et al., 2019).

In conclusion, it is important that governments as well as HCPs ensure that digital therapeutics do not increase the digital divide, but instead lead to new forms of health equity (Burr, Morley, Taddeo, & Floridi, 2019; Doraiswamy et al., 2019).

CASE STUDY II Robo-trainer*

Robo-trainer offers flexible, digital online courses for depression that are based on the methods of cognitive-behavioral therapy. It uses online content to train users to deal with mental conditions such as stress, depression, eating disorders, or chronic pain. It is often used as an interim solution for patients waiting for a traditional therapy slot, although patients can access a medical professional on demand. Robo-friend was developed by trained psychologists, is reviewed regularly by external scientists, and is a CE-certified class-1 medical product. As one of the first mental health apps, Robo-trainer is covered by health insurance in its domestic market, provided that a doctor has diagnosed depression or a similar mental condition and has prescribed the course. This renders digital apps reimbursable, provided that their safety and data security have been validated by the government, and their utility proven during a one-year trial period. Thus, it is possible for patients to profit from innovative therapy tools.

What HCPs can do about it

- *Ensure fair access:* In many healthcare systems, HCPs have considerable scope to ensure that patients gain adequate access to either traditional and digital therapeutics, regardless of status or wealth. Where differences in availability exist, HCPs should ensure that they are aware of free or low-cost alternatives, and make patients aware of the most relevant options.
- *Build digital skills:* HCPs can address issues such as fairness and the digital divide by seeing that patients gain the necessary skills to participate in digital therapeutics. This may involve educating them directly or connecting them with relevant patient education resources.

16.2.3 Health care professionals

As outlined above, digital therapeutics raises major ethical issues affecting patients and society, and imply new roles, responsibilities, and deliberations for HCPs. This section discusses the changing roles and responsibilities of HCPs, and the ethical foundation for these changes, specifically patient trust and HCP duty to care.

16.2.3.1 HCPs' new role: control, augmentation, and competition for trust

With the rise of digital therapeutics, the role of HCPs is undoubtedly changing. The popularity of digital therapeutics can be partly attributed to the fact that "human resources" are not adequate to meet the demand. Digital therapeutics offer an almost infinitely scalable supplement, available 24/7, that HCPs cannot match. As a result, patients interact more frequently with digital therapy solutions than with their HCP (even when interactions are limited to avoid addiction). In addition to higher availability, digital therapeutics have other "competitive advantages" when compared to HCPs. Take *Ellie*, a digital therapeutic that was developed by the US Defense Advanced Research Projects Agency to help members of the US military cope with post-traumatic stress disorder. Ellie was designed as a conversational agent (i.e., an algorithm that can understand and respond using human language). Studies on *Ellie* have found that patients trust "artificial" doctors more than "human" HCPs. Patients reported that they felt less judged or stigmatized by a robot than by an HCP when they shared their deeper experiences and feelings (Lucas et al., 2017).

Despite reduced interaction and potential "trust competition" from digital therapeutics, HCPs retain considerable responsibility throughout the entire patient journey of prescribing, undergoing, and completing digital therapeutics. These responsibilities raise several issues, in turn relying on both established and novel ethical foundations.

A. *Recommending digital therapeutics*

As described above, HCPs carry an increased ethical responsibility when recommending digital therapeutics, and should thus exercise ample care. This is based on HCPs' occupational *duty to care* based both in the law, as well as morally implied by patients' trust in them (Bending, 2015). The importance of such trust is aggravated in the case of innovative digital health applications. Patients have been shown to lack trust for novel technologies in medicine such as AI (Gille, Jobin, & Ienca, 2020). As patients lack the tools to validate digital offerings, they rely heavily on recommendations of trusted instances, such as their HCP's (Adjekum et al., 2018).

HCP direct recommendation thus tends to lead to a reduction in patient caution and skepticism. Given the wide array of unresolved ethical issues, HCPs should a) only recommend apps they have validated and of whose data ethics and practices they are aware, and b) educate patients about moral and ethical issues surrounding the use of these apps and/or devices and about which precautions to take (Bauer et al., 2017).

B. *Monitoring digital therapies—complex and unresolved responsibilities*

The recency of digital therapeutics implies that even for relatively well-validated solutions, adverse, or unintended outcomes such as dependence or addiction may arise. When these algorithms make a wrong diagnosis or lead to adverse effects, it is unclear who is responsible (Martinez-Martin & Kreitmair, 2018). The same question arises for potential discrimination caused by bias (e.g., with regard to gender or ethnicity), which some algorithms have displayed (Martinez-Martin & Kreitmair, 2018). Many private app and consumer device providers state in their T&Cs that they are not a professional therapeutic or diagnostic tool and view these consequences as supererogatory, thereby sidestepping responsibility (Burr, Morley, Taddeo, & Floridi, 2019; Ienca et al., 2018). This allows them to often bypass medical device regulation.

This issue is further complicated by the fact that when digital therapeutics are prescribed as the primary form of therapy, HCPs are less involved. While it is technologically possible to digitally monitor patients more extensively than before (e.g., smartphone interaction, Facebook profile) no comprehensive guidance delineating the line between patient privacy and HCP oversight exists thus far (Bauer et al., 2017). To complicate matters further, technology is effectively blurring the responsibility line between therapist and technology developers (Burr, Morley, Taddeo, & Floridi, 2019).

While it is beyond the remit of individual HCPs to resolve these issues, it is highly recommended that HCPs continue to exercise their responsibility. This includes educating patients about possible (and potentially unknown) side effects, and maintaining regular contact with patients, while refraining from infringing on patient privacy.

C. *Completing digital therapies*

HCPs should also perform several novel functions when patients complete digital therapies. For one, HCPs should validate whether a patient has developed dependency or addiction. Public health systems' lack of assessment and validation for different digital therapies further means that these could benefit significantly from the insights HCPs have gained. HCPs should therefore share their experience and qualified view on the efficacy of therapeutics with their health systems and the wider scholarly debate.

16.3 Conclusion

A growing number of digital therapeutics, in particular offerings for mental health and addiction, are becoming available to citizens and patients. However, digital therapeutics, in general, and offerings for mental health and addiction, in particular, raise a plethora of ethical challenges. This chapter clusters issues into those affecting individuals directly, concerning society at large, and especially arising for HCPs. On an individual level, privacy, transparency, and autonomy are of major relevance. At the societal level, lack of clinical validation as well as fairness and equality play a major role.

Privacy is a major concern as highly sensitive data have been resold by therapeutics providers to third-party private actors that may use this to influence or even manipulate individuals. Missing transparency on how data and insights are collected, analyzed, and shared further complicates this issue and deteriorates patient trust and thus intervention effectiveness. Digital therapeutics may also reduce individuals' autonomy. Widespread lack of clinical validation for many digital therapeutics makes it difficult to verify the real medical benefit. This is especially problematic in the context of mental health and addiction, where patients require the best help available, but might even end up harmed. Digital therapeutics also raise major issues around equality and fairness. On the one hand, not all patients may have access to these technologies. On the other hand, digital therapeutics' lower cost may in some instances be used to supplement more expensive therapies for less advantaged groups.

Healthcare professionals (HCPs) gain additional roles and enhanced responsibility. We argue that it is important that HCP's have a well-founded knowledge of the opportunities as well as the ethical challenges and risks of digital therapeutics. They should educate patients of the ethical issues involved and provide them practical guidance on how to use digital therapeutics safely. They should also only recommend digital therapeutics whose efficacy has been validated. HCPs should cautiously monitor and interact with their patients while they use digital therapeutics and ensure that they themselves remain aware of major ethical and legal developments in the field.

16.4 Appendix

When using the case studies in Sections 16.2.1.3 and 16.2.2.2 for further reflection or classroom discussion, you could use the following questions as guidelines:

- Which ethical issues arise from each intervention?
- Which intervention would you be comfortable recommending to a patient? Why?
- What could other interventions learn from Robo-trainer and Robo-friend mitigation and management of ethical challenges?

Bibliography

Adjekum, A., Blasimme, A., & Vayena, E. (2018). Elements of trust in digital health systems: Scoping review. *Journal of Medical Internet Research, 20*(12), e11254. doi:10.2196/11254.

American Medical Association. (2016). Digital Health Study—physicians' motivations and requirements for adopting digital clinical tools. https://www.ama-assn.org/sites/ama-assn.org/files/corp/media-browser/specialty%20group/washington/ama-digital-health-report923.pdf. Accessed on March 21, 2021.

Ananny, M., & Crawford, K. (2018). Seeing without knowing: Limitations of the transparency ideal and its application to algorithmic accountability. *New Media & Society, 20*(3), 973–989. doi:10.1177/1461444816676645.

Bauer, M., Glenn, T., Monteith, S., Bauer, R., Whybrow, P. C., & Geddes, J. (2017). Ethical perspectives on recommending digital technology for patients with mental illness. *International Journal of Bipolar Disorders, 5*(1), 6. doi:10.1186/s40345-017-0073-9.

Beauchamp, T. L., & J. F. , C. (2001). *Principles of Biomedical Ethics* (5 ed). New York, NY: Oxford University Press.

Bending, Z. J. (2015). Reconceptualising the doctor-patient relationship: Recognising the role of trust in contemporary health care. *Journal of Bioethical Inquiry, 12*(2), 189–202. doi:10.1007/s11673-014-9570-z.

Berensmann, M., & Gratzfeld, M. (2018). [Requirements for CE-marking of apps and wearables]. Bundesgesundheitsblatt, Gesundheitsforschung, Gesundheitsschutz, 61 (3), 314–320. https://doi.org/10.1007/s00103-018-2694-2.

Burr, C., Morley, J., Taddeo, M., & Floridi, L. (2019). Digital Psychiatry: Risks and Opportunities for Public Health and Wellbeing. *IEEE Transactions on Technology and Society, 1*(1), 21–33. doi:10.1109/TTS.2020.2977059.

Campbell, A. N., Nunes, E. V., Matthews, A. G., Stitzer, M., Miele, G. M., Polsky, D., & Ghitza, U. E. (2014). Internet-delivered treatment for substance abuse: A multisite randomized controlled trial. *American Journal of Psychiatry, 171*(6), 683–690. doi:10.1176/appi.ajp.2014.13081055.

Doraiswamy, P. M., London, E., Varnum, P., Harvey, B., Saxena, S., Tottman, S., ... Candeias, V. (2019). Empowering 8 Billion Minds: Enabling Better Mental Health for All via the Ethical Adoption of Technologies. *NAM Perspectives, 2019*(10). doi:10.31478/201910b.

Ferreri, F., Bourla, A., Mouchabac, S., & Karila, L. (2018). e-Addictology: An overview of new technologies for assessing and intervening in addictive behaviors. *Frontiers in Psychiatry, 9*, 51. doi:10.3389/fpsyt.2018.00051.

Gille, F., Jobin, A., & Ienca, M. (2020). What we talk about when we talk about trust: Theory of trust for AI in healthcare. *Intelligence-Based Medicine, 1-2*, 100001. https://doi.org/10.1016/j.ibmed.2020.100001.

Glenn, T., & Monteith, S. (2014). New measures of mental state and behavior based on data collected from sensors, smartphones, and the Internet. *Current Psychiatry Reports, 16*(12), 523. doi:10.1007/s11920-014-0523-3.

Haug, S., Castro, R. P., Kwon, M., Filler, A., Kowatsch, T., & Schaub, M. P. (2015). Smartphone use and smartphone addiction among young people in Switzerland. *Journal of Behavioral Addictions, 4*(4), 299–307. doi:10.1556/2006.4.2015.037.

Ienca, M., & Andorno, R. (2017). Towards new human rights in the age of neuroscience and neurotechnology. *Life Sciences, Society and Policy, 13*(1), 5. doi:10.1186/s40504-017-0050-1.

Ienca, M., Fabrice, J., Elger, B., Caon, M., Scoccia Pappagallo, A., Kressig, R. W., & Wangmo, T. (2017). Intelligent assistive technology for Alzheimer's disease and other dementias: A systematic review. *Journal of Alzheimer's Disease, 56*(4), 1301–1340. doi:10.3233/jad-161037.

Ienca, M., Haselager, P., & Emanuel, E. J. (2018). Brain leaks and consumer neurotechnology. *Nature Biotechnology, 36*(9), 805–810. doi:10.1038/nbt.4240.

Ienca, M., Shaw, D. M., & Elger, B. (2019). Cognitive enhancement for the ageing world: Opportunities and challenges. *Ageing and Society, 39*(10), 2308–2321. doi:10.1017/S0144686X18000491.

Jones, D. J., Anton, M., Gonzalez, M., Honeycutt, A., Khavjou, O., Forehand, R., & Parent, J. (2015). Incorporating mobile phone technologies to expand evidence-based care. *Cognitive and Behavioral Practice, 22*(3), 281–290. doi:10.1016/j.cbpra.2014.06.002.

Khoury, M. J., & Ioannidis, J. P. A. (2014). Medicine. Big data meets public health. *Science (New York, N.Y.), 346*(6213), 1054–1055. doi:10.1126/science.aaa2709.

Lattie, E. G., Adkins, E. C., Winquist, N., Stiles-Shields, C., Wafford, Q. E., & Graham, A. K. (2019). Digital mental health interventions for depression, anxiety, and enhancement of psychological well-being among college students: Systematic review. *Journal of Medical Internet Research, 21*(7), e12869. doi:10.2196/12869.

Lucas, G. M., Rizzo, A., Gratch, J., Scherer, S., Stratou, G., Boberg, J., & Morency, L.-P. (2017). Reporting mental health symptoms: Breaking down barriers to care with virtual human interviewers. *Frontiers in Robotics and AI, 4*(51). doi:10.3389/frobt.2017.00051.

Mares, M.-L., Gustafson, D. H., Glass, J. E., Quanbeck, A., McDowell, H., McTavish, F., & Ward, V. (2016). Implementing an mHealth system for substance use disorders in primary care: A mixed methods study of clinicians' initial expectations and first year experiences. *BMC Medical Informatics and Decision Making, 16*(1), 126. doi:10.1186/s12911-016-0365-5.

Marshall, J. M., Dunstan, D. A., & Bartik, W. (2019). Clinical or gimmickal: The use and effectiveness of mobile mental health apps for treating anxiety and depression. *Australian & New Zealand Journal of Psychiatry, 54*(1), 20–28. doi:10.1177/0004867419876700.

Martinez-Martin, N., Insel, T. R., Dagum, P., Greely, H. T., & Cho, M. K. (2018). Data mining for health: Staking out the ethical territory of digital phenotyping. *NPJ Digital Medicine, 1*. doi:10.1038/s41746-018-0075-8.

Martinez-Martin, N., & Kreitmair, K. (2018). Ethical issues for direct-to-consumer digital psychotherapy apps: Addressing accountability, data protection, and consent. *JMIR Mental Health, 5*(2), e32. doi:10.2196/mental.9423.

McFarland, M. S. (2012). What is Privacy?. https://www.scu.edu/ethics/focus-areas/internet-ethics/resources/what-is-privacy/. Accessed on December 18, 2020.

Meurk, C., Hall, W., Carter, A., & Chenery, H. (2014). Collecting real-time data from substance users raises unique legal and ethical issues: Reply to Kuntsche & Labhart. *Addiction, 109*(10), 1760. doi:10.1111/add.12640.

O'Hara, D. (2019). Lisa A. Marsch develops digital interventions for the treatment of opioid addiction. https://www.apa.org/members/content/digital-interventions-opioid-addiction. Accessed on: March 21, 2021.

Olsen, J. A. (2011). Concepts of equity and fairness in health and health care. The Oxford Handbook of Health Economics. Oxford University Press, doi:10.1093/oxfordhb/9780199238828.013.0034.

Parker, L., Halter, V., Karliychuk, T., & Grundy, Q. (2019). How private is your mental health app data? An empirical study of mental health app privacy policies and practices. *International Journal of Law and Psychiatry, 64*, 198–204. doi:10.1016/j.ijlp.2019.04.002.

Robillard, J. M., Feng, T. L., Sporn, A. B., Lai, J. A., Lo, C., Ta, M., & Nadler, R. (2019). Availability, readability, and content of privacy policies and terms of agreements of mental health apps. *Internet Interventions, 17*, 100243. doi:10.1016/j.invent.2019.100243.

Shah, P., Thornton, I., Turrin, D., & Hipskind, J. E. (2020). Informed consent. https://europepmc.org/books/n/statpearls/article-23518/?extid=29083707&src=med. Accessed on March 22, 2021.

Tofighi, B., Leonard, N., Greco, P., Hadavand, A., Acosta, M. C., & Lee, J. D. (2019). Technology use patterns among patients enrolled in inpatient detoxification treatment. *Journal of Addiction Medicine, 13*(4), 279–286. doi:10.1097/adm.0000000000000494.

Torous, J., Andersson, G., Bertagnoli, A., Christensen, H., Cuijpers, P., Firth, J., & Arean, P. A. (2019). Towards a consensus around standards for smartphone apps and digital mental health. *World Psychiatry, 18*(1), 97–98. doi:10.1002/wps.20592.

Torous, J., Firth, J., Huckvale, K., Larsen, M. E., Cosco, T. D., Carney, R., & Christensen, H. (2018). The emerging imperative for a consensus approach toward the rating and clinical recommendation of mental health apps. *Journal of Nervous and Mental Disease, 206*(8), 662–666. doi:10.1097/nmd.0000000000000864.

Vayena, E., Haeusermann, T., Adjekum, A., & Blasimme, A. (2018). Digital health: Meeting the ethical and policy challenges. *Swiss Medical Weekly, 148,* w14571. https://doi.org/10.4414/smw.2018.14571.

Vayena, E., & Madoff, L. (2019). Navigating the ethics of big data in public health. *The Oxford Handbook of Public Health Ethics,* edited by Anna C. Mastroianni, Jeffrey P. Kahn, and Nancy E. Kass p. 354. Oxford Univeristy Press.

Wykes, T., Lipshitz, J., & Schueller, S. M. (2019). Towards the design of ethical standards related to digital mental health and all its applications. *Current Treatment Options in Psychiatry, 6*(3), 232–242. doi:10.1007/s40501-019-00180-0.

Chapter 17

A look forward to digital therapeutics in 2040 and how clinicians and institutions get there

Donald M. Hilty[a], Christina M. Armstrong[b], Amanda Edwards-Stewart[c] and David D. Luxton[d]

[a] Northern California Veterans Administration Health Care System Professor, Department of Psychiatry & Behavioral Sciences, UC Davis 10535 Hospital Way, Mather, CA, [b] Office of Connected Care, U.S. Department of Veterans Affairs, Washington D.C., [c] AES Psychological Service, LLC, Tukwila, Washington, [d] Department of Psychiatry and Behavioral Sciences, University of Washington School of Medicine, Seattle, Washington

17.1 Introduction

The digital revolution over the last three decades has changed how services are provided to consumers, and nearly all industries, from higher education, retail, banking, and even to dating firms, had to adjust to people's preferences for electronic and online modalities to remain viable and competitive (Christensen, 2007; Kotter, 1996). These changes were largely driven by consumers and were reinforced by increased adoption of online options over time. Modern healthcare, with its focus on well-being, prevention, and integrated care approaches, has transitioned from patient-centered care to person-centered care (deBronkart, 2015).

The COVID-19 pandemic provided an opportunity to advance the use of technology solutions in person-centered healthcare, and patients, clinicians, and healthcare staff are expecting more from technology-based tools than ever before. This current period in history builds on several tipping points in recent decades—the effectiveness of care delivered through synchronous telehealth (Hilty et al., 2015c); the uptake of smartphones and mobile health applications (apps) (Pew Research Center, 2019) social media in healthcare (Naslund et al., 2019), wearable sensors (Hilty et al., 2021b), and informatics reshaping of clinical workflow in healthcare institutions worldwide (Luxton, 2016). These shifts align with theories in technology adoption and innovation advancement, such as Roger's Diffusion of Innovation (Rogers, 2010), Moore's Law (Gustafson, 2011), as well as technology advancement in past centuries (e.g., telegraph) (Mermelstein et al., 2017). More recently, digital natives (i.e., persons born or brought up during the age of the Internet and digital technology) are being contrasted with digital immigrants using the concept of digital fluency rather than by generation and age (Wang et al., 2013). Learning from decades of prior research and experience, hybrid solutions that blend in-person, video, and other technology treatments will be the most effective solution (Torous et al., 2020). While technology continues to advance, the current era of rapid technology adoption represents a paradigm shift for patients and clinicians from in-person, cross-sectional, and care to future in time, longitudinal, technologically based care interventions (Hilty et al., 2021d).

Driven by both the shift toward person-centered care and the vast expansion of an e-spectrum of care using technology, we can expect healthcare technology to transition from an appendage or facilitator to an organizer of care in over the next two decades (Hilty et al., 2019c). This is a potentially dramatic shift in the experience for all healthcare users and healthcare institutions. The new vision of patient and clinician engagement will require digital literacy assessment, training (competencies) for healthcare staff, and evaluation/process improvement to move forward. Multidirectional decision support for users will be based on ongoing, 24 hours × 7 days/week data collection, integration, and analysis. The daily workflow

Digital Therapeutics for Mental Health and Addiction: The State of the Science and Vision for the Future. DOI: https://doi.org/10.1016/B978-0-323-90045-4.00014-9

will move from structural "visits" to teams used for collaborative, stepped functional in-time data analysis, automated interventions, and communication. Current and integrated care models will have additional team members—technology *de novo* as one and digital navigators—in the move from primary care and behavioral health settings to community settings including the home.

Public health and healthcare institutions, while still using traditional approaches of population-centered health, team-based care, and public health surveillance must adapt to advancements in integrated workflow across private, public, and other settings. Artificial intelligence (AI), predictive modeling (PM), and other informatics capabilities are rapidly transforming information flow, insights, and ultimately decision-making. Institutional, population health, and administrative technological competencies should, in principle, guide this transformation. If, however, financing, reimbursement, and regulatory bodies in countries, provinces/states, and organizations consider more than basic adjustments, or adapt inroads via COVID-19 and apply them to other technologies, a redesign of financing, reimbursement, and regulatory systems could create a "pull" that will be even more challenging.

This chapter will explore: (1) how care will be delivered, moving from current in-person, cross-sectional, and care to future in time, longitudinal, technologically based care interventions; (2) how shift patients and clinician engagement, experience, and participation via technology; and (3) shifts for systems and institutions based on person- and population-centered health, implementation science, effectiveness, and team-based care.

17.2 Technology's role in shifting care: market pull from users and the push from artificial intelligence

17.2.1 People empowered by mobile technologies

A paradigm shift is taking place for persons/patients and clinicians from current in-person, cross-sectional care to future in-time, longitudinal, technologically based interventions. Technologies applied to behavioral health are being added to in-person care and synchronous video to help people, patients, families, caregivers, and primary care providers (Hilty et al., 2019a; Hilty et al., 2020a; Hilty et al., 2015a; Yellowlees et al., 2018). On a spectrum of low to high engagement and technology requirements, participants use website information, support & chat groups, social media, resources for self-directed assessment and care, asynchronous patient-clinician communication (i.e., apps, text, or e-mail), STP, and hybrid care models (i.e., in-person and/or technology combinations) (Chan et al., 2014; Hilty et al., 2015a). Many of these technologies are used for patient self-management of symptoms, but others aim to support primary care provider and specialist delivery of care and/or collaboration, via low- (e.g., e-mail, phone), mid- (e.g., disease management, consultation care), and high-intensity (e.g., collaborative, hybrid care) models of care (Hilty et al., 2018b).

17.2.2 The research and planning enabling the shift

Health technology research continues to expand on the effectiveness of asynchronous communication (i.e., text, secure messaging), and synchronous communication (video telehealth, e-consultation), as well as technologies that facilitate the management of healthcare across a wide range of use cases (i.e., electronic health record (EHR), patient portals, apps, social media) (Hilty et al., 2015a; Hilty et al., 2018b; Yellowlees et al., 2018). Definitions for these technologies are as follows:

- Asynchronous video: refers to the "store-and-forward" technique, whereby a patient, primary care staff or behavioral health clinician collects psychiatric history and other medical information by video and then sends it to a psychiatrist for diagnostic and treatment recommendations.
- E-mail: a method of exchanging messages ("mail") between people using electronic devices.
- E-consultation: involves remote communication between patients and clinicians, or between clinicians and specialists, using e-mail, text, and/or documented notes; using a video to complement using e-mail, text, and/or documented notes is more dynamic.
- Sensors and wearables: refer to wearable hardware or textiles (i.e., wristbands, smartwatches, smart glasses, shirts, patches, bandages) that include sensors that allow for the passive collection of biological and behavioral data, and can include analytics and machine learning providing the user with feedback.
- Text/app: messages typically consisting of alphabetic, numeric characters, and ideograms known as emoji (happy faces, sad faces, and other icons) sent between two or more users of mobile devices or desktops/laptops (previously short message service or SMS and now multimedia messages or MMS) with digital images, videos, and sound content.

Wearable sensors use AI and ML techniques to collect behavioral data, alert, communicate/give feedback, detect changes, monitor symptoms, access information, and provide preventive and therapeutic interventions (Hilty et al., 2021b; Luxton, 2016). This enables behavior change via algorithms that provide customized feedback to the user (Cho et al., 2019), enables clinicians to make decisions "in-time," and helps healthcare organizations perform robust analyses of clinical data (Watson et al., 2020). These developments have come at an opportune time, as there is significant interest in and commitment toward investment in analytics to improve care delivery and bend the cost curve.

Sensors and wearables, cognitive computing (Fraccaro et al., 2019; McIntyre et al., 2014), technology miniaturization, and placement, communication architectures and fifth-generation (5G) cellular network technology, data analytics and algorithms (Luxton, 2016), and evolution of cloud and edge computing architectures (Hobson et al., 2019) are exciting fronts in healthcare technology and clinical decision-making. The future state with cognitive computing emphasizes temporal modeling and perception of deltas (i.e., change over time). Bioinformatics integrate vast and complex phenotypic, anamnestic, behavioral, family, and personal "omics" profile data, by using a variety of processing approaches with cloud- and grid-enabled computing. This could fundamentally revolutionize how psychiatric disorders are predicted, prevented, and treated (McIntyre et al., 2014). Additional options like hearables (in-ear devices) and nearables (neighboring devices) will interact with wearables to transform future healthcare and lifestyles. Fortunately, much of the workflow in this substantial process shift in care does not fall on patients and clinicians.

This dramatic shift in healthcare also creates a need for a collaboration between healthcare, clinical informatics, and computer science. This includes behind the scenes work and collaboration by partners from diverse areas of expertise, such as informaticians, health system administrators, quality/process improvement teams, clinicians, software developers, and engineers. PM, AI, and ML techniques have the potential to improve care and decrease costs through a variety of mechanisms, such as early identification of patients requiring more intensive follow-up through readmission and postoperative complication risk models, and automation of diagnostic interpretation previously completed by humans (Edgcomb & Zima, 2019). PM supplements clinical judgment when identifying risks associated with mental or behavioral illness. It employs computing power and statistical techniques to model complex processes with higher accuracy—for depression (Nemesure et al., 2021; Pigoni et al., 2019), schizophrenia (Mikolas et al., 2018)—and other disorders like delirium (Jauk et al., 2020). AI, ML, and PM also further diagnostic "bench" research with functional magnetic resonance imaging applied to the bedside (de Filippis et al., 2019; Kalmady et al., 2019).

17.2.2.1 The shift from 2020 to 2040: new technologies entering life and healthcare

We can expect a shift from current technologies on the e-spectrum of care—used in varying degrees by patients, clinicians, and health systems—to what may be standard or readily elective by 2040 (Table 17.1). There are a variety of technological options, which may be organized or stratified into levels of care based on technology access (i.e., easy to less available), the type of interface (e.g., a human–computer vs human–human interface via computer), capacity of the technology to engage (e.g., chatbot), other participants in care (e.g., caregiver, primary care provider), specialist time (i.e., psychiatrist or behavioral health clinician), model of care and use of informatics or health system resources (Hilty et al., 2018b). The contemporary spectrum of care *within* health systems typically involves telephone, secure e-mail, in-person, video, EHR, patient portals, and a few other options (e.g., use of apps). The range of low- (e.g., e-mail, phone), mid- (e.g., disease management, consultation care), and high-intensity (e.g., collaborative, hybrid care) options provides systems flexibility (Hilty et al., 2018b).

People have been using many technological options for health and wellness purposes–health information websites, texting/chat, social media, and other options on their own– that many clinicians may see as *outside* of care. Yet, social media, for example, is integral to everyday life of many people, especially adolescents and young adults, and therefore clinicians would be remiss to not consider its role in health and wellness of people. Clinicians should systematically screen technology use and for what purpose(s) (e.g., entertainment, health or mental health), as well as exposure to risks (e.g., self-disclosure, cyberbullying, privacy) (Luxton et al., 2015; Zalpuri et al., 2018). The field of data science has emerged as a way of addressing the growing scale of data, and the analytics and computational power it requires (Wongkoblap et al., 2017). ML techniques allow researchers to extract information from complex datasets in order to interpret data, detect patterns, and create predictive models in various domains, such as finance, economics, and health. A specific segment of this work has focused on analyzing and detecting symptoms of mental disorders through status updates in social networking websites (e.g., risk for postpartum depression). More research is needed on what methods are best used to collect data from online social network sites and for what populations, as well as which techniques for predictive analytics of work best with regard to social network data in mental health. However, as technology use increases, ethical issues arise. For example, one of the main ethical concerns for patients and clinicians is related to privacy and other safeguards when using health technology.

TABLE 17.1 An e-behavioral health continuum of interventions for patients, families, and clinicians.

Source	User goals/aims	Liabilities	Suggestions
User-friendly—simple technology—less interaction—less regulated—variable quality			
Publically available options			
Website information	Gain perspective, obtain standard and updated health information Good for somatic symptom disorders	Quality of information and lack of regulation Ease of access	Refer to sites by psychiatric and medical organizations Keep information simple
Support/chat groups	Patient: education and perspective Caregivers: tips and support on coping	Peer compatibility Information quality	Verify quality and those with a professional facilitator
Social media (SM) passive and active	Patient: education and perspective Caregivers: tips and support on coping	Privacy Little time for clinicians	Active participation is more therapeutic than passive surfing
Natural language processing (computational linguistics or statistical text classification)	Patient: use for immediate crisis intervention tool Clinician: have patient diaries, statements, clinical notes, and background data analyzed	Coarse or nonspecific outcomes that require more human time Errors Adaptation to each new area often required	Customize in a few areas for clinical use Generalize with artificial intelligence and use advanced machine learning techniques
Clinical care screening and engagement			
Education program for self-assessment	Person/patient: education, tips Caregiver: education, support, and advice	Some learners prefer in-person Ease of access Quality of information	Give assignments Clarify if knowledge versus skill development
Telephone-based diagnostic systems	Person/patient/caregiver: additional options, speed up diagnostic process Clinician: reduces in-person time	Not user-friendly—as not particularly warm—but effective	Prepare by setting expectations and emphasizing how it helps
e-base self-report questionnaires fed into electronic health record	Person/patient/caregiver: additional option, taking action Clinician: reduces in-person time	False positives and negatives, but nonetheless helpful Clinicians/team members may need training to score	Use common, easy-to-score ones like the Patient Health Questionnaire Integrate automatic scoring
Patient-computer virtual, augmented, and mixed reality approaches (XR for all)			
XR applications to clinical populations	Digitally-recreated simulations of real-world activities, often involving dynamic social and emotionally engaging stimuli for interaction and assessment	Lab setting design is not always able to create psychological realism Need a sensorially rich and perceptually realistic environment	Design with lab and field settings to facilitate natural interaction between the user, objects, and the environment
Autonomous virtual human (VH) agents for clinical interviewing	Start the early steps for creating a treatment plan that would be followed on by a live clinician. Kiosk-based VH agent can leverage advances in sensing technology (e.g., cameras and microphones) Support and train social and safety skills	Impersonal, automated Gaps requiring human judgment, errors Creation and use of highly interactive, intelligent VHs for such clinical purposes is still in progress	Comparative evaluation across interviews Participants reported willingness to disclose, willingness to recommend and general satisfaction

(continued on next page)

TABLE 17.1 An e-behavioral health continuum of interventions for patients, families, and clinicians—cont'd

Source	User goals/aims	Liabilities	Suggestions
User-friendly—simple technology—less interaction—less regulated—variable quality			
Virtual effective agents and therapeutic games	Capable of effective interaction, such as expression of "emotions" and recognition of human emotions, based on emotional and social intelligence (e.g., awareness of social cues and user goals),	Computing diagnostic features require solving all sorts of open problems in respective fields (e.g., computer vision, acoustic signal processing, digital signal processing, time series modeling, natural language understanding)	Engage people in provide highly immersive learning, training and therapeutic environments that can be customized to the user's specific learning needs or therapeutic goals
Chatbot and other robotic technologies	Affective human-computer interaction: respond to the user's emotions and other stimuli; assessment of facial recognition	Working out autonomy, information exchange, how people and robots adapt/learn, and how the task shapes interaction	Ensure respect, dignity, privacy, and consent Avoid racist, sexist, and ableist morphologies
Automated mental state detection	Read-out" a person's mental state via verbal, behavior, body, and other data	Challenges from sensor inputs to diagnostic patterns Need supervised learning with supplemental annotations	Affective states and attentional lapses (i.e., lack of mindfulness) prototypes
Human behavior modeling and simulation (e.g., agent-based modeling and simulation) using motivation, state, and action	Used for serious mental illness and addressing the needs of the most vulnerable populations in our society, based on comprehension, manageability, and meaningfulness	A person's behavior is not always determined by their intention to perform the behavior and this intention is, in turn, may not be a function of their attitude toward the behavior and subjective norms	Improvement on discriminative ability of traditional risk prediction models, which are little better than coin-flipping at predicting readmissions
Brain computer interface (BCI): system that translates "brain signals into new kinds of outputs": implantable and noninvasive	Exercise training devices for neurological disorders, usually: wearable powered braces for daily use; range of motion and strength training devices for musculoskeletal disorders; assistance (e.g., exercise) with activities of daily living; and social or telepresence robots	Professionals believe it may interfere with, or be perceived by patients to interfere with, in-person communication between the provider and patient. Cultural acceptance: some have more positive attitudes toward some robot morphologies compared to other cultures	The interface determines: (1) what degree the user is separated from the physical world; (2) what stimuli are introduced and how; (3) artificial intelligence versus human control; and (4) the proportion of objective, quantifiable data collected
Customized human–human care interactions			
Consult with e-services clinician for self-care decisions	Person/patient/caregiver: additional options, perspective, context Clinician: helps independent learners and requires less time with patient	Complexity: not often as simple as do A or do B Scope of practice questions	Link with social work, hotline and/or clinic, if needed Context with input from own clinician is better
Assisted self-care assessment and decision-making	Person/patient/caregiver: empowering as customized and supported Clinician: effective to distribute skills	Asynchronous communication has limitations Context for decisions is lacking	Use clinicians in health system Integrate in electronic health record (EHR) and care team

(continued on next page)

TABLE 17.1 An e-behavioral health continuum of interventions for patients, families, and clinicians—cont'd

Source	User goals/aims	Liabilities	Suggestions
User-friendly—simple technology—less interaction—less regulated—variable quality			
Asynchronous video or one-time synchronous consultation	Person/patient/caregiver: good step to enable primary provider Clinician: efficient use of time, help from others and empowers patients	Can primary provider use tips? They will work for which patients? Learning curve takes some time?	Build into the regular care continuum Provide training on how to do each step
Synchronous, telepsychiatry (TP)	Person/patient: effective and is much more convenient Clinician: if patients like it, it is good	Has to be scheduled (and paid for) Workflow sometimes a little more demanding on clinician/team	Have technology, clerical, and administrative support Training on clinical skills
Hybrid care: in-person & e-option (TP & e-option)	Person/patient: options to connect Clinician: preferred if gives options and workflow not cumbersome	Systems vary in quality, training, prioritization and quality improvement; time and $ costs	Develop integrated workflow for clinicians and administrators Facilitate positive e-culture
Systems for patients and clinicians via informatics system designs for care			
eConsult between primary care provider (PCP) and specialist in EHR	PCP (geriatrician, family medicine): timely to visit and sent "in time" Specialist: simple questions (e.g., facts, steps to do) can be answered	May not work for difficult patient cases that require integration Time to clarify question, review chart and communicate plan	Monitor timeliness, follow up and quality; build into care workflow and culture of care Use video: effective and easier
Asynchronous, between-session, patient-clinician contact (e.g., app, text)	Person/patient/caregiver has minor question or needs a detail → e-mail/text; tracking symptoms → app Clinician: e-mail/text for quick, simple advice; apps good for monitoring	Align 1–2 apps with 1–2 purposes to focus Errors, miscommunications Time, documentation, and privacy issues	Training for faculty and team Integrate into EHR for clinical decision support Use evidence-based app and evidence-based approach
24 X 7 ecological, in-time and multidirectional collection, integration, communication, and decision support based on artificial intelligence, machine learning and predictive modeling	Person/patient: options to reflect and make decisions in real-time (e.g., behavior to improve mood or reduce stress via electrodermal activity) Clinician: adds new workflow but better data for decision-making System: benefits from context-aware systems that automatically detect when patients require assistance or intervention	Culture shift from traditional visit or TP care to in-time care Work in teams with technology as a new "team member" or digital navigators "steer" Functional > structural	Train professionals in teams and shift faculty beyond health "knowledge" experts to synthetic data leaders Define roles in teams and system for workflow
Digital phenotyping/biomarkers and public/population health surveillance	Public health surveillance and population health integration Clinical and nonclinical populations to correlate multimodal sensor data, cognitions and depressive mood Early signs of relapse and intervention	Privacy concerns Quality and safety barriers to adoption Specificity to state and trait	Use large multicentric databases is an essential element for development and validation
Workflow investment—complex technology—more interaction—high quality—high clinician time			

17.2.2.2 Health information on the Internet, support/chat groups, and educational programming

The users of the Internet (Hilty et al., 2015a) are slightly more female (86% vs 73% of men) and hold a higher level of educational attainment (89% of those with a college degree vs 70% with a high school degree vs 38% without a high school diploma). These people, patients, and caregivers (a term used for adults who provide unpaid care to a parent, child, friend, or other loved ones). They seek information on diseases or medical problems, treatments or procedures, doctors or other health professionals, hospitals or other medical facilities, food safety or recalls, drug safety or recalls, and pregnancy and childbirth. Caregivers in about two-thirds of studies who use Internet-based services have significantly reduced stress and improved quality of life, according to a review of medical and mental health disorders (Hu et al., 2015). Internet-based interventions range from interactive communities to bulletin boards to therapy groups. Many patients migrate to sites like *PatientsLikeMe* (http://www.patientslikeme.com/) is a consumer driven site where individuals log-on to connect with others in the community who are experiencing similar medical issues. Health promotion strategies are typically at free-standing websites (Siemer et al., 2011).

Young people with developmental challenges may have limitations in access to traditional care options and feel more comfortable anonymously or at a distance, to share experiences and try to learn new behaviors (Berger et al., 2005). Comfortable with Internet-based chats and groups, they may even express ideas of self-harm, negative affective states, or pessimistic cognitions of other peers (Griffiths et al., 2009). Moreover, young persons may also seek out means to suicide on chats and groups, and also become victims to cyberbullying, increasing risk for suicide (Luxton et al., 2012)—concerning, though, if these things are not also shared with parents and/or professionals. Those with anxiety (phobias or panic disorder) and trauma (e.g., military personnel) patients may be avoidant of in-person care and choose the Internet (Hilty et al., 2015a). Likewise, those with barriers to access to care based on geographic location (i.e., rural or otherwise long travel), or those with little/no access to care (e.g., waiting time), or those that dislike of local services (e.g., seeing options as ineffective or of limited quality) also tend to use the Internet.

Most support groups are for consumers and patients. These Internet interventions are developed around theoretically grounded concepts and offer good prospects for improving health-related behaviors, are based on the following premises: (1) knowledge affects changes in behaviors, (2) peer support/feedback may induce such changes (or in some cases, the opposite), and (3) even informal contact by e-mail, chat or telephone with a healthcare provider feels personalized and affects such changes. Internet-mediated support groups can include specialized groups for individuals with disabilities or unique modes of experience (Antze, 2010).

17.2.2.3 Virtual reality and augmented reality (AR) experimental approaches in social science and neuroscience

The fields of behavioral health and social psychology can make use of these technologies as they contribute four total physical response factors/determinants: (1) control (i.e., degree, immediacy and mode of it; anticipation of events); (2) sensory experience (i.e., environmental richness, multimodal presentation, consistency of information, degree of movement perception); (3) attention (i.e., focus or distraction); and (4) realism (e.g., scene, information, meaningfulness, separation anxiety/disorientation) (Witmer & Singer, 1998). To focus a clinical intervention, augmented reality (AR) is not technically limited to introducing virtual elements into the physical world. It may also inhibit the perception of physical objects by overlaying them with virtual representations, such as a virtual objects or even virtual empty spaces—this reduces distractions (Hilty et al., 2020c).

Assessment of interpersonal behavior can be piloted in synthetic and applied to real environments. Social scientists studying interpersonal behavior are interested in investigating why one person behaves differently from another person (i.e., study of individual difference). So, they are creating, exploring, and testing environments used for learning, engagement, and reflection by simulating day-to-day life. This includes social situations with objective stimuli and reactions to unusual stimuli (e.g., disasters, moral dilemmas, stress response) (Bombari et al., 2015; Parsons et al., 2017; Smith, 2015). Behavioral or performance measures examine natural behavior (e.g., facial expressions and reflex, postural responses, or social responses) exhibited by the user in responses to objects or events in the virtual environment. These are considered objective measures and are collected without involving participants' conscious deliberation (van Baren and Ijsselsteijn, 2004). These are considered most helpful with when used as adjuncts to subjective measures. Technology can improve the approach to environmental assessment—which consists of a wide range of cognitive, affective, and behavioral responses—by standardizing conditions, measurement instruments, and outcome targets (i.e., responses).

Within the public health domain, substantive interactive virtual environment advances have been made in several specific areas such as obesity prevention and maintenance, dementia, genomics research, and health communication. A turning event in the delivery of care has occurred: this technology can be used to simulate *distal consequences* of individuals' *immediate decisions* related daily choices (e.g., soft drink consumption, which leads to weight gain can be "experienced" in-time, whereas in the past the patient was "educated" with information). Indeed, many factors may influence the adoption of preventive behaviors in-time (e.g., print vs radio vs television; depicting celebrities' opinion).

17.2.2.4 Human behavior modeling and simulation for mental health conditions

Health reform is ultimately concerned with social justice and social change. These goals require a sophisticated understanding of the contexts that give rise to social problems and the use of research methods and change strategies that attend to the complexities of social settings. Although researchers dedicate considerable attention to these concerns, the ability to understand the intricate and dynamic relationship at individual, organizational, and societal levels still lags behind the considerable need in our society for transformative change in healthcare delivery. Over the last 30 years, minimal progress has been made on the discriminative ability of traditional risk prediction models, and they are little better than coin-flipping at predicting readmissions. Agent-based modeling and simulation may be used for understanding individual, organizational, and societal levels of a hospitalization (Silverman et al., 2015).

The basis for this approach is comprehension, manageability, and meaningfulness. The theory of reasoned action suggests that a person's behavior is determined by their intention to perform the behavior and that this intention is, in turn, a function of their attitude toward the behavior and subjective norms (Fishbein & Ajzen, 1977). The best predictor of behavior is intention or instrumentality (belief that the behavior will lead to the intended outcome). Instrumentality is determined by their attitude toward the specific behavior, their subjective norms, and their perceived behavioral control—the more favorable these are, then the stronger the person's intention to perform the behavior. These concepts are often applied in healthcare by stratifying people into low mental and physical health problems (Quadrant I), high mental health problems and low physical health problems (Quadrant II), low mental health and high physical health problems (Quadrant III), and high mental and physical health problems (Quadrant IV).

17.2.2.5 Robotics technology

The use of robotics technology in mental healthcare is nascent, but represents a potentially useful tool. The central issues involve the level and behavior of a robot's autonomy, the nature of information exchange between human and robot, how people and robots adapt and learn from one another, and how the task shapes interaction. Robots currently used in mental healthcare vary greatly in their morphology and include zoomorphic, mechanistic, cartoon-like, and humanoid representations. These robots have been used for helping treat people with dementia, autism, and cognitive impairments and in rehabilitation (Luxton & Riek, 2019). Robots are quite engaging, due to people's innate tendency to anthropomorphize anything with animacy (Waytz et al., 2010). Many modern robotic systems have adjustable autonomy, where a person can change how they interact with a robot in real-time, which is particularly useful in mental healthcare scenarios, where a professional may wish to have certain robot behaviors directly controlled and others autonomous. Robotics technology has also been used in novel ways to study people with schizophrenia, based on interactions to evaluate social rapport and group participation (Lavelle et al., 2014).

17.2.2.6 Ethical issues with new technologies (e.g., robotics)

There are many issues to evaluate and the example of robotics is a good case example.

One is the prime directive or the principle of respect for human persons, including respect for human autonomy, respect for human bodily, and mental integrity, and the affordance of all rights and protections ordinarily assumed in human–human interactions. The human's right to privacy shall always be respected to the greatest extent consistent with reasonable design objectives; likewise, data privacy, security, confidentiality, and integrity are important. Legal considerations include relevant laws and regulations concerning individuals' rights and protections (e.g., food and drug administration, federal trade commission, European Union) and human informed consent. It is prudent to avoid turing deceptions (i.e., a person is unable to determine whether they are interacting with a machine or not) and the tendency for humans to form attachments to and anthropomorphize robots should be carefully considered during design. Finally, it is important to avoid racist, sexist, and ableist morphologies (i.e., discrimination and social prejudice against people with disabilities and/or people who are perceived to be disabled) and behaviors in robot design.

17.3 A look at how users' experience of care and clinical workflow will change by 2040

17.3.1 Overview

In response to COVID-19, many health systems have been moving toward the integration of in-person and video services (Torous et al., 2020), due to patient requests and attempts to offer more points-of-service, while trying to more efficiently use providers at the top of their license (Hilty et al., 2015c). Normally, change with technology takes time, competes with other demands and requires investment—but buy-in has come from all participants of care and technological assistance rapidly moved up the priority list. What are the features of the potential 2040 model of care? At a minimum, it will include: customer-based health, wellness and prevention; joint ownership of care; 24 hours by 7 days/week data collection, monitoring, and flow; and multidirectional decision support for users. There will be many new signposts of care, including patient-initiated reflection, identification of needs/preferences, and participation in available technologies. For clinicians, less are doing inpatient and more outpatient services, and many who did only in-person are doing only video and telephone services. The inroads in mobile health, wearable sensors, patient portals, and use of mobile phones for life activities could usher in a broader shift in healthcare. Normally, there is a need for training to use new technologies, but attitudes have been shifted and workflows can be simplified or intuitive, clinicians are more likely to adopt technology and see the benefits if it goes smoother than with the EHR implementation (Cowan et al., 2019; Hilty et al., 2021d).

Furthermore, the person receiving care may be an older adult patient with a caregiver and/or be an adolescent with a family—anyone can face homelessness, suicide, substance use, and mood and anxiety disorders (Hilty et al., 2021e; Hu et al., 2015). Youth and their parents/guardians often have no or very little guidance about vetting apps associated with evidence-supported behavioral interventions, compared with nonevidence-based apps (Grist et al., 2017). A review of asynchronous technologies found suicide prevention apps found disparity in app quality, content used/developed, and research methodology, and that participants using an app for homelessness reported receiving the most benefit from the daily motivational tips (e.g., overcoming struggles, making progress in life, well-being), daily surveys, up-to-date resources, and an automated self-help system (Hilty et al., 2021e). Indeed, women's health approaches have used family based care with attachment-based and include skill-based treatments delivered to individuals, dyads, family, and groups, and in locales ranging from hospitals to clinics to community centers to homes (Dossett & Shoemaker, 2015).

In some current systems (e.g., US Department of Veterans Health Administration, private sector) patients are assessed for interest in using technology—video telehealth, apps and others—at the beginning of care. This occurred before but was certainly enhanced/furthered by COVID-19. Modest examples include video versus in-person visits and use of video on demand instead of a telephone care for inter-visit spontaneous appointments—the latter is often preferred by the clinicians to better "see" the patient for evaluation and treatment for patients with behavioral health conditions. In the future, though, patients could choose from additional options besides the Internet, chat rooms, educational programs and/or advice nurses. Some diagnosed Veterans will be triaged to very specific technologies like virtual or AR for trauma or a phobia. Others may be lined up for an autonomous virtual human agents for a clinical interview or more specifically with a virtual affective agent if a mood disorder is likely—the high incidence may have this built into clinical workflow.

17.3.2 Evolution of the healthcare team

Changes are required from individual care to teamwork, adaptation of members' roles, and a substantial change in team membership are needed for technology to be integrated into mental healthcare (Hilty et al., 2019a; Hilty et al., 2018c). Related to primary care, customary members of the team are the patient/family, peer provider, social work, primary care, psychologist, nurse, and psychiatrist. Technology itself can be considered a team member (Hilty et al., 2018c) or a "practice extender" by performing some of the tasks others did to integrate care (Raney et al., 2017). A relatively new digital navigator team member has also been suggested, as an entry healthcare role to provide systems flexibility. The role could be filled by an established clinician, nurse case manager, peer specialist, or other who gains specific expertise in digital health and serves in the role of a clinical champion to support integration into the clinical workflow (Wisniewski et al., 2020).

Regardless of the navigator's background experience and skills, the three core responsibilities focus on mobile health: (1) selecting apps; (2) troubleshooting technology; and (3) reviewing and quality checking digital data to facilitate care. Much like a radiology or pathology technologist became an essential role as each of these fields by developed new imaging and lab tests/workflow, so behavioral health must now with digital navigators. However, the nature of mental healthcare and a focus on the therapeutic alliance demands additional steps in the approach (e.g., communication with patients and teams, engagement within reasonable times and boundaries). While the digital navigator role and training is not designed to develop medical thinking or decision-making, it can bring a valuable skillset to the interprofessional team via a potential 10-hour training, certification, and recertification (Wisniewski et al., 2020).

17.3.3 Interprofessional teamwork

Synchronous and asynchronous technologies, in particular, may promote a patient-centered approach and emphasize interprofessional teamwork. Efficient clinical operations match clinician expertise (i.e., at the "top of one's license") and teamwork to meet patient needs at the point-of-service. For example, care coordinators/managers can manage secure mail, nurse practitioners/physician assistants can initiate e-consults, and mid-level behavioral health professionals may evaluate less complex cases—each of these options preserves physician time for analysis of data, complex cases, and supervision. Team-based care with technology ideally offers a variety of options: learning by patients and clinicians (e.g., curricula); levels for low- to high-experienced members; development of attitudes and skills in addition to knowledge outcomes; explicit activities for teams to communicate (e.g., huddles, chat rooms, group texts); and perhaps most-importantly, supervision for feedback, reflection, and developing good habits (e.g., text to supervisor in time for help).

Teamwork is facilitated by a shared mental model of expectation, roles, and outcomes (Ross & Allen, 2012). Physical (e.g., schedules, huddles), virtual (i.e., on-site and distant member), and other training interventions may substantially improve team-based care—coordination, communication, and teamwork—and lead to decreased length of stay, fewer emergency room visits/readmissions, and better quality and safety (Agency for Healthcare Research and Quality, 2019; Will et al., 2019). Specifically, technology can organize workflow, as mobile health architecture with data monitoring may alert participants to take action (Silva et al., 2015). It may also serve as a virtual team member by performing tasks previously done by others (Hilty et al., 2019a; Hilty et al., 2018c). This may help organizations offer flexible work schedules without lowering quality of service and raising the frequency of errors (Stimpfel et al., 2012; Will et al., 2019).

Clinicians who spend more time reviewing data from asynchronous technologies will also be in more communication with patients. This requires awareness of some of the complexities related to trust and the therapeutic relationship studied by the field of computer-mediated communication (i.e., the exchange of text, images, audio, and video). These communications have a sense of immediacy and interaction that builds "trust," but transcend physical materiality (or proximity), time, and space. Direct consequences of action may seem a step away and asymmetrical communication limits social negotiation of meaning, yet provides a sense of control regarding what messages are received and when (Hilty et al., 2020c). Text-based communication poses some additional challenging issues, as keyboard characters (e.g., symbols, emoticons) used since 1982 enrich communication and comprehension (Aldunate & Gonzalez-Ibanez, 2016). Studies of brain regions involved in the emotional processing of emoticons have found these unnatural, iconic, and static representations of human facial expressions may not reflect a sentiment accurately—as users employ emoticons for different reasons and meanings—nor are they adequate to replace words. Since miscommunications appear more frequent when conveyed using English as the second language, using emoticons to text may also cause confusion across cultures and languages (Hilty et al., 2020c).

17.3.4 Technology and active learning

"Data rounds" for interprofessional teams will focus on triage, screening, diagnosis, and treatment. Algorithms would signify response to treatment or failure to benefit, based on AI, ML, and PM. Pathways for treatment-resistant, chronic patient, and other high-risk populations will be evidence-based. Modeled after problem and team-based learning principles—as well as traditional clinical rounds—learning facilitators can be programmed into workflow and barriers managed (Lane et al., 2013). In recent years, active learning curricula and methods increasingly incorporate web-based and video technology, partly to appeal to a generation of learners who habitually teach themselves via digital sources such as YouTube tutorials (Kamei et al., 2012) and who are experiential learners.

Technology has traditionally not been considered, in and of itself, an educational method (Hilty et al., 2019c). It has been seen rather an alternative vehicle for delivering educational methods. However, technologies can be used to explore more innovative approaches that harness the unique abilities of digital technology. A broader perspective could consider integrative practices of the US Department of Education (US Department of Education, 2017a, 2017b) and business (e.g., IBM) (Levy & Murnane, 2004) with learning theory (Matzen & Edmunds, 2007). This suggests technology has much more impact, perhaps on a spectrum, to: affirm/engage, accentuate, augment, complement existing practices—to help people explore and experiment with ideas in new ways—and to advance, engineer, innovate and create new ideas and practices (Hilty et al., 2019c).

17.3.5   Workflow practices on the rise

e-consultation (i.e., e-consult or eConsult) is commonly used in many healthcare systems in Canada and the United States (Agency for Healthcare Research and Quality, 2017; Liddy et al., 2016). e-consults generally involve a primary care provider referral for a consultation related to questions about a patient's care that is outside of their expertise. The goal of psychiatric e-consults is to help the consultee to be able to more quickly diagnose and treat behavioral health or other medical conditions. Typically, e-consults exist in text notes within the EHR between consultees and specialists, with meaningful review of patient information, data (e.g., results of tests, images), and other system information. Outcome measures focus on access to care (e.g., time first appointment), timeliness of consultation (e.g., wait times), and impact (e.g., depression scores) (Agency for Healthcare Research and Quality, 2017; Archibald et al., 2018; Liddy et al., 2016). Specialist text e-consultations may be replaced by video, as it is more engaging, memorable, synthetic, efficient, and user-friendly (Hilty et al., 2020c).

Research is increasing on sensor, wearable, and remote patient monitoring technologies used by people, patients, and clinicians. Initially, these were *added to* in-person care and synchronous video, but in select instances they have been *replacing* these modes as a primary assessment and/or treatment modality (Hilty et al., 2018b; Hilty et al., 2021d). These technologies are a key part of mobile health, connected health, and technology-enabled care. They have the capacity to integrate health, lifestyle, and clinical care, and they contextually change the culture of care and training—with more "in time" engagement, continuity of experience, and dynamic data for decision-making. There are a number of devices, software, and platforms for collecting behavioral data, alerting, communicating/giving feedback, detecting change, monitoring symptoms, accessing information, and providing preventive and therapeutic interventions (Luxton et al., 2011; Silva et al., 2015). These options bring patients together with clinical teams for communication, support and intervention (Kumari et al., 2017; Torous & Roberts, 2017).

Sensors, wearables, and remote patient monitoring technologies transform care by moving from cross-sectional, manual transfer of data at a healthcare appointment or from daily diary methods by hand or via time-, signal-, and event-dependent reporting—to an 24 hours by 7 days a week, longitudinal, integrated approach facilitated by AI methods (e.g., ML, integrated computer sensing technologies) and nanotechnology (Arean et al., 2016; Ariga et al., 2019; Kirchner & Shiffman, 2016; Luxton, 2016). The process is minimally intrusive, allows interactive sharing of data, and provides real-time feedback to monitor, intervene and follow trajectories. This approach has advantages to simply doing in-person, video and/or home visits, as it is based on the ecology of a person or patient in their natural settings (health, lifestyle, social). Researchers are exploring the link between objectively measured behavioral features (e.g., phone usage, mood rating, short message service text messages), location, and social interaction data for depression (Faurholt-Jepsen et al., 2018; Saeb et al., 2015). More broadly, digital phenotyping or behavioral markers are being developed for both clinical and nonclinical populations to correlate multimodal sensor data, cognitions, and depressive mood symptoms.

These technologies provide opportunities—and challenges—for patients, clinicians (e.g., keeping up, education/training, skills), and healthcare organizations (e.g., technology integration into workflow, privacy). Studies typically describe methods, interventions, technologies used, and care outcomes rather what clinical skills are needed and how they are developed or acquired. Education/training appears either narrowly limited to use of a specific technology for an intervention, or about the importance of patient buy-in, usability, and engagement. Academic health centers and health systems may assume that clinicians and systems are adapting to technology—and they are using video in the COVID-19 era—but clinical, technological, and administrative workflow is still in progress and many clinicians struggle to use new technologies, and they may not see this technology as a part of care (Hilty et al., 2021d).

17.4   Using technology for integrating care

While mental healthcare can be added to services, it is better to integrate it into primary care using a stepped, collaborative, or integrated care model (Hilty et al., 2018c). There are many other delivery settings and models that are juxtaposed to and overlap with integrated care, and video may or may not be used, but technology offers flexibility and efficiency. These include the patient-centered medical home model, which was founded to increase access, continuity and quality of care via team, coordination and safety practices. Home-based tele-interventions are on the rise due to patient nonattendance in clinics, which results in care delays and costs to healthcare systems. Even with automated reminders, inconvenience, and travel time are associated with nonadherence. Data are encouraging efforts in using telepsychiatric consultation to the home, resulting in reduced depression and disability, increased medication use, and promotion of self-efficacy. Stepped care in-person models are efficient for depression in primary care and common mental disorders, including via telehealth (Hilty et al., 2018c).

Integrated care will be the expectation rather than the exception and it requires many participants. There were five core requirements of integrated care: (1) responsibility, decision-making, and oversight of patient care by the interprofessional team; (2) colocation of services, both literally and/or virtually; (3) integrated funding; (4) evaluation; and (5) outcome measurement (Substance Abuse and Mental Health Services Administration, 2018). Two additional suggestions are (6) an e-platform; and (7) the alignment of reimbursement (Hilty et al., 2018c). While not included in the study of integrated, traditionally, mobile health has remarkable options for integrating care across settings, offering connectivity to meet patient needs, and facilitating clinical decision-making (i.e., clinical decision support (CDS)) (Hilty et al., 2018c).

Mobile health is using patient-to-clinician, patient-to-system, clinician-to-clinician, and system-user workflow for low and high acuity care (Hilty et al., 2021b). The 24 hours × 7 days/week collection of intelligently filtered data support CDS for clinicians "in-time" (Greenes et al., 2018)—this improves patient outcomes, reduces unnecessary mistakes and expenses, and increases efficiency. This collection of data in the patient's day-to-day life is called ecological momentary assessment and it involves repeated sampling of naturalistic behaviors and experiences (Carlson et al., 2016). This moves from paper-and-pencil diary methods (e.g., medication calendars) to signal- and event-dependent reporting—an expectation or pull demand from patients with a level of engagement and motivation that may exceed the capacity of some participants, but made possible by newer smartphones and wearable sensors. Additional automatic monitoring systems have been suggested by the National Academy of Sciences, engineering, and medicine (National Academies of Sciences Engineering & Medicine, 2019; Rohani et al., 2018).

17.4.1 Technology training (competencies) and evaluation

Clinical skill (i.e., competencies) for synchronous care parallel in-person care, but asynchronous technologies use a variety of options like apps, e-consultation, sensors, store-and-forward, text, wearables, and social media (Hilty et al., 2019a; Hilty et al., 2020a; Hilty et al., 2021d). Previous research has shown a technology gap between patients and clinicians (e.g., service members and providers, with service members having higher ownership and use of smartphones than military providers) (Armstrong, 2019; Armstrong et al., 2010). The competency-based medical education movement has focused on skill and attitudinal development in addition to knowledge acquisition. Outcomes are optimal via systematic curricular planning and evaluation (Kirkpatrick J. & Kirkpatrick W., 2009) and linking the learner, the teacher, the patient, and the desired outcome (i.e., skill, attitude, and/or knowledge).

Telepsychiatry video (2015, 2018) (Crawford et al., 2016; Hilty et al., 2015b), telebehavioral health (2017) (Maheu et al., 2018), social media (2018) (Luxton et al., 2015; Zalpuri et al., 2018), and mobile health (2019, 2020) (Hilty et al., 2019a; Hilty et al., 2020a) competency sets. These have included suggestions for training, faculty development, and institutional change to shift from in-person to synchronous and asynchronous care (Hilty et al., 2015b). Synchronous video has been gradually integrated into clinical care and curricula through seminar, supervision, grand rounds, and rotations (Hilty et al., 2019d). As digital life and healthcare evolve, the role of cultural factors cannot be emphasized enough. A conceptual framework has been suggested for video cultural competencies to improve patient-centered care, inclusion of the family and community, and training (Hilty et al., 2018a; Hilty et al., 2013; Hilty et al., 2020b) (Table 17.2).

Fortunately, mobile health (Hilty et al., 2019a; Hilty et al., 2020a) and broader asynchronous (Hilty et al., 2021d) competencies have been developed in addition to those described above. Mobile health, social media, and informatics practices require more substantial planning and organizational change. For example, rather than spontaneous use of many apps, a clinician and patient could purposely decide to use one to monitor habits (e.g., smoking), mood changes (i.e., depression), level of activity and vital signs (e.g., blood pressure). The data flow into the EHR and may be dispensed to the patient for a behavioral change or the clinician for the same or other pathways. Specifically, mobile health competencies should include CDS, prudent selection of technology, and management of flow of information across an EHR platform (Hilty et al., 2019a; Hilty et al., 2020a) (Table 17.3). Educational (e.g., using resources for searches, publishing) and administrative/practice management skills (e.g., licensure, jurisdictional, liability, and prescribing requirements) have also been suggested.

The competencies for asynchronous technology for adults (Hilty et al., 2021d) and children, adolescents, and families (Hilty et al., 2021e) include teaching and assessment methods and descriptive clinical application examples for programs to address the specific asynchronous competencies. Training subdomains are didactic teaching; case-based learning; clinical care with patients; professionalism, supervision, and practice-based improvement and learning; systems-based practice, quality improvement (QI), evaluation and research; and role as an educator.

TABLE 17.2 Overview of the intersection between culture and video, mobile health, and wearable sensor competencies.

Area/topic	Novice/advanced beginner (e.g., learn clinical and technology-based skills)	Competent/proficient (e.g., apply "good" in-person skill to technology-based care with appreciation of context)	Expert (or authority) (e.g., has advanced knowledge, skill and experience in VIDEO care, research, administration and/or policy)
General			
Royal College of Physicians and Surgeons Canada, 2015	Health and technology roles that all physicians play: medical expert, communicator, collaborator, manager, health advocate, scholar, and professional		
Patient care			
History-taking—engagement and interpersonal skills—assessment and physical examination—management and treatment planning—documentation—billing			
Maheu et al. (2018) *JTIBS* (Clinical evaluation and care) Assesses for cultural factors influencing care	Identifies obvious cultural factors; considers this theme if a dilemma arises in care and adjusts assessment and treatment strategies. Explores with an attitude of cultural humility and interest in learning. Seeks appropriate consultation to address cultural considerations/challenges	Systematically screens for, differentiates between regular and technology-specific cultural factors, for example, preference for telephone rather than video. Appropriately involves "cultural facilitators," for example, interpreters and members of the cultural community to assist with assessment and care. Uses culturally sensitive and evidence-based approaches, for example, assessment instruments like the cultural formulation interview.	Researches, trains and teaches peer-reviewed and when possible, evidence-based methods for problem-solving obstacles related to VIDEO and culture, for example, identifying implicit biases, opting out of a preferred technology if it is not working; obtaining a cultural consultant; researching trends across (and via mixing) technologies; multi-site care of a client/patient.
Maheu et al. (2018) *JTIBS* (Clinical evaluation and care) Creates a climate that encourages reflection and discussion of cultural issues in an ongoing manner	Contributes to a climate of humility and learning by identifying implicit biases as well as commonalities and differences between the client/patient and professionals.	Promotes a climate of humility and learning by identifying implicit biases and weighing client/patient and professional commonalities and differences as well as how these affect the therapeutic relationship. Facilitates reflection, manages complexities and uses VIDEO to optimize "fit" between client/patient and professional based on cultural identity, belief system(s), help-seeking behaviors, and preferences for care.	Researches, disseminates, and delivers evidence-based training related to public and population health data regarding the impact of these factors on outcomes related to in-person and/or VIDEO, for example, integrates these factors with data on the adoption of technology related to geographic mapping/trending, patterns of technology and resource availability of mobile technologies.
Hilty et al. (2021) *JTIBS* wearable sensors Assessment	Educate patients on technology and inquire about preferences for use Screen use of mobile phone, apps, and for what (e.g., exercise, music, healthcare)	Evaluate digital literacy and interest in app, text (i.e., generic, motivational or tailored) or wearable Screen use of apps, Bluetooth, wearable (e.g., activity tracker, holter monitor sensor-imbedded textile) and for what (e.g., glucometer for diabetes)	Use/teach/develop evidence-based strategies for clinical, technical, and administrative tasks ● Consent, policy, and procedures ● Populations and settings ● Medico-legal considerations Train/supervise/consult to optimize assessment and flexibility

(continued on next page)

TABLE 17.2 Overview of the intersection between culture and video, mobile health, and wearable sensor competencies—cont'd

Area/topic	Novice/advanced beginner (e.g., learn clinical and technology-based skills)	Competent/proficient (e.g., apply "good" in-person skill to technology-based care with appreciation of context)	Expert (or authority) (e.g., has advanced knowledge, skill and experience in VIDEO care, research, administration and/or policy)
Hilty et al. (2021) *JTIBS* culture and mobile health Assessment Cultural, diversity, language/interpreter use, and social determinants of health	Ask how culture, language and other factors affect life, family, and community (e.g., customs) Consider patient's culture, values, behaviors, and preferences Discuss if culture affects use of apps, texts, e-mail, and wearables Learn how social determinants affect health and healthcare	Check/adjust options in consideration of patient culture and preference Elicit cultural meaning of terminology used (e.g., illness/wellness, medical/spiritual) Educate on use of asynchronous technology and how it works or doesn't (e.g., emoticon confusion with nonprimary language) Assess how social determinants affect health and use of technology	Follow cultural formulation frameworks and ask if culture explains illness and affects use of technology Consider patient–doctor relationship in the context of culture, values, behaviors, and preferences Promote generalizations not stereotypes Assess if social determinants affect access to health care and technology
Hilty et al. (2021) *JTIBS* culture and mobile health Management and treatment planning	Integrate mobile technologies into biopsychosocial (BPS) approach Consider pros/cons of the decision support tool or app Monitor ongoing mobile technologies use, as well as documenting memorable and problematic events as they occur Clarify/spell out brief communications Pilot 1 or 2 mobile technologies to learn communication options	Select mobile technology for one treatment goal based on patient preference, skill and need (i.e., purpose) within BPS Cultural (BPSC) approach • App to monitor mood • Capture day-to-day accurate accounts of a patient's emotions, functioning, and activity (i.e., EMA) Blend mobile technologies with regular clinical discussions, facilitate reflection, and assess effect on relationship Weigh pros/cons of mH versus other technologies for informed consent	Use BPS outline with prioritization, with adjustments for technology Teach best practices, such as an evidence-based app with an evidence-based approach Select "best" mode for a given task: mobile technologies, e-mail/text, telephone, and/or in-person Be aware of legal, billing, and jurisdictional issues for medication Research and disseminate procedures to prevent problems and manage clinical and administrative issues to ensure quality (e.g., hard/software; accessories; troubleshooting)
Hilty et al. (2021) *JTIBS* wearable sensors Management and treatment planning	Make clear communications (e.g., text, notifications, symbols) Pilot/monitor/evaluate a technology for a specific purpose	Assess comfort with wearable in daily life (e.g., ease, fit with contact/water sports) Select option based on patient preference, feasibility, and purpose	Tailor recommendations to resources, culture, and patient preference Research/disseminate procedures to prevent problems and manage clinical and administrative issues

(continued on next page)

TABLE 17.2 Overview of the intersection between culture and video, mobile health, and wearable sensor competencies—cont'd

Area/topic	Novice/advanced beginner (e.g., learn clinical and technology-based skills)	Competent/proficient (e.g., apply "good" in-person skill to technology-based care with appreciation of context)	Expert (or authority) (e.g., has advanced knowledge, skill and experience in VIDEO care, research, administration and/or policy)
Interpersonal and communication skills			
Hilty et al. (2021) *JTIBS* culture and mobile health Communication	Create nonjudgmental, safe environment to engage others to share information and perspectives Ask advice/consultation to promote relationships and resolve problems Seek advice on merit and method of responses, if any, to patient's communications	Discuss the scope of communication with mobile technologies, clarify expectations and anticipate problems Build rapport in a continuous manner that fosters trust, respect, and understanding, including the ability to manage conflict Identify physical, cultural, psychological, and social barriers to communication and use of technology; avoid stereotypes (e.g., one group is more technologically savvy than another group)	Identify and trouble-shoot communication issues related to technology and other Educate and provide consultation to colleagues about asynchronous technology use Clarify expectations and potential ambiguous (i.e., multiple) meanings of acronyms, abbreviations, and such communication Role model effective, continuous, personal and professional relationships that optimize well-being
Hilty et al. (2015) *IRP* Cultural, diversity and social determinants of health	Consider the diversity of patients, families, and communities related to language fluency, and customs Consider one's culture, values, behaviors, and preferences Learn how social determinants affect in-person care	Adjust in consideration of patient culture and preference Check language fluency to confirm Elicit cultural meaning of illness/wellness Be aware that social determinants may affect interest in, using of, and experience with telemedicine	Follow cultural formulation frameworks Ask if culture affects using video (general exploration) or explanation of illness Consider patient–doctor relationship in context of culture, values, behaviors, and preferences Adjust interview, assessment, and treatment per social determinants; consider in-person care if critical need
Hilty et al. (2021) JTIBS culture and mobile health Cultural, diversity, and social determinants of health; attend to language issues	Consider cultural issues related to technologies ● How social determinants affect synchronous and asynchronous healthcare ● Access to mobile technologies ● Sentinel events Notices positive and negative trends in patient populations (e.g., generation Y, autism spectrum)	Ask patient if and how culture impacts use and preferences for mobile technologies and other technologies Promote reflection and awareness of how social determinants and mobile technologies intersect Observe, adjust and manage language and communication issues (e.g., emoticon use) Consider preferences of mobile technologies use (e.g., adolescent, Veteran with posttraumatic stress disorder)	Include mobile technologies use in cultural formulation interview, if applicable Instruct on generalizations (and how to avoid stereotypes) of how culture may affect mobile technologies use and impact treatment/patient care Instruct on how to adapt assessment and management approaches according to cultural differences

(continued on next page)

TABLE 17.2 Overview of the intersection between culture and video, mobile health, and wearable sensor competencies—cont'd

Area/topic	Novice/advanced beginner (e.g., learn clinical and technology-based skills)	Competent/proficient (e.g., apply "good" in-person skill to technology-based care with appreciation of context)	Expert (or authority) (e.g., has advanced knowledge, skill and experience in VIDEO care, research, administration and/or policy)
Hilty et al. (2015) *IRP* Language/interpreter ability	Use the interpreter	Time management and options (e.g., professionals preferred to medical staff and family)	Verbal and nonverbal dimensions
Hilty et al. (2018) *AP* Cultural, diversity, and social determinants 2018 AP	Consider participants' needs and preferences	Adjust to patient culture/preferences for therapeutic relationship Ensure language fluency and preferences	Teach on cultural formulation, generalizations for practice, and approach with humility
Maheu et al. (2018) *JTIBS* (Clinical evaluation and care) Ensures communication with a reasonable language option	Assesses primary/native language and client/patient preference for use of interpreter or certified language assistance service, when needed.	Ensures primary and/or preferred language is operational. Explores how language affects the story/narrative and level of intimacy. If an interpreter is used, explores ethnicity, interpersonal communication style and skill, that is, not analyzing, shaping story/narrative. Manages two-site complexities (client/patient site "a" and professional and/or interpreter site "b")	Teaches and consults regarding the therapeutic relationship's cultural issues, comparing similarities and differences between in-person and video-based communication. Disseminates evidence-based information about communication trends related to cultural values, practices, preferences, and language in video.
Hilty et al. (2018) *AP* video Language/interpreter ability	Use the interpreter as best as possible with supervision	Manage time, pick best option (e.g., professionals > staff and family) and use interpreters on either site or on telephone	Verbal and nonverbal dimensions Teach differences of relationship when using interpreter and quality thereof (e.g., nurse vs certified professional)
Hilty et al. (2018) *JTIBS* social media Cultural, diversity, and social determinants of health	How do these affect asynchronous methods?What is the impact of: • Technology fluency? • Idioms, "shorthand" expressions, and acronyms? Generational differences?	Show interest and flexibility in discussing diversity and technology issues Be aware of how social determinants affect in-person care and apply this information to use of SM/N	Ask about the impact of culture and diversity on preferences related to SM/N and other technology use Promote reflection, discussion, and awareness of how social determinants affect interest in, use of, and experience with technology Ask about immigrant/assimilation, generational, and other cultural factors that impact family

(continued on next page)

TABLE 17.2 Overview of the intersection between culture and video, mobile health, and wearable sensor competencies—cont'd

Area/topic	Novice/advanced beginner (e.g., learn clinical and technology-based skills)	Competent/proficient (e.g., apply "good" in-person skill to technology-based care with appreciation of context)	Expert (or authority) (e.g., has advanced knowledge, skill and experience in VIDEO care, research, administration and/or policy)
Zalpuri et al. (2018) *AP* social media Cultural, diversity, and social determinants of health; language access	Show interest and flexibility Be aware of how social determinants affect in-person care and SM Identify communication issues that may affect in-person care and SM	Ask about culture and use/preferences Promote reflection and awareness of how social determinants affect technology use Anticipate issues, make adjustments and ensure language flexibility	Include in cultural formulation interview and how culture may affect SM use and impact treatment/patient care Provide consultation
Hilty et al. (2019) *JTIBS* mobile health Cultural, diversity, and social determinants of health; attend to language issues	Consider culture and diversity issues, related to SP/device and/or apps and other technologies • How social determinants affect synchronous and asynchronous healthcare • Access to SP/device and/or apps • Sentinel events	Ask patient if/how culture impacts use and preferences for SP/device and/or apps and other technologies Promote reflection and awareness of how social determinants and SP/device and/or apps intersect Observe, adjust, and manage language and communication issues (e.g., emoticon use)	Include SP/device and/or apps use in cultural formulation interview, if applicable Instruct on generalizations (and how to avoid stereotypes) of how culture may affect SP/device and/or apps use and impact treatment/patient care Consider consultation
Hilty et al. (2019) *Psych Clinics* mobile health Culture, diversity and special populations	Recognize culture impacts use and other trends in populations (e.g., generation Y, autism spectrum)	Consider preferences and other implications of use and preference (e.g., adolescent, Veteran with posttraumatic stress disorder)	Instruct on cultural variations and how to adapt assessment and management approaches according to differences
Hilty et al. (2020) *JMIR* Cultural, diversity, and social determinants of health; attend to language issues	Consider culture and diversity issues, related to mobile technologies and other technologies • How social determinants affect synchronous and asynchronous healthcare • Access to mobile technologies • Sentinel events	Ask patient if/how culture impacts use and preferences for mobile technologies and other technologies Promote reflection and awareness of how social determinants and mobile technologies intersect Observe, adjust and manage language and communication issues (e.g., emoticon use)	Include mobile technologies use in cultural formulation interview, if applicable Instruct on generalizations (and how to avoid stereotypes) of how culture may affect mobile technologies use and impact treatment/patient care Consider consultation

(continued on next page)

TABLE 17.2 Overview of the intersection between culture and video, mobile health, and wearable sensor competencies—cont'd

Area/topic	Novice/advanced beginner (e.g., learn clinical and technology-based skills)	Competent/proficient (e.g., apply "good" in-person skill to technology-based care with appreciation of context)	Expert (or authority) (e.g., has advanced knowledge, skill and experience in VIDEO care, research, administration and/or policy)
Hilty et al. (2021) *JTIBS* wearable sensors Cultural, diversity, and social determinants of health; attend to language issues	Identify technology utilization patterns across demographic variables (age, ethnicity, race, national origin, literacy level, disability and identify inaccurate biases that are not supported by data Promote reflection and awareness how cultural, diversity issues, and social determinants of health might impact readiness to adopt, availability of, and access to health technology	Utilize a cultural competency framework to understand how experiences with technology may impact integration into care Utilize strategies to increase technological cultural competency and decrease biases and/or stereotypes Determine patient regarding readiness of use, and if/how culture impacts use Adjust to preferences based on patient need/interest Observe, adjust and manage language and communication issues (e.g., emoticon use)	Instruct/use components of a cultural formulation interview, if applicable Instruct on utilization of cultural competency frameworks to shift from ethnocentric to ethno-relative perceptive for use of health technology Instruct on how to adapt assessment and management approaches according to differences

Systems-base practice

Outreach to community—interprofessional education issues—care models—licensure regulations for vide and model used

Hilty et al. (2018) AP Rural health	Learn rural health basics related to care	Learn about rural access, epidemiology, and barriers	Teach, practice, and role model
Hilty et al. (2015) IRP Rural health	Learn about rural access, epidemiology, $, and other	Learn rural health basics	Practice and role model
Hilty et al. (2018) AP Special populations	Adjust to a difference (e.g., child/adolescent vs adult)	Recognize differences and adapts assessment and management (e.g., veterans, child/adolescent/parent, culture, geriatric) 2018 AP	Teach, practice, and role model

Professionalism

Attitude—integrity and ethical behavior—scope

Practice-based learning

Administration—safety and quality improvement (QI)—teaching and learning

Knowledge

Hilty et al. (2019) JTIBS mobile health Patient care	Ability to answer questions, discuss and adjust mH, SP/device, and/or apps in comparison to in-person care, including consent, privacy, data protection/integrity and security safety, and documentation Aware of SP/device security measures (i.e., password protection)	Answer questions/teach, discuss/clarify and adjust/develop options for mH, SP/device, and/or apps in comparison to in-person care in additional areas of scope of practice, communication, culture and diversity, ethics, and care models Aware of SP/device security measures (i.e., password protection bypassed if incoming call "goes around" security measure)	Demonstrate extensive knowledge of mH, SP/device, and/or apps to advise colleagues on practical knowledge of how to mitigate them

(continued on next page)

TABLE 17.2 Overview of the intersection between culture and video, mobile health, and wearable sensor competencies—cont'd

Area/topic	Novice/advanced beginner (e.g., learn clinical and technology-based skills)	Competent/proficient (e.g., apply "good" in-person skill to technology-based care with appreciation of context)	Expert (or authority) (e.g., has advanced knowledge, skill and experience in VIDEO care, research, administration and/or policy)
Patient care	Ability to answer questions, discuss and adjust mobile technologies in comparison to in-person care, including consent, privacy, data protection/integrity, and security safety and documentation Aware of mobile technologies security measures (i.e., password protection)	Answer questions/teach, discuss/clarify and adjust/develop options for mobile technologies in comparison to in-person care in additional areas of scope of practice, communication, culture and diversity, ethics, and care models Aware of mobile technologies security measures (i.e., password protection bypassed if incoming call "goes around" security measure)	Demonstrate extensive knowledge of mobile technologies to advise colleagues on practical knowledge of how to mitigate them
Technology (not in ACGME)			
Adapt to technology—remote site design—technology operation			
Telepresence & virtual environment (not in ACGME)			
Maheu et al. (2018) JTIBS Identifies obvious cultural factors; considers this theme if a dilemma arises in care and adjusts assessment and treatment strategies. Explores with an attitude of cultural humility and interest in learning. Seeks appropriate consultation to address cultural considerations/challenges.	Screens for client/patient communication styles/habits including culture/language and how technology alters them Adapts in-person skills to communicate with clients/patients, families, and other healthcare professionals via video Identifies ways in which communication is different, based on technology used; inquiries about the use of abbreviations, nouns, emoticons and thought fragments Names three ways that technology adds complexity between the client/patient and professional	Anticipates differences in client/patient styles/habits related to technology and encourages reflection Demonstrates and applies in-person skills to communicate with clients/patients, families, and other healthcare professionals Identifies ways in which in-person and technology-based communication are similar/different, for example, use of emoticons, "textese" and abbreviations Considers flow of conversation and the impact of the medium as well as related to language and culture, for example, East versus West coast in the United States Adjusts communication specifically to modality, for example, engages differently when client/patient is visibly anxious on video	

TABLE 17.3 An outline of mobile technology competencies in accreditation council for graduate medical education (i.e., Milestones).

Area/topic	Novice/advanced beginner (ACGME milestone Levels 1–2)	Competent/proficient (ACGME milestone Levels 3–4)	Advanced/Expert (ACGME milestone Level 5)
Patient care			
History-taking	Inquire about experience with and exposure to technology tScreen with questions such as: ● Are you using a mobile technologies and for what? Exercise Entertainment Health care Social	Evaluate patient digital literacy and recommend information/resources to learnScreen systematically/encourage reflection ● SP/device and/or apps versus e-mail, Internet, social media or other ● Health care with doctor, nurse or other ● Mental health issues with friends, clinicians, or others ● Risks (e.g., privacy, self-disclosure)?	Include informed consent Integrate details of personal and healthcare mobile technologies use ● Evaluate types of personal versus professional use ● Screen for the patient use of privacy settings for mobile technologies and show/teach skills
Engagement and interpersonal skills	Inquire about the use of technology and comfort/openness and trustDiscuss impact on ● Relationships with others ● Professional life ● Healthcare Generally, avoid humor, self-deprecatory remarks, and jokes	Ask preferences and assess impact on relationship (i.e., family, peers, vocation) ● Effect on intimacy and emotion Assess impact on therapeutic relationship ● Communication ● Intimacy ● Boundaries Compare to other technologies and in-person care (e.g., how to express empathy)	Guide effective communication Instruct on the best ways to use mH ● Simplify with purpose ● Match use with goal/intention Teach about asynchronous versus synchronous impact on relationshipsDiscuss expectations of participantsDetermine/teach alliance-building practices, responses, and barriers
Cultural, diversity, language/interpreter use, and social determinants of health	Ask how culture, language, and customs affect choices Consider one's culture, values, behaviors, and preferences Discuss if culture affects use of apps, texts, e-mail and wearables Consider social determinants	Adjust options in consideration of patient culture, preference & social determinants Elicit cultural meaning of terminology used (e.g., illness/wellness, medical/spiritual) Educate on use of asynchronous technology and how it works/doesn't (e.g., emoticon confusion with nonprimary language)	Follow cultural formulation frameworks Consider patient–doctor relationship in context of culture, values, behaviors, and preferences Promote generalizations not stereotypes Assess if social determinants affects access to health care and technology
Assessment	Assess how mH should be used or not and document Learn how to administer tools (e.g., MMSE) with supervision Ask for advice/consultation if question/problem arises	Consider the need for collateral info from in-person care or others Integrate mobile technology components with overall in-person assessments Demonstrate flexibility and share decisions to meet patient's needs and preferences	Synthesize information from in-person TP, mH, and other Optimize assessment and demonstrate flexibility (i.e., shared decisions) Identify pros/cons of using mobile technologies and for what purpose(s)

(continued on next page)

TABLE 17.3 An outline of mobile technology competencies in accreditation council for graduate medical education (i.e., Milestones)—cont'd

Area/topic	Novice/advanced beginner (ACGME milestone Levels 1–2)	Competent/proficient (ACGME milestone Levels 3–4)	Advanced/Expert (ACGME milestone Level 5)
Management and treatment planning	Integrate mobile technologies into BPS approach (see also Decision Support in Knowledge) Consider pros/cons of the decision support tool or app (see Decision Support in Knowledge) Monitor and document ongoing mobile technologies use If indicated, focus part of a visit on the use of mobile technologies and other technologies to talk in-depth Pilot 1 or 2 mobile technologies to learn communication options	Select mobile technology option based on patient preference, skill, and need (i.e., purpose); focus on one treatment goal ● App to monitor mood ● Data on a patient's emotions and activity (i.e., ecological momentary assessment (EMA)) Blend mobile technologies with regular BPS clinical discussions, facilitate reflection and assess effect on relationship Identify safety/risk factors of mobile technologies and triage complex, urgent/emergent issues	Use BPS outline with prioritization, with adjustments for technologyTeach best practices ● An evidence-based app with ● An evidence-based approach Select "best" mode for a given task: mobile technologies, e-mail/text, telephone, and/or in-personBe aware of legal, billing, workflow, jurisdictional and administrative issues for treatments Weigh populations impact and relative/absolute contraindications
Clinical decision support (CDS)	Use mobile technologies within for decision-making and care Review examples with learner/supervisor	Adjust mobile technologies within parameter(s) to for decision-making Help patients, learners, and staff use decision support tools based on evidence Prioritize mobile technologies options, e-mail, and tools that integrate into the EHR	Instruct on how to use pre- and intra-platform data feeds (e.g., questionnaire upload) into EHR to improve quality of care and be efficient Evaluate and adjust workflow for users and system
Administration and documentation	Adhere to clinic, health system and professional requirements for mobile and other technologies Document in informed consent and key events and seek supervision	Develop standard language for consent form, treatment plan and sentinel events Adapt current practices and develop new policies/procedures for mobile technologies and other technologies	Instruct on in-person, telepsychiatry, and mobile technologies related to documentation, privacy, and billing Consider/attend to business & financial issues (e.g., pros/cons of time used)
Medico-legal issues; privacy, confidentiality, safety, data protection/integrity, and security	Identify and adhere to regulations in the jurisdiction(s) of practice and of that of the patient Advise patients to communicate and send data privately (e.g., secure email within EHR not Gmail)	Apply in-person relevant regulations in any/all jurisdiction(s) to mobile technologies, and adjust clinical care Educate patient/team about mobile technologies and adapt existing laws Obtain clinical and/or legal advice	Teach/consult on in-person laws and regulations for mobile technologies Adapt legal and regulatory principles from in-person to mH Work with authorities, professional organizations & regulatory boards
Medical knowledge			
Define mobile technologies	Define mH, smartphone, app, and other mobile technologies	Compare mobile technologies for uses/purposes, risks/benefits & platforms	Teach on options, resources for patients and clinicians
Evidence-base	Know basic evidence base and how to evaluate asynchronous care	Know the outcome data, concepts, and principles (e.g., standards, guidelines, if an app is evidence-based and how to use it in an evidence-based clinical approach)	Teach/consult to healthcare colleagues Develop best practice guidelines Teach evidence-base, clinical guidelines, and competency sets

(continued on next page)

TABLE 17.3 An outline of mobile technology competencies in accreditation council for graduate medical education (i.e., Milestones)—cont'd

Area/topic	Novice/advanced beginner (ACGME milestone Levels 1–2)	Competent/proficient (ACGME milestone Levels 3–4)	Advanced/Expert (ACGME milestone Level 5)
Technical operations	Provide support on technologies for assessment and treatment Explain ways to learn how to use a mobile technologies product Recognize and know systems for reporting problems/errors	Assess performance of current options and evaluate new products and their pros/cons Assess user requirements and determine best match for patients/participants to options Serve as a resource, request technical assistance and help with QI processes	Research and disseminate information on mobile technologies platforms (e.g., feeds into EHR, integration of data across community systems) Teach QI and process improvement, including challenges and solutions
Clinical care and administration	Answer questions, discuss and apply mobile technologies to care Know basic privacy, data protection/integrity/security, and safety measures/steps	Adjust/customize for communication, cultural factors, teams, care model, EHR, and challenging cases/disorders Learn privacy, documentation, and reimbursement regulations for systems	Optimize hardware, software, and accessories and minimize distraction Develop prevention, mitigation, and management policies (e.g., data security, workflow, reimbursement)
Decision support	Learn the role in supporting decisions with supervision Pilot/review examples with supervisor and/or teacher Compare manual and decision support option	Find/evaluate the role in initiating, enhancing and monitoring decisions Compare versus manual operations (e.g., error identification, duplicative processes) for a variety of acute/chronic conditions Help patients, learners and staff use decision support tools based on pros/cons	Develop/research/evaluate decision support tools Guide others to make adjustments on parameter(s) to aide decision-making (e.g., fundamentals of adjustments) Review/research aggregate data on current and target practices
Systems-based, problem-based, and professionalism	Learn approach, logistics, safety, and outcomes for patient and system Identify potential risks (e.g., privacy violation) Be accountable, share concerns, and obtain feedback	Identify potential user risks and prevent, mitigate them (e.g., use privacy settings) Adapt regulations, adjust policy and procedures with administration Develop an approach to establish, maintain and promote ethical, accountable care	Know risks and advises colleagues Develop prevention, mitigation, and management policies (e.g., data security, professionalism lapses) Teach practices for efficient outreach and care across systems
Systems-based practice			
IPE and teamwork	Learn about technologies and share information with others	Work with interprofessional team apply mobile technologies to enhance care Weigh pros/cons related to communication, privacy, and clinical productivity	Incorporate IPE team point-of-view into system plan and teach strategies Adjust assignments/roles based on effectiveness and efficiency
Safety (see Patient care and professionalism)	Educate patient to call and/or set up additional appointment for emergencies	Prevent, identify and risk stratify potential problems based on past history in case mobile technologies	Adjust/manage risk of care model Instruct others in pitfalls of mobile technologies use in healthcare
Safety and QI	Learn systematic assessment with supervision and triage urgent issues to in-person care Follow policies in using reporting systems for problems/errors Discuss quality gaps and measures' pros/cons with supervisor	Outline factors/causal chains contributing to quality gaps in team and system process Recognize system error, participate in conferences and use QI tools to prevent adverse events Educate participants on technology-specific principles (e.g., emergencies)	Manage problems, participate in QI solutions, and instruct others of pitfalls and error prevention Analyze aggregate data, and lead root cause analyze & change management Weigh/adjust risk to manage clinical, administrative, and legal matters

(continued on next page)

TABLE 17.3 An outline of mobile technology competencies in accreditation council for graduate medical education (i.e., Milestones)—cont'd

Area/topic	Novice/advanced beginner (ACGME milestone Levels 1–2)	Competent/proficient (ACGME milestone Levels 3–4)	Advanced/Expert (ACGME milestone Level 5)
Models, practices and systems of care	Learn consultation, evaluation, triage and management clinician roles Incorporate workflow into EHR	Give input on (in)efficiencies and opportunities to integrate data in workflow/EHR for decision-making (e.g., apply sensors, remote monitors devices)	Consider what part, if any, of the "therapeutic hour" is used for mH Analyze mH options for timely, available, affordable and quality care
Practice-based learning			
Evaluation approach	Learn from/participate in global evaluations from patients, IP team and clinic/hospital	Develop/promote attitudes and skills for consistency, quality/specificity and stability of evaluation	Teach practice standards of evaluation & adjust across professions for states/provinces/countries
Quality improvement (QI)	Participate in chart review, case/M&M conference and other activities related to in-person and technology-based care	Apply/adapt in-person QI principles to mobile technologies in order to adjust assessment and/or care Educate participants on technology-specific principles and measures	Develop QI strategies to adhere to and adapt legal, regulatory, and ethical standards (e.g., privacy, access) Teach/consult on how to analyze, select and evaluate QI options
Professionalism			
Attitude	Show interest and reflect about use of mobile technologies	Express interest, be nonjudgmental and be spontaneous in discussing technology	Leadership organizational policy, curricula, and workflow
Integrity and ethical behavior	Maintain integrity by adhering to professional and governmental guidelines Recognize boundary, privacy, and confidentiality issues with mobile technologies communication	Reflect on personal versus professional contexts and potential micro- and macro-boundary violations (e.g., texting patient after clinical hours as "convenient") Recognize that personal information (e.g., health) may be accessible and monitor	Role model, teach/consult others to manage complicated ethical issues and maintain professional identity Research/develop approaches uphold quality of the therapeutic relationship and communication for care
Scope and therapeutic objective(s)	Practice within scope(s) and discuss expectations with patient Keep focus on shared primary objective of care	Attend to and evaluate scope issues Assess if mobile technologies are licensed and reputable; avoid fraudulent practices	Teach/consult use to adjust for patient populations (e.g., age, illness/disorder) Evaluate and advise on complex cases (e.g., risk populations, legal matters)
Interpersonal and communication skills			
Communication	Discuss problems and arrange alternative options Seek advice on merit and method of responses, if any, to patient's communication	Clarify expectations and anticipate problems (e.g., feasibility of checking mobile technology at other sites, clinics) Make brief, clear mobile technologies communications	Develop policies for asynchronous technology use Clarify expectations and potential ambiguous (i.e., multiple) meanings of acronyms and abbreviations
Special populations	Notices positive and negative trends in patient populations (e.g., generation Y or Z, autism spectrum)	Consider preferences of mobile technologies use (e.g., adolescent, Veteran with posttraumatic stress disorder) Be aware of trends across asynchronous technologies (e.g., e-mail/text, apps)	Instruct on how to adapt assessment and management approaches according to differences

17.4.2 Digital and cultural literacy

Digital literacy has an increasing, central role in our society and has become an important focus of institutions and policy makers (European Commission, 2018; US Department of Education, 2017a; Vial, 2019). One challenge is a unified definition since terms are used like e-skills, digital literacy, information and communication technology literacy, digital skills, digital competence, and so on (Spante et al., 2018). Most definitions go back to the original definition of digital literacy as "the ability to understand and use information in multiple formats from a wide range of sources when it is presented via computers" (Gilster, 1998). Over time, technical skills, cognitive skill (e.g., critical-thinking), and socio-emotional dimensions which integrate online behaviors and the sensibility that is required to behave appropriately. Digital competence is broadly defined by as "a set of different skills for achieving a good performance on digital society," which emphasizes the notion of performance and may infer knowledge production (Picatoste et al., 2018). Lastly, digital skills are the "set of skills that users need to operate computers and their networks, to search and select information, and the ability to use them for the fulfillment of one's goals" (van Dijk, 2006). This includes technical, information-seeking, and strategic skills (i.e., using technical and information skills in order to achieve something). Communication and media dimensions are also important.

The shift to technology-based platforms like mobile health may better avail services to ethnic minorities (e.g., African Americans with congestive heart failure, women with HIV) if they have access to technology and an approach (i.e., attitudes, skills, and knowledge) to use it. Twitter and Facebook are common among the Digital Native (Z; 1998-present), Millennial (Y; 1981-97), and X (1965-80) Generations. Users need sufficient technology access and user experience specific to an illness (e.g., app or Internet-based interventions for type 2 diabetes management). Some patients still prefer in-person care and technology literacy and language capacity may determine responsivity to mobile intervention. Some patients embrace technology (e.g., Hispanic/Latino participants who used Spanish text messages) (Arora et al., 2014), but there are many factors with affect the level of interactive engagement (Jang et al., 2018). Successful e-health programs involve frequent communication, bidirectional information flow, feedback, multicomponent interventions, and multimodal delivery (Hilty et al., 2021c). Technological fluency interfaces with health literacy, as well as language, social, economic, and other cultural issues for both behavioral and primary care settings (Hilty et al., 2021c; Hilty et al., 2018c).

For underserved and disadvantaged populations, clinicians need training for culturally competent care to learn attitudes, awareness, and skills in addition to knowledge (Benuto et al., 2018). There are numerous healthcare professional organization publications on cultural competency, cultural humility, and cultural safety for clinical care, including the role of experiences, resilience, trauma, and other factors. Existing cultural competencies are consensus-based, yet are not operationalized in terms of skills that can be measured (Indigenous Physicians Association of Canada, 2009; Kirmayer et al., 2012). Fortunately, video competencies with culturally based adaptation have been published (Hilty et al., 2020b) and the intersection of cultural competencies and mobile health was mapped out with synthetic, multidimensional cultural competencies (Hilty et al., 2021c); free-standing cultural competencies in a medicine framework was also organized for medical students, residents, faculty, and other members of interprofessional teams.

17.5 System and institutional shifts for 2030–2040

17.5.1 Overview

The US healthcare system will struggle to adapt to new, postpandemic norms and each health system will have to decide the role of technology. Institutional leadership is needed to support technological interventions so that clinical, technological, and administrative operations ensure the health of patients, the well-being of providers, and the health of communities. As important, is that clinicians, learners, teachers, and administrators "think together" about quality of care and clinical practices (Armstrong et al., 2010; Chan et al., 2017; Hilty et al., 2021d; Schueller et al., in press). Health systems could benefit from assessing providers' skills and implementing training with competencies—which exist for video telehealth (Maheu et al., 2018), wearable sensors and remote patient monitoring (Hilty et al., 2021a), social media (Zalpuri et al., 2018), mobile health (Hilty et al., 2019a; Hilty et al., 2020a), and asynchronous healthcare (Hilty et al., 2021d). Telehealth may also afford opportunities to reach new populations, help underserved ones, and build relationships with community partners (Hilty et al., 2021d). Building a positive e-culture for learning includes help for patients, clinicians, and other team members to learn skills, promote teamwork, and adapt to the new age of technology (Fig. 17.1).

17.5.2 Human factors

Recent research suggests the most common barrier to the successful implementation of technology in care is not the technology itself, but rather workflow issues related to training—true for video, mobile health, social media, and other

FIGURE 17.1   Steps for building a positive E-culture for synchronous and asynchronous technologies in healthcare.

asynchronous technologies (Hilty et al., 2021d). There are clinical, technical, and administrative workflow issues. For video integration, "delegating a clinician champion and a teaching faculty to learn and implement clinical competencies to steer training and faculty development" makes sense (Hilty et al., 2019d). But it is even more necessary than ever to address the rapid pace of technology change combined with unique implementation challenges. For patients seen, the encounters typically may be more similar to integrated care models, with visits/contacts that are briefer and more problem-focused than traditional specialty mental health encounters—the focus will be on access, quality, user satisfaction, and efficiency.

As technology has traditionally been added to care, there are significant physician/clinician and workflow barriers for video and mobile health. For video, barriers relate to engagement, technology workflow, legal (e.g., licensure, credentialing), organizational fitness, program/system evaluation, QI, and finance/reimbursement (Cowan et al., 2019; Institute for Healthcare Improvement, 2018). Barriers regarding clinician implementation of *video* have been characterized into three categories (1) personal barriers; (2) clinical workflow and technology barriers; and (3) licensure, credentialing, and reimbursement. Personal barriers include concerns that they will have difficulty establishing rapport, establishing a good clinical relationship, and be able to assess for nonverbal signs of psychiatric illness (e.g., initial greeting, poor hygiene, alcohol on breath). This is best handled by technology training, sitting in on others' video clinics and experience, and visiting another's practice with a clinic manager and staff.

Clinical workflow and technology barriers include the additional time to plan and organize operations for a video visit, which are not needed for an in-person encounter (e.g., room preparation, different location, equipment). An instrumental step is the creation of a culture in which in-person and video care is part of workflow—this can work well in video due to the regular patient appointment schedule (e.g., 15-, 30-, or 60-minute visits), unusual timeliness of the therapeutic "hour" and generalizable care across private, clinic, and academic health center practices. Workflow barriers of video include orienting patients to it, staff flexibility and dependability and provider-distant site coordination—video has more demands than in-person care. If video is combined with in-person visits for other patients and/or additional technologies, the transitions take coordination. The EHR may help with integration and coordination of the components of care. The use of patient questionnaires and other data (e.g., mobile health apps) may reduce interview and decision-making time and documentation.

The evaluation of video telehealth research has moved beyond general satisfaction and feasibility—to the issues of validity, reliability, cost/economics, and impact on clinical outcomes (Hilty et al., in press). A standard framework for economic cost analysis should include: an economist for planning, implementation and evaluation; a toolkit or guideline; comprehensive analysis (e.g., cost-effectiveness or cost-benefit) with an incremental cost-effectiveness ratio; measures for health, quality of life, and utility outcomes for populations; methods to convert outcomes into economic benefits (e.g., monetary, quality of adjusted life year); broad perspective (e.g., societal perspective); sensitivity analysis for uncertainty in modeling; and adjustments for differential timing (e.g., discounting and future costs). Process-based evaluation with iterative feedback from patients, providers, and staff is very important as programs start, maintain, and improve. For learners, the evaluation suggests that it should include four different levels: (1) reaction, (2) learning, (3) behavior, and (4) results (Kirkpatrick J. & Kirkpatrick W., 2009).

For mobile health, a systematic review found eight social and organizational factors that affect adoption, including workflow-related, patient-related, policy and regulations, culture or attitude or social influence, monetary factors, evidence base, awareness, and user engagement (Jacob et al., 2020). Technological factors, for example, included usefulness, ease of use, design, compatibility, technical issues, content, personalization, and convenience. A qualitative assessment simplified clinician-perceived barriers to adoption into three themes: (1) personal (clinician); (2) patient; and (3) organizational (i.e., time, uniformity, policy/direction) (Weichelt et al., 2019).

17.5.3 Financing, reimbursement, and licensing

The Affordable Care Act (ACA) and other trends in healthcare are reshaping services significantly. Significant money is spent on healthcare in the gross national product (GNP), trouble balancing budgets, and recession herald changes. While costs could increase in the future (Keehan et al., 2016), the rate of growth in the GNP has slowed since 2008 and is lower than before the Great Depression (Aaron, 2015). Consumers and payers expect more accountability of clinicians, clinics, and health systems. The focus is on access, safety, and quality care. Basic tenets are waste elimination, efficiency, and integration. Americans are plugged into the ACA, which has increased access but is not fully efficient (e.g., administration, exchanges) (Aaron, 2015). Healthcare organizations and many providers focus on payment issues, which have moved from fee-for-service, bundled payments, relative value units, managed care, cost-plus reimbursement, and prospective reimbursement. Legislation and financing are focusing value and quality/performance improvement.

For video, long-term costs have been the primary problem throughout the United States and reimbursement barriers still exist. Start-up grants generally pay for technology, but not for ongoing staff coordination and clinician service. Insurance or third-party payors have fallen into line, though they often require preliminary educational and administrative interventions. Patient care may or may not covered in the medical sector: (1) behavioral health service is carved out or poorly reimbursed; and (2) state transfers the responsibility to county mental health systems. Patients sometimes prefer this option due to less stigmatization and the ongoing relationship with the primary care provider. Federal programs were established with high specialist reimbursement for rural patients in federally qualified health clinics and rural health clinics, but telehealth services have sometimes not qualified because of inexplicably being viewed as provided "outside" the clinic. Bypassing barriers to services provided for indigent and rural patients remain a top priority, as these populations remain underserved, though some statewide telehealth programs have increased access(e.g., South Carolina).

The most significant, necessary paradigm shift in the US model of healthcare related to telehealth and culture may be related to reimbursement. There is trending toward greater telehealth and limited telephone care (e.g., hematology and thromboembolic diseases) reimbursement. Mobile health, telehealth, remote monitoring, and data infrastructure may add efficiency and not increase costs (Frakt & Pizer, 2016). US Department of Veterans Health Affairs and other capitated or managed care organizations see the utility and efficiency of low-end technology. But technology is not high on the radar of value-based care and accountable care organizations that are driven by the CMS and the ACA (Hilty et al., 2018d).

Licensing and credentialing may be significant, but with planning these can partially be overcome. Unless COVID-19 changes are continued long-term, unless there are interstate agreements, licenses are needed in all the states in which patients are located, unless they are doing a consultation or one-time assessment. This process is both complex and expensive for those drawing patients in/around many small states and additional exploration of state/federal laws may be necessary. Credentialling is easier if providers have had telehealth training or experience, when there are existing video proctors (who may teach/train at the same time), and there is reciprocity between health systems (e.g., rural hospital accepts an AHC's credentialing). For mobile health, reimbursement and interstate medical licensure rules remain problematic (Weinstein & Stason, 1977).

17.5.4   Institutional competencies

Many health systems are still working on basic clinical, technological, and administrative workflow, and may benefit from institutional level competencies to prepare to implement new technologies (Hilty et al., 2021d; Hilty et al., 2019d) (Table 17.4). Competencies for clinical care can help ensure quality, and faculty should role-model placing the patient's needs first and embracing technology for healthcare reform. Further student and trainee learning opportunities should be developed in order support the practice of new skills. Competencies, such as digital professionalism, should be included into training opportunities, such as during internships, residencies, postdoctoral fellowships, and clerkships (Hilty et al., 2021d).

Medical students, graduate education and graduate medical education councils, and professional organizations/boards are providing guidance on use of technology, but have not yet put forward competencies to ensure quality care (Hilty et al., 2021d; Maheu et al., 2018). Integrating training on health technology competencies in graduate training will facilitate implementation, training, and evaluation. Because technological innovation moves quickly, annual recertification may be needed for quality control of skills and to adjust to relevant technological advances for the curricula. The recertification process could also explore the digital navigator role, in general, and in relation to service delivery (e.g., interprofessional teams, stepped and/or integrated care models). As it becomes clear who the digital navigator serves and types of technology used, subtypes of digital navigators may bare consideration with additional/adjusted training requirements. Recertification processes are typically short and consist of one–two modules that are administered online.

17.6   Discussion

The shift from digital medicine to digital health is in progress, with the latter of which is comprised of a broad range of technologies, platforms, and systems to engage consumers for lifestyle, wellness, and health-related purposes. Digital, health, and cultural literacy for participants will be a key foundation for clinical care and administration. Many organizations are still reluctant to change, in general, and specifically related to technology despite telemedicine's impact (Hilty et al., 2021d). Progressive businesses have integrated core business divisions—research and development, operations, marketing, and finance—with an information technology (IT) division (i.e., a shared IT-business framework) to leverage knowledge and capital (Ray et al., 2007). This paradigm has been applied to telepsychiatry as a way to organize/integrate healthcare rather than adding technology to existing systems (Hilty et al., 2019c). This suggests that *organizing* care with technology will have better outcomes than *appending* it to healthcare. An example would be systems of building "in" user friendliness and workflow rather than adding steps for patients and clinicians in existing systems. Leveraging established models to increase the usability of health technologies is important as we look ahead to increase adoption. For example, the People at the Centre of Mobile Application Development usability model emphasizes seven factors: effectiveness; efficacy; satisfaction; learnability; memorability; errors; and cognitive load (Harrison et al., 2013).

Asynchronous technologies help systems improve access, reduce costs and complement other care options via clinical operating efficiency with care at multiple points-of-service, practice extension, and virtual team care (Hilty et al., 2021d; O'Keefe et al., 2021). Clinical, training, faculty development, and administrative missions should be evidence-based, but can also be consolidated and supported by system policy and procedures (Hilty et al., 2019b; Hilty et al., 2021d). This suggests a further shift toward patient- and learner-centered approaches, based on effectiveness, implementation science, and model of assessment of technology paradigms (Curran et al., 2012; Gargon et al., 2019; Kidholm et al., 2018; Marcolino et al., 2020; Proctor et al., 2011) – to be for sustainable and feasible across cultures (Hilty et al., 2020b). Patient–clinician interactions and measurement of empathy, positive regard, and genuineness related to therapeutic effectiveness will be instrumental (Grekin et al., 2019; Hilty et al., 2021d).

TABLE 17.4 Competencies for institutions/academic health centers for synchronous and asynchronous telehealth.

Competency focus	General technology approach	Shift to include synchronous care	Shift to include asynchronous care tips
Patient-centered care	Offer multiple points-of-entry Employ interprofessional teams and care coordination Data warehouse, analysis, and health information exchange Screen for technology use	Educate on in-person and synchronous similarities Use as one of many care models/treatment options Use templates and adjust policies and procedures	Offer options (e.g., apps, sensors/wearables) Design clinical technology workflows Import social science, health behavior, and business ideas
Evaluation and outcomes	Assess readiness for change Link behavior to outcomes for a patient or program Use evidence-based measures Use accreditation principles: Goal, Measure, Benchmark, Target, and Data	Use disease state measures and adapt, if applicable, aligned with in-person accreditation Use 360 evaluation Employ quality measures	Organize care on a technology platform (e.g., EHR, pre- and post-loads) Use technology-specific measures, evidence-based if available Use 360 evaluation
Trainee/student needs/roles	Patient- and learner-centered outcomes Prepare as Resource Manager Clarify personal versus professional technology use Use technology as a lifelong learner/teacher	Integrate skill development, care, teaching, and supervision Monitor well-being and professionalism Adjust curricula (e.g., part-time rotations, supervision)	Use quantitative and qualitative approaches Use observation, video, and simulation Role model healthy behaviors Capitalize on personal expertise
Faculty clinical, teaching, and leadership roles	Emphasize communication, well-being, and professionalism Emphasize resource manager technology leadership role Use social science, health service, and business constructs to shift attitudes	Monitor technology impact on care, well-being/fatigue Integrate part-time use for care with teaching by champions	Use sustainable, longitudinal approaches Remember that "less is more" and evidence base is key Use technology for portfolio, curricula, dissemination, networking, and other purposes
Teams, professions, and, systems within institutions	Assess structure/function of social groups that govern behavior of a community Use faculty development to build skills and shift culture	Define success based on populations Foster alignment across systems Organize goals and outcomes for success Employ team-based care and virtual teams	Align shared outcomes: Patient/clinician, Learner/teacher, Clinic/system, and Institution/community Use stepped care and interprofessional principles
AHC organizational structure, process, and finance	Evaluate/manage governance structure and change Weigh human resources, Technology, and cost issues Market technology delivery of care competitively Build AHC-community partnerships to share resources and integrate care Align clinical, educational, and research missions and values Integrate (not add or append) information technology into organizational structure	Use faculty development projects for existing/new leaders, as a gateway to others (e.g., mobile health) Measure technology in performance evaluations and provide feedback Add research and funding infrastructure for pilot and full-scale projects, to impact health service delivery and training programs	Assess context, pace, scope, and drive of/for change Monitor private, federal, state, and other sectors for best practices, partner agencies, and grant funding Strive for incremental, sustainable solutions Use/adapt others' evidence-based system approaches Develop strategies for promoting adoption/optimization of clinical information systems

One of the most pressing challenges facing countries is the increasing cost of healthcare. Organizations are seeking strategies to slow the growth of healthcare spending without compromising access, effectiveness, and safety. A research gap exists in reliable, comparative economic data for policy makers, health system administrators, and other stakeholder decision-making—a key obstacle to the widespread adoption, proliferation, and funding of telehealth programs. For technology to be embraced, decision-makers need to be sure that it adds to/helps rather than causes problems (Hilty et al., 2021d). For any technology to be adapted, it must be medically feasible and effective, as well as economically viable, that is, adaptability = health + economic feasibility. An overarching goal with telehealth and healthcare is to maximize the population-level effectiveness at a reasonable cost. This may be defined as how many patients are reached (i.e., treated), the clinical effectiveness of the treatment, and the return on investment. Since tensions unavoidably arise among the parties to medical decisions—patients, families and friends, clinicians, third-party payers, and others—economic approaches with a societal perspective are suggested to be fair to all parties (Drummond et al., 2015). Rigorous economic cost evaluation (e.g., cost-benefit or effective) and population-level analysis of telehealth programs, and health technologies more broadly, remain rare, and the costs of technology continue to change rapidly.

Research opportunities and challenges for sensors, wearables, and remote patient monitoring can be broadly organized into four categories: (1) clinical health outcomes; (2) medico-legal, professional, and privacy policy issue; (3) outcome, evaluation, and other models; and (4) human–computer interaction. Overall, the challenges that remain include the need for larger samples and even better-standardized methods, interventions, and evaluation measures—both in traditional randomized/efficacy *and* implementation/effectiveness designs—with longitudinal, quality of life, economic cost analysis, and other dimensions (Cornet & Holden, 2018). It is still too early to know which of the automatically generated objective sensor and wearable data will correlate, predict, and/or biomark/digital phenotype patients' outcomes (e.g., levels of depressive or manic symptoms or diagnostic classification of new affective states) (Insel, 2017; Onnela & Rauch, 2016). At a minimum, however, monitoring diagnosed patients and evaluating those under stress (e.g., combat) is helpful (Bourla et al., 2018).

While advancements in technology-based healthcare solutions accelerated during the COVID-19 pandemic, continued focus on research and development is needed to meet the healthcare needs of the future. Studies designed to investigate the impact of technologies on health outcomes, and how to best customize them based on biological and other clinical data to identify markers of risk, diagnosis, state, stage, a treatment response, and prognosis in different populations, could provide advancements needed to deliver patient and person-centered care. Analyzing large amounts of data will require close collaboration between partners from diverse areas of expertise, such as researchers, clinicians, statisticians, software developers, and engineers. Furthermore, considering, planning, and documenting the statistical analyses in advance (e.g., accounting for confounding factors), multiple comparisons, and un/adjusted analyses should be a priority (Faurholt-Jepsen et al., 2018). Controlled trials are needed of efficacy and comparative effectiveness, with adequate validation of measures and integration into the HER (Burnham et al., 2018). Temporal modeling, which leverages dynamical information, extends methods across multiple decision time points (i.e., sequential decision-making). One challenge is figuring out how it fits into the real-world clinical process. Relative differences over time are particularly important for chronic illnesses, both physical and behavioral symptoms (e.g., related to blood glucose readings, depression scales) or functional measures of the patient's ability to engage in normal daily activities (e.g., quality-of-life). Similarly, while treatment recommendations for each patient appear individualized, based on their characteristics, they may be affected by clinician characteristics and preferences, which can also be captured.

17.7 Conclusions

Widespread use of telehealth, mobile health, and remote patient monitoring during the COVID-19 pandemic has given users a vision of what is possible in the future to reshape the delivery of healthcare worldwide. The shift from digital medicine to digital health is in progress, with the latter employing technologies, platforms, and systems to engage consumers for lifestyle, wellness, and health-related purposes. Digital, health, and cultural literacy for participants will be a key foundation for clinical care and administration. Countries and organizations vary in level of technology integration in clinical care, research, and administration—some using it to organize work while others are adding it. AI and PM capitalize on in-time, continuous data collection, and analytics to support patient and clinician decision-making. Institutional competencies are needed for the application of technology to clinical services, professional development, and administrative leadership. Countries, provinces/states, and organizations may also have to redesign financing, reimbursement, regulation, training, and infrastructure needed for a new vision of healthcare. Research is needed in the areas of user-centered design, implementation and effectiveness approach, models of care, human–computer interaction, ethical and legal issues, and economic cost analyses.

Acknowledgments

American Telemedicine Association and the Telemental Health Interest Group; Department of Psychiatry and Behavioral Sciences, University of California, Davis School of Medicine; Veteran Affairs Northern California Health Care System and Mental Health Service.

Conflicts of Interest

None.

References

Aaron, J. (2015). *Five Years Old, Going on Ten: The Future of the Affordable Care Act*. Brookings. https://www.brookings.edu/blog/usc-brookings-schaeffer-on-health-policy/2015/03/27/five-years-old-going-on-ten-the-future-of-the-affordable-care-act/. (Accessed date: August 2, 2022).

Agency for Healthcare Research and Quality. (2017). *2017 National healthcare quality and disparities report*. https://www.ahrq.gov/research/findings/nhqrdr/nhqdr17/index.html. (Accessed date: August 2, 2022).

Agency for Healthcare Research and Quality. (2019). *Agency for healthcare research and quality team STEPPS*. https://www.ahrq.gov/teamstepps/index.html. (Accessed date: August 2, 2022).

Aldunate, N., & Gonzalez-Ibanez, R. (2016). An integrated review of emoticons in computer-mediated communication. *Front Psychol, 7*, 2061. https://doi.org/10.3389/fpsyg.2016.02061.

Antze, P. (2010). On the pragmatics of empathy in the neurodiversity movement. In M. Lambek (Ed.), *Ordinary Ethics* (pp. 310–327). New York, NY: Fordham University Press.

Archibald, D., Stratton, J., Liddy, C., Grant, R. E., Green, D., & Keely, E. J. (2018). Evaluation of an electronic consultation service in psychiatry for primary care providers. *BMC Psychiatry, 18*(1), 119. https://doi.org/10.1186/s12888-018-1701-3.

Arean, P. A., Hoa Ly, K., & Andersson, G (2016). Mobile technology for mental health assessment. *Dialogues Clin Neurosci, 18*(2), 163–169. https://www.ncbi.nlm.nih.gov/pubmed/27489456.

Ariga, K., Makita, T., Ito, M., Mori, T., Watanabe, S., & Takeya, J. (2019). Review of advanced sensor devices employing nanoarchitectonics concepts. *Beilstein J Nanotechnol, 10*, 2014–2030. https://doi.org/10.3762/bjnano.10.198.

Armstrong, C. M. (2019). Mobile health provider training: Results and lessons learned from year four of training on core competencies for mobile health in clinical care. *Journal of Technology in Behavioral Science, 4*(2), 86–92. https://doi.org/10.1007/s41347-019-00089-8.

Armstrong, C. M., McGee-Vincent, P., Juhasz, K., Owen, J., Avery, T., Jaworski, B., Jamison, A. L., Cone, W., Gould, C., Ramsey, K., Mackintosh, M. A., & Hilty, D. M. (2010). *VA Mobile Health Practice Guide (1st ed.)*. Washington, DC: U.S. Department of Veterans Affairs.

Arora, S., Peters, A. L., Burner, E., Lam, C. N., & Menchine, M. (2014). Trial to examine text message-based mHealth in emergency department patients with diabetes (TExT-MED): a randomized controlled trial. *Ann Emerg Med, 63*(6), 745–754e746. https://doi.org/10.1016/j.annemergmed.2013.10.012.

Benuto, L. T., Casas, J. B., & O'donohue, W. T. (2018). Training culturally competent psychologists: a systematic review of the training outcome literature. *Training and Education in Professional Psychology, 12*(12U), 134.

Berger, M., Wagner, T. H., & Baker, L. C. (2005). Internet use and stigmatized illness. *Soc Sci Med, 61*(8), 1821–1827. https://doi.org/10.1016/j.socscimed.2005.03.025.

Bombari, D., Schmid Mast, M., Canadas, E., & Bachmann, M. (2015). Studying social interactions through immersive virtual environment technology: virtues, pitfalls, and future challenges. *Front Psychol, 6*, 869. https://doi.org/10.3389/fpsyg.2015.00869.

Bourla, A., Mouchabac, S., El Hage, W., & Ferreri, F. (2018). e-PTSD: An overview on how new technologies can improve prediction and assessment of Posttraumatic Stress Disorder (PTSD). *Eur J Psychotraumatol, 9*(sup1), 1424448. https://doi.org/10.1080/20008198.2018.1424448.

Burnham, J. P., Lu, C., Yaeger, L. H., Bailey, T. C., & Kollef, M. H. (2018). Using wearable technology to predict health outcomes: a literature review. *J Am Med Inform Assoc, 25*(9), 1221–1227. https://doi.org/10.1093/jamia/ocy082.

Carlson, E. B., Field, N. P., Ruzek, J. I., Bryant, R. A., Dalenberg, C. J., Keane, T. M., & Spain, D. A. (2016). Advantages and psychometric validation of proximal intensive assessments of patient-reported outcomes collected in daily life. *Qual Life Res, 25*(3), 507–516. https://doi.org/10.1007/s11136-015-1170-9.

Chan, S., Godwin, H., Gonzalez, A., Yellowlees, P. M., & Hilty, D. M. (2017). Review of use and integration of mobile apps into psychiatric treatments. *Curr Psychiatry Rep, 19*(12), 96. https://doi.org/10.1007/s11920-017-0848-9.

Chan, S., Torous, J., Hinton, L., & Yellowlees, P. (2014). Mobile tele-mental health: increasing applications and a move to hybrid models of care. *Healthcare (Basel), 2*(2), 220–233. https://doi.org/10.3390/healthcare2020220.

Cho, G., Yim, J., Choi, Y., Ko, J., & Lee, S. H. (2019). Review of machine learning algorithms for diagnosing mental illness. *Psychiatry Investig, 16*(4), 262–269. https://doi.org/10.30773/pi.2018.12.21.2.

Christensen, C. M. (2007). Disruptive innovation: can health care learn from other industries? A conversation with Clayton M. Christensen. Interview by Mark D. Smith. *Health Aff (Millwood), 26*(3), w288–w295. https://doi.org/10.1377/hlthaff.26.3.w288.

Cornet, V. P., & Holden, R. J. (2018). Systematic review of smartphone-based passive sensing for health and wellbeing. *J Biomed Inform, 77*, 120–132. https://doi.org/10.1016/j.jbi.2017.12.008.

Cowan, K. E., McKean, A. J., Gentry, M. T., & Hilty, D. M. (2019). Barriers to use of telepsychiatry: clinicians as gatekeepers. *Mayo Clin Proc, 94*(12), 2510–2523. https://doi.org/10.1016/j.mayocp.2019.04.018.

Crawford, A., Sunderji, N., Lopez, J., & Soklaridis, S. (2016). Defining competencies for the practice of telepsychiatry through an assessment of resident learning needs. *BMC Med Educ, 16*, 28. https://doi.org/10.1186/s12909-016-0529-0.

Curran, G. M., Bauer, M., Mittman, B., Pyne, J. M., & Stetler, C. (2012). Effectiveness-implementation hybrid designs: combining elements of clinical effectiveness and implementation research to enhance public health impact. *Med Care, 50*(3), 217–226. https://doi.org/10.1097/MLR.0b013e3182408812.

de Filippis, R., Carbone, E. A., Gaetano, R., Bruni, A., Pugliese, V., Segura-Garcia, C., & De Fazio, P. (2019). Machine learning techniques in a structural and functional MRI diagnostic approach in schizophrenia: a systematic review. *Neuropsychiatr Dis Treat, 15*, 1605–1627. https://doi.org/10.2147/NDT.S202418.

deBronkart, D. (2015). From patient centred to people powered: Autonomy on the rise. *BMJ, 350*. h148 https://doi.org/10.1136/bmj.h148.

Dossett, E., & Shoemaker, E. (2015). Integrated care for women, mothers, children and newborns: Approaches and models for mental health, pediatric and prenatal care settings. *Journal of Womens Health Care, 04*. doi:10.4172/2167-0420.1000223.

Drummond, M., Sculpher, M. J., Claxton, K., Stoddart, G. L., & Torrance, G. W. (2015). *Methods for the Economic Evaluation of Health Care Programmes, 4th ed.* Oxford: Oxford University Press. https://books.google.com/books?id=lvWACgAAQBAJ.

Edgcomb, J. B., & Zima, B. (2019). Machine learning, natural language processing, and the electronic health record: innovations in mental health services research. *Psychiatr Serv, 70*(4), 346–349. https://doi.org/10.1176/appi.ps.201800401.

European Commission. (2018). *The European digital strategy. shaping Europe's digital future.* https://digital-strategy.ec.europa.eu/en. (Accessed date: August 2, 2022).

Faurholt-Jepsen, M., Bauer, M., & Kessing, L. V. (2018). Smartphone-based objective monitoring in bipolar disorder: Status and considerations. *Int J Bipolar Disord, 6*(1), 6. https://doi.org/10.1186/s40345-017-0110-8.

Fishbein, M., & Ajzen, I. (1977). Belief, attitude, intention, and behavior: an introduction to theory and research. *Philosophy and Rhetoric, 10*(2), 21–52.

Fraccaro, P., Beukenhorst, A., Sperrin, M., Harper, S., Palmier-Claus, J., Lewis, S., Van der Veer, S. N., & Peek, N. (2019). Digital biomarkers from geolocation data in bipolar disorder and schizophrenia: a systematic review. *J Am Med Inform Assoc, 26*(11), 1412–1420. https://doi.org/10.1093/jamia/ocz043.

Frakt, A. B., & Pizer, S. D. (2016). The promise and perils of big data in healthcare. *Am J Manag Care, 22*(2), 98–99. https://www.ncbi.nlm.nih.gov/pubmed/26885669.

Gargon, E., Gorst, S. L., & Williamson, P. R. (2019). Choosing important health outcomes for comparative effectiveness research: 5th annual update to a systematic review of core outcome sets for research. *PLoS One, 14*(12), e0225980. https://doi.org/10.1371/journal.pone.0225980.

Gilster, P. (1998). *Digital Literacy.* Wiley https://books.google.com/books?id=ppVx7pHr07kC.

Greenes, R. A., Bates, D. W., Kawamoto, K., Middleton, B., Osheroff, J., & Shahar, Y. (2018). Clinical decision support models and frameworks: Seeking to address research issues underlying implementation successes and failures. *J Biomed Inform, 78*, 134–143. https://doi.org/10.1016/j.jbi.2017.12.005.

Grekin, E. R., Beatty, J. R., & Ondersma, S. J. (2019). Mobile health interventions: exploring the use of common relationship factors. *JMIR mHealth and uHealth, 7*.

Griffiths, K. M., Calear, A. L., & Banfield, M. (2009). Systematic review on Internet Support Groups (ISGs) and depression (1): Do ISGs reduce depressive symptoms? *J Med Internet Res, 11*(3), e40. https://doi.org/10.2196/jmir.1270.

Grist, R., Porter, J., & Stallard, P. (2017). Mental health mobile apps for preadolescents and adolescents: a systematic review. *J Med Internet Res, 19*(5), e176. https://doi.org/10.2196/jmir.7332.

Gustafson, J. L. (2011). Moore's law. In D. Padua (Ed.), *Encyclopedia of Parallel Computing* (pp. 1177–1184). Springer US. https://doi.org/10.1007/978-0-387-09766-4_81.

Harrison, R., Flood, D., & Duce, D. (2013). Usability of mobile applications: literature review and rationale for a new usability model. *Journal of Interaction Science, 1*(1), 1. https://doi.org/10.1186/2194-0827-1-1.

Hilty, D. M., Crawford, A., Teshima, J., Nasatir-Hilty, S. E., Luo, J., Chisler, L. S. M., Gutierrez Hilty, Y. S. M., Servis, M. E., Godbout, R., Lim, R. F., & Lu, F. G. (2021). Mobile health and cultural competencies as a foundation for telehealth care: scoping review. *J Tech Behav Sci, 6*, 197–230.

Hilty, D. M., Armstrong, C. M., Edwards-Stewart, A., Gentry, M. T., Luxton, D. D., & Krupinski, E. A. (2021a). Sensor, wearable, and remote patient monitoring competencies for clinical care and training: Scoping review. *J Technol Behav Sci*, 1–26. doi:10.1007/s41347-020-00190-3.

Hilty, D. M., Armstrong, C. M., Luxton, D. D., Gentry, M. T., & Krupinski, E. A. (2021b). A scoping review of sensors, wearables, and remote monitoring for behavioral health: Uses, outcomes, clinical competencies, and research directions. *Journal of Technology in Behavioral Science, 6*(2), 278–313. https://doi.org/10.1007/s41347-021-00199-2.

Hilty, D. M., Maheu, M., Drude, K., & Hertlein, K. (2018). The need to implement and evaluate telehealth competency frameworks to ensure quality care across behavioral health professions. *Acad Psychiatry, 42*(6), 818–824.

Hilty, D. M., Chan, S., Torous, J., Luo, J., & Boland, R. J. (2019). A competency-based framework for psych/behavioral health apps for trainees, faculty, programs and health systems. *Psych Clin N Amer, 42*, 513–534.

Hilty, D. M., Chan, S., Torous, J., Luo, J., & Boland, R. (2019a). A telehealth framework for mobile health, smartphones, and apps: competencies, training, and faculty development. *Journal of Technology in Behavioral Science, 4*(2), 106–123. https://doi.org/10.1007/s41347-019-00091-0.

Hilty, D. M., Chan, S., Torous, J., Luo, J., & Boland, R. J. (2020). A framework for competencies for the use of mobile technologies in psychiatry and medicine. *JMIR Uhealth Mobile Health, 8*(2). http://mhealth.jmir.org/2020/2/e12229/.

Hilty, D. M., Chan, S., Torous, J., Luo, J., & Boland, R. (2020a). A framework for competencies for the use of mobile technologies in psychiatry and medicine: Scoping review. *JMIR Mhealth Uhealth, 8*(2), e12229. https://doi.org/10.2196/12229.

Hilty, D. M., Crawford, A., Teshima, J., Chan, S., Sunderji, N., Yellowlees, P. M., & Li, S. T. (2015). A framework for telepsychiatric training and e-health: Competency-based education, evaluation and implications. *Int Rev Psychiatry, 27*(6), 569–592.

Hilty, D. M., Chan, S., Torous, J. B., Matmahur, J., & Mucic, D. (2015a). New frontiers in healthcare and technology: Internet-and web-based mental options emerge to complement in-person and telepsychiatric care options. *Journal of Health and Medical Informatics, 6,* 1–14.

Hilty, D. M., Crawford, A., Teshima, J., Chan, S., Sunderji, N., Yellowlees, P. M., Kramer, G., O'Neill, P., Fore, C., Luo, J., & Li, S. T. (2015b). A framework for telepsychiatric training and e-health: competency-based education, evaluation and implications. *Int Rev Psychiatry, 27*(6), 569–592. https://doi.org/10.3109/09540261.2015.1091292.

Hilty, D. M., Crawford, A., Teshima, J., Nasatir-Hilty, S. E., Luo, J., Chisler, L. S. M., Gutierrez Hilty, Y. S. M., Servis, M. E., Godbout, R., Lim, R. F., & Lu, F. G. (2021c). Mobile health and cultural competencies as a foundation for telehealth care: scoping review. *Journal of Technology in Behavioral Science, 6*(2), 197–230. https://doi.org/10.1007/s41347-020-00180-5.

Hilty, D. M., Evangelatos, G., Valasquez, G. A., Le, C., & Sosa, J. (2018a). Telehealth for rural diverse populations: cultural and telebehavioral competencies and practical approaches for clinical services. *Journal of Technology in Behavioral Science, 3*(3), 206–220. https://doi.org/10.1007/s41347-018-0054-6.

Hilty, D. M., Ferrer, D. C., Parish, M. B., Johnston, B., Callahan, E. J., & Yellowlees, P. M. (2013). The effectiveness of telemental health: A 2013 review. *Telemed J E Health, 19*(6), 444–454. https://doi.org/10.1089/tmj.2013.0075.

Hilty, D. M., Gentry, M. T., McKean, A. J., Cowan, K. E., Lim, R. F., & Lu, F. G. (2020b). Telehealth for rural diverse populations: Telebehavioral and cultural competencies, clinical outcomes and administrative approaches. *Mhealth, 6,* 20. https://doi.org/10.21037/mhealth.2019.10.04.

Hilty, D. M., Liu, H. Y., Stubbe, D., & Teshima, J. (2019b). Defining professional development in medicine, psychiatry, and allied fields. *Psychiatr Clin North Am, 42*(3), 337–356. https://doi.org/10.1016/j.psc.2019.04.001.

Hilty, D. M., Rabinowitz, T., McCarron, R. M., Katzelnick, D. J., Chang, T., Bauer, A. M., & Fortney, J. (2018b). An update on telepsychiatry and how it can leverage collaborative, stepped, and integrated services to primary care. *Psychosomatics, 59*(3), 227–250. https://doi.org/10.1016/j.psym.2017.12.005.

Hilty, D. M., Randhawa, K., Maheu, M. M., McKean, A. J. S., Pantera, R., Mishkind, M. C., & Rizzo, A. S. (2020c). A review of telepresence, virtual reality, and augmented reality applied to clinical care. *Journal of Technology in Behavioral Science, 5*(2), 178–205. https://doi.org/10.1007/s41347-020-00126-x.

Hilty, D. M., Serhal, E., & Crawford, A. (in press). A telehealth and telepsychiatry economic cost analysis framework: scoping review. Telemed J E-Health.

Hilty, D. M., Sunderji, N., Suo, S., Chan, S., & McCarron, R. M. (2018c). Telepsychiatry and other technologies for integrated care: evidence base, best practice models and competencies. *Int Rev Psychiatry, 30*(6), 292–309. https://doi.org/10.1080/09540261.2019.1571483.

Hilty, D. M., Torous, J., Parish, M. B., Chan, S. R., Xiong, G., Scher, L., & Yellowlees, P. M. (2021d). A literature review comparing clinicians' approaches and skills to in-person, synchronous, and asynchronous care: Moving toward competencies to ensure quality care. *Telemed J E Health, 27*(4), 356–373. https://doi.org/10.1089/tmj.2020.0054.

Hilty, D. M., Turvey, C., & Hwang, T. (2018d). Lifelong learning for clinical practice: How to leverage technology for telebehavioral health care and digital continuing medical education. *Curr Psychiatry Rep, 20*(3), 15. https://doi.org/10.1007/s11920-018-0878-y.

Hilty, D. M., Uno, J., Chan, S., Torous, J., & Boland, R. J. (2019c). Role of technology in faculty development in psychiatry. *Psychiatr Clin North Am, 42*(3), 493–512. https://doi.org/10.1016/j.psc.2019.05.013.

Hilty, D. M., Unutzer, J., Ko, D. G., Luo, J., Worley, L. L. M., & Yager, J. (2019d). Approaches for departments, schools, and health systems to better implement technologies used for clinical care and education. *Acad Psychiatry, 43*(6), 611–616. https://doi.org/10.1007/s40596-019-01074-2.

Hilty, D. M., Yellowlees, P. M., Parrish, M. B., & Chan, S. (2015c). Telepsychiatry: effective, evidence-based, and at a tipping point in health care delivery? *Psychiatr Clin North Am, 38*(3), 559–592. https://doi.org/10.1016/j.psc.2015.05.006.

Hilty, D. M., Zalpuri, I., Torous, J., & Nelson, E. L. (2021e). Child and adolescent asynchronous technology competencies for clinical care and training: scoping review. *Fam Syst Health, 39*(1), 121–152. https://doi.org/10.1037/fsh0000536.

Hobson, G. R., Caffery, L. J., Neuhaus, M., & Langbecker, D. H. (2019). Mobile health for first nations populations: systematic review. *JMIR Mhealth Uhealth, 7*(10), e14877. https://doi.org/10.2196/14877.

Hu, C., Kung, S., Rummans, T. A., Clark, M. M., & Lapid, M. I. (2015). Reducing caregiver stress with internet-based interventions: a systematic review of open-label and randomized controlled trials. *J Am Med Inform Assoc, 22*(e1), e194–e209. https://doi.org/10.1136/amiajnl-2014-002817.

Indigenous Physicians Association of Canada. (2009). *First nations, inuit, metis health core competencies.* http://www.afmc.ca/sites/default/files/pdf/IPAC-AFMC_Core_Competencies_EN.pdf. (Accessed date: August 2, 2022).

Insel, T. R. (2017). Digital phenotyping: technology for a new science of behavior. *JAMA, 318*(13), 1215–1216. https://doi.org/10.1001/jama.2017.11295.

Institute for Healthcare Improvement. (2018). *Quality improvement essentials toolkit.* http://www.ihi.org/resources/Pages/Tools/Quality-Improvement-Essentials-Toolkit.aspx. (Accessed date: August 2, 2022).

Jacob, C., Sanchez-Vazquez, A., & Ivory, C. (2020). Social, organizational, and technological factors impacting clinicians' adoption of mobile health tools: Systematic literature review. *JMIR Mhealth Uhealth, 8*(2), e15935. https://doi.org/10.2196/15935.

Jang, M., Johnson, C. M., D'Eramo-Melkus, G., & Vorderstrasse, A. A. (2018). Participation of racial and ethnic minorities in technology-based interventions to self-manage type 2 diabetes: a scoping review. *J Transcult Nurs, 29*(3), 292–307. https://doi.org/10.1177/1043659617723074.

Jauk, S., Kramer, D., Grossauer, B., Rienmuller, S., Avian, A., Berghold, A., Leodolter, W., & Schulz, S. (2020). Risk prediction of delirium in hospitalized patients using machine learning: an implementation and prospective evaluation study. *J Am Med Inform Assoc, 27*(9), 1383–1392. https://doi.org/10.1093/jamia/ocaa113.

Kalmady, S. V., Greiner, R., Agrawal, R., Shivakumar, V., Narayanaswamy, J. C., Brown, M. R. G., Greenshaw, A. J., Dursun, S. M., & Venkatasubramanian, G. (2019). Towards artificial intelligence in mental health by improving schizophrenia prediction with multiple brain parcellation ensemble-learning. *NPJ Schizophr, 5*(1), 2. https://doi.org/10.1038/s41537-018-0070-8.

Kamei, R. K., Cook, S., Puthucheary, J., & Starmer, C. F. (2012). 21st century learning in medicine: traditional teaching versus team-based learning. *Medical Science Educator, 22*(2), 57–64. https://doi.org/10.1007/BF03341758.

Keehan, S. P., Poisal, J. A., Cuckler, G. A., Sisko, A. M., Smith, S. D., Madison, A. J., Stone, D. A., Wolfe, C. J., & Lizonitz, J. M. (2016). National health expenditure projections, 2015–25: economy, prices, and aging expected to shape spending and enrollment. *Health Affairs, 35*(8), 1522–1531. https://doi.org/10.1377/hlthaff.2016.0459.

Kidholm, K., Jensen, L. K., Kjolhede, T., Nielsen, E., & Horup, M. B. (2018). Validity of the model for assessment of telemedicine: a Delphi study. *J Telemed Telecare, 24*(2), 118–125. https://doi.org/10.1177/1357633 × 16686553.

Kirchner, T. R., & Shiffman, S. (2016). Spatio-temporal determinants of mental health and well-being: advances in geographically-explicit ecological momentary assessment (GEMA). *Soc Psychiatry Psychiatr Epidemiol, 51*(9), 1211–1223. https://doi.org/10.1007/s00127-016-1277-5.

Kirkpatrick, J., & Kirkpatrick, W. (2009). The Kirkpatrick four levels: a fresh look after 50 years, 1959-2009. *Training Magazine.* (Accessed date: August 2, 2022).

Kirmayer, L., Fung, K., Rousseau, C., Lo, H.-T., Menzies, P., Guzder, J., Ganesan, S., Andermann, L., & McKenzie, K. (2012). Guidelines for training in cultural psychiatry-position paper. *Canadian Journal of Psychiatry, 57.* Insert 1-16 https://doi.org/10.1177/0706743720907505.

Kotter, J. P. (1996). Leading Change. McGraw-Hill Companies. https://books.google.com/books?id=wNSmPwAACAAJ.

Kumari, P., Mathew, L., & Syal, P. (2017). Increasing trend of wearables and multimodal interface for human activity monitoring: a review. *Biosens Bioelectron, 90*, 298–307. https://doi.org/10.1016/j.bios.2016.12.001.

Lane, D., Ferri, M., Lemaire, J., McLaughlin, K., & Stelfox, H. T. (2013). A systematic review of evidence-informed practices for patient care rounds in the ICU. *Crit Care Med, 41*(8), 2015–2029. https://doi.org/10.1097/CCM.0b013e31828a435f.

Lavelle, M., Healey, P. G., & McCabe, R. (2014). Participation during first social encounters in schizophrenia. *PLoS One, 9*(1), e77506. https://doi.org/10.1371/journal.pone.0077506.

Levy, F., & Murnane, R. (2004). A role for technology in professional development? lessons from IBM. *Phi Delta Kappan, 85*, 728–734. https://doi.org/10.1177/003172170408501005.

Liddy, C., Drosinis, P., & Keely, E. (2016). Electronic consultation systems: Worldwide prevalence and their impact on patient care-a systematic review. *Fam Pract, 33*(3), 274–285. https://doi.org/10.1093/fampra/cmw024.

Luxton, D. D. (2016). *Artificial Intelligence in Behavioral and Mental Health Care.* Boston, MA: Elsevier Science. https://books.google.com/books?id=INvUBQAAQBAJ.

Luxton, D. D., June, J. D., & Chalker, S. A. (2015). Mobile health technologies for suicide prevention: feature review and recommendations for use in clinical care. *Current Treatment Options in Psychiatry, 2*(4), 349–362. https://doi.org/10.1007/s40501-015-0057-2.

Luxton, D. D., June, J. D., & Fairall, J. M. (2012). Social media and suicide: a public health perspective. *Am J Public Health, 102*(Suppl 2), S195–S200. https://doi.org/10.2105/AJPH.2011.300608.

Luxton, D. D., McCann, R. A., Bush, N. E., Mishkind, M. C., & Reger, G. M. (2011). mHealth for mental health: Integrating smartphone technology in behavioral healthcare. *Professional Psychology: Research and Practice, 42*(6), 505–512. https://doi.org/10.1037/a0024485.

Luxton, D. D., & Riek, L. D. (2019). Artificial intelligence and robotics in rehabilitation. *Handbook of rehabilitation psychology* (3rd ed., pp. 507–520). Washington DC: American Psychological Association. https://doi.org/10.1037/0000129-031.

Maheu, M. M., Drude, K. P., Hertlein, K. M., & Hilty, D. M. (2018). A framework of interprofessional telebehavioral health competencies: implementation and challenges moving forward. *Acad Psychiatry, 42*(6), 825–833. https://doi.org/10.1007/s40596-018-0988-1.

Marcolino, M. S., Alkmim, M. B., Pessoa, C. G., Maia, J. X., & Cardoso, C. S. (2020). Development and implementation of a methodology for quality assessment of asynchronous teleconsultations. *Telemed J E Health, 26*(5), 651–658. https://doi.org/10.1089/tmj.2019.0049.

Matzen, N., & Edmunds, J. (2007). Technology as a catalyst for change. *Journal of Research on Technology in Education, 39*, 417–430. https://doi.org/10.1080/15391523.2007.10782490.

McIntyre, R. S., Cha, D. S., Jerrell, J. M., Swardfager, W., Kim, R. D., Costa, L. G., Baskaran, A., Soczynska, J. K., Woldeyohannes, H. O., Mansur, R. B., Brietzke, E., Powell, A. M., Gallaugher, A., Kudlow, P., Kaidanovich-Beilin, O., & Alsuwaidan, M. (2014). Advancing biomarker research: utilizing 'Big Data' approaches for the characterization and prevention of bipolar disorder. *Bipolar Disord, 16*(5), 531–547. https://doi.org/10.1111/bdi.12162.

Mermelstein, H., Guzman, E., Rabinowitz, T., Krupinski, E., & Hilty, D. M. (2017). The application of technology to health: The evolution of telephone to telemedicine and telepsychiatry: a historical review and look at human factors. *Journal of Technology in Behavioral Science, 2*(1), 5–20. https://doi.org/10.1007/s41347-017-0010-x.

Mikolas, P., Hlinka, J., Skoch, A., Pitra, Z., Frodl, T., Spaniel, F., & Hajek, T. (2018). Machine learning classification of first-episode schizophrenia spectrum disorders and controls using whole brain white matter fractional anisotropy. *BMC Psychiatry, 18*(1), 97. https://doi.org/10.1186/s12888-018-1678-y.

Naslund, J. A., Aschbrenner, K. A., McHugo, G. J., Unutzer, J., Marsch, L. A., & Bartels, S. J. (2019). Exploring opportunities to support mental health care using social media: A survey of social media users with mental illness. *Early Interv Psychiatry, 13*(3), 405–413. https://doi.org/10.1111/eip.12496.

National Academies of Sciences Engineering & Medicine (2019). Taking action against clinician burnout: A systems approach to professional well-being. Washington, DC: National Academies Press. https://nam.edu/systems-approaches-to-improve-patient-care-by-supporting-clinician-well-being/?gclid=CjwKCAjw2P-KBhByEiwADBYWChB818zZC1PzDVuR9MKjU8ThnChfUeuctyVtenux0W84tBeCBY4WQBoCPpcQAvD_BwE.

Nemesure, M. D., Heinz, M. V., Huang, R., & Jacobson, N. C. (2021). Predictive modeling of depression and anxiety using electronic health records and a novel machine learning approach with artificial intelligence. *Sci Rep, 11*(1), 1980. https://doi.org/10.1038/s41598-021-81368-4.

O'Keefe, M., White, K., & Jennings, J. C. (2021). Asynchronous telepsychiatry: a systematic review. *J Telemed Telecare, 27*(3), 137–145. https://doi.org/10.1177/1357633 × 19867189.

Onnela, J. P., & Rauch, S. L. (2016). Harnessing smartphone-based digital phenotyping to enhance behavioral and mental health. *Neuropsychopharmacology, 41*(7), 1691–1696. https://doi.org/10.1038/npp.2016.7.

Parsons, T. D., Gaggioli, A., & Riva, G. (2017). Virtual reality for research in social neuroscience. *Brain Sci, 7*(4). https://doi.org/10.3390/brainsci7040042.

Pew Research Center. (2019). *Pew Research internet project: mobile technology fact sheet* (Pew Research Center Internet and Technology. http://www.pewinternet.org/fact-heets/mobiletechnology-fact-sheet/. (Accessed date: August 2, 2022).

Picatoste, J., Pérez-Ortiz, L., & Ruesga-Benito, S. M. (2018). A new educational pattern in response to new technologies and sustainable development. Enlightening ICT skills for youth employability in the European Union. *Telematics and Informatics, 35*(4), 1031–1038. https://doi.org/10.1016/j.tele.2017.09.014.

Pigoni, A., Delvecchio, G., Madonna, D., Bressi, C., Soares, J., & Brambilla, P. (2019). Can machine learning help us in dealing with treatment resistant depression? A review. *J Affect Disord, 259*, 21–26. https://doi.org/10.1016/j.jad.2019.08.009.

Proctor, E., Silmere, H., Raghavan, R., Hovmand, P., Aarons, G., Bunger, A., Griffey, R., & Hensley, M. (2011). Outcomes for implementation research: Conceptual distinctions, measurement challenges, and research agenda. *Adm Policy Ment Health, 38*(2), 65–76. https://doi.org/10.1007/s10488-010-0319-7.

Raney, L., Bergman, D., Torous, J., & Hasselberg, M. (2017). Digitally driven integrated primary care and behavioral health: How technology can expand access to effective treatment. *Curr Psychiatry Rep, 19*(11), 86. https://doi.org/10.1007/s11920-017-0838-y.

Ray, G., Muhanna, W. A., & Barney, J. B. (2007). Competing with IT: The role of shared IT-business understanding. *Communications of the ACM, 20*(12), 87–91. https://doi.org/10.1145/1323688.1323700.

Rogers, E. (2010). *Diffusion of Innovations* (4th ed.). New York, NY: Simon & Schuster [Original work published 1962].

Rohani, D. A., Faurholt-Jepsen, M., Kessing, L. V., & Bardram, J. E. (2018). Correlations between objective behavioral features collected from mobile and wearable devices and depressive mood symptoms in patients with affective disorders: systematic review. *JMIR Mhealth Uhealth, 6*(8), e165. https://doi.org/10.2196/mhealth.9691.

Ross, S., & Allen, N. (2012). Examining the convergent validity of shared mental model measures. *Behav Res Methods, 44*(4), 1052–1062. https://doi.org/10.3758/s13428-012-0201-5.

Saeb, S., Zhang, M., Kwasny, M. M., Karr, C. J., Kording, K., & Mohr, D. C. (2015). The relationship between clinical, momentary, and sensor-based assessment of depression. *Int Conf Pervasive Comput Technol Healthc, 2015*. doi:10.4108/icst.pervasivehealth.2015.259034.

Schueller, S. M., Armstrong, C. M., & Neary, M. e. a. (in press). Identifying and using mobile apps in clinical practice article type: Special series ABCT digital intervention.

Siemer, C. P., Fogel, J., & Van Voorhees, B. W. (2011). Telemental health and web-based applications in children and adolescents. *Child Adolesc Psychiatr Clin N Am, 20*(1), 135–153. https://doi.org/10.1016/j.chc.2010.08.012.

Silva, B. M., Rodrigues, J. J., de la Torre Diez, I., Lopez-Coronado, M., & Saleem, K. (2015). Mobile-health: a review of current state in 2015. *J Biomed Inform, 56*, 265–272. https://doi.org/10.1016/j.jbi.2015.06.003.

Silverman, B. G., Hanrahan, N., Bharathy, G., Gordon, K., & Johnson, D. (2015). A systems approach to healthcare: agent-based modeling, community mental health, and population well-being. *Artif Intell Med, 63*(2), 61–71. https://doi.org/10.1016/j.artmed.2014.08.006.

Smith, J. W. (2015). Immersive virtual environment technology to supplement environmental perception, preference and behavior research: a review with applications. *Int J Environ Res Public Health, 12*(9), 11486–11505. https://doi.org/10.3390/ijerph120911486.

Spante, M., Hashemi, S. S., Lundin, M., & Algers, A. (2018). Digital competence and digital literacy in higher education research: systematic review of concept use. *Cogent Education, 5*(1), 1519143. https://doi.org/10.1080/2331186X.2018.1519143.

Stimpfel, A. W., Sloane, D. M., & Aiken, L. H. (2012). The longer the shifts for hospital nurses, the higher the levels of burnout and patient dissatisfaction. *Health Aff (Millwood), 31*(11), 2501–2509. https://doi.org/10.1377/hlthaff.2011.1377.

Substance Abuse and Mental Health Services Administration. (2018). *Health resources and services integrated health solutions.* https://www.integration.samhsa.gov/integrated-care-models/A_Standard_Framework_for_Levels_of_Integrated_Healthcare.pdf. (Accessed date: August 2, 2022).

Torous, J., Jan Myrick, K., Rauseo-Ricupero, N., & Firth, J (2020). Digital mental health and COVID-19: using technology today to accelerate the curve on access and quality tomorrow. *JMIR Ment Health, 7*(3), e18848. https://doi.org/10.2196/18848.

Torous, J., & Roberts, L. W. (2017). The ethical use of mobile health technology in clinical psychiatry. *J Nerv Ment Dis, 205*(1), 4–8. https://doi.org/10.1097/NMD.0000000000000596.

US Department of Education. (2017a). *Reimagining the role of technology in education: 2017 national education technology plan update.* https://tech.ed.gov/files/2017/01/NETP17.pdf. (Accessed date: August 2, 2022).

US Department of Education. (2017b). *Reimagining the role of technology in higher education: a supplement to the 2017 national education technology plan.* https://tech.ed.gov/files/2017/01/Higher-Ed-NETP.pdf. (Accessed date: August 2, 2022).

van Baren J., Ijsselsteijn W. (2004). OmniPres project IST-2001-39237 deliverable 5 measuring presence: a guide to current measurement approaches. IST FET OMNIPRES project. https://www8.informatik.umu.se/~jworth/PresenceMeasurement.pdf.

van Dijk, J. (2006). *The Network Society: Social Aspects of New Media, 2nd ed.* New York: SAGE Publications. https://books.google.com/books?id=b7ktTPViIYMC.

Vial, G. (2019). Understanding digital transformation: a review and a research agenda. *The Journal of Strategic Information Systems, 28*(2), 118–144. https://doi.org/10.1016/j.jsis.2019.01.003.

Wang, Q., Myers, M. D., & Sundaram, D. (2013). Digital natives and digital immigrants. *Business & Information Systems Engineering, 5*(6), 409–419. https://doi.org/10.1007/s12599-013-0296-y.

Watson, J., Hutyra, C. A., Clancy, S. M., Chandiramani, A., Bedoya, A., Ilangovan, K., Nderitu, N., & Poon, E. G. (2020). Overcoming barriers to the adoption and implementation of predictive modeling and machine learning in clinical care: What can we learn from US academic medical centers? *JAMIA Open, 3*(2), 167–172. https://doi.org/10.1093/jamiaopen/ooz046.

Waytz, A., Cacioppo, J., & Epley, N. (2010). Who sees human? The stability and importance of individual differences in anthropomorphism. *Perspect Psychol Sci, 5*(3), 219–232. https://doi.org/10.1177/1745691610369336.

Weichelt, B., Bendixsen, C., & Patrick, T. (2019). A model for assessing necessary conditions for rural health care's mobile health readiness: qualitative assessment of clinician-perceived barriers. *JMIR Mhealth Uhealth, 7*(11), e11915. https://doi.org/10.2196/11915.

Weinstein, M. C., & Stason, W. B. (1977). Foundations of cost-effectiveness analysis for health and medical practices. *N Engl J Med, 296*(13), 716–721. https://doi.org/10.1056/NEJM197703312961304.

Will, K. K., Johnson, M. L., & Lamb, G. (2019). Team-based care and patient satisfaction in the hospital setting: a systematic review. *J Patient Cent Res Rev, 6*(2), 158–171. https://doi.org/10.17294/2330-0698.1695.

Wisniewski, H., Gorrindo, T., Rauseo-Ricupero, N., Hilty, D. M., & Torous, J. (2020). The role of digital navigators in promoting clinical care and technology integration into practice. *Digit Biomark, 4*(Suppl 1), 119–135. https://doi.org/10.1159/000510144.

Witmer, B. G., & Singer, M. J. (1998). Measuring presence in virtual environments: a presence questionnaire. *Presence: Teleoperators and Virtual Environments, 7*(3), 225–240. https://doi.org/10.1162/105474698565686.

Wongkoblap, A., Vadillo, M. A., & Curcin, V. (2017). Researching mental health disorders in the era of social media: systematic review. *J Med Internet Res, 19*(6), e228. https://doi.org/10.2196/jmir.7215.

Yellowlees, P., Burke Parish, M., Gonzalez, A., Chan, S., Hilty, D. M., Iosif, A. M., McCarron, R., Odor, A., Scher, L., Sciolla, A., Shore, J., & Xiong, G. (2018). Asynchronous telepsychiatry: a component of stepped integrated care. *Telemed J E Health, 24*(5), 375–378. https://doi.org/10.1089/tmj.2017.0103.

Zalpuri, I., Liu, H. Y., Stubbe, D., Wrzosek, M., Sadhu, J., & Hilty, D. M. (2018). Social media and networking competencies for psychiatric education: skills, teaching methods, and implications. *Acad Psychiatry, 42*(6), 808–817. https://doi.org/10.1007/s40596-018-0983-6

Index

Page numbers followed by "*f*" and "*t*" indicate, figures and tables respectively.